Lecture Notes in Computer Science 10029

Commenced Publication in 1973
Founding and Former Series Editors:
Gerhard Goos, Juris Hartmanis, and Jan van Leeuwen

More information about this series at http://www.springer.com/series/7412

Antonio Robles-Kelly · Marco Loog
Battista Biggio · Francisco Escolano
Richard Wilson (Eds.)

Structural, Syntactic, and Statistical Pattern Recognition

Joint IAPR International Workshop, S+SSPR 2016
Mérida, Mexico, November 29 – December 2, 2016
Proceedings

 Springer

Editors
Antonio Robles-Kelly
Data 61 - CSIRO
Canberra
Australia

Marco Loog
Pattern Recognition Laboratory
Technical University of Delft
CD Delft
The Netherlands

Battista Biggio
Electrical and Electronic Engineering
University of Cagliari
Cagliari
Italy

Francisco Escolano
Computación e IA
Universidad de Alicante
Alicante
Spain

Richard Wilson
Computer Science
University of York
York
UK

ISSN 0302-9743   ISSN 1611-3349   (electronic)
Lecture Notes in Computer Science
ISBN 978-3-319-49054-0   ISBN 978-3-319-49055-7   (eBook)
DOI 10.1007/978-3-319-49055-7

Library of Congress Control Number: 2016956001

LNCS Sublibrary: SL6 – Image Processing, Computer Vision, Pattern Recognition, and Graphics

Printed on acid-free paper

This Springer imprint is published by Springer Nature
The registered company is Springer International Publishing AG
The registered company address is: Gewerbestrasse 11, 6330 Cham, Switzerland

Preface

This volume contains the proceedings of the joint IAPR International Workshops on Structural and Syntactic Pattern Recognition (SSPR 2016) and Statistical Techniques in Pattern Recognition (SPR 2016). S+SSPR 2016 is a joint biannual event organized by Technical Committee 1 (Statistical Pattern Recognition Technique) and Technical Committee 2 (Structural and Syntactical Pattern Recognition) of the International Association of Pattern Recognition (IAPR). This year, S+SSPR was held in Mérida, Mexico, from November 29 to December 2, 2016. Merida is the largest city in the Yucatan Peninsula and boasts the second-largest historic center in Mexico after Mexico City.

All submitted full papers were reviewed by at least two members of the Technical Committee. The 51 accepted papers span a wide variety of topics in structural, syntactical, and statistical pattern recognition. Indeed, since its inception, S+SSPR has attracted participants working in a wide variety of fields that make use of statistical, structural or syntactic pattern recognition techniques as well as researchers who make methodological contributions to the field. This has traditionally resulted in a rich mix of papers with applications in image processing, computer vision, bioinformatics, chemo-informatics, machine learning, and document analysis. We are pleased to say that S+SSPR 2016 is no exception to this trend, featuring up-to-date research on areas such as spatio-temporal pattern recognition, texture and shape analysis, semantic segmentation, clustering and classification. S+SSPR also had invited talks by two prominent speakers: Professor Sudeep Sarkar from the University of South Florida and Professor Hamid Krim from North Carolina State University. In addition to parallel oral sessions for SPR and SSPR, there were also joint oral sessions covering topics of interest to both communities and the Pierre Devijver Award Lecture delivered by Professor Mario A.T. Figueiredo.

We would like to thank our sponsors, the British Machine Vision Association (BMVA) and the International Association of Pattern Recognition (IAPR), who helped us to achieve the goals of the conference. We wish to express our appreciation to Springer for publishing this volume. Would also like to thank all the members of the Technical Committee. Without the enthusiasm of the Technical Committee members, their time, and expertise, the reviewing process that has made S+SSPR 2016 possible could never have been completed. Last, but not least, we would like to also thank the authors for submitting their work and making S+SSPR 2016 a success.

November 2016

Antonio Robles-Kelly
Marco Loog
Battista Biggio
Francisco Escolano
Richard Wilson

Organization

Program Committee

Ethem Alpaydin	Bogazici University, Turkey
Silvia Biasotti	CNR — IMATI, Italy
Manuele Bicego	University of Verona, Italy
Gavin Brown	University of Manchester, UK
Umberto Castellani	University of Verona, Italy
Veronika Cheplygina	Erasmus Medical Center, The Netherlands
Francesc J. Ferri	Universitat de Valencia, Spain
Pasi Fränti	University of Eastern Finland
Giorgio Fumera	University of Cagliari, Italy
Edel Garcia	CENATAV, Cuba
Marco Gori	Universita di Siena, Italy
Michal Haindl	Institute of Information Theory and Automation of the CAS, Czech Republic
Barbara Hammer	Bielefeld University, Germany
Edwin Hancock	University of York, UK
Laurent Heutte	Université de Rouen, France
Tin Kam Ho	IBM Watson Research Center, USA
Atsushi Imiya	IMIT Chiba University, Japan
Xiaoyi Jiang	University of Münster, Germany
Tomi Kinnunen	University of Eastern Finland, Finland
Jesse Krijthe	Leiden University, The Netherlands
Mineichi Kudo	Hokkaido University, Japan
Arjan Kuijper	Fraunhofer IGD and TU Darmstadt, Germany
Ludmila Kuncheva	Bangor University, Wales
Xuelong Li	Chinese Academy of Sciences, China
Richard Nock	CEREGMIA-UAG, France
Mauricio Orozco-Alzate	Universidad Nacional de Colombia Sede Manizales, Colombia
Nikunj Oza	NASA
Giorgio Patrini	Australian National University, Australia
Marcello Pelillo	University of Venice, Italy
Adrian Peter	Florida Institute of Technology, USA
Filiberto Pla	University Jaume I, Spain
Marcos Quiles	Federal University of Sao Paulo, Brazil
Mariano Rivera	Centro de Investigacion en Matematicas AC, Mexico
Fabio Roli	University of Cagliari, Italy
Samuel Rota Bulò	Fondazione Bruno Kessler, Italy

Contents

Graph-Theoretic Methods

Model Selection, Classification and Clustering

Semi and Fully Supervised Learning Methods

Shape Analysis

Spatio-temporal Pattern Recognition

Structural Matching

Text and Document Analysis

Dimensionality Reduction, Manifold Learning and Embedding Methods

P2P Lending Analysis Using the Most Relevant Graph-Based Features

Lixin Cui[1], Lu Bai[1]([✉]), Yue Wang[1], Xiao Bai[2], Zhihong Zhang[3],
and Edwin R. Hancock[4]

[1] School of Information, Central University of Finance and Economics,
Beijing, China
bailucs@cufe.edu.cn, bailu69@hotmail.com
[2] School of Computer Science and Engineering, Beihang University, Beijing, China
[3] Software School, Xiamen University, Xiamen, Fujian, China
[4] Department of Computer Science, University of York, York, UK

Abstract. Peer-to-Peer (P2P) lending is an online platform to facilitate borrowing and investment transactions. A central problem for these P2P platforms is how to identify the most influential factors that are closely related to the credit risks. This problem is inherently complex due to the various forms of risks and the numerous influencing factors involved. Moreover, raw data of P2P lending are often high-dimension, highly correlated and unstable, making the problem more untractable by traditional statistical and machine learning approaches. To address these problems, we develop a novel filter-based feature selection method for P2P lending analysis. Unlike most traditional feature selection methods that use vectorial features, the proposed method is based on graph-based features and thus incorporates the relationships between pairwise feature samples into the feature selection process. Since the graph-based features are by nature completed weighted graphs, we use the steady state random walk to encapsulate the main characteristics of the graph-based features. Specifically, we compute a probability distribution of the walk visiting the vertices. Furthermore, we measure the discriminant power of each graph-based feature with respect to the target feature, through the Jensen-Shannon divergence measure between the probability distributions from the random walks. We select an optimal subset of features based on the most relevant graph-based features, through the Jensen-Shannon divergence measure. Unlike most existing state-of-the-art feature selection methods, the proposed method can accommodate both continuous and discrete target features. Experiments demonstrate the effectiveness and usefulness of the proposed feature selection algorithm on the problem of P2P lending platforms in China.

1 Introduction

Online Peer-to-Peer (P2P) lending has recently emerged as an useful financing alternative where individuals can borrow and lend money directly through an online trading platform without the help of institutional intermediaries such

A. Robles-Kelly et al. (Eds.): S+SSPR 2016, LNCS 10029, pp. 3–14, 2016.
DOI: 10.1007/978-3-319-49055-7_1

as banks [24]. Despite its explosive development, recent years have witnessed several acute problems such as high default rate of borrowers and bankruptcy of a large number of P2P lending platforms, etc [15]. To prevent personal investors from economic losses and ensure smooth and effective operations of the P2P lending industry, it is of great necessity to develop efficient credit risk assessment methods. Indeed, state-of-the-art P2P lending platforms such as Prosper and Lending Club, have utilized credit rating models for evaluating risk for each loan [6].

Credit risk evaluation decisions are inherently complex due to the various forms of risks and the numerous influencing factors involved [4]. Along this line, tremendous efforts have been devoted to developing quantitative credit rating methods due to their effectiveness. These methods can be broadly divided into two groups: traditional statistical methods [5]and machine learning approaches [20]. Using statistical methods is difficult because of the complexities of dependencies between various factors that influence the final credit risk evaluations. On the other hand, machine learning approaches, such as tree-based classifiers, support vector machines (SVM), and neural networks (NN), etc., do not require the factors to be independent and identically distributed (i.i.d.), and are capable of tackling computationally intensive credit risk evaluation problems. However, along with emergence of the Internet and E-Commerce, data sets of P2P lending are getting larger and larger. In addition, these raw data are often high-dimensional, highly-correlated, and unstable. These characteristics of the P2P lending data present new challenges for traditional machine learning algorithms, which may consume large amount of computational time and can not process information effectively.

To mitigate this problem, one potent way is to use feature selection in the data preprocessing process before implementing the learning algorithms [7]. By choosing a small subset of the most informative features that ideally is necessary and sufficient to describe the target concept [10], feature selection is capable of solving data mining and pattern recognition problems with data sets involving large number of features. Some have attempted to explore the advantages of feature selection for credit risk evaluation for P2P lending. For instance, Malekipir-bazari and Aksakalli [15] proposed a random forest based classification method for predicting borrower status. To reduce data dimensionality, they proposed a feature selection method based on the information gain of each individual feature. In Jin and Zhu [21], a random forest method is used to evaluate the significance of each feature, and a feature subset is selected based on this measure. By comparing the performance of decision tree, SVM, and NN on a dataset from Lending Club, the authors demonstrated the effectiveness of using feature selection method for credit risk analysis for P2P lending. Despite its usefulness in solving credit risk evaluation problems in P2P lending, most existing feature selection methods accommodate each feature as a vector, and thus ignore the relationship between pairwise samples in each feature. This drawback lead to significant information loss. Thus, developing effective feature selection method still remains a challenge.

The aim of this paper is to address the aforementioned shortcoming, by developing a new feature selection method. We commence by transforming each vectorial feature into a graph-based feature, that not only incorporates relationships between pairwise feature samples but also reflects richer characteristics than the original vectorial features. We also transform the target feature into a graph-based target feature. Furthermore, we use the steady state random walk and compute a probability distribution of the walk visiting the vertices. With the probability distribution for the graph-based features and the graph-based target feature to hand, we measure the discriminant power of each graph-based feature with respect to the graph-based target feature, through the Jensen-Shannon divergence measure between the probability distributions from the random walks. We select an optimal subset of features based on the most relevant graph-based features, through the Jensen-Shannon divergence measure. Unlike most existing state-of-the-art feature selection methods, the proposed method can accommodate both continuous and discrete target features. Experiments demonstrate the effectiveness and usefulness of the proposed feature selection algorithm on the problem of P2P lending platforms in China.

This paper is organized as follows. Section 2 briefly reviews the related works of feature selection methods. Section 3 presents preliminary concepts that will be used in this work. Then Sect. 4 define the proposed feature selection method. Section 5 presents the experimental evaluation of the proposed approach on a dataset collected from a large P2P lending portal. Section 6 concludes this work.

2 Literature Review

Feature selection has been a fundamental research topic in data mining and machine learning [8]. By choosing from the input data a subset of features that maximizes a generalized performance criterion, feature selection reduces the high dimensionality of the original data, improves learning performance, and provides faster and more cost-effective predictors [7].

Feature selection methods can be broadly divided into two categories, depending on their interaction with the classifier [18]. Filter-based methods [23] are independent from the classifier and focuses on the intrinsic properties of the original data. It usually provides a feature weighting or ranking based on some evaluation criteria and outputs a subset of selected features. By contrast, wrapper approaches [13] perform a search for an optimal subset of features using the outcome of a classifier as guidance. Often, the results obtained by wrapper methods are better than those obtained by filter methods, but the computational cost is also much higher.

Generally, evaluation criteria are of great significance for feature selection and a great variety of effective evaluation criteria have been proposed to locate informative features. These methods include distance [12], correlation [9], information entropy [14], rough set theory [3], etc. Among them, the correlation criterion and its extensions are probably one of the most widely used criteria to characterize the relevance between features, due to its good performance and ease of implementation. For instance, Hall [9] employed some correlation measures to evaluate the optimal feature subsets based on the assumption that a good subset contains features which are highly correlated to the class, yet uncorrelated to each other. In [19], a supervised feature selection method which builds a dissimilarity space by hierarchical clustering with conditional mutual information is developed.

Broadly speaking, there are two types of methods to measure the correlation between the same type of features. One is traditional linear correlation and the other kind is based upon information theory. For the first type, the most well-known similarity measure between two features x and y is the linear correlation measure $\text{sim}(x, y)$ and

$$\text{sim}(x, y) = \text{cov}(x, y)/\sqrt{\text{var}(x)\text{var}(y)}.$$

Here $\text{var}(\cdot)$ represents the variance of a feature and $\text{cov}(x, y)$ denotes the covariance between feature x and y. Other measures in this category are basically variations of this measure, including the maximal information compression index and least square regression error. Although this type of similarity measure can reduce redundancy among relevant features, it has several shortcomings. First, the linear correlation assumption between features are often not reasonable because many real world data such as P2P lending and finance, have very complex non-linear relationships. Second, the linear correlation measure is not applicable in the cases when discrete data are involved.

To address these shortcomings, various information theory based correlation measures such as information gain [22] and symmetrical uncertainty [17] have been proposed. The amount by which the entropy of x decreases reflects additional information about X provided by Y, and is called information gain, which is expressed as

$$IG(x, y) = H(x) - H(x|y) = H(y) - H(y|x).$$

Here $H(\cdot)$ denotes the entropy of a feature X and $H(x|y)$ refers to the entropy of X after observing values of another discrete feature Y. Despite its efficiency, information gain is biased towards features with more values. On the other hand, symmetrical uncertainty normalizes its value to the range of $[0, 1]$. It can be defined as

$$SU(x, y) = 2[IG(x, y)/H(x) + H(y)].$$

If $SU(x, y) = 1$, it indicates that features x and y are completely related. Otherwise, if SU takes the value of zero, it suggests that x and y are totally independent.

Although many robust correlation-based evaluation criteria have been developed in literature, there is no existing method that can incorporate relationships between pairwise samples of each feature dimension into the feature selection process. We also notice that a number of existing feature selection criteria implicitly select features that preserve sample relationship, which can be inferred from either a predefined distance metric or label information [25]. This indicates that it would be beneficial to incorporate sample relationship into the feature selection algorithm.

3 Preliminary Concepts

3.1 Probability Distributions from the Steady State Random Walk

Assume $G(V, E)$ is a graph with vertex set V, edge set E, and a weight function $\omega : V \times V \to \mathbb{R}^+$. If $\omega(u, v) > 0$ $(\omega(u, v) = \omega(v, u))$, we say that (u, v) is an edge of G, i.e., the vertices $u \in V$ and $v \in V$ are adjacent. The vertex degree matrix of G is a diagonal matrix D whose elements are given by

$$D(v, v) = d(v) = \sum_{u \in V} \omega(v, u). \tag{1}$$

Based on [2], the probability of the steady state random walk visiting each vertex v is

$$p(v) = d(v) / \sum_{u \in V} d(u). \tag{2}$$

Furthermore, from the probability distribution $P = \{p(1), \ldots, p(v), \ldots, p(|V|)\}$, we can straightforwardly compute the Shannon entropy of G as

$$H_S(G) = - \sum_{v \in V} p(v) \log p(v). \tag{3}$$

3.2 Jensen-Shannon Divergence

In information theory, the JSD is a dissimilarity measure between probability distributions over potentially structured data, e.g., trees, graphs, etc. It is related to the Shannon entropy of the two distributions [2]. Consider two (discrete) probability distributions $P = (p_1, \ldots, p_m, \ldots, p_M)$ and $Q = (q_1, \ldots, q_m, \ldots, q_M)$, then the classical Jensen-Shannon divergence between P and Q is defined as

$$D_{JS}(P, Q) = H_S\left(\frac{P + Q}{2}\right) - \frac{1}{2}H_S(P) - \frac{1}{2}H_S(Q)$$

$$= - \sum_{m=1}^{M} \frac{p_m + q_m}{2} \log \frac{p_m + q_m}{2} + \frac{1}{2} \sum_{m=1}^{M} p_m \log p_m + \frac{1}{2} \sum_{m=1}^{M} q_m \log q_m, \tag{4}$$

where $H_S(.)$ is the Shannon entropy of a probability distribution. Note that, the JSD measure is used as a means of measuring the information theoretic dissimilarity of graphs. However, in this work, we are more interested in the similarity measure between features. Thus, we define the JSD based similarity measure by transforming the JSD into its negative form and obtaining the corresponding exponential value, i.e.,

$$S(P, Q) = \exp\{-D_{JS}(P, Q)\}. \tag{5}$$

4   The Feature Selection Method on Graph-Based Features

4.1   Construction of Graph-Based Features

In this subsection, we transform each vectorial feature into a new graph-based feature, that is a complete weighted graph. The main advantage of using the new feature representation is that the graph-based feature can incorporate the relationship between samples of each original vectorial feature, and thus leading to less information loss. Given a dataset having N features denoted as $\mathcal{X} = \{\mathbf{f}_1, \ldots, \mathbf{f}_i, \ldots, \mathbf{f}_N\} \in \mathbb{R}^{M \times N}$, $\mathbf{f}_i$ represents the i-th vectorial feature that has M samples. We transform each vectorial feature $\mathbf{f}_i$ into a graph-based feature $\mathbf{G}_i(V_i, E_i)$, where each vertex $v_a \in V_i$ indicates the a-th sample f_a of $\mathbf{f}_i$, each pair of vertices v_a and v_b is connected by a weighted edge $(v_a, v_b) \in E_i$, and the weight $w(v_a, v_b)$ is the Euclidean distance as

$$w(v_a, v_b) = \sqrt{(f_a - f_b)(f_a - f_b)^T}. \tag{6}$$

Similarly, if the sample of the target feature $\mathbf{Y} = \{y_1, \ldots, y_a, \ldots, y_b, \ldots, y_M\}^T$ are continuous, its graph-based feature $\hat{\mathbf{G}}(\hat{V}, \hat{E})$ can also be computed using Eq. (6) and each vertex $\hat{v}_a \in \hat{V}$ represents the a-th sample y_a. However, for some instances, the sample y_a of the target feature Y may take discrete values $c = 1, 2, \ldots, C$. For this instance, we first compute the graph-based target feature $\hat{\mathbf{G}}_{\mathbf{i}}(\hat{V}_i, \hat{E}_i)$ for each feature $\mathbf{f}_i$, where the weight $w(\hat{v}_{ia}, \hat{v}_{ib})$ of each edge $(\hat{v}_{ia}, \hat{v}_{ib}) \in \hat{E}_i$ is

$$w(\hat{v}_a, \hat{v}_b) = \sqrt{(\mu_{ia} - \mu_{ib})(\mu_{ia} - \mu_{ib})^T}, \tag{7}$$

where μ_{ia} is the mean value of all samples in $\mathbf{f}_i$ that are corresponded by the same discrete value c of the target feature samples and $c = y_a$. Moreover, based on [11], we also compute the Fisher score $F(\mathbf{f}_i)$ for each feature $\mathbf{f}_i$ as

$$F(\mathbf{f}_i) = \frac{\sum_{c=1}^{C} n_c(\mu_c - \mu)^2}{\sum_{c=1}^{C} n_c\sigma_c^2}, \tag{8}$$

where μ_c and σ_c^2 are the mean and variance of the samples corresponded by the same discrete value c, μ is the mean of feature $\mathbf{f}_i$, and n_c is the number of the samples corresponded by c-th in feature $\mathbf{f}_i$. From Eq. (8), we observe that the Fisher score $S(\mathbf{f}_i)$ reveal the quality of the graph-based target feature $\hat{\mathbf{G}}_i$ for $\mathbf{f}_i$. In other words, a higher Fisher score means a better target feature graph. As a result, the graph-based target feature $\hat{\mathbf{G}}(\hat{V}, \hat{E})$ can be identified by

$$\hat{\mathbf{G}}(\hat{V}, \hat{E}) = \hat{\mathbf{G}}(\hat{V}_i^*, \hat{E}_i^*), \tag{9}$$

where

$$i^* = \arg\max_i F(\mathbf{f}_i). \tag{10}$$

4.2 Feature Selection Based on Relevant Graph-Based Features

We aim to select an optimal subset of features. Specifically, by measuring the Jensen-Shannon divergence between graph-based features, we compute the discriminant power of each vectorial feature with respect to the target feature. For a set of N features $\mathbf{f}_1, \ldots, \mathbf{f}_i, \ldots, \mathbf{f}_j, \ldots, \mathbf{f}_N$ and the associated continuous or discrete target feature $\mathbf{Y}$, the relevance degree or discriminant power of the feature $\mathbf{f}_i$ with respect to $\mathbf{Y}$ is

$$R_{\mathbf{f}_i, \mathbf{Y}} = S(\mathbf{G}_i, \hat{\mathbf{G}}), \tag{11}$$

where $\mathbf{G}_i$ and $\hat{\mathbf{G}}$ are the graph-based features of $\mathbf{f}_i$ and $\mathbf{Y}$, S is the JSD based similarity measure defined in Eq. (5). Based on the relevance degree of each feature $\mathbf{f}_i$ with respect to the target feature $\mathbf{Y}$ computed by Eq. (11) (for the continuous target feature) or Eq. (9) (for the discrete target feature), we can rank the original vectorial features in descending order and then select a subset of the most relevant features.

5 Experiments

We evaluate the effectiveness of the proposed graph-based feature selection algorithm on the problem of P2P lending platforms in China. This is of great significance for the credit risk analysis of the P2P platforms because the P2P lending industry has developed rapidly since the year of 2007, and many have suffered from severe problems such as default of borrowers and bankruptcy. More specifically, we use a data of 200 P2P platforms collected from a famous P2P lending portal in China (http://www.wdzj.com/). For each platform, we use 19 features including: (1) transaction volume, (2) total turnover, (3) total number of borrowers, (4) total number of investors, (5) online time, which refers to the foundation year of the platform, (6) operation time, i.e., number of months since the foundation of the platform, (7) registered capital, (8) weighted turnover, (9) average term of loan, (10) average full mark time, i.e., tender period of a loan raised to the required full capital, (11) average amount borrowed, i.e., average loan amount of each successful borrower, (12) average amount invested, which is the average investment amount of each successful investor, (13) loan dispersion, i.e., the ratio of the repayment amount to the total capital, (14) investment dispersion, the ratio of the invested amount to the total capital, (15) average times of borrowing, (16) average times of investment, (17) loan balance, (18) popularity, and (19) interest rate.

5.1 The Most Influential Features for Credit Risks (Continuous Target Features)

We first use the proposed feature selection algorithm to identify the most influential features which are most relevant to the interest rate of P2P platforms. In finance, the interest rates of P2P lending can also be interpreted as the rate of return on a loan (for investors), and the higher the rate of return, the greater the likelihood of default. Identifying the most relevant features to the interest rate can help investors effectively manage the credit risks involved in P2P lending [24]. Therefore, in our experiment, we set the interest rate as our **continuous target feature**. Our purpose is to identify the features that are most influential for the credit risks of the P2P platforms. To realize this goal, we use the proposed feature selection algorithm to rank the remaining 18 features according to their similarities to the target label in descending order. The results are shown in Table 1.

Table 1. Influential factors for bankruptcy problems for P2P lending platforms in China

Ranking	Feature score	Name of feature	Ranking	Feature score	Name of feature
1#	0.975	Registered capital	10#	0.935	Popularity
2#	0.967	Operation time	11#	0.927	Total number of borrowers
3#	0.966	Average amount invested	12#	0.926	Weighted turnover
4#	0.965	Loan dispersion	13#	0.919	Loan balance
5#	0.965	Average times of investment	14#	0.916	Total turnover
6#	0.963	Online time	15#	0.908	Average times of borrowing
7#	0.950	Average term of loan	16#	0.903	Average full mark time
8#	0.949	Total number of investors	17#	0.903	Average amount borrowed
9#	0.939	Investment dispersion	18#	0.902	Transaction volume

Results and Discussions: It is shown that registered capital, operation time, average amount invested, loan dispersion, and average times of investment are the top five features which are most relevant to the interest rate (target feature). These results are in consistent with the finance theory. For instance, the registered capital indicates stronger financial stability of the platform. In addition, a longer operation time of the platform usually implies that the platform accumulates abundant risk management knowledge and skills, which are helpful to maintain a lower credit risk level. Moreover, a more dispersed loan rate often indicates a higher degree of security for the platform, which implies a relatively lower interest rate. The average amount invested and average times of investment indicate investors' preferences for the less risky platforms. On the contrary, features such as average times of borrowing and average amount borrowed are of less relevance because these features reflect the financing needs of the borrowers and are less relevant to the credit risks of the platforms.

Comparisons: In this section, we compare the proposed feature selection (FS) method with two widely used methods including correlation analysis (CA) and multiple linear regression (MLR). Table 2 presents a comparison of the results obtained via these methods. Each method identifies 10 features which have higher correlation to the interest rate. It can be noticed that the most influential factors identified by the proposed method tend to be more in consistent with the factors selected by MLR, whereas CA ranks different features higher. For example, among the top five most influential factors, both FS and MLR select operation

Table 2. Comparison of three methods

Ranking	Feature selection	Correlation analysis	Multiple linear regression
1#	Registered capital	Popularity	Loan dispersion
2#	Operation time	Loan balance	Investment dispersion
3#	Average amount invested	Average times of investment	Online time
4#	Loan dispersion	Average times of borrowing	Popularity
5#	Average times of investment	Investment dispersion	Operation time
6#	Online time	Loan dispersion	Average times of borrowing
7#	Average term of loan	Average amount invested	Total number of borrowers
8#	Total number of investors	Average amount borrowed	Loan balance
9#	Investment dispersion	Average full mark time	Transaction volume
10#	Popularity	Average term of loan	Weighted turnover

time and loan dispersion. This is reasonable because a more dispersed loan rate often indicates a higher degree of security for the platform, which implies a relatively lower interest rate. Also, a longer operation time of the platform often indicates that the platform accumulates abundant risk management knowledge and skills, which are helpful to maintain a lower credit risk level. These results are in consistent with the finance theory and demonstrate the effectiveness and usefulness of the proposed method for the identification of the most influential factors for credit risk analysis of P2P lending platforms.

5.2 Classification for the Credit Rating (Discrete Target Features)

We further evaluate the performance of the proposed method when the target features are discrete. We set the credit rating (taking discrete values) as the target feature, and our purpose is to identify the most influential features for the credit rating of the P2P lending platforms in China. These rating values are collected from the "Report on the Development of the P2P lending industry in China, 2014–2015", issued by the Financial Research Institute of the Chinese Academy of Social Sciences. Due to the strict evaluation criteria involved, only 104 P2P platforms are included in this report, among which only 42 platforms belong to the 200 P2P platforms used in the above data set. Therefore, we take these 42 platforms as samples for evaluation.

In our experiment, we set the discrete credit rating targets as the classification labels. Because the 42 platforms are categorized into four classes according to their credit rating values, we set the number of classes as four. We randomly select 50 % of the 42 samples as training data, and use the other half for testing. By repeating this selection process 10 times, we obtain 10 random partitions of the original data. For each of the 10 partitions of the original data, we perform a 10-fold cross-validation using a C-Support Vector Machine (C-SVM) to evaluate the classification accuracy associated with the selected features located via different feature selection methods. These methods include: (1) the proposed feature selection method (GS), (2) the Fisher Score method (FS) [11], and (3) the

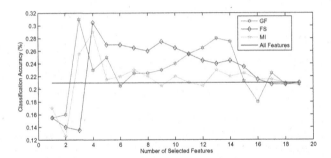

Fig. 1. Accuracy vs. number of selected features for different feature selection methods.

Mutual Information based method (MI) [16]. We perform cross-validation on the testing samples taken from the feature selection process. Specifically, the entire sample is randomly partitioned into 10 subsets and then we choose one subset for testing and use the remaining 9 subsets for training, and this procedure is repeated 10 times. The final accuracy is computed by averaging the accuracies from each of the random subsets, over all the 10 partitions. The final accuracy is computed by averaging the accuracies from each of the random subsets, over all the 10 partitions. The classification accuracy of each feature selection method based on different number of the most influential features is shown in Fig. 1.

Figure 1 indicates that the proposed method (GS) achieves the best classification accuracy (31.50 %) while requiring the lowest number of features, i.e., 3 features. In contrast, the FS and MI methods request 3 and 4 features respectively for their best classification accuracies 30.50 % and 29.00 %, respectively. The reasons for this effectiveness is that only the proposed method incorporates the sample relationship into the feature selection process, and thus encapsulates more information. Although the classification accuracy is 31.50 %, it is very promising because dividing 42 samples into four different classes is a very challenging classification task. Thus, the classification accuracy demonstrates the effectiveness of the proposed method.

6 Conclusion

In this paper, we have developed a novel feature selection algorithm to conduct credit risk analysis for the P2P lending platforms. Unlike most existing feature selection methods, the proposed method is based on graph-based feature and encapsulate global topological information of features into feature selection process. The proposed method thus avoid information loss between feature samples that arises in traditional feature selection methods. Using a dataset collected from a famous P2P portal in China, we demonstrate the effectiveness of our method.

The proposed feature selection method ignores the redundancy between pairwise features. As a result, the optimal subset of selected features may include redundant features. Furthermore, the proposed method cannot adaptively select the most informative feature subset. To address these problems, future work will be aimed at proposing a new framework that can adaptively select the most informative and less redundant graph-based feature subset. Furthermore, it is also interesting to propose new approaches of establishing graph-based features from original vectorial features. Finally, note that, the similarity measure between a graph-based feature and the target graph-based feature defined by Eq. (11) is the Jensen-Shannon diffusion graph kernel [1,2] over probability distributions. In fact, one can also adopt other alternative graph kernels. In other words, the proposed framework provides a way of developing feature selection methods associated with graph kernels. It is interesting to explore the performance of the proposed method associated with different graph kernels in future works.

Acknowledgments. This work is supported by the National Natural Science Foundation of China (Grant nos. 61602535, 61503422 and 61402389), and the Open Projects Program of National Laboratory of Pattern Recognition. Lu Bai is supported by the program for innovation research in Central University of Finance and Economics. Edwin R. Hancock is supported by a Royal Society Wolfson Research Merit Award. Lixin Cui is supported by the Young Scholar Development Fund of Central University of Finance and Economics, No. QJJ1540.

References

1. Bai, L., Bunke, H., Hancock, E.R.: An attributed graph kernel from the Jensen-Shannon divergence. In: Proceedings of ICPR, pp. 88–93 (2014). DBLP:conf/icpr/2014
2. Bai, L., Rossi, L., Bunke, H., Hancock, E.R.: Attributed graph kernels using the Jensen-Tsallis q-differences. In: Calders, T., Esposito, F., Hüllermeier, E., Meo, R. (eds.) ECML PKDD 2014. LNCS (LNAI), vol. 8724, pp. 99–114. Springer, Heidelberg (2014). doi:10.1007/978-3-662-44848-9_7
3. Chen, Y., Miao, D., Wang, R.: A rough set approach to feature selection based on ant colony optimization. Pattern Recogn. Lett. **31**(3), 226–233 (2010)
4. Crook, J.N., Edelman, D., Thomas, L.C.: Recent developments in consumer credit risk assessment. Eur. J. Oper. Res. **183**(3), 1447–1465 (2007)
5. Hand, D.J., Henley, W.E.: Statistical classification methods in consumer credit scoring: a review. J. R. Stat. Soc. Ser. A **160**(3), 523–541 (1997)
6. Guo, Y., Zhou, W., Luo, C., Liu, C., Xiong, H.: Instance-based credit risk assessment for investment decisions in P2P lending. Eur. J. Oper. Res. **249**(2), 417–426 (2016)
7. Guyon, I., Elisseeff, A.: An introduction to variable and feature selection. J. Mach. Learn. Res. **3**, 1157–1182 (2003)
8. Hájek, P., Michalak, K.: Feature selection in corporate credit rating prediction. Knowl.-Based Syst. **51**, 72–84 (2013)
9. Hall, M.A.: Correlation-based feature selection for discrete and numeric class machine learning. In: Proceedings of the ICML, pp. 359–366 (2000)
10. Han, J., Sun, Z., Hao, H.: Selecting feature subset with sparsity and low redundancy for unsupervised learning. Knowl.-Based Syst. **86**, 210–223 (2015)

11. He, X., Cai, D., Niyogi, P.: Laplacian score for feature selection. In: Advances in Neural Information Processing Systems 18 [Neural Information Processing Systems, NIPS 2005, Vancouver, British Columbia, Canada, 5–8 December 2005], pp. 507–514 (2005)
12. Huang, Y., McCullagh, P.J., Black, N.D.: An optimization of relieff for classification in large datasets. Data Knowl. Eng. **68**(11), 1348–1356 (2009)
13. Kohavi, R., John, G.H.: Wrappers for feature subset selection. Artif. Intell. **97**(1–2), 273–324 (1997)
14. Last, M., Kandel, .A., Maimon, O.: Information-theoretic algorithm for feature selection. Pattern Recogn. Lett. **22**(6/7), 799–811 (2001)
15. Malekipirbazari, M., Aksakalli, V.: Risk assessment in social lending via random forests. Expert Syst. Appl. **42**(10), 4621–4631 (2015)
16. Pohjalainen, J., Räsänen, O., Kadioglu, S.: Feature selection methods and their combinations in high-dimensional classification of speaker likability, intelligibility and personality traits. Comput. Speech Lang. **29**(1), 145–171 (2015)
17. Press, W.H., Teukolsky, S.A., Vetterling, W.T., Flannery, B.P.: Numerical Recipes in C, 2nd edn. Cambridge University Press, Cambridge (1992)
18. Saeys, Y., Inza, I., Larrañaga, P.: A review of feature selection techniques in bioinformatics. Bioinformatics **23**(19), 2507–2517 (2007)
19. Sotoca, J.M., Pla, F.: Supervised feature selection by clustering using conditional mutual information-based distances. Pattern Recogn. **43**(6), 2068–2081 (2010)
20. Yeh, I.-C., Lien, C.-H.: The comparisons of data mining techniques for the predictive accuracy of probability of default of credit card clients. Expert Syst. Appl. **36**(2), 2473–2480 (2009)
21. Jin, Y., Zhu, Y.D.: A data-driven approach to predict default risk of loan for online Peer-to-Peer (P2P) lending. In: Proceedings of Fifth International Conference on Communication Systems and Network Technologies, pp. 609–613 (2015)
22. Yu, L., Liu, H.: Efficient feature selection via analysis of relevance and redundancy. J. Mach. Learn. Res. **5**, 1205–1224 (2004)
23. Zhang, D., Chen, S., Zhou, Z.-H.: Constraint score: a new filter method for feature selection with pairwise constraints. Pattern Recogn. **41**(5), 1440–1451 (2008)
24. Zhao, H., Le, W., Liu, Q., Ge, Y., Chen, E.: Investment recommendation in P2P lending: a portfolio perspective with risk management. In: Proceedings of ICDM, pp. 1109–1114 (2014)
25. Zhao, Z., Wang, L., Liu, H., Ye, J.: On similarity preserving feature selection. IEEE Trans. Knowl. Data Eng. **25**(3), 619–632 (2013)

Simultaneous Nonlinear Label-Instance Embedding for Multi-label Classification

Keigo Kimura[✉], Mineichi Kudo, and Lu Sun

Graduate School of Information Science and Technology, Hokkaido University,
Sapporo 060-0814, Japan
{kkimura,mine,sunlu}@main.hokudai.ac.jp

Abstract. In this paper, unlike previous many linear embedding methods, we propose a non-linear embedding method for multi-label classification. The algorithm embeds both instances and labels into the same space, reflecting label-instance relationship, label-label relationship and instance-instance relationship as faithfully as possible, simultaneously. Such an embedding into two-dimensional space is useful for simultaneous visualization of instances and labels. In addition linear and nonlinear mapping methods of a testing instance are also proposed for multi-label classification. The experiments on thirteen benchmark datasets showed that the proposed algorithm can deal with better small-scale problems, especially in the number of instances, compared with the state-of-the-art algorithms.

Keywords: Multi-label classification · Nonlinear embedding · Visualization

1 Introduction

Multi-Label Classification (MLC), which allows an instance to have more than one label at the same time, has been recently received a surge of interests in a variety of fields and applications [10,15]. The main task of MLC is to learn the relationship between a F-dimensional feature vector x and an L-dimensional binary vector y from N training instances $\{(x^{(1)}, y^{(1)}), \ldots, (x^{(N)}, y^{(N)})\}$, and to predict a binary vector $\hat{y} \in \{0,1\}^L$ for a test instance $x \in \mathbb{R}^F$. To simplify the notation, we use a matrix $\mathbf{X} = [x^{(1)}, x^{(2)}, \ldots, x^{(N)}]^T \in \mathbb{R}^{N \times F}$ and a matrix $\mathbf{Y} = [y^{(1)}, y^{(2)}, \ldots, y^{(N)}]^T \in \{0,1\}^{N \times L}$ for expressing the training set.

A key of learning in MLC is how to utilize dependency between labels [10]. However, an excessive treatment of label dependency causes over-learning and brings larger complexity, sometimes, even intractable. Thus, many algorithms have been proposed to model the label dependency efficiently and effectively. Embedding is one of such methods for MLC. This type of methods utilizes label dependency through dimension reduction. The label dependency is explicitly realized by reducing the dimension of the label space from L to K ($\ll L$). Embedding methods in general learn relationships instances in F-dimensional

A. Robles-Kelly et al. (Eds.): S+SSPR 2016, LNCS 10029, pp. 15–25, 2016.
DOI: 10.1007/978-3-319-49055-7_2

space and latent labels in K-dimensional space, then, linearly transform the relationship to those in F-dimensional and real labels in L-dimensional space [4–6,8,12,16].

In this paper, we propose a novel method of a nonlinear embedding. Usually, either a set of labels or a set of instances is embedded [4–6,8,16], but in our method, both are embedded in the same time. We realize a mapping into a low-dimensional Euclidean space keeping three kinds of relationships between instance-instance, label-label and label-instance as faithfully as possible. In addition, for classification, a linear and a non-linear mappings of a testing instance are realized.

2   The Proposed Embedding

2.1   Objective Function

In contrast to traditional embedding methods, we explicitly embed both labels and instances into the same K-dimensional space ($K < F$) while preserving the relationships among labels and instances.[1] To preserve such relationships, we use a manifold learning method called Laplacian eigen map [1]. It keeps the distance or the degree of similarity between any pair of points or objects even in a low-dimensional space. For example, given similarity measure $\mathbf{W}_{ij}$ between two objects indexed by i and j, we find $z^{(i)}$ and $z^{(j)}$ in $\mathbb{R}^K$ so as to minimize $\sum_{i,j} \mathbf{W}_{ij} \| z^{(i)} - z^{(j)} \|_2^2$ under an appropriate constraint for scaling.

Now, we consider to embed both instances and labels at once. Let $g^{(i)} \in \mathbb{R}^K$ be the low-dimensional representation of ith instance $x^{(i)}$ on the embedding space and $h^{(l)} \in \mathbb{R}^K$ be the representation of lth label on the same space as well. In this embedding, we consider three types of relationships: instance-label, instance-instance and label-label relationships. In this work, we quantify the above relationships by focusing on their localities. In more detail, we realize a mapping to preserve the following three kinds of properties in the training set:

1. Instance-Label (IL) relationship: Explicit relationship given by $(x^{(i)}, y^{(i)})$ ($i = 1, \ldots, N$) should be kept in the embedding as closeness between $g^{(i)}$ and $h^{(l_i)}$ where l_i is one label of value one in $y^{(i)}$
2. Label-Label (LL) relationship: Frequently co-occurred label pairs should be placed more closely in the embedded space $\mathbb{R}^K$.
3. Instance-Instance (II) relationship: Instances close in $\mathbb{R}^F$ should be placed closely even in $\mathbb{R}^K$.

Let us denote them by $\mathbf{W}^{(IL)} \in \mathbb{R}^{N \times L}$, $\mathbf{W}^{(LL)} \in \mathbb{R}^{L \times L}$ and $\mathbf{W}^{(II)} \in \mathbb{R}^{N \times N}$, respectively. Then our objective function of $\{g^{(i)}, h^{(l)}\}$ become, with α, β (>0),

---

[1] Note that labels do not have their representations explicitly before embedding.

$$O = 2O_{IL} + \alpha O_{II} + \beta O_{LL}$$

$$= 2 \sum_{i,l} \mathbf{W}_{il}^{(IL)} \|g^{(i)} - h^{(l)}\|_2^2 + \alpha \sum_{ij} \mathbf{W}_{i,j}^{(II)} \|g^{(i)} - g^{(j)}\|_2^2 \qquad (1)$$

$$+ \beta \sum_{l,m} \mathbf{W}_{l,m}^{(LL)} \|h^{(l)} - h^{(m)}\|_2^2$$

$$= \sum_{s,t} \mathbf{W}_{st} \|e^{(s)} - e^{(t)}\|_2^2 \quad (s,t = 1,2,\ldots,(N+L)),$$

where $e^{(s)} = g^{(s)}$ or $h^{(s)}$, and $\mathbf{W}_{st} = \mathbf{W}_{st}^{(IL)}$, $\mathbf{W}_{st}^{(II)}$ or $\mathbf{W}_{st}^{(LL)}$ depending on the values of s and t. As their matrix representation, let us use $\mathbf{G} = [g^{(1)},\ldots,g^{(N)}]^T \in \mathbb{R}^{N \times K}$ and $\mathbf{H} = [h^{(1)},\ldots,h^{(L)}]^T \in \mathbb{R}^{L \times K}$. Then using

$$\mathbf{W} = \underbrace{\begin{pmatrix} \alpha \mathbf{W}^{(II)} & \mathbf{W}^{(IL)} \\ \mathbf{W}^{(IL)T} & \beta \mathbf{W}^{(LL)} \end{pmatrix}}_{N \qquad L} \text{ and } \mathbf{E} = \underbrace{\begin{pmatrix} \mathbf{G} \\ \mathbf{H} \end{pmatrix}}_{K} \begin{matrix} \}N \\ \}L \end{matrix},$$

our objective function is rewritten as

$$O = \sum_{s,t} \mathbf{W}_{st} \|e^{(s)} - e^{(t)}\|_2^2 = 2\text{Tr}(\mathbf{E}^T \mathbf{L} \mathbf{E}), \quad \text{s.t. } \mathbf{E}^T \mathbf{D} \mathbf{E} = \mathbf{I} \qquad (2)$$

where $\mathbf{L} = \mathbf{D} - \mathbf{W}$ and $\mathbf{D}$ is a diagonal matrix with elements $\mathbf{D}_{ii} = \sum_j \mathbf{W}_{ij}$ [1]. The constraint $\mathbf{E}^T \mathbf{D} \mathbf{E} = \mathbf{I}$ is imposed to remove an arbitrary scaling factor in the embedding. This formulation is that of the Laplacian eigen map. Next, let us explain how to determine the similarity matrix $\mathbf{W}$.

Instance-Label Relationship: For the instance-label relationship $\mathbf{W}^{(IL)}$, we use $\mathbf{W}^{(IL)} = \mathbf{Y}$. In this case, $\mathbf{W}^{(IL)}$ has elements of zero or one. The corresponding objective function of Instance-Label relationship becomes:

$$O_{IL} = \min_{g^{(i)}, h^{(l)}} \sum_{\substack{i=1,\ldots,N \\ l=1,\ldots,L}} \mathbf{W}_{il}^{(IL)} \|g^{(i)} - h^{(l)}\|_2^2,$$

where $\mathbf{W}_{il}^{(IL)} = \mathbf{Y}_{il} \in \{0,1\}$.

Instance-Instance Relationship: We use the symmetric k-nearest neighbor relation in $\mathbb{R}^F$ for constructing $\mathbf{W}^{(II)}$ as seen in [3]. Thus, our second objective function becomes

$$O_{II} = \min_{g^{(i)}, g^{(j)}} \sum_{i,j=1,\ldots,N} \mathbf{W}_{ij}^{(II)} \|g^{(i)} - g^{(j)}\|_2^2,$$

where

$$\mathbf{W}_{ij}^{(II)} = \begin{cases} 1 & (i \in \mathcal{N}_k(x^{(j)}) \vee j \in \mathcal{N}_k(x^{(i)})), \\ 0 & (\text{otherwise}), \end{cases}$$

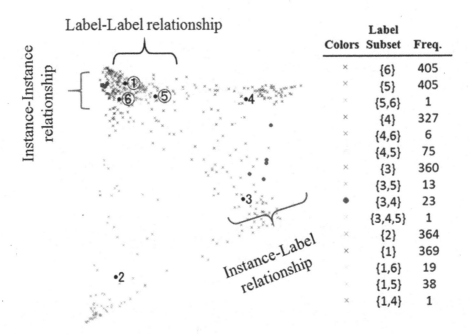

Fig. 1. The result of the proposed embedding in *Scene* dataset. Only 20 % of instances are displayed. The numbers indicate the labels $1, \ldots, 6$, and small crosses show the instances. (Color figure online)

where $\mathcal{N}_k(\boldsymbol{x}^{(i)})$ denotes the index set of k nearest neighbors of the ith instance. It is worth noting that we can construct $\mathbf{W}^{(II)}$ on the basis of the similarity between $\boldsymbol{y}^{(i)}$ and $\boldsymbol{y}^{(j)}$ as seen in [3] instead of that between $\boldsymbol{x}^{(i)}$ and $\boldsymbol{x}^{(j)}$ above.

Label-Label Relationship: We construct $\mathbf{W}^{(LL)}$ in such a way that $\mathbf{W}^{(LL)}_{lm}$ takes a large positive value when labels l and m co-occur frequently in $\mathbf{Y}$, otherwise a small positive value. We also use the symmetric k-nearest neighbor relation in the frequency. The corresponding third objective function becomes

$$O_{LL} = \min_{\boldsymbol{h}^{(l)}, \boldsymbol{h}^{(m)}} \sum_{l,m=1,\ldots L} \mathbf{W}^{(LL)}_{lm} \|\boldsymbol{h}^{(l)} - \boldsymbol{h}^{(m)}\|_2^2,$$

where

$$\mathbf{W}^{(LL)}_{lm} = \begin{cases} 1 & \text{(if } l \text{ is one of top-}k \text{ co-occurrence labels of } m \text{ and vice versa)}, \\ 0 & \text{(otherwise)}. \end{cases}$$

Note that $\mathbf{W}^{(LL)}$ is symmetric as well as $\mathbf{W}^{(II)}$. The symmetricity of those guarantees the existence of a solution in (2).

The solution of (2) is obtained by solving the following generalized eigen problem:

$$\mathbf{LE} = \lambda \mathbf{DE}. \tag{3}$$

Algorithm 1. MLLEM (Training)

1: **Input:** Label matrix $\mathbf{Y}$, Feature matrix $\mathbf{X}$, The number of dimension K, wighting parameters α and β;
2: **Output:** K-dimensional representation of labels $\mathbf{H}$ and instances $\mathbf{G}$;
3: Construct $\mathbf{W}^{(II)}$ for instances (Section 2);
4: Construct $\mathbf{W}^{(LL)}$ for labels (Section 2);
5: $\mathbf{W}^{(IL)} = \mathbf{Y}$;
6: $\mathbf{W} = \begin{bmatrix} \alpha\mathbf{W}^{(II)} & \mathbf{W}^{(IL)} \\ \mathbf{W}^{(IL)T} & \beta\mathbf{W}^{(LL)} \end{bmatrix}$;
7: Solve the generalized eigen problem $\mathbf{LE} = \lambda\mathbf{DE}$ where $\mathbf{L} = \mathbf{D} - \mathbf{W}$ and $\mathbf{D}_{ii} = \sum_j \mathbf{W}_{ij}$ and obtain the bottom K eigen vectors excluding an eigen vector with zero eigen value;
8: $\begin{bmatrix} \mathbf{G} \\ \mathbf{H} \end{bmatrix} = \mathbf{E}$;

Hence, the optimal solution $\mathbf{E}$ of the objective function is the bottom K eigenvectors excluding an eigenvector with zero eigenvalue [1].

An example of this embedding is shown in Fig. 1. This is the result of mapping for *Scene* dataset [11] where $N = 2407$, $F = 294$, $L = 6$ and $K = 2$. In Fig. 1, we can see that the instance-label, instance-instance and label-label relations are fairly preserved. First, for instance-label relationship, four instances that share a label subset $\{3, 4\}$ (large brown dots) are mapped between labels 3 and 4. Second, for label-label relationship, highly co-occurred labels 1, 5 and 6 are closely mapped (highlighted by a circle). Finally, for instance-instance relationship, an instance and its k nearest neighbors ($k = 2$) in the original F-dimensional space (a blue square and 2 blue diamonds) are closely placed.

2.2   Embedding Test Instances

For assigning labels for a testing instance, we need to embed it into the same low-dimensional space constructed from the training instances with multiple labels. Unfortunately above embedding is not functionally realized, we do not have an explicit way of mapping. Therefore, we propose two different ways of a linear mapping and a nonlinear mapping.

In the linear mapping, we simulate the nonlinear mapping from $\mathbf{X}$ to $\mathbf{G}$ (the former part of $\mathbf{E}$) by a linear mapping $\mathbf{V}$ so as to $\mathbf{G} \simeq \hat{\mathbf{G}} = \mathbf{XV}$. We use *Ridge regression* to find such a $\mathbf{V}$:

$$\min_{\mathbf{V}} \|\mathbf{XV} - \mathbf{G}\|_2^2 + \lambda\|\mathbf{V}\|_2^2.$$

where λ is a parameter. A test instance $\boldsymbol{x}$ is mapped to $\boldsymbol{g}$ such as $\boldsymbol{g} = \boldsymbol{x}^T\mathbf{V}$.

Algorithm 2. MLLEM (Testing)

1: **Input:** Test instance x, Feature matrix $\mathbf{X}$, K-dimensional representation of labels $\mathbf{H}$ and instances $\mathbf{G}$;

2: **Output:** Predicted multi-label (score) $\hat{y} \in \mathbb{R}^L$;

   {**Linear Embedding**}

3: Minimize $\|\mathbf{X}\mathbf{V} - \mathbf{G}\|_2^2 + \lambda\|\mathbf{V}\|_2^2$ in $\mathbf{V}$;

4: Embed test instance by $g = x^T\mathbf{V}$;

   {**Nonlinear Embedding**}

5: Find k-nn $\mathcal{N}_{k(t)}$ of test instance x in training set $\mathbf{X}$;

6: $g = \underset{g'}{\operatorname{argmin}} \frac{1}{k} \sum_{i \in \mathcal{N}_k(t)} \|g' - g^{(i)}\|_2^2$;

   {After **Linear Embedding** or **Nonlinear Embedding**}

7: Calculate the score for each label $\hat{y}_i = \|g - h^{(i)}\|_2^2$;

In the nonlinear mapping, we use again the k-nearest neighbor relation to the testing instance x. We map x into g by the average point of its k-nearest neighbors in the training instances.

$$g = \underset{g'}{\operatorname{argmin}} \frac{1}{K} \sum_{i \in \mathcal{N}_k(x)} \|g' - g^{(i)}\|_2^2 = \frac{1}{K} \sum_{i \in \mathcal{N}_k(x)} g^{(i)}.$$

Since the objective function (2) is solved by Laplacian Eigen Map [1], we name the proposed method **Multi-Label** classification using **Laplacian Eigen Map** (shortly, **MLLEM**). The combined pseudo-code of **MLLEM-L** (for linear mapping of a testing instance) and **MLLEM-NL** (for nonlinear mapping of a test instance) is described in Algorithms 1 and 2.

2.3 Computational Complexity

The training procedure of the proposed algorithm (Algorithm 1) can be divided into two parts. The first part constructs k-nn graphs for both labels and instances (Step 3 and Step 4), in $O(NL^2)$ for labels and in $O(FN^2)$ for instances, respectively. The second part solves the generalized eigen problem (Step 6). This part takes $O((N + L)^3)$. However, it is known that this complexity can be largely reduced when the matrix $\mathbf{W}$ is sparse and only a small number K of eigen vectors are necessary [9]. Therefore, the complexity of the proposed algorithm can be estimated as $O(NL^2 + FN^2)$. This complexity is the same to those of almost all embedding methods including the compared methods on the experiments.

In the testing phase, the linear embedding needs $O(F^2N)$ for the ridge regression. In contrast, nonlinear embedding needs only $O(FN)$ for each test instance that is faster than linear embedding.

3   Related Work

Label embedding methods for MLC are employed to utilize label-dependency via the low-rank structure of an embedding space. Recently, several methods based on traditional factorizations [4,6,8] and based on regressions with various loss functions [12,13] have been proposed. Canonical Correlation Analysis based method [16] is also one of them. This method conceptually embeds both instance and labels at the same time like the proposed **MLLEM** does. However, it conducts only one-side embedding in the actual classification process. This is because the linear regression after embedding includes the other-side embedding. Although all methods utilizes low-rank structure and succeeded to improve classification accuracy, they are limited to linear transformation.[2] In contrast to these methods, our **MLLEM** utilizes label dependency in a nonlinear way so that it is more flexible for mapping. On the other hand, we have to be careful for overfitting when we use nonlinear mappings. In **MLLEM**, the nonlinear mappings rely only on the similarity measures $\mathbf{W}^{(IL)}$, $\mathbf{W}^{(II)}$ and $\mathbf{W}^{(LL)}$. Therefore, overfitting is limited to some extent.

Bhatia *et al.* proposed linear embedding method for instances [3]. In their embedding, only instance locality on the label space is considered and ML-KNN [14] is conducted on the low-dimensional space. In the sense of using locality, the proposed **MLLEM** is close to theirs, but the proposed **MLLEM** is different from their approach in the sense that label-instance relationship, label-label relationship and instance-instance relationship are all taken into consideration at the same time.

4   Experiments

4.1   Setting

We conducted experiments on thirteen benchmark datasets [11] (Table 1). Each dataset was separated into 67 % of training instances and 33 % of test instances at random. On large datasets (*i.e. delicious*, *bookmarks* and *mediamill*), we sampled randomly 5000 instances (4000 samples for training and 1000 samples for testing) according to [6].

Since all embedding methods return scores of labels, not a label subset, we used Area Under ROC-Curve (AUC) and top-k precision to evaluate the results [13].[3] AUC is a popular criteria to evaluate the ranking of all labels. We used ROC-curve between true-positive rate and false-positive rate for AUC and Top-1 precision [3].[4]

We compared the following three state-of-the-art embedding methods to ours:

1. Low-rank Empirical risk minimization for MLC (**LEML**) [13]
2. Feature-aware Implicit Encoding(**FaIE**) [6]

---

[2] Several methods can utilize kernel regressions instead a liner regression, however, after regression, they linearly transform the latent labels into the original labels. This means that the way of utilizing label dependency is still limited to be linear.

[3] All embedding methods use a threshold to obtain a label subset.

[4] We only show the result of top-1 precision since the ordering was not changed in top-3 or top-5.

Table 1. Dataset used in the experiment. F_{nnz} and L_{nnz} are the average of number non-zero features and labels among instances in the corresponding set, respectively.

Dataset	F	L	Training set			Test set		
			N	F_{nnz}	L_{nnz}	N	F_{nnz}	L_{nnz}
CAL500	68	174	335	67.84	26.09	167	68	25.94
enron	1001	53	387	79.17	3.37	192	93.30	3.33
emotions	72	6	396	71.79	1.88	197	71.71	1.83
birds	258	19	431	158.00	1.10	214	158.40	1.10
genbase	1186	27	442	2.52	1.20	220	2.55	1.34
medical	1449	45	653	13.04	1.24	325	14.10	1.25
scene	294	6	1606	290.54	1.07	801	290.78	1.07
yeast	103	14	1613	103.00	4.21	804	102.99	4.27
corel5k	499	374	3335	8.28	3.53	1665	8.23	3.49
bibtex	1836	159	4933	69.13	2.41	2462	67.71	2.38
delicious	500	983	4000	18.08	19.02	1000	18.71	19.00
bookmarks	2150	208	4000	125.27	2.02	1000	126.17	2.03
mediamill	120	101	4000	120.00	4.38	1000	120.00	4.36

3. Sparse Local Embedding for Extreme Multi-label Classification (**SLEEC**) [3]
4. Proposal with linear embedding (**MLLEM-L**) and nonlinear embedding (**MLLEM-NL**)

The proposed **MLLEM** has five parameters, the number of nearest neighbors $k^{(I)}$ for instances, $k^{(L)}$ for labels, weighting parameters α for $\mathbf{W}^{(II)}$ and β for $\mathbf{W}^{(LL)}$ and the dimension K of the embedding space. On all datasets, we used $K = 20$. All the other parameters were tuned by five-cross validation on training dataset. The code is available at the authors' web site.[5] For **CPLST** and **FaIE**, we set their numbers of dimension for labels space to the 80 % of their numbers of labels following the setting in [3]. The other parameters were tuned as well. We used the implementations provided by the authors [13][6], [6][7]. For **SLEEC**, we set the number of dimension as $K = 100$ following the their setting [3]. We tuned best the number of k-nn and the number of neighborhoods for ML-KNN after embedding. The other parameters such as smoothing parameter in their regression is set to the default setting they used. We used the implementations provided by the authors too.

---

[5] https://dl.dropboxusercontent.com/u/97469461/MLLEM.zip.

[6] http://www.cs.utexas.edu/~rofuyu/exp-codes/leml-icml14-exp.zip.

[7] The code is available at the authors' site (https://sites.google.com/site/linzijia72/).

4.2   Results

Table 2 shows the averaged AUC and Top-1 precision on all thirteen benchmark datasets (the larger, the better). **SLEEC** was the best in AUC. The proposed **MLLEM** follows. In Top-1 **MLLEM** was the best, especially in relatively small-scale datasets. This difference is possibly explained from the difference between objective functions of **MLLEM**. **MLLEM** ignores the distance between two objects which do not have a local similarity relation to each other. Therefore, occasionally, such two objects are placed closely in the embedding space in spite that they are not similar. This affects the result measured by AUC which takes overall ranking into evaluation. On the other hand, on top-k labels, such an indicated bad effect problem seldom occurs.

Table 2. Results

Dataset	Averaged-AUC					Top-1 precision				
	LEML	FaIE	SLEEC	MLLEM		LEML	FaIE	SLEEC	MLLEM	
				L	NL				L	NL
CAL500	.7592	.7738	.8187	.8206	**.8211**	.7365	.7605	.8503	**.8623**	.8563
enron	.7929	.7710	.8748	.8857	**.8902**	.6458	.5260	.5677	.6146	**.7292**
emotions	.8073	.8106	.8362	**.8372**	.7565	7107	.7107	.7107	**.7563**	.6599
birds	.7604	.7240	.7394	**.8265**	.7240	.2850	.3271	.2617	**.3645**	.2196
genbase	.9944	.9950	.9950	**.9985**	.9979	1.000	1.000	1.000	1.000	1.000
medical	.9620	.9502	**.9736**	.9697	.9342	.8246	.8245	.8246	**.8646**	.7446
scene	.8852	.8683	**.9134**	.9114	.9074	.6667	.6442	.7378	.7466	**.7528**
yeast	.6318	.8188	**.8299**	.8288	.8247	.6457	.7550	.7836	**.7886**	.7724
corel5k	.7805	.7862	**.8906**	.8700	.8717	.3495	.3610	**.3928**	.3538	.2505
bibtex	.8895	.8868	**.9480**	.9066	.7563	**.6409**	.6186	.6255	.5902	.3298
delicious	.8097	.8768	**.8921**	.8463	.8416	.5940	.6274	**.6830**	.6291	.6253
bookmarks	.8000	.7551	**.8662**	.8023	.7700	.3370	.3040	**.4110**	.3020	.2950
mediamill	.9451	.9425	**.9475**	.9286	.9283	.8380	**.8440**	.8210	.8010	.7801

Table 3. Training time ($K = 20$).

Dataset	LEML	FaIE	SLEEC	MLLEM
CAL500	2.62	0.36	0.48	0.61
bibtex	3.62	5.51	9.44	12.72
enron	3.53	0.39	0.49	0.43
genbase	1.00	0.32	0.36	0.66
medical	2.78	0.38	1.21	0.96
corel5k	6.24	1.46	5.55	11.62
delicious	3.53	2.91	10.54	19.15
bookmarks	7.19	5.56	7.86	8.40
mediamill	8.14	3.92	8.69	7.66

MLLEM was superior to **SLEEC** when the number N of instances is relatively small even in AUC. This is probably because **SLEEC** considers only locality of instances (instance-instance relationship). When the number of available instances is limited, instance-instance relationship is not enough to capture the relationship between features and labels. From this viewpoint, **MLLEM** is the best choice for small- to medium-sample size problem (Table 3).

5   Discussion

Since the proposed **MLLEM** uses Laplacian eigen map for the nonlinear embedding, there are several ways to increase the scalability of **MLLEM** such as an incremental method [2], Nyström approximation or column sampling [9] and efficient k-NN constructors [7]. Note that the framework used in **MLLEM** is very general. It comes from the freedom of choice of matrix **W** (*e.g.*, using a heat kernel). It is also able to handle categorical features. This generalization is not shared with **SLEEC** [3].

6   Conclusion

In this paper, we have proposed an embedding based approach for multi-label classification. The proposed algorithm takes into consideration three relationships: label-instance relationship, label-label relationship and instance-instance relationship, and realized a nonlinear mapping. All these three relationships are preserved in the embedded low-dimensional space as the closeness between instances and individual labels. We have shown that the algorithm is useful to visualize instances and labels at the same time, which helps us to understand a given multi-label problem, especially, how strongly those labels are related to each other. Linear and nonlinear mapping have been also proposed for classification. On experiments, the proposed algorithm outperformed the other state-of-the-art methods in small-scale datasets in sample number.

Acknowledgment. We would like to thank Dr. Kush Bhatia for providing the code of SLEEC and large-scale datasets. This work was partially supported by JSPS KAKENHI Grant Number 14J01495 and 15H02719.

References

1. Belkin, M., Niyogi, P.: Laplacian eigenmaps for dimensionality reduction and data representation. Neural Comput. **15**(6), 1373–1396 (2003)
2. Bengio, Y., Paiement, J.F., Vincent, P., Delalleau, O., Le Roux, N., Ouimet, M.: Out-of-sample extensions for LLE, Isomap, MDS, Eigenmaps, and spectral clustering. Adv. Neural Inf. Process. Syst. **16**, 177–184 (2004)
3. Bhatia, K., Jain, H., Kar, P., Varma, M., Jain, P.: Sparse local embeddings for extreme multi-label classification. Adv. Neural Inf. Process. Syst. **28**, 730–738 (2015)

4. Chen, Y.N., Lin, H.T.: Feature-aware label space dimension reduction for multi-label classification. In: Advances in Neural Information Processing Systems, pp. 1529–1537 (2012)
5. Hsu, D., Kakade, S., Langford, J., Zhang, T.: Multi-label prediction via compressed sensing. Adv. Neural Inf. Process. Syst. **22**, 772–780 (2009)
6. Lin, Z., Ding, G., Hu, M., Wang, J.: Multi-label classification via feature-aware implicit label space encoding. In: Proceedings of the 31st International Conference on Machine Learning, pp. 325–333 (2014)
7. Liu, T., Moore, A.W., Yang, K., Gray, A.G.: An investigation of practical approximate nearest neighbor algorithms. In: Advances in Neural Information Processing Systems, pp. 825–832 (2004)
8. Tai, F., Lin, H.T.: Multilabel classification with principal label space transformation. Neural Comput. **24**(9), 2508–2542 (2012)
9. Talwalkar, A., Kumar, S., Rowley, H.: Large-scale manifold learning. In: IEEE Conference on Computer Vision and Pattern Recognition, pp. 1–8. IEEE (2008)
10. Tsoumakas, G., Katakis, I.: Multi-label classification: an overview. Department of Informatics, Aristotle University of Thessaloniki, Greece (2006)
11. Tsoumakas, G., Spyromitros-Xioufis, E., Vilcek, J., Vlahavas, I.: Mulan: a Java library for multi-label learning. J. Mach. Learn. Res. **12**, 2411–2414 (2011)
12. Weston, J., Bengio, S., Usunier, N.: Wsabie: scaling up to large vocabulary image annotation. IJCAI **11**, 2764–2770 (2011)
13. Yu, H.f., Jain, P., Kar, P., Dhillon, I.: Large-scale multi-label learning with missing labels. In: Proceedings of the 31st International Conference on Machine Learning, pp. 593–601 (2014)
14. Zhang, M.L., Zhou, Z.H.: ML-KNN: a lazy learning approach to multi-label learning. Pattern Recogn. **40**(7), 2038–2048 (2007)
15. Zhang, M.L., Zhou, Z.H.: A review on multi-label learning algorithms. IEEE Trans. Knowl. Data Eng. **26**(8), 1819–1837 (2014)
16. Zhang, Y., Schneider, J.G.: Multi-label output codes using canonical correlation analysis. In: International Conference on Artificial Intelligence and Statistics, pp. 873–882 (2011)

Least-Squares Regression with Unitary Constraints for Network Behaviour Classification

Antonio Robles-Kelly[1,2(✉)]

[1] DATA61 - CSIRO, Tower A, 7 London Circuit, Canberra ACT 2601, Australia
antonio.robles-kelly@data61.csiro.au
[2] College of Engineering and Computer Science, Australian National University,
Canberra, Australia

Abstract. In this paper, we propose a least-squares regression method [2] with unitary constraints with applications to classification and recognition. To do this, we employ a kernel to map the input instances to a feature space on a sphere. In a similar fashion, we view the labels associated with the training data as points which have been mapped onto a Stiefel manifold using random rotations. In this manner, the least-squares problem becomes that of finding the span and kernel parameter matrices that minimise the distance between the embedded labels and the instances on the Stiefel manifold under consideration. We show the effectiveness of our approach as compared to alternatives elsewhere in the literature for classification on synthetic data and network behaviour log data, where we present results on attack identification and network status prediction.

1 Introduction

The literature on classification is vast, comprising approaches such as Linear Discriminant Analysis, Projection Pursuit and kernel methods [20]. All these algorithms treat the input data as vectors in high dimensional spaces and look for linear or non-linear mappings to the feature space, often with reduced dimensionality, leading to statistically optimal solutions.

For computer network behaviour, misuse strategies are often use to guarantee security and privacy, whereby breaches and intrusions are recognised using network states or traffic patterns. This can be viewed as a classification over the network behaviour data. Hence, classification methods, such as support vector machines (SVMs) [19] have been traditionally adapted and applied to network behaviour analysis.

One of the most popular classification methods is Linear Discriminant Analysis (LDA) [16]. LDA is a classical method for linear dimensionality reduction which can utilise label information for purposes of learning a lower dimensional space representation suitable for feature extraction, supervised learning and classification. Both LDA and the closely related Fisher's Linear Discriminant (FDA) [13] are concerned with learning the optimal projection direction for binary classes. Further, various extensions and improvements to LDA have been

© Springer International Publishing AG 2016
A. Robles-Kelly et al. (Eds.): S+SSPR 2016, LNCS 10029, pp. 26–36, 2016.
DOI: 10.1007/978-3-319-49055-7_3

proposed in the literature. For instance, non-parametric discriminant analysis (NDA) [16] incorporates boundary information into between-class scatter. Mika *et al.* [23] and Boudat and Anour [3] have proposed kernel versions of LDA that can cope with severe non-linearity of the sample set. In a related development, Maximum Margin Criterion (MMC) [21] employs an optimisation procedure whose constraint is not dependent on the non-singularity of the within-class scatter matrix.

Indeed, these methods are closely related to manifold learning techniques [22], where a distance metric is used to find a lower dimensional embedding of the data under study. For instance, ISOMAP [29], focuses on the recovery of a lower-dimensional embedding of the data which is quasi-isometric in nature. Related algorithms include locally linear embedding [25], which is a variant of PCA that restricts the complexity of the input data using a nearest neighbour graph and the Laplacian eigenmap method [4] which constructs an adjacency weight matrix for the data-points and projects the data onto the principal eigenvectors of the associated Laplacian matrix. In [15], a pairwise discriminant analysis method with unitary constraints is minimised making use of unconstrained optimisation over a Grassmann manifold.

Note that LDA, PCA and their kernel and regularised versions correspond to particular cases of least-squares reduced rank regression [2, 10]. Here, we present a least-squares regression method for classification. We note that the classification labels can be embedded in a Stiefel manifold making use of a matrix of random rotations. This treatment allows us to arrive to a formulation in which the feature vector for an instance to be classified is mapped onto the Stiefel manifold making use of a kernel mapping. The matching of these features and the labels mapped onto the Stiefel manifold can then be effected via a regularised least-squares formulation. This yields a least-squares approach where the aim of computation are given by the parameter matrix for the kernel function and the span of the resulting vector on the Stiefel manifold.

2 Stiefel Manifolds and Least Squares

In this section, we provide some background on Stiefel manifolds [1, 6, 31] and review the concepts used throughout the paper. With these formalisms, we introduce, later on in the paper, our least-squares approach. To this end, we depart from a linear regression setting and view the jth coefficient $\mathbf{z}_j$ of target vector $\mathbf{z}$ arising from the label $\mathbf{y}$ corresponding to the instance $\mathbf{x}$ in the training set Ψ as a linear combination of k kernel functions $K(\mathbf{A}_i, \mathbf{x})$ such that

$$\mathbf{z}_j = \sum_{i=1}^{k} b_{j,i} K(\mathbf{A}_i, \mathbf{x}) + \epsilon_{\mathbf{x}} \tag{1}$$

where $\epsilon_{\mathbf{x}}$ is the residual for $\mathbf{x}$, $\mathbf{A}_i$ are the parameters for the ith kernel function K and $b_{j,i}$ are mixture coefficients.

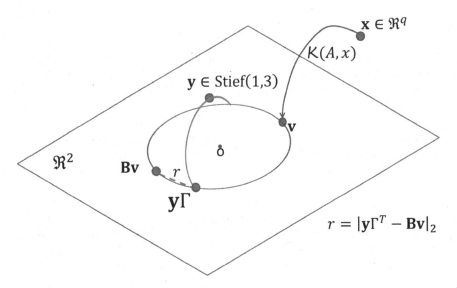

Fig. 1. Graphical illustration of our least-squares approach. In the figure, we have denoted the origin of the example $\Re^2$ space as o and used the notation introduced in Sect. 2.

Here, we aim at minimising the residual by considering the case where the targets $\mathbf{z}$ are p-dimensional vectors in a real Stiefel manifold [18]. Recall that a Stiefel manifold Stief(p, n) is the set of all orthonormal p-dimensional frames in $\Re^n$, where n and p be positive integers such that $p \leq n$. This formalism is important since, with these ingredients, we can define the span Span$(\mathbf{v})$, where $\mathbf{v} \in$ Stief(p, n), in a straightforward manner as a linear operation on a p-dimensional unit vector in $\Re^n$. Moreover, in practice, we can consider the span Span$(\mathbf{v})$ to be a matrix $\mathbf{B}$ and denote the set of vectors spanned by all the possible choices of $\mathbf{B}$ as $\mathcal{V}_\mathbf{B}$.

To formulate our least-squares cost function, we consider the Stiefel manifold Stief$(1, n)$, *i.e.* the manifold comprised by the 1-frames in $\Re^n$, and view it as the frame of a unit ball of dimension n. In Fig. 1, we use the case where $n = 2$ to elaborate further on the approach taken here and provide a better intuition of the concepts presented previously in the section. In the figure, we show a vector $\mathbf{v}$ in the manifold Stief$(1, 2)$. Note that we have used the kernel map $\mathcal{K} : \Re^n \times \Re^m \times \Re^m \to \mathcal{S}^n$ to embed a vector $\mathbf{x}$ in $\Re^m$ onto the unit n-sphere $\mathcal{S}^n$ such that $\mathbf{v} = \mathcal{K}(\mathbf{A}, \mathbf{x})$, where $\mathbf{A}$ is a set of parameters in $\Re^n \times \Re^m$. It is worth noting in passing that we have chosen to define the parameters of the function $\mathcal{K}$ in this manner to be consistent with the linear case in Eq. 1, *i.e.* $\mathbf{v} = \frac{\mathbf{A}\mathbf{x}}{|\mathbf{A}\mathbf{x}|_2}$ where $\mathbf{A}$ is a matrix whose i^{th} column is $\mathbf{A}_i$ and $|\cdot|_2$ denotes the L-2 norm.

Later on, we comment further on our choice of function $\mathcal{K}$. For now, we continue our discussion on Figure 1. Note that, in the figure, we also show a label vector $\mathbf{y} \in$ Stief$(1, 3)$ and the target induced by the matrix Γ onto

Stief$(1, 2)$. Here, we focus on matrices Γ that correspond to rotations about the origin, *i.e.* $\Gamma\Gamma^T = \mathbf{I}$, where $\Gamma^{-1} = \Gamma^T$ and $\mathbf{I}$ is the identity matrix. As can be observed from the figure, this treatment leads in a straightforward manner to a geometric interpretation of the distances in the ambient space and the consequent formulation of the least-squares cost function. With the notation and concepts introduced above, we can write the cost function as follows

$$f = \sum_{\mathbf{x} \in \Psi} \epsilon_{\mathbf{x}}^2 = \sum_{\mathbf{x} \in \Psi} |\mathbf{y}\Gamma - \mathbf{B}\mathcal{K}(\mathbf{A}, \mathbf{x})|_2^2 \tag{2}$$

where the target for the instance $\mathbf{x}$ is $\mathbf{z} = \mathbf{y}\Gamma$ and the entries of the matrix $\mathbf{B}$ are given by $b_{i,j}$.

3 Implementation and Discussion

3.1 Optimisation and Initialisation

As a result, the aim of computation becomes that of recovering $\mathbf{A}$ and $\mathbf{B}$ such that f is minimum subject to $|\mathbf{B}\mathcal{K}(\mathbf{A}, \mathbf{x})|_2 = 1$, where $|\cdot|_2$ denotes the L-2 norm. Note that, the minimisation of the cost function in Eq. 2 can be rewritten as a maximisation by using the inner product and adding an L-2 norm term so as to induce sparsity. Thus, here we solve

$$\Theta = \max_{\mathbf{A}, \mathbf{B}} \left\{ \sum_{\mathbf{x} \in \Psi} \langle \mathbf{y}\Gamma, \mathbf{B}\mathcal{K}(\mathbf{A}, \mathbf{x}) \rangle - \frac{\lambda}{|\Psi|} |\mathbf{B}|_2^2 \right\} \tag{3}$$

where λ is a constant that controls the influence of the regularisation term in the maximisation process, as before $|\cdot|_2$ denotes the matrix L-2 norm and we have used Θ as a shorthand for the set of parameters comprised by the maximisers $\mathbf{A}^*$ and $\mathbf{B}^*$ of the cost function over the space of solutions for $\mathbf{A}$ and $\mathbf{B}$, *i.e.* $\Theta = \{\mathbf{A}^*, \mathbf{B}^*\}$.

In practice, the ith coefficient of the label vector for the feature $\mathbf{x}$ can be defined as

$$\mathbf{y}_i = \begin{cases} 1 & \text{if } \mathbf{x} \text{ belongs to the class indexed } i \text{ in } \Psi \\ 0 & \text{otherwise} \end{cases} \tag{4}$$

whereas the rotation matrix Γ can be computed exploiting the group structure of all rotations about the origin. We do this by using the algorithm in [11] as follows. We commence by constructing a 2×2 rotation matrix, *i.e.* a two-dimensional rotation. With this matrix in hand, we recursively increase the dimensionality from q to $q + 1$ by augmenting the current $q \times q$ rotation matrix making use of the embedding of a vector in an $q+1$ dimensional sphere. The idea underpinning this procedure is that a vector on an q-dimensional sphere should correspond to the embedding induced by the rotation matrix operating on an analogous vector on a sphere in $q + 1$ dimensions.

For the optimisation, we employ a coordinate descent scheme [14]. The idea is to optimise the multivariable cost function in Eq. 3 by maximising it along one coordinate direction at a time while keeping all other coordinates fixed. Thus, the multivariable problem is in fact solved by computing a sequence of two optimisation subproblems, *i.e.* one for $\mathbf{A}$ and another one for $\mathbf{B}$. For each subproblem, we perform line search optimisation in the steepest gradient descent direction.

Our choice of this optimisation scheme resides in the notion that, when maximising in the coordinate direction spanned by $\mathbf{A}$, the problem in hand turns into a standard non-linear least-squares one which can be solved, in a straightforward manner, using an interior point trust region method [8]. Once the matrix $\mathbf{A}^*$ for the current iteration is in hand, computing the iterate of $\mathbf{B}^*$ can be done by viewing the problem as a standard regularised linear least squares one. We iterate until convergence is reached, which is when the norm difference between $\mathbf{A}^*$ and $\mathbf{B}^*$ for consecutive iterations is less or equal to a predefined tolerance τ.

In practice, for our gradient descent approach above, we require an initial estimate of $\mathbf{A}$ prior to the recovery of $\mathbf{B}$ for our first iteration. For our initial estimate of $\mathbf{A}$, we have used randomly generated values in the interval $[-1, 1]$ and normalised the columns of $\mathbf{A}$ to unity.

3.2 Classification

Note that, so far, we have not constrained the dimensionality of the feature vector or the vectors given by $\mathbf{y}\Gamma$. Its clear, however, that if the feature vector dimensionality is m and that of the vector $\mathbf{A}\mathbf{x}$ is n, the matrix $\mathbf{A}$ should be of size $n \times m$. Similarly, for a training set containing k classes, the vector $\mathbf{y}$ is expected to be of length k and, hence, the matrix Γ is of size $k \times k$. Further, since we have imposed the constraint $\mathbf{B}\mathbf{B}^T = \mathbf{I}$, $\mathbf{B}$ should also be $k \times k$.

Nonetheless at first glance this fixes the dimensionality of the matrices $\mathbf{A}$ and $\mathbf{B}$ to correspond to $k \times m$ and $k \times k$, respectively, note the length of the vector $\mathbf{y}$ can be set to any length $m \geq k$. This implies that those coefficients whose indexes are greater than k will be identical to zero. The main advantage of this treatment strives in the fact that the dimensionality of both matrices, $\mathbf{A}$ and $\mathbf{B}$, can be rendered independent of the number of classes.

To classify a testing instance $\hat{\mathbf{x}}$, we employ a nearest-neighbour classifier [12] with respect to the vectors $\mathbf{y}\Gamma$ as yielded by $\mathbf{B}\mathcal{K}(\mathbf{A}, \hat{\mathbf{x}})$. This is consistent with the notion that the cost function being minimised at training is, in effect, the Euclidean distance of the feature vectors with respect to their label vectors embedded in the Stiefel manifold under consideration. For the kernel function, here we use both, a Radial Basis function (RBF) [5] and a linear kernel. For the linear kernel, the ith element of the vector $\mathcal{K}(\mathbf{A}, \mathbf{x})$ is given by

$$K_L(\mathbf{A}_i, \mathbf{x}) = \frac{\langle \mathbf{A}_i, \mathbf{x} \rangle}{|\mathbf{A}\mathbf{x}|_2} \qquad (5)$$

where $\mathbf{A}_i$ denotes the ith column of the matrix $\mathbf{A}$. Similarly, for a matrix $\mathbf{A}$ with n columns the RBF kernel is given by

$$K_{RBF}(\mathbf{A}_i, \mathbf{x}) = \frac{\exp\left(-\rho|\mathbf{A}_i - \mathbf{x}|_2^2\right)}{\left|\sum_{i=1}^{n} \exp\left(-\rho|\mathbf{A}_i - \mathbf{x}|_2^2\right)\right|_2} \qquad (6)$$

where, as before, $| \cdot |_2$ denotes the vector L-2 norm, ρ^{-1} is the bandwidth of the kernel function and $\langle \cdot, \cdot \rangle$ accounts for the dot product.

3.3 Relation to Neural Networks and Support Vector Machines

As mentioned above, here we use RBF and linear kernels. These are kernels commonly used in neural networks [17] and Support Vector Machines (SVMs) [9]. Further, the nearest neighbour classifier as used here is reminiscent of the mean-squared error soft-max linear classifier in feed forward neural networks [24]. Despite these apparent similarities, there are a number of differences between our method and single-layer neural networks and SVMs.

Regarding SVMs, we commence by noting that, for binary classification, the dual form of the support vector machine depends on a single set of parameters, *i.e.* the alpha-weights, where the dividing hyperplane is given by $\langle \mathbf{w}, \mathbf{x} \rangle + b = 0$ where b is the intersect of the hyperplane and $\mathbf{w}$ is its normal vector.

Amongst those variants of the SVMs elsewhere in the literature, the least-squares support vector machine [30] imposes a quadratic penalty function which is probably the one that most closely resembles our formulation. Here, we limit our discussion to the linear case. We do this without any loss of generality and for the sake of clarity. Further, extending the following to the "kernelised" setting is a straightforward task. For non-linearly separable training data, the minimisation in hand for the linear least-squares SVM is given by [27].

$$\min_{\mathbf{w}} \left\{ \langle \mathbf{w}, \mathbf{w} \rangle + \varsigma \sum_{\mathbf{x} \in \Psi} \left(y - (\langle \mathbf{w}, \mathbf{x} \rangle + b) \right)^2 \right\} \tag{7}$$

where $\langle \cdot, \cdot \rangle$ denotes, as usual, the dot product, $y \in -1, 1$ is the instance label and the hyperplane normal vector is

$$\mathbf{w} = \sum_{\mathbf{x} \in \Psi} y \alpha \mathbf{x} \tag{8}$$

ς is a hyperparameter and α is a positive scalar, *i.e.* a slack variable, such that

$$\sum_{\mathbf{x} \in \Psi} y \alpha = 0 \tag{9}$$

In the equations above, we have written y instead of $\mathbf{y}$ to reflect the fact that $\mathbf{y}$ is a vector, which in the binary case where the length of the vector equals the number of classes in the training set is given by $[1, 0]^T$ or $[0, 1]^T$ whereas y is a scalar in $\{-1, 1\}$. In addition to this, from inspection, its clear the two cost functions, ours in Eq. 3 and that of the least-squares SVM in Eq. 7 are quite distinct. Also, note that there is no constraint on the number of support vectors in Eq. 7, *i.e.* the number of feature vectors $\mathbf{x}$ contributing to the hyperplane normal vector $\mathbf{w}$. This contrasts with our method, where the complexity of evaluating the kernel is given by the number of rows in $\mathbf{A}$. This is a major difference with respect to SVMs, where non-linear kernels can be quite cumbersome to evaluate when the number of support vectors is large.

Now we turn our attention to the multiple class case and comment on the similarities of our method with single-layer feed forward networks [17]. Note that the matrix **A** can be viewed as the weights of the hidden layer of the network and the **B** as the coefficients of the all-connected layer. That said, there are two major differences. The first of these pertains the cost function itself, whereby, there is no evident analogue of our rotation matrix Γ in neural networks. The second relates to the optimisation scheme used here. Recall that, in neural networks, the training is often achieved by back propagation [26]. In contrast, our approach is an iterative coordinate descent one which imposes unitary constraints on $\mathbf{B}\mathcal{K}(\mathbf{A}, \mathbf{x})$. For our approach, the final classification step via a nearest-neighbour classifier is akin to the classification layer often used in neural networks. This opens-up the possibility of using well known techniques on soft-max and entropy-based classification layers as an alternative to our nearest-neighbour classifier.

4 Experiments

In this section, we show classification results yielded by our approach and provide comparison with alternative methods elsewhere in the literature. To this end, we commence by presenting qualitative results using synthetic data and, later on in the section, we focus our attention on intrusion detection using a widely available network behaviour data set.

4.1 Synthetic Data

To illustrate the utility of our method for classification, we have used two synthetically generated point clouds in 2D. The first of these accounts for a binary classification setting comprising 750 points arranged in an annulus where the centre corresponds to one class and the periphery to another one. The second of these data sets is given by 1500 points describing two annuli distributed into four well defined clusters.

For both points clouds, we have used an **A** matrix with 10 rows, *i.e.* $n = 10$, and a rotation Γ and **B** matrices of size 10×10. For purposes of comparison we have used a least-squares SVM (LS-SVM) [30] for binary classification and, for the multiclass annuli data, an stochastic feed forward neural network [28] with a single hidden layer composed of 10 neurons.

In Fig. 2, we show the results yielded by our method and the alternatives. In the left-hand column, we show the input point clouds. In the middle and right-hand columns we show the results yielded by our method and the alternatives, respectively. Note that, from the figure, we can note that our method delivers very similar results to those obtained by the LS-SVM. Nonetheless these classification boundaries are very similar, the LS-SVM yields 116 support vectors. This contrasts with our approach, which employs a matrix **A** with 10 rows. Thus, for

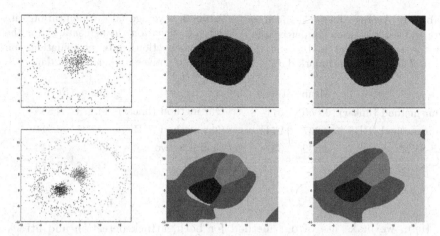

Fig. 2. Classification results on synthetic data. Left-hand column: input point clouds; Middle column: results yielded by our approach; Right-hand column: results delivered by a LS-SVM (top panel) and an RBF neural network (bottom panel).

testing, the evaluation of the kernel for our method is approximately an order of magnitude less computationally costly. Regarding multiclass classification, note the boundaries yielded by the neural network are smoother than those obtained by our method. As a result, the region comprising the small group of points next to the centre of the second annulus are completely misclassified. In contrast, the decision boundary yielded by our algorithm does classify these points correctly.

4.2 Network Behaviour Classification

We now turn our attention to the task of cyber attack discrimination in network behaviour data. Here, we view this task as a classification one where the aim is to discern attacks from legitimate connections in the network. For our experiments, we have used the sanitized Kyoto University benchmark data[1]. These benchmark dataset is comprised of the traffic net logs for the Kyoto University's honeypots from November 2006 to August 2009[2]. The data contains 24 features including the attack label and the connection state identifier and is arranged by day per calendar month.

For our experiments, we have used the data corresponding to November 2006 and trained our method on the connections established on the even days of the month. We have then tested our method and the alternatives on the data corresponding to each of the odd days. For our method, we have used a matrix **A** with 30 rows and, for the neural network, whenever applicable, have set to 30 the number of hidden neurons.

---

[1] The data is accessible at http://www.takakura.com/Kyoto_data/.

[2] The description of the data can be found at http://www.takakura.com/Kyoto_data/ BenchmarkData-Description-v5.pdf.

Table 1. Average false classification rate and std. deviation (in percentage) for the attack detection task on the Kyoto University benchmark data.

	Mean	St. dev
Our method (linear)	1.056	1.011
Our method (RBF)	0.967	0.974
LS-SVM	4.839	3.331

Table 2. Average false classification rate and std. deviation (in percentage) for the attack-connection state combinations on the Kyoto University benchmark data.

	Mean	St. dev
Our method (linear)	14.273	3.893
Our method (RBF)	9.027	2.941
Neural network	11.746	4.342
LDA	33.656	19.379

Here, we make use of our method for both, attack detection and attack-connection state classification. The former of these tasks is a binary classification one whereas the latter is a multiple class one. The data contains a label entry which indicates whether the connection under study was an attack. It also contains a connection state variable that indicates the status of the connection, generally when it ends. Since the dataset provides 13 different connection states, this yields 26 classes, *i.e.* for each state, there is the possibility of the connection being an attack or not.

In Tables 1 and 2 we show the mean false classification rate and the corresponding std. deviation for our method and the alternatives for both, attack detection and attack-connection state classification. Since our method employs a random initialisation for the matrix **A**, we have effected ten trails and averaged accordingly before computing the mean and std. deviation results shown in the tables. In our experiments, for the attack detection, we have again compared our results to those yielded by the LS-SVM. For the attack-connection state classification task, we have compared our results with those delivered by the neural network in [28] and those yielded by linear discriminant analysis (LDA) [7]. From the tables, its clear that our method outperforms the alternatives. Moreover, when an RBF is used, our method can correctly predict an attack 99.033% of the time and can account for the status of the network at termination of the connection with 90.973% accuracy. It also delivers a lower standard deviation and is quite competitive even when a linear kernel is used.

5   Conclusions

In this paper, we have proposed a least-squares reduced rank regression method on with applications to classification. Departing from the notion that a kernel function can be used to map an input instance to a Steifel manifold, we have used a rotation matrix to pose the problem of recovering the span and kernel parameter matrices in a least-squares minimisation setting. We have discussed the link between our approach and support vector machines and feed forward neural networks and shown qualitative results on synthetic data. We have also performed experiments on a widely available network behaviour dataset and compared our approach to alternatives elsewhere in the literature.

References

1. Absil, P.A., Mahony, R., Sepulchre, R.: Riemannian geometry of Grassmann manifolds with a view on algorithmic computation. Acta Applicandae Math. **80**(2), 199–220 (2004)
2. Anderson, T.W.: Estimating linear restrictions on regression coefficients for multivariate normal distributions. Ann. Math. Stat. **22**(3), 327–351 (1951)
3. Baudat, G., Anouar, F.: Generalized discriminant analysis using a kernel approach. Neural Comput. **12**(10), 2385–2404 (2000)
4. Belkin, M., Niyogi, P.: Laplacian eigenmaps and spectral techniques for embedding and clustering. In: Neural Information Processing Systems, vol. 14, pp. 634–640 (2002)
5. Bishop, C.M.: Pattern Recognition and Machine Learning. Springer, Heidelberg (2006)
6. Boothby, W.M.: An Introduction to Differentiable Manifolds and Riemannian Geometry. Academic Press, Cambridge (1975)
7. Borg, I., Groenen, P.: Modern Multidimensional Scaling, Theory and Applications. Springer Series in Statistics. Springer, Heidelberg (1997)
8. Coleman, T.F., Li, Y.: An interior, trust region approach for nonlinear minimization subject to bounds. SIAM J. Optim. **6**, 418–445 (1996)
9. Cristianini, N., Shawe-Taylor, J.: An Introduction to Support Vector Machines. Cambridge University Press, Cambridge (2000)
10. De la Torre, F.: A least-squares framework for component analysis. IEEE Trans. Pattern Anal. Mach. Intell. **34**(6), 1041–1055 (2012)
11. Diaconis, P., Shahshahani, M.: The subgroup algorithm for generating uniform random variables. Probab. Eng. Inf. Sci. **1**, 15–32 (1987)
12. Duda, R.O., Hart, P.E.: Pattern Classification. Wiley, Hoboken (2000)
13. Fisher, R.A.: The use of multiple measurements in taxonomic problems. Ann. Eugenics **7**, 179–188 (1936)
14. Friedman, J., Hastie, T., Tibshirani, R.: Regularization paths for generalized linear models via coordinate descent. J. Stat. Softw. **33**(1), 1–22 (2010)
15. Fu, Z., Robles-Kelly, A., Tan, R.T., Caelli, T.: Invariant object material identification via discriminant learning on absorption features. In: Object Tracking and Classification in and Beyond the Visible Spectrum (2006)
16. Fukunaga, K.: Introduction to Statistical Pattern Recognition, 2nd edn. Academic Press, Cambridge (1990)
17. Hassoun, M.H.: Fundamentals of Artificial Neural Networks. MIT Press, Cambridge (1995)
18. James, I.M.: The Topology of Stiefel Manifolds. Cambridge University Press, Cambridge (1976)
19. Khan, L., Awad, M., Thuraisingham, B.: A new intrusion detection system using support vector machines and hierarchical clustering. Int. J. Very Large Data Bases **16**(4), 507–521 (2007)
20. Landgrebe, D.: Hyperspectral image data analysis. IEEE Sig. Process. Mag. **19**, 17–28 (2002)

21. Li, H., Jiang, T., Zhang, K.: Efficient and robust feature extraction by maximum margin criterion. In: Neural Information Processing Systems, vol. 16 (2003)
22. Ma, Y., Fu, Y.: Manifold Learning Theory and Applications. CRC Press, Inc., Boca Raton (2011)
23. Mika, S., Ratsch, G., Weston, J., Scholkopf, B., Muller, K.: Fisher discriminant analysis with kernels. In: IEEE Neural Networks for Signal Processing Workshop, pp. 41–48 (1999)
24. Neal, R.M.: Bayesian Learning for Neural Networks. Springer, New York (1996)
25. Roweis, S.T., Saul, L.K.: Nonlinear dimensionality reduction by locally linear embedding. Science **290**, 2323–2326 (2000)
26. Rumelhart, D.E., Hinton, G.E., Williams, R.J.: Learning representations by back-propagating errors. Nature **323**, 533–536 (1986)
27. Suykens, J.A.K., Van Gestel, T., De Brabanter, J., De Moor, B., Vandewalle, J.: Least Squares Support Vector Machines. World Scientific, Singapore (2002)
28. Tang, Y., Salakhutdinov, R.R.: Learning stochastic feedforward neural networks. In: Advances in Neural Information Processing Systems, pp. 530–538 (2013)
29. Tenenbaum, J.B., de Silva, V., Langford, J.C.: A global geometric framework for nonlinear dimensionality reduction. Science **290**(5500), 2319–2323 (2000)
30. Van Gestel, T., Suykens, J.A.K., De Brabanter, J., Lambrechts, A., De Moor, B., Vandewalle, J.: Bayesian framework for least-squares support vector machine classifiers, Gaussian processes, and kernel fisher discriminant analysis. Neural Comput. **14**(5), 1115–1147 (2002)
31. Wong, Y.C.: Differential geometry of Grassmann manifolds. Proc. Natl. Acad. Sci. United States Am. **57**(3), 589–594 (1967)

Mathematical Aspects of Tensor Subspace Method

Hayato Itoh[1], Atsushi Imiya[2(✉)], and Tomoya Sakai[3]

[1] School of Advanced Integration Science, Chiba University, 1-33 Yayoi-cho,
Inage-ku, Chiba 263-8522, Japan
hayato-itoh@graduate.chiba-u.jp
[2] Institute of Media and Information Technology, Chiba University, Yayoi-cho 1-33,
Inage-ku, Chiba 263-8522, Japan
imiya@faculty.chiba-u.jp
[3] Graduate School of Engineering, Nagasaki University,
Bunkyo-cho 1-14, Nagasaki 852-8521, Japan
tsakai@cis.nagasaki-u.ac.jp

Abstract. The mathematical and computational backgrounds of pattern recognition are the geometries in Hilbert space for functional analysis and applied linear algebra for numerical analysis, respectively. Organs, cells and microstructures in cells dealt with in biomedical image analysis are volumetric data. We are required to process and analyse these data as volumetric data without embedding vector space from the viewpoints of object oriented data analysis. Therefore, sampled values of volumetric data are expressed as three-way array data. These three-way array data are expressed as the third order tensor. This embedding of the data leads us to the construction of subspace method for higher-order tensors expressing multi-way array data.

1 Introduction

The aim of the paper is to clarify mathematical properties of pattern recognition of tensor data from the viewpoints of object oriented data analysis for volumetric data, which dealt with in biomedical image analysis, retrievals and recognition. In traditional pattern recognition, sampled patterns for numerical computation are embedded in an appropriate-dimensional Euclidean space as vectors. The other way is to deal with sampled patterns as three-way array data. These three-way array data are expressed as tensors [6–10] to preserve the linearity of the original pattern space.

The subspace method based on Karhunen-Loève transform is a fundamental technique in pattern recognition. Modern pattern recognition techniques for sampled value of patterns are described using linear algebra for sampled value embedded in vector space. Organs, cells and microstructures in cells dealt with in biomedical image analysis are volumetric data. We are required to process and analyse these data as volumetric data without embedding sampled values in vector space from the viewpoints of object oriented data analysis [5]. We express sampled values of volumetric data as three-way array data. These three-way

© Springer International Publishing AG 2016
A. Robles-Kelly et al. (Eds.): S+SSPR 2016, LNCS 10029, pp. 37–48, 2016.
DOI: 10.1007/978-3-319-49055-7_4

array data are processed as the third order tensor. This expression of data requires to develop subspace method for tensor data.

We derive and clarify the mutual subspace and constrained mutual subspace methods for tensors using tensor PCA based on the Tucker-3 tensor decomposition. The mutual subspace method is stable against geometric perturbation of queries for pattern recognition, since the method assumes that a query is an element of a low-dimensional subspace. Furthermore, since the constrained subspace method eliminates the common parts of subspaces among categories, the method confirms robust recognition against global deformation of queries in Hilbert space. Since the constrained mutual subspace method is a combination of the mutual subspace method and constrained subspace method the method is stable and robust both against geometric perturbations and global deformations of queries for pattern recognition.

2   Pattern Recognition in Vector Space

A volumetric pattern is assumed to be a square integrable function in a linear space and to be defined on a finite support in three-dimensional Euclidean space [1–3] such that $\int_{\Omega} |f|^2 dx < \infty$ for $\Omega \subset \mathbb{R}^3$. Furthermore, we assume $\int_{\Omega} |\nabla f|^2 dx < \infty$ and $\int_{\Omega} tr\{(\nabla\nabla^\top f)^\top (\nabla\nabla^\top f)\} dx < \infty$, where $\nabla\nabla^\top f$ is the Hessian matrix of f. For an orthogonal projection $\boldsymbol{P}_\perp = \boldsymbol{I} - \boldsymbol{P}$, $f^\| = \boldsymbol{P}f$ and $f^\perp = \boldsymbol{P}_\perp f$ are the canonical element and canonical form of f with respect to $\boldsymbol{P}$ and $\boldsymbol{P}_\perp$, respectively. If $\boldsymbol{P}$ is the projection to the space spanned by the constant element, the operation $\boldsymbol{P}_\perp f$ is called the constant canonicalisation. Let $\boldsymbol{P}_i$ be the orthogonal projection to the linear subspace corresponding to the category C_i. For a pattern f, if $|\boldsymbol{P}_{i^*}(f/|f|)| \leq \delta$ for an appropriately small positive number δ, we conclude that $f \in C_{i^*}$.

Setting $\boldsymbol{\delta}$ and ε to be a small vector and a small positive number, we have the relation

$$|f(\boldsymbol{x} + \boldsymbol{\delta}) - (f(\boldsymbol{x}) + \boldsymbol{\delta}^\top \nabla f + \frac{1}{2}\boldsymbol{\delta}^\top (\nabla\nabla^\top f)\boldsymbol{\delta})| < \varepsilon, \qquad (1)$$

for local geometric perturbations. All f, f_x, f_y, f_z, f_{xx}, f_{yy}, f_{zz}, f_{xy}, f_{yz} and f_{zx} are independent, if f is not sinusoidal in each direction. Therefore, Eq. (1) implies that, for a pattern defined on three-dimensional Euclidean space, the local dimensions of a pattern are four and ten, if local geometric perturbations and local bending deformation of the pattern are assumed as local transformations to the pattern. This property of the local dimensionality allows us to establish the mutual subspace method, which deals with a query as a pattern in a subspace.

Setting (f, g) to be the inner product in Hilbert space $\mathfrak{H}$, the relation $|f|^2 = (f, f)$ is satisfied. Let θ be the canonical angle between a pair of linear subspaces L_1 and L_2. Setting $\boldsymbol{P}_1$ and $\boldsymbol{P}_2$ to be the orthogonal projections to L_1 and L_2, respectively, $\cos^2 \theta$ is the maximiser of $(\boldsymbol{P}_1 f, \boldsymbol{P}_2 g)^2$ with respect to the conditions

$|f| = 1$, $|g| = 1$ $P_1 f = f$ and $P_2 g = g$. The relation $\cos^2 \theta = \lambda_{\max}^2$ is satisfied, where $\lambda_{\max}$ is the maximal singular value of $P_2 P_1$.

Since, in mutual subspace method, a query f is expressed by using a set of local bases, we set that Q_f is the orthogonal projection to linear subspace expressing the query f. Then, if the canonical angle between Q_f and P_i satisfies the relation $\angle(Q_f, P_i) < \angle(Q_f, P_i^*)$ for all C_i, we conclude that $f \in C_{i^*}$.

Setting P_i to be the orthogonal projection to linear subspace $\mathcal{L}_i$ corresponding to the category C_i, the orthogonal projection which maximises the criterion $J = \sum_{i=1}^n |QP_i|_2^2$ with respect to the condition $Q^* Q = I$ where Q^* is the conjugate of Q and $|A|$ is the trace norm of the operator A in Hilbert space $\mathfrak{H}$. Though operation Qf removes common part for all categories from f, $(I - Q)f$ preserves essentially significant parts for pattern recognition of f.

For f and g in $\mathfrak{H}$, we define the metric d for $\mu(f)$ and $\mu(g)$, such that $d(\mu(f), \mu(g))$, using an appropriate transform μ from $\mathfrak{H}$ to its subset. Furthermore, using an appropriate mapping Φ, we define a measure

$$s(f, g) = \Phi(d(\mu(f), \mu(g))). \tag{2}$$

If we set $\mu(f) = \frac{f}{|f|}$ and set d and Φ the geodesic distance on the unit sphere in $\mathfrak{H}$ and $\Phi(x) = \cos x$, respectively, $s(f, g)$ becomes the similarity measure based on the angle between f and g. For $f' = f + \delta_f$ and $g' = g + \delta_g$, setting

$$\min(|f|, |g|) = \Lambda, \quad \max(\delta_f, \delta_g) = \Delta, \tag{3}$$

we have the relation

$$\left| \left(\frac{f'}{|f'|}, \frac{g'}{|g'|} \right) - \left(\frac{f}{|f|}, \frac{g}{|g|} \right) \right| = c\frac{\Delta}{\Lambda}, \tag{4}$$

for a positive constant c. Therefore, $s(f, g)$ is stable and robust against perturbations and noises for f and g.

For patterns in $\mathfrak{H}$, we have the following property.

Property 1. For $|f| = 1$ and $|g| = 1$, assuming $|f - g| \leq \frac{1}{3} \cdot \frac{\pi}{2}$ the geodesic distance $\theta = d_S(f, g)$ between f and g satisfies the relation $|\theta - |f - g|| < \varepsilon$ for a positive small number ε.

In traditional pattern recognition, these sampled patterns are embedded in an appropriate-dimensional Euclidean space as vectors. For $x \in \mathbb{R}^n$ and $X \in \mathbb{R}^{n \times n}$, $|x|_2$ and $|X|_F$ are the vector norm and Frobenius norm of x and X, respectively.

Setting the data matrix X to be $X = (f_1, f_2, \cdots, f_m)$ for data vectors $\{f_i\}_{i=1}^m$ in $\mathbb{R}^N$, whose mean is zero, the Karhunen-Loève transform is established by computing $\hat{f}_i = U f_i$ for U which minimises $J_1 = |UX|_F^2$ with the condition $U^\top U = I_N$. The orthogonal matrix U is the minimiser of

$$J_{11} = |UX|_F^2 + \langle (U^\top U - I)\Lambda \rangle \tag{5}$$

where
$$\Lambda = Diag(\lambda_1, \lambda_2, \cdots, \lambda_N) \tag{6}$$

for
$$\lambda_1 \geq \lambda_2 \geq \lambda_2 \geq \cdots \geq \lambda_N \geq 0. \tag{7}$$

The minimiser of Eq. (5) is the solution of the eigenmatrix problem

$$MU = U\Lambda, \ M = XX^\top \tag{8}$$

The row vectors of U are the principal components.

The compression of f_i to a low-dimensional linear subspace is achieved by computing the transform $P_n U f$, where P_n, for $n < N$, is the orthogonal projection such that

$$P_n = \begin{pmatrix} I_n & O \\ O^\top & O \end{pmatrix}. \tag{9}$$

3 Pattern Recognition in Multi-linear Forms

For the triplet of positive integers I_1, I_2 and I_3, the third-order tensor $\mathbb{R}^{I_1 \times I_2 \times I_3}$ is expressed as $\mathcal{X} = ((x_{ijk}))$ Indices i, j and k are called the 1-mode, 2-mode and 3-mode of $\mathcal{X}$, respectively. The tensor space $\mathbb{R}^{I_1 \times I_2 \times I_3}$ is interpreted as the Kronecker product of three vector spaces $\mathbb{R}^{I_1}$, $\mathbb{R}^{I_2}$ and $\mathbb{R}^{I_3}$ such that $\mathbb{R}^{I_1} \otimes \mathbb{R}^{I_2} \otimes \mathbb{R}^{I_3}$. We set $I = \max(I_1, I_2, I_3)$.

For a square integrable function $f(\boldsymbol{x})$, which is zero outside of a finite support Ω in three-dimensional Euclidean space, the sample $Sf(\Delta \mathbf{z})$ for $\mathbf{z} \in \mathbf{Z}^3$ and $|\mathbf{z}|_\infty \leq I$ defines an $I \times I \times I$ three-way array $\mathbf{F}$. To preserve the multi-linearity of the function $f(\boldsymbol{x})$, we deal with the array $\mathbf{F}$ as a third-order tensor $\mathcal{F}$. The operation $vec\mathcal{F}$ derives a vector $\boldsymbol{f} \in \mathbb{R}^{I_{123}}$ for $I_{123} = I_2 \cdot I_2 \cdot I_3$. We can reconstruct f from $\mathcal{F}$ using an interpolation procedure.

For $\mathcal{X}$, the n-mode vectors, $n = 1, 2, 3$, are defined as the I_n-dimensional vectors obtained from $\mathcal{X}$ by varying this index i_n while fixing all the other indices.

The unfolding of $\mathcal{X}$ along the n-mode vectors of $\mathcal{X}$ is defined as matrices such that

$$\mathcal{X}_{(1)} \in \mathbb{R}^{I_1 \times I_{23}}, \ \mathcal{X}_{(2)} \in \mathbb{R}^{I_2 \times I_{13}}, \ \mathcal{X}_{(3)} \in \mathbb{R}^{I_3 \times I_{12}} \tag{10}$$

for $I_{12} = I_1 \cdot I_2$, $I_{23} = I.I_3$ and $I_{13} = I_1 \cdot I_3$, where the column vectors of $\mathcal{X}_{(j)}$ are the j-mode vectors of $\mathcal{X}$ for $i = 1, 2, 3$. We express the j-mode unfolding of $\mathcal{X}_i$ as $\mathcal{X}_{i,(j)}$.

For matrices

$$\boldsymbol{U} = ((u_{ii'})) \in \mathbb{R}^{I_1 \times I_1}, \ \boldsymbol{V} = ((v_{jj'})) \in \mathbb{R}^{I_2 \times I_2}, \ \boldsymbol{W} = ((w_{kk'})) \in \mathbb{R}^{I_3 \times I_3}, \tag{11}$$

the n-mode products for $n = 1, 2, 3$ of a tensor $\mathcal{X}$ are the tensors with entries

$$
(\mathcal{X} \times_1 U)_{ijk} = \sum_{i'=1}^{I_1} x_{i'jk} u_{i'i},
$$

$$
(\mathcal{X} \times_2 V)_{ijk} = \sum_{j'=1}^{I_2} x_{ij'k} v_{j'j}, \tag{12}
$$

$$
(\mathcal{X} \times_3 W)_{ijk} = \sum_{k'=1}^{I_3} x_{ijk'} w_{k'k},
$$

where $(\mathcal{X})_{ijk} = x_{ijk}$ is the ijk-th element of the tensor $\mathcal{X}$. The inner product of two tensors $\mathcal{X}$ and $\mathcal{Y}$ in $\mathbb{R}^{I_1 \times I_2 \times I_3}$ is

$$
\langle \mathcal{X}, \mathcal{Y} \rangle = \sum_{i=1}^{I_1} \sum_{j=1}^{I_2} \sum_{k=1}^{I_3} x_{ijk} y_{ijk}. \tag{13}
$$

Using this inner product, we have the Frobenius norm of a tensor $\mathcal{X}$ as $|\mathcal{X}|_F = \sqrt{\langle \mathcal{X}, \mathcal{X} \rangle}$. The Frobenius norm $|\mathcal{X}|_F$ of the tensor $\mathcal{X}$ satisfies the relation $|\mathcal{X}|_F = |f|_2$, where $|f|_2$ is the Euclidean norm of the vector f.

To project a tensor $\mathcal{X}$ in $\mathbb{R}^{I_1 \times I_2 \times I_3}$ to the tensor $\mathcal{Y}$ in a lower-dimensional tensor space $\mathbb{R}^{P_1 \times P_2 \times P_3}$, where $P_n \leq I_n$, three projection matrices $\{U^{(n)}\}_{n=1}^3$, for $U^{(n)} \in \mathbb{R}^{I_n \times P_n}$ are required for $n = 1, 2, 3$. Using these three projection matrices, we have the tensor orthogonal projection such that

$$
\mathcal{Y} = \mathcal{X} \times_1 U^{(1)\top} \times_2 U^{(2)\top} \times_3 U^{(3)\top}. \tag{14}
$$

This projection is established in three steps, where in each step, each n-mode vector is projected to a P_n-dimensional space by $U^{(n)}$ for $n = 1, 2, 3$.

For a collection of matrices $\{F_i\}_{i=1}^N \in \mathbb{R}^{m \times n}$ satisfying $\mathrm{E}_i(F_i) = 0$, the orthogonal-projection-based data reduction $\hat{F}_i = U^\top F_i V$ is performed by maximising

$$
J_2(U, V) = \mathrm{E}_i \left(|U \hat{F}_i V^\top|_F^2 \right) \tag{15}
$$

with respect to the conditions $U^\top U = I_m$ and $V^\top V = I_n$. The solutions are the minimiser of the Euler-Lagrange equation

$$
J_{22}(U, V) = \mathrm{E} \left(|U \hat{F}_i V^\top|_F^2 \right) + \langle (I_m - U^\top U), \Sigma \rangle + \langle (I_n - V^\top V), \Lambda \rangle \tag{16}
$$

For diagonal matrices Λ and Σ.

Setting

$$
\frac{1}{N} \sum_{i=1}^N F_i^\top F_i = M, \quad \frac{1}{N} \sum_{i=1}^N F_i F_i^\top = N, \tag{17}
$$

U and V are the solutions of the eigendecomposition problems

$$
MV = V\Lambda, \quad NU = U\Sigma, \tag{18}
$$

where $\Sigma \in \mathbb{R}^{m \times m}$ and $\Lambda \in \mathbb{R}^{n \times n}$ are diagonal matrices satisfying the relationships $\lambda_i = \sigma_i$ for

$$\Sigma = \text{diag}(\sigma_1, \sigma_2, \ldots, \sigma_K, 0, \ldots, 0), \tag{19}$$
$$\Lambda = \text{diag}(\lambda_1, \lambda_2, \ldots, \lambda_K, 0, \ldots, 0). \tag{20}$$

The equation

$$(P_1 U)^\top X (P_2 V) = Y \tag{21}$$

is rewritten as

$$(P_2 V \otimes P_1 U)\text{vec} X = (P_2 \otimes P_1)(V \otimes U)X = P\text{vec}X = \text{vec}Y. \tag{22}$$

Using three projection matrices $U^{(i)}$ for $i = 1, 2, 3$, we have the tensor orthogonal projection for a third-order tensor as

$$\mathcal{Y} = \mathcal{X} \times_1 U^{(1)\top} \times_2 U^{(2)\top} \times_3 U^{(3)\top}. \tag{23}$$

For a collection $\{\mathcal{X}_k\}_{k=1}^m$ of the third-order tensors, the orthogonal-projection-based dimension reduction procedure is achieved by maximising the criterion

$$J_3 = E_k(|\mathcal{X}_k \times_1 U^{(1)} \times_2 U^{(2)} \times_3 U^{(3)}|_F^2) \tag{24}$$

with respect to the conditions $U^{(i)\top} U^{(i)} = I$ for $i = 1, 2, 3$. The Euler-Lagrange equation of this conditional optimisation problem is

$$J_{33}(U^{(1)}, U^{(2)}, U^{(3)}) = E_k(|\mathcal{X}_k \times_1 U^{(1)} \times_2 U^{(2)} \times_3 U^{(3)}|_F^2)$$
$$+ \sum_{i=1}^3 \langle (I - U^{(i)\top} U^{(i)}), \Lambda^{(i)} \rangle. \tag{25}$$

This minimisation problem is solved by the iteration procedure [11].

Algorithm 1. Iterative method for the tensor PCA

Data: $0 < \varepsilon \ll 1$
Data: $\alpha := 0$
Data: For $i = 1, 2, 3$, $U_{(0)}^i := Q^{(i)}$, where $Q^{(i)\top} Q^{(i)} = I$
Result: Orthogonal matrices $U^{(i)}$ for $i = 1, 2, 3$
while $|U_{\alpha+1}^{(i)} - U_\alpha^{(i)}| \gg \varepsilon$ **do**

> maximise $E_i | U_{(\alpha+1)}^{(1)\top} \mathcal{X}_{i1}^\top \mathcal{X}_{i1} U_{(\alpha+1)}^{(1)} |_F^2 + \langle (U_{\alpha+1}^{(1)} U_{(\alpha+1)}^{(1)} - I)\Lambda_{(\alpha)}^{(1)} \rangle$;
> for $\mathcal{X}_{i1} = \mathcal{X}_i \times_2 U_{(\alpha)}^{(2)\top} \times_3 U_{(\alpha)}^{(3)\top}$;
> maximise $E_i | U_{(\alpha+1)}^{(2)\top} \mathcal{X}_{i2}^\top \mathcal{X}_{i2} U_{(\alpha+1)}^{(2)} |_F^2 + \langle (U_{(\alpha+1)}^{(2)} U_{(\alpha+1)}^{(2)} - I)\Lambda_{(\alpha)}^{(2)} \rangle$;
> for $\mathcal{X}_{i2} = \mathcal{X}_i \times_1 U_{(\alpha+1)}^{(1)\top} \times_3 U_{(\alpha)}^{(3)\top}$;
> maximise $E_i | U_{(\alpha+1)}^{(3)\top} \mathcal{X}_{i3}^\top \mathcal{X}_{i3} U_{(\alpha+1)}^{(3)} |_F^2 + \langle (U_{(\alpha+1)}^{(3)} U_{(\alpha+1)}^{(3)} - I)\Lambda_{(\alpha)}^{(3)} \rangle$;
> for $\mathcal{X}_{i3} = \mathcal{X}_i \times_1 U_{(\alpha+1)}^{(1)\top} \times_2 U_{(\alpha+1)}^{(2)\top}$;
> $\alpha := \alpha + 1$

end

As an relaxation of the iteration algorithm, we define the system of optimisation problems

$$J_j = E(|U^{(j)\top}\mathcal{X}_{i,(j)}U^{(j)}|_F^2) + \langle(U^{(j)\top}U^{(j)} - I_j), \Lambda^{(j)}\rangle \tag{26}$$

for $i = 1, 2, 3$, where $\mathcal{X}_{i,(j)}$ is the ith column vector of the unfolding matrix $\mathcal{X}_{(j)}$. These optimisation problems derive a system of eigenmatrix problems

$$M^{(j)}U^{(j)} = U^{(j)}\Lambda^{(j)}, \quad M^{(j)} = \frac{1}{N}\sum_{i=1}^{N}\mathcal{X}_{i,(j)}\mathcal{X}_{i,(j)}^{\top} \tag{27}$$

for $j = 1, 2, 3$.

Setting $P^{(j)}$ to be an orthogonal projection in the linear space $\mathcal{L}(\{u_i^{(j)}\}_{i=1}^{I_j})$ spanned by the column vectors od $U^{(j)}$, data reduction is computed by

$$\mathcal{Y} = \mathcal{X} \times_1 P^{(1)}U^{(1)} \times_2 P^{(2)}U^{(2)} \times_3 P^{(3)}U^{(3)}. \tag{28}$$

This expression is equivalent to the vector form as

$$vec\mathcal{Y} = (P^{(3)} \otimes P^{(2)} \otimes P^{(1)})(U^{(3)} \otimes U^{(2)} \otimes U^{(1)})vec\mathcal{X}, \tag{29}$$

The eigenvalues of eigenmatrices of Tucker-3 orthogonal decomposition satisfy the following theorem.

Theorem 1. *The eigenvalues of* $U = U^{(1)} \otimes U^{(2)} \otimes U^{(3)}$ *define a semi-order.*

(*Proof*). For the eigenvalues $\lambda_i^{(1)}, \lambda_j^{(2)}, \lambda_k^{(3)}$ of the 1-, 2- and 3-modes of tensors, the inequalities $\lambda_i^{(1)}\lambda_j^{(2)}\lambda_k^{(3)} \geq \lambda_i^{(1)}\lambda_j^{(2)}\lambda_{k+1}^{(3)}$, $\lambda_i^{(1)}\lambda_j^{(2)}\lambda_k^{(3)} \geq \lambda_i^{(1)}\lambda_{j+1}^{(2)}\lambda_k^{(3)}$, $\lambda_i^{(1)}\lambda_j^{(2)}\lambda_k^{(3)} \geq \lambda_{i+1}^{(1)}\lambda_j^{(2)}\lambda_k^{(3)}$ define the semi-orders among eigenvalues as

$$\lambda_i^{(1)}\lambda_j^{(2)}\lambda_k^{(3)} \succeq \left\langle \lambda_i^{(1)}\lambda_j^{(2)}\lambda_{k+1}^{(3)}, \lambda_i^{(1)}\lambda_{j+1}^{(2)}\lambda_k^{(3)}, \lambda_{i+1}^{(1)}\lambda_j^{(2)}\lambda_k^{(3)}\right\rangle$$

□

Regarding the selection of the dimension of tensor subspace, Theorem 1 implies the following theorem.

Theorem 2. *The dimension of the subspace of the tensor space for data compression is* $\frac{1}{6}n(n+1)(n+2)$ *if we select n principal components in each mode of three-way array data.*

(*Proof*). For a positive integer n, the number s_n of eigenvalues $\lambda_i^{(1)}\lambda_j^{(2)}\lambda_k^{(3)}$ is

$$s_n = \sum_{i+j+k=0, i,j,k\geq 0}^{n-1}(i+j+k) = \sum_{l=1}^{n}\sum_{m=1}^{l}m$$

$$= \frac{1}{2}\left(\frac{1}{6}n(n+1)(2n+1) + \frac{1}{2}n(n+1)\right) = \frac{1}{6}n(n+1)(n+2) \tag{30}$$

(*Q.E.D*)

If $n = 1, 2, 3, 4$, we have $N = 1, 4, 10, 20$, respectively, for tensors $mathcalX = ((x_{ijk}))$ in $\mathbb{R}^{I \times I \times I}$.

Since the discrete cosine transform (DCT) [4] is asymptotically equivalent to the matrix of K-L transform [13] for data observed from a first-order Markov model [12,13], the dimension reduction by PCA is performed using DCT as

$$f_{ijk}^n = \sum_{i'j'k'=0}^{n-1} g_{i'j'k'}\varphi_{i'i}\varphi_{j'j}\varphi_{k'k}, \quad g_{ijk} = \sum_{i'j'k'=0}^{N-1} f_{i'j'k'}\varphi_{ii'}\varphi_{jj'}\varphi_{kk'} \tag{31}$$

for $n \leq N$, where

$$\boldsymbol{\Phi}_{(N)} = ((\epsilon \cos \frac{(2j+1)i}{2\pi N})) = ((\varphi_{ij})), \quad \epsilon = \begin{cases} 1 & \text{if } j = 0 \\ \frac{1}{\sqrt{2}} & \text{otherwise} \end{cases} \tag{32}$$

is the DCT-II matrix of the order N. If we apply the fast cosine transform to the computation of the 3D-DCT-II matrix, the computational complexity is $\mathcal{O}(3n \log n)$.

Since

$$vec(\boldsymbol{u} \circ \boldsymbol{v} \circ \boldsymbol{w}) = \boldsymbol{u} \otimes \boldsymbol{v} \otimes \boldsymbol{w} \tag{33}$$

outer products of vectors redescribes the DCT-based transform as

$$\mathcal{F} = \sum_{i,j,k=0}^{N-1} a_{ijk}\varphi_i \circ \varphi_j \circ \varphi_k, \quad a_{ijk} = \langle \mathcal{F}, (\varphi_i \circ \varphi_j \circ \varphi_k) \rangle \tag{34}$$

where

$$\boldsymbol{\Phi}^{(N)} = (\varphi_0, \varphi_1, \cdots, \varphi_{N-1}). \tag{35}$$

4   Mutual Tensor Subspace Method

The mutual subspace method is stable against geometric perturbation of queries for pattern recognition. Furthermore, the constrained subspace method confirms robust recognition against global perturbations of queries. Therefore, as the combination of two methods, the constrained mutual subspace method is stable and robust both against geometric and global perturbations of queries.

The angle $\theta = \angle(\mathcal{A}, \mathcal{B})$ between two tensor $\mathcal{A}$ and $\mathcal{B}$ is computed as $\cos \theta = \frac{\langle \mathcal{A}, \mathcal{B} \rangle}{|\mathcal{A}|_F |\mathcal{B}|_F}$. The angle between two spaces defined by $\{\boldsymbol{U}^{(i)}, \boldsymbol{V}^{(i)}\}_{i=1}^3$ is the extremal of the criterion

$$\cos^2 \theta = |\langle \mathcal{A} \times_1 \boldsymbol{U}^{(1)} \times_2 \boldsymbol{U}^{(2)} \times_3 \boldsymbol{U}^{(3)}, \mathcal{B} \times_1 \boldsymbol{V}^{(1)} \times_2 \boldsymbol{V}^{(2)} \times_3 \boldsymbol{V}^{(3)} \rangle|^2 \tag{36}$$

with respect to the conditions $|\mathcal{A}|_F = 1$ and $|\mathcal{B}|_F = 1$. Therefore, the minimiser of Eq. (36) is the solution of the Euler-Lagrange equation

$$T_1 = \cos^2 \theta + \lambda(1 - |\mathcal{A}|_F^2) + \mu(1 - |\mathcal{B}|_F^2) \tag{37}$$

Since, we have the system of equations

$$\mathcal{A} \times_1 \boldsymbol{U}^{(1)} \times_2 \boldsymbol{U}^{(2)} \times_3 \boldsymbol{U}^{(3)} = \mu \mathcal{B} \tag{38}$$

$$\mathcal{B} \times_1 \boldsymbol{V}^{(1)} \times_2 \boldsymbol{V}^{(2)} \times_3 \boldsymbol{V}^{(3)} = \lambda \mathcal{A}, \tag{39}$$

for the tensor singular value problem

$$\mathcal{A} \times_1 \boldsymbol{P}_1 \times_2 \boldsymbol{P}_2 \times_3 \boldsymbol{P}_3 = \lambda \mu \mathcal{A} \tag{40}$$

$$\mathcal{B} \times_1 \boldsymbol{Q}_1 \times_2 \boldsymbol{Q}_2 \times_3 \boldsymbol{Q}_3 = \lambda \mu \mathcal{B}, \tag{41}$$

where $\boldsymbol{P}_i = \boldsymbol{U}^{(1)} \boldsymbol{V}^{(i)}$ and $\boldsymbol{Q}_i = \boldsymbol{V}^{(i)} \boldsymbol{U}^{(i)}$, the maximiser of Eq. (36) is

$$T_{\max} = \lambda_{\max} \mu_{\max} \tag{42}$$

with the condition $\lambda = \mu$. This mathematical property implies the following theorem.

Theorem 3. *The canonical angle between a pair of linear subspaces spanned by triples of tensors $\{\boldsymbol{U}^{(i)}\}_{i=1}^3$ and $\{\boldsymbol{V}^{(i)}\}_{i=1}^3$ is $\cos^{-1} \sigma$ where σ is the maximum eigenvalue of tensor $\boldsymbol{P}_1 \times_2 \boldsymbol{P}_2 \times_3 \boldsymbol{P}_3$.*

To shrink the size of linear problem, we evaluate the difference between two subspaces using the perpendicular length of the normalised tensors.

For a collection of tensors $\{\mathcal{X}_i\}_{i=1}^M$ with the condition and $\mathrm{E}(\mathcal{X}_i) = 0$, we define a collection of categories of volumetric data. We assume that we have N_{C} categories For the kth category, we compute a system of orthogonal projections $\{\boldsymbol{U}_{k,j}\}_{j=1,k=1}^{3,N_c}$ to define tensor subspaces, that is, a tensor subspace corresponds to a category of volumetric data.

Setting $\{\mathcal{G}_{i'}\}_{i'=1}^{M'}$ to be queries, we compute projected tensor

$$\mathcal{A}_{i'} = \mathcal{G}_{i'} \times_1 \boldsymbol{U}_k^{(1)\top} \times_2 \boldsymbol{U}_k^{(2)\top} \times_3 \boldsymbol{U}_k^{(3)\top}. \tag{43}$$

Furthermore, assuming that queries belong a category from N_{C} categories, we have orthogonal projection matrices $\{\boldsymbol{V}_j\}_{j=1}^3$, from a tensor subspace of queries. This system of orthogonal projections yields the projected tensor

$$\mathcal{B}_{i'} = \mathcal{G}_{i'} \times_1 \boldsymbol{V}^{(1)\top} \times_2 \boldsymbol{V}^{(2)\top} \times_3 \boldsymbol{V}^{(3)\top}. \tag{44}$$

Using a tensor subspace $\mathcal{C}_k$ corresponding to a category and a tensor subspace $\mathcal{C}_{\mathrm{q}}$ computed from queries, we define the dissimilarity of subspaces $d(\mathcal{C}_k, \mathcal{C}_{\mathrm{q}})$ by

$$\mathrm{E}\left(|\mathcal{A}_{i'} \times_1 \boldsymbol{P} \boldsymbol{U}_k^{(1)} \times_2 \boldsymbol{P} \boldsymbol{U}_k^{(2)} \times_3 \boldsymbol{P} \boldsymbol{U}_k^{(3)} - \mathcal{B}_{i'} \times_1 \boldsymbol{P} \boldsymbol{V}^{(1)} \times_2 \boldsymbol{P} \boldsymbol{V}^{(2)} \times_3 \boldsymbol{P} \boldsymbol{V}^{(3)}|_{\mathrm{F}}^2 \right), \tag{45}$$

for $\mathcal{A}_{i'} := \mathcal{A}_i / |\mathcal{A}_i|_F$ and $\mathcal{B}_{i'} := \mathcal{B}_i / |\mathcal{B}_i|_F$, where a unitary matrix $\boldsymbol{P}$ selects bases for each mode of tensors. For the dissimilarity of (45), the condition

$$\arg \left(\min_l d(\mathcal{C}_l, \mathcal{C}_{\mathrm{q}}) \right) = \mathcal{C}_k, \tag{46}$$

leads to the property that $\{\mathcal{G}_{i'}\}_{i'=1}^{M'} \in \mathcal{C}_k(\delta)$ for $k, l = 1, 2, \dots, N_{\mathrm{C}}$.

For the construction of the common space spanned by $\{P_1^{(j)} \otimes P_2^{(j)} \otimes P_2^{(j)}\}_{j=1}^{N_C}$, the orthogonal projection $Q = Q^{(1)} \otimes Q^{(2)} \otimes Q^{(3)}$ which minimises criterion

$$J_{\text{CMS1}} = \sum_{i=1}^{N_C} |QP_i|_F^2 \qquad (47)$$

with the conditions

$$P_i = P_i^{(1)} \otimes P_i^{(2)} \otimes P_i^{(3)} \qquad (48)$$

and

$$Q_i^{(j)\top} Q_i^{(j)} = I_{n_i}, \quad j = 1, 2, 3. \qquad (49)$$

Therefore, we minimise the Euler-Lagrange equation

$$J_{\text{CMS2}} = \sum_{i=1}^{N_C} |QP_i|_F^2 + \sum_{j=1}^{3} \langle \Lambda^{(j)} (I_{m_j} - Q^{(j)\top} Q^{(j)}) \rangle \qquad (50)$$

Same with the minimisation for the tensor PCA, we solve the system of optimisation problems.

In each mode, the orthogonal projection $Q^{(j)}$, which maximises the criterion

$$J_{\text{CMSM}} = \sum_{i=1}^{N_j} |Q^{(j)} U_i^{(j)}|_F^2, \qquad (51)$$

for the collection of orthogonal projections $\{U_i^{(j)}\}_{i=1}^{N_j}$, approximates the common space of spaces spanned by $\{U_i^{(j)}\}_{i=1}^{N_j}$.

The projection $Q^{(j)}$ is the solution of the variational problem

$$J_{\text{CMSM-EL}} = \sum_{i=1}^{N} |QU_i^{(j)}|_F^2, + \langle (I_{m_j} - Q^{(j)\top} Q^{(j)}), \Sigma \rangle. \qquad (52)$$

Therefore, the $Q^{(j)}$ is the eigenmatrix of

$$\sum_{i=1}^{N_j} U_i^{(j)} Q^{(j)} = Q^{(j)} \Sigma^{(j)}. \qquad (53)$$

Let $\{\mathcal{X}_i^j\}_{i=1}^{j(n)}$ be elements of category $\mathcal{C}_j$ and For $\Pi_i \in 2^N$, where $N = \{0, 1, 2, \cdots, N-1\}$, we define the minimisation criterion

$$J(\Pi_k, \Pi_m, \Pi_n) = E_i |\mathcal{X}_i^j - \mathcal{X}_i^j \times_1 P_{\Pi_k} \Phi_{(N)} \times_2 P_{\Pi_m} \Phi_{(N)} \times_3 P_{\Pi_n} \Phi_{(N)}|_F^2 \quad (54)$$

for the selection of the ranges of orthogonal projections P_{Π_k}, P_{Π_m} and P_{Π_n}. The minimisation of Eq. (54) defines the base of the category of $\mathcal{C}_i$ as

$$\mathcal{L}(\mathcal{C}_i) = \{\varphi_p\}_{p \in \Pi_k} \otimes \{\varphi_q\}_{q \in \Pi_m} \otimes \{\varphi_r\}_{r \in \Pi_n} \qquad (55)$$

using row vectors of DCT-II matrix.

Setting

$$\bar{F}^j_{pqr} = E_i(F^j_{ipqr}), \tag{56}$$

where F^j_{pqr} is the DCT coefficient of f^j_{ipqr}, the ranges are selected

$$\Pi_k \times \Pi_m \times \Pi_n = \{(p,q,r) \,|\, \bar{F}^j_{pqr} \geq T\} \tag{57}$$

for an appropriate threshold T. Since the low frequency parts are common for almost all data, the canonicalisation of patterns is achieved by separating data to low and high frequency parts. The separation of the constrain subspace $\cap^n_{i=1} \mathcal{C}_i$ for $\{\mathcal{C}_i\}^n_{i=1}$ is achieved by estimating common low frequency parts to categories.

5 Conclusions

We have reviewed several fundamental and well-established methodologies in pattern recognition in vector space to unify data procession for higher-dimensional space using tensor expressions and multi-way principal component analysis.

In traditional method in medical image analysis outline shapes of objects such as organs and statistical properties of interior textures are independently extracted using separate methods. However, the tensor PCA for volumetric data allows us to simultaneously extract both outline shapes of volumetric objects and statistical properties of interior textures of volumetric data from data projected onto a low-dimensional linear subspace spanned by tensors. We also showed frequency-based pattern recognition methodologies for tensor subspace method.

This research was supported by the "Multidisciplinary Computational Anatomy and Its Application to Highly Intelligent Diagnosis and Therapy" project funded by a Grant-in-Aid for Scientific Research on Innovative Areas from MEXT, Japan, and by Grants-in-Aid for Scientific Research funded by the Japan Society for the Promotion of Science.

References

1. Iijima, T.: Pattern Recognition. Corona-sha, Tokyo (1974). (in Japanese)
2. Otsu, N.: Mathematical studies on feature extraction in pattern recognition. Res. Electrotech. Lab. **818**, 1–220 (1981). (in Japanese)
3. Grenander, U., Miller, M.: Pattern theory: from representation to inference. In: OUP (2007)
4. Strang, G., Nguyen, T.: Wavelets and Filter Banks. Wellesley-Cambridge Press, Wellesley (1996)
5. Marron, J.M., Alonso, A.M.: Overview of object oriented data analysis. Biom. J. **56**, 732–753 (2014)
6. Cichocki, A., Zdunek, R., Phan, A.-H., Amari, S.: Nonnegative Matrix and Tensor Factorizations: Applications to Exploratory Multi-way Data Analysis and Blind Source Separation. Wiley, Hoboken (2009)
7. Itskov, M.: Tensor Algebra and Tensor Analysis for Engineers. Springer, Berlin (2013)

8. Mørup, M.: Applications of tensor (multiway array) factorizations and decompositions in data mining. Wiley Interdiscipl. Rev.: Data Min. Knowl. Disc. **1**, 24–40 (2011)
9. Malcev, A.: Foundations of Linear Algebra, in Russian edition 1948, (English translation W.H, Freeman and Company (1963)
10. Kolda, T.G., Bader, B.W.: Tensor decompositions and applications. SIAM Rev. **51**, 455–500 (2008)
11. Kroonenberg, P.M.: Applied Multiway Data Analysis. Wiley, Hoboken (2008)
12. Hamidi, M., Pearl, J.: Comparison of the cosine Fourier transform of Markov-1 signals. IEEE ASSP **24**, 428–429 (1976)
13. Oja, E.: Subspace Methods of Pattern Recognition. Research Studies Press, Letchworth (1983)

Thermodynamic Characterization of Temporal Networks

Giorgia Minello[1(⊠)], Andrea Torsello[1], and Edwin R. Hancock[2]

[1] DAIS, Università Ca' Foscari Venezia, Venice, Italy
giorgia.minello@unive.it
[2] Department of Computer Science, University of York, York, UK

Abstract. Time-evolving networks have proven to be an efficient and effective means of concisely characterising the behaviour of complex systems over time. However, the analysis of such networks and the identification of the underlying dynamical process has proven to be a challenging problem, particularly trying to model the large-scale properties of graphs. In this paper we present a novel method to characterize the behaviour of the evolving systems based on a thermodynamic framework for graphs. This framework aims at relating the major structural changes in time evolving networks to thermodynamic phase transitions. This is achieved by relating the thermodynamics variables to macroscopic changes in network topology. First, by considering a recent quantum-mechanical characterization of the structure of a network, we derive the network entropy. Then we adopt a Schrödinger picture of the dynamics of the network, in order to obtain a measure of energy exchange through the estimation of a hidden time-varying Hamiltonian from the data. Experimental evaluations on real-world data demonstrate how the estimation of this time-varying energy operator strongly characterizes the different states of time evolving networks.

Keywords: Complex networks · Quantum thermodynamics · Graphs

1 Introduction

Complex systems can be regarded as a collection of in homogeneously and generically interacting units and are ubiquitous both in nature and man-made systems. They appear in a wide range of scenarios, varying from biological and ecological to social and technological fields and can refer to any phenomena properties, from the molecular level to the scale of large communications infrastructures [4]. Notable examples are the World Wide Web, metabolic reaction networks, financial market stock correlations, scientific collaboration, coauthorship and citation relations, and social interactions [9].

Therefore, the study of the dynamics of these structures has become increasingly important and a main subject for interdisciplinary research. Especially, network evolution mechanisms play a central role in science. This follows a crucial change of viewpoint in network analysis, from a rather unnatural static view

© Springer International Publishing AG 2016
A. Robles-Kelly et al. (Eds.): S+SSPR 2016, LNCS 10029, pp. 49–59, 2016.
DOI: 10.1007/978-3-319-49055-7_5

of the networks to a more realistic characterization of the dynamics of the system, in order to predict their behavior and explain processes acting over the networks. Indeed now the majority of efforts aims at identifying relations between the system structure and network performance, *e.g.*, how the network evolves with time to respond to structural needs, whereas previously efforts were aimed at representing the problem, *i.e.* the characterization of network structure [3].

An excellent framework for the study of complex networks relies on statistical physics and thermodynamics, connecting the macroscopic properties of a system to the behavior of microscopic particles [6–8]. In particular, thermodynamics defines the macroscopic properties of a system through three variables, subject to constraints imposed by the four laws of thermodynamics. For instance, in the case of graph representation of complex networks, Escolano et al. [2] provide a thermodynamic characterization based on the variation of local histories over graphs.

In this paper we present a novel method to characterize the behaviour of the evolving systems based on a thermodynamic framework for graphs. Specifically, the graph Laplacian of each time slice is seen as a quantum mixed state undergoing free evolution - through the Schrödinger equation - under an unknown time-dependent Hamiltonian representing the change in potential due to external factors, and entropy and energy changing direct interaction with the environment.

In this way, the evolution of the network allows to estimate the hidden time-varying Hamiltonian and consequently the Energy-exchange at each time interval - and the variation in entropy of the underlying structure as well. From these we derive all the thermodynamic variables of networks, including the free energy and temperature.

The consequent characterization is utilized to represent two real-world time-varying networks: the price correlation of selected stocks in the New York Stock Exchange (NYSE) [12], and the gene expression of the life cycle of the Fruit Fly (*Drosophila melanogaster*) [1,13].

2   Quantum Thermodynamics of the Network

Let $G(V, E)$ be an undirected graph with node set V and edges set $E \subseteq V \times V$ and let $A = a_{ij}$ be the adjacency matrix, where

$$a_{ij} = \begin{cases} 1, & v_i \sim v_j, \\ 0, & \text{otherwise.} \end{cases}$$

The degree d of a node is the number of edges incident to the node and it can be represented through the degree matrix $D = (d_{ij})$ which is a diagonal matrix with $d_{ii} = \sum_i a_{ij}$. The graph Laplacian is then defined as $L = D - A$, and it can be interpreted as a combinatorial analogue of the discrete Laplace-Beltrami operator. The normalized Laplacian matrix $\tilde{L}$ is defined as

$$\tilde{L} = D^{-1/2}(D - A)D^{-1/2} \tag{1}$$

If we divide the normalized Laplacian by the number of vertices in the graph we obtain a unit-trace positive semidefinite matrix that Passerini and Severini [11] suggest can be seen as a density matrix in a quantum system representing a quantum superposition of the transition steps of quantum walk over the graph.

The *continuous-time quantum walk* is the quantum counterpart of the continuous-time random walk, and it is similarly defined as a dynamical process over the vertices of the graph [5]. Here the classical state vector is replaced by a vector of complex amplitudes over V, and a general state of the walk is a complex linear combination of the basis states $|v\rangle$, $v \in V$, such that the state of the walk at time t is defined as

$$|\psi_t\rangle = \sum_{u \in V} \alpha_u(t) |u\rangle \tag{2}$$

where the amplitude $\alpha_u(t) \in \mathbb{C}$ and $|\psi_t\rangle \in \mathbb{C}^{|V|}$ are both complex. Moreover, we have that $\alpha_u(t)\alpha_u^*(t)$ gives the probability that at time t the walker is at the vertex u, and thus $\sum_{u \in V} \alpha_u(t)\alpha_u^*(t) = 1$ and $\alpha_u(t)\alpha_u^*(t) \in [0,1]$, for all $u \in V$, $t \in \mathbb{R}^+$.

The evolution of the walk is then given by the Schrödinger equation, where we denote the time-independent Hamiltonian as $\mathcal{H}$.

$$\frac{\partial}{\partial t} |\psi_t\rangle = -i\mathcal{H} |\psi_t\rangle . \tag{3}$$

Given an initial state $|\psi_0\rangle$, we can solve Eq. (3) to determine the state vector at time t

$$|\psi_t\rangle = e^{-i\mathcal{H}t} |\psi_0\rangle . \tag{4}$$

The *density operator* (or *density matrix*) is introduced in quantum mechanics to describe a system whose state is an ensemble of pure quantum states $|\psi_i\rangle$, each with probability p_i. The density operator of such a system is a positive unit-trace matrix defined as

$$\rho = \sum_i p_i |\psi_i\rangle \langle\psi_i| . \tag{5}$$

The *von Neumann entropy* [10] H_N of a mixed state is defined in terms of the trace and logarithm of the density operator ρ

$$H_N = -\operatorname{Tr}(\rho \log \rho) = -\sum_i \xi_i \ln \xi_i \tag{6}$$

where $\xi_1, \ldots, \xi_n$ are the eigenvalues of ρ. The von Neumann entropy is related to the distiguishability of the states, *i.e.*, the amount of information that can be extracted from an observation on the mixed state.

The observation process for a quantum system is defined in terms of projections onto orthogonal subspaces associated with operators on the quantum state-space called *observables*. Let O be an observable of the system, with spectral decomposition

$$O = \sum_i a_i P_i \tag{7}$$

where the a_i are the (distinct) eigenvalues of O and the P_i the orthogonal projectors onto the corresponding eigenspaces. The outcome of an observation, or projective measurement, of a quantum state $|\psi\rangle$ is one of the eigenvalues a_i of O, with probability

$$P(a_i) = \langle \psi | P_i | \psi \rangle \tag{8}$$

After the measurement, the state of the quantum systems becomes

$$|\bar{\psi}\rangle = \frac{P_i |\psi\rangle}{|| P_i |\psi\rangle ||}, \tag{9}$$

where $|| |\psi\rangle || = \sqrt{\langle \psi | \psi \rangle}$ is the norm of the vector $|\psi\rangle$.

Density operators play an important role in the quantum observation process. The observation probability of a_i is $P(a_i) = \mathrm{Tr}(\rho P_i)$, with the mixed state being projected by the observation process onto the state represented by the modified density matrix $\rho' = \sum_i P_i \rho P_i$. The expectation of the measurement is $\langle O \rangle = \mathrm{Tr}(\rho O)$. The projective properties of quantum observation means that an observation actively modifies the system, both by altering its entropy and forcing an energy exchange between quantum system and observer.

Thermodynamics describes the behavior of a composite system in terms of macroscopic variables such as energy, entropy and temperature. These are linked together by the thermodynamic identity

$$dU = TdS - PdV \tag{10}$$

where U is the internal energy, S the entropy, V the volume, T the temperature, and P the pressure.

Following Passerini and Severini [11] in their use of the normalized Laplacian matrix as a density operator defining the current state of the network, we derive the network entropy in terms of the von Neumann entropy

$$S_{VN} = -\sum_{i=1}^{|V|} \frac{\tilde{\lambda}_i}{|V|} \ln \frac{\tilde{\lambda}_i}{|V|} \tag{11}$$

With this we can measure dS the change in entropy as the network evolves. Previous work used similar entropic measure to define thermodynamic variables on networks, but linked energy to the number of edges in the graph [15] or derived it through the Boltzmann partition function of the network [14]. However, in these approaches the structure of the graph has the dual function of state (determining the density operator) and operator. Here we opt for a different approach that does away with this duality, assuming that the energy operator is unknown and estimated from the evolution. We assume that the dynamics of the network is governed by a free evolution following the Schrödinger equation under an unknown time-varying Hamitonian $\mathcal{H}_t$, and an interaction with the outside world which acts as an observer. The free evolution does not change the thermodynamic variables, while the cause of the variation in Entropy has to be sought from the interaction process which also causes an energy exchange.

To measure the energy exchange we need to recover the potential term expressed by the unknown Hamiltonian. In fact, the Hamiltonian acts as an energy operator, resulting the following expression for the change in energy between state ρ_t and ρ_{t+1}

$$dU = \mathrm{Tr}(\mathcal{H}_t \rho_{t+1}) - \mathrm{Tr}(\mathcal{H}_t \rho_t) \tag{12}$$

We estimate the Hamiltonian $\mathcal{H}_t$ as the one that minimizes the exchange of energy through the interaction with the environment. To this end we assume that the interaction intervenes at the end of the free evolution, where ρ_t is transformed by the Schrödinger equation into

$$\hat{\rho}_{t+1} = \exp(-i\mathcal{H}_t)\rho_t \exp(i\mathcal{H}_t) \tag{13}$$

The exchange of energy in the interaction is then

$$\mathcal{H} = \arg\min_{H} \mathrm{Tr}(H\rho_{t+1}) - \mathrm{Tr}(H\hat{\rho}_{t+1}) \tag{14}$$
$$= \arg\min_{H} \mathrm{Tr}\left(H\left(\rho_{t+1} - \exp(-iH)\rho_t \exp(iH)\right)\right)$$

Let $\rho_t = \Phi_t \Lambda_t \Phi_t^T$ be the spectral decomposition of the state of the network at time t, Eq. (14) can be solved by noting that the minimum energy exchange intervenes when the interaction changes the eigenvalues of the density matrices, and with them the entropy, but does not change the corresponding eigenspaces. In other words, the Hamiltonian is the cause of the eigenvector rotation and can be recovered by it:

$$\mathcal{H}_t \approx i \log(\Phi_{t+1}\Phi_t^T) \tag{15}$$

It is worth noting that we have computed a lower bound of the Hamiltonian, since we cannot observe components on the null spaces of ρs. Furthermore, we have

$$\underbrace{\Phi_{t+1}\Phi_t^T}_{\mathcal{U}} \rho_0 \underbrace{\Phi_t\Phi_{t+1}^T}_{\mathcal{U}} = \hat{\rho}_{t+1}, \tag{16}$$

where $\mathcal{U} = \Phi_{t+1}\Phi_t^T$ is the unitary evolution matrix. The final change in internal energy is then

$$dU = Tr(\mathcal{H}_t \rho_{t+1}) - Tr(\mathcal{H}_t \rho_t) \tag{17}$$

The thermodynamic temperature T can then be recovered through the fundamental thermodynamic relation $dU = TdS - PdV$ but where we assume that the volume is constant, i.e. $dV = 0$ (isochoric process). As a result, the reciprocal of the temperature T is the rate of change of internal energy with entropy

$$T = \frac{dU}{dS} \tag{18}$$

This definition can be applied to evolving complex networks which do not change the number of nodes during their evolution.

3   Experimental Evaluation

In this section we evaluate the ability of the thermodynamic variables to describe the overall dynamics of a system and to characterize significant changes of network's state. Especially, we will investigate how the estimated Energy-exchange describes the temporal trend of the network and whether the approach turns out efficient to detect critical events of a complex phenomena (*e.g.* financial crises or crashes). To this aim, we focused on two real-world time-evolving networks, representing the stock price correlation of the New York Stock Exchange (NYSE) and the gene expression of the Fruit Fly (*Drosophila melanogaster*).

3.1   Datasets

NYSE: The dataset is extracted from a database containing the daily prices of 3799 stocks traded on the New York Stock Exchange (NYSE). The dynamic network is built by selecting 347 stocks with historical data from May 1987 to February 2011 [12]. To obtain an evolving network, a time window of 28 days is used and moved along time to obtain a sequence (from day 29 to day 6004); so doing every temporal window becomes a subsequence of the daily return stock values over a 28 day period. Then, to set trades among the different stocks in the form of a network, for each time window, the cross correlation coefficient between the time-series for each pair of stocks is computed. We create connections between them if the absolute value of the correlation coefficient exceeds a threshold and in this way we construct a stock market network which changes over the time, with a fixed number of 347 nodes and varying edge structure for each of trading days.

Drosophila: The dataset belongs to the biology field and collects interactions among genes of Fruit Fly - *Drosophila melanogaster* - during its life cycle. The fruit fly life cycle is divided into four stages; data is sampled at 66 sequential developmental time points. Early embryos are sampled hourly and adults are sampled at multiday intervals, according to the speed of the morphological changes. Each stage gathers a set of samples: the embryonic phase contains samples from time point 1 to time point 30, larval has samples 31–40, pupal 41–58 and the remaining samples concerns the adulthood. To represent data using a time evolving network, the following steps are followed [13]. At each developmental point the 588 genes that are known to play an important role in the development of the *Drosophila* are selected. These genes are the nodes of the network, and edges are established based on the microarray gene expression measurements reported in [1]. To make more tractable the normalized Laplacian any self-loop in the obtained undirect graph - at each time - has been removed. This dataset yields a time-evolving network with a fixed number of 588 nodes, sampled at 66 developmental time points.

3.2 Experiments

To carry out our analysis, firstly we computed the normalized Laplacian of the network at each step (*e.g.* the time interval in the NYSE is a day) and then the thermodynamic variables, entropy and Energy-exchange (*i.e.* the change in internal energy), as shown in Eqs. (6) and (17), respectively. Then, by means of the entropy variation dS, we computed the temperature (Eq. (18)) and finally we derived the Energy, from the energy variation. Initial investigations were oriented towards a general analysis of three main variables' behaviour and afterward we shifted the focus on the one with the best (qualitative) performance.

We commenced by examining the energy variation dU, the entropy variation dS and the temperature T, as fluctuation indicators for the NYSE dataset (more suitable at this exploratory level since presenting many phase oscillations to be detected). Figures 1 and 2 are of help to compare the three quantities, throughout two slices of the time series including well-distinct occurrences. We can see that both signals tend to exhibit clear alterations in proximity of some major events even if the entropy variation appears slightly noisier than the energy. For instance, in Fig. 2, the Asian financial crisis is well-defined within boundaries (for the energy variation) as well as the Persian Gulf War in Fig. 1, while the entropy's signal lightly errs in terms of precision, still remaining acceptable. Consequently the temperature, strongly affected by the entropy variation, sometimes oscillates even if none financial incident influences the system. An example of these unjustified swings is in Fig. 1, after January 1995.

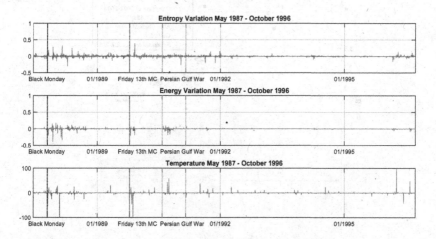

Fig. 1. Up-Bottom: entropy variation, energy variation and temperature versus time (May 1987 - October 1996), for the dynamic stock correlation network. The vertical colored lines refer to the most important and devasting events for the trade market. Left-Right: Black Monday (19th October 1987), Friday the 13th Mini-Crash (13rd October 1989), Persian Gulf War (2nd August 1990 - 17th January 1991). (Color figure online)

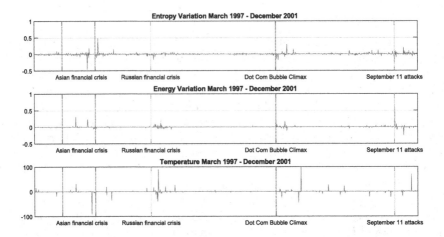

Fig. 2. Up-Bottom: entropy variation, energy variation and temperature versus time (March 1997 - December 2001), for the NYSE dataset. The vertical colored lines signal important events. Left-Right: Asian Financial Crisis (July 1997 – October 1997), Russian Financial Crisis – Ruble devaluation (17th August 1998), Dot-com bubble - climax (10th March 2000), September 11 attacks. (Color figure online)

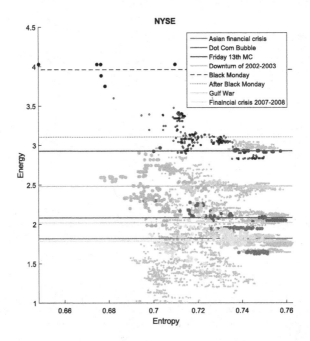

Fig. 3. Scatter plot of Energy vs Entropy (New York Stock Exchange data). Each dot is a day and grey dots are the background. Dots of the same color belong to the same network phase. Horizontal lines represent cluster centroids for the energy dimension. (Color figure online)

Now we turn our attention to the energy dimension, which has proven to be the one with the lowest volatility from the preliminary examinations. In Fig. 3, we show the scatter plot of the Entropy over Energy for the NYSE dataset. Exploiting this kind of representation, we were able to assess the effectiveness of the Energy-exchange in characterizing the network state. Indeed, from the chart, an interesting feature of the network emerges: it exists a clustering-like behavior of the market when the system endures strong modifications. However, each pattern presents a wide entropy variation but a low energy variation. Thus, we conclude network's states are better identified by the energy, which effectively catches cluster compactness, rather than the entropy, more dispersive. A further evidence of such energetic typifying comes from Fig. 5, concerning the *Drosophila melanogaster* data. Here again the entropy over energy plot modality was adopted; we can observe that stages of the fruit fly life cycle, seen as phase transitions, are being recognized by the Energy-exchange, in a more succinct way than the entropy dimension. Qualitative comparisons with other approaches adopting thermodynamic characterizations, such as in [14], confirm that a clear distinction is not always straightforward, above all when the amount of data is scarce (*e.g.*, time epochs in the time-series).

Finally, in Fig. 4, the temporal trend of the energy (recovered from Energy-exchange), distinctly proves how the estimation of the hidden time-varying Hamiltonial successfully extracts information from data and how the energy can be considered a decisive state function.

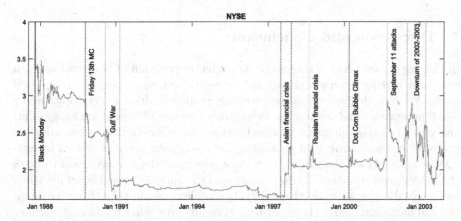

Fig. 4. Energy-exchange versus time of the NYSE network (May 1987 - December 2003). Left-Right: Black Monday (19th October 1987), Friday the 13th Mini-Crash (13rd October 1989), Persian Gulf War (2nd August 1990 - 17th January 1991), Asian Financial Crisis (July 1997 – October 1997), Russian Financial Crisis – Ruble devaluation (17th August 1998), Dot-com bubble - climax (10th March 2000), September 11 attacks, Downturn of 2002–2003. (Color figure online)

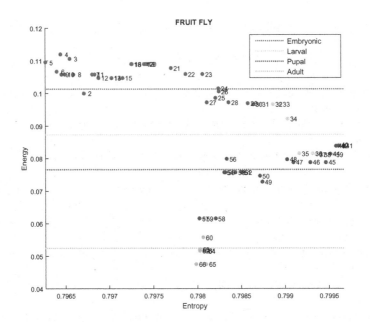

Fig. 5. Scatter plot of Energy vs Entropy of the *Drosophila melanogaster* network. Each dot represents a sample. Numbers and colors are used to identify samples and life cycle stages. Horizontal lines represent cluster centroids for the energy dimension. (Color figure online)

4 Discussion and Conclusion

In this paper, we adopt a thermodynamic characterization of temporal network structure in order to represent and understand the evolution of time-varying networks. We provide expressions for thermodynamic variables, *i.e.* entropy, energy and temperature, and in addition we derive a measure of Energy-exchange. This analysis is based on quantum thermodynamics and connected to recent works on the von Neumann entropy of networks. The Energy-exchange is derived by estimating an unknown Hamiltonian operator governing the free evolution through the Schrödinger equation. We have evaluated the approach experimentally using real-world data, representing time-varying complex systems taken from the financial and biological fields. The experimental results prove that the Energy-exchage is a convenient and efficient measure for analyzing the evolutionary properties of dynamic networks, able to detect phase transitions and abrupt changes occurring in complex phenomena.

References

1. Arbeitman, M.N., Furlong, E.E., Imam, F., Johnson, E., Null, B.H., Baker, B.S., Krasnow, M.A., Scott, M.P., Davis, R.W., White, K.P.: Gene expression during the life cycle of drosophila melanogaster. Science **297**(5590), 2270–2275 (2002)

2. Escolano, F., Hancock, E.R., Lozano, M.A.: Heat diffusion: thermodynamic depth complexity of networks. Phys. Rev. E **85**(3), 036206 (2012)
3. Estrada, E.: The Structure of Complex Networks: Theory and Applications. OUP, Oxford (2011)
4. Estrada, E.: Introduction to complex networks: structure and dynamics. In: Banasiak, J., Mokhtar-Kharroubi, M. (eds.) Evolutionary Equations with Applications in Natural Sciences. LNM, vol. 2126, pp. 93–131. Springer, Heidelberg (2015). doi:10.1007/978-3-319-11322-7_3
5. Farhi, E., Gutmann, S.: Quantum computation and decision trees. Phys. Rev. A **58**(2), 915 (1998)
6. Huang, K.: Statistical Mechanics. Wiley, New York (1987)
7. Javarone, M.A., Armano, G.: Quantum-classical transitions in complex networks. J. Stat. Mech.: Theor. Exp. **2013**(04), P04019 (2013)
8. Mikulecky, D.C.: Network thermodynamics and complexity: a transition to relational systems theory. Comput. Chem. **25**(4), 369–391 (2001)
9. Newman, M.E.: The structure and function of complex networks. SIAM Rev. **45**(2), 167–256 (2003)
10. Nielsen, M., Chuang, I.: Quantum Computation and Quantum Information. Cambridge University Press, Cambridge (2010)
11. Passerini, F., Severini, S.: Quantifying complexity in networks: the Von Neumann entropy. Int. J. Agent Technol. Syst. (IJATS) **1**(4), 58–67 (2009)
12. Peron, T.D., Rodrigues, F.A.: Collective behavior in financial markets. EPL (Europhys. Lett.) **96**(4), 48004 (2011)
13. Song, L., Kolar, M., Xing, E.P.: Keller: estimating time-varying interactions between genes. Bioinformatics **25**(12), i128–i136 (2009)
14. Ye, C., Comin, C.H., Peron, T.K.D., Silva, F.N., Rodrigues, F.A., Costa, L.D.F., Torsello, A., Hancock, E.R.: Thermodynamic characterization of networks using graph polynomials. Phys. Rev. E **92**(3), 032810 (2015)
15. Ye, C., Torsello, A., Wilson, R.C., Hancock, E.R.: Thermodynamics of time evolving networks. In: Liu, C.-L., Luo, B., Kropatsch, W.G., Cheng, J. (eds.) GbRPR 2015. LNCS, vol. 9069, pp. 315–324. Springer, Heidelberg (2015). doi:10.1007/978-3-319-18224-7_31

Dissimilarity Representations

A Multi-Stage Approach for Fast Person Re-identification

Bahram Lavi$^{(\boxtimes)}$, Giorgio Fumera, and Fabio Roli

Department of Electrical and Electronic Engineering,
University of Cagliari Piazza d'Armi, 09123 Cagliari, Italy
{lavi.bahram,fumera,roli}@diee.unica.it
http://pralab.diee.unica.it

Abstract. One of the goals of person re-identification systems is to support video-surveillance operators and forensic investigators to find an individual of interest in videos taken by a network of non-overlapping cameras. This is attained by sorting images of previously observed individuals for decreasing values of their similarity with the query individual. Several appearance-based descriptors have been proposed so far, together with ad hoc similarity measures, mostly aimed at improving ranking quality. We address instead the issue of the processing time required to compute the similarity values, and propose a multi-stage ranking approach to attain a trade-off with ranking quality, for any *given* descriptor. We give a preliminary evaluation of our approach on the benchmark VIPeR data set, using different state-of-the-art descriptors.

1 Introduction

Person re-identification is a computer vision task consisting in recognizing an individual who had previously been observed over a network of video surveillance cameras with non-overlapping fields of view [1]. One of its applications is to support surveillance operators and forensic investigators in retrieving videos where an individual of interest appears, using an image of that individual as a query (*probe*). To this aim, the video frames or tracks of all the individuals recorded by the camera network (*template gallery*) are sorted by decreasing similarity to the probe, allowing the operator to find the occurrences (if any) of the individual of interest, ideally, in the top positions. This task is challenging in typically unconstrained surveillance settings, due to low image resolution, unconstrained pose, illumination changes, and occlusions, which do not allow to exploit strong biometrics like face. Clothing appearance is therefore one of the most widely used cues, although cues like gait and anthropometric measures have also been investigated. Most of the existing person re-identification techniques are based on a specific descriptor of clothing appearance (typically including color and texture), and a specific similarity measure between a pair of descriptors which can be either manually defined or learnt from data [1,3,4,6,12]. Their focus is to improve recognition accuracy, i.e., ranking quality.

© Springer International Publishing AG 2016
A. Robles-Kelly et al. (Eds.): S+SSPR 2016, LNCS 10029, pp. 63–73, 2016.
DOI: 10.1007/978-3-319-49055-7_6

In this work we address the complementary issue of the processing time required to compute the similarity measure (*matching score*). Many of the similarity measures defined so far are indeed rather complex, and require a relatively high processing time (e.g., [3,16]). Moreover, in real-world application scenarios the template gallery can be very large, and even when a single matching score is fast to compute (e.g., the Euclidean distance between fixed-length feature vectors [12]), computing it for all templates is time-consuming. This issue has been addressed so far only by a few works [2,9,15].

Inspired by the multi-stage approach for generic classification problems of [14], and by the object detection approach based on a cascade of classifiers of [19], we propose a multi-stage ranking approach specific to person re-identification, aimed at attaining a trade-off between ranking quality and processing time. Both the approaches of [14,19] consist in a cascade of classifiers, where each stage uses features that are increasingly more discriminant but also more costly [14] or slower to compute [19]. The goal of [14] is to assign an input instance (e.g., a medical image) to one of the classes (e.g., the outcome of a diagnosis) with a predefined level of confidence, using features (e.g., medical exams) with the lowest possible cost; if a classifier but the last one does not reach the desired confidence level, it *rejects* the input instance (i.e., withholds making a decision), and sends it to the next stage. This approach has later been exploited to attain a trade-off between classification accuracy and processing time, e.g., in handwritten digit classification [8,17,18]. The similar approach of [19] focuses on designing fast object detectors: its goal is to detect background regions of the input image as quickly as possible, using classifiers based on features fast to compute, and to focus the attention on regions more likely to contain the object of interest, using classifiers based on more discriminant features that also require a higher processing time.

The above approaches cannot be directly applied to person re-identification, which is a *ranking* problem, not a classification one. In this paper we adapt it to person re-identification, to attain a trade-off between ranking quality and processing time, for a *given* descriptor and similarity measure. To this aim, we build a multi-stage re-identification system in which the chosen descriptor is used in the last stage, whereas "reduced" versions of the *same* descriptor are used in previous stages, characterized by a decreasing processing time and a lower recognition accuracy; the first stage ranks *all* templates, whereas each subsequent stage re-ranks a subset of the top-ranked templates by the previous stage. After summarizing in Sect. 2 related re-identification approaches, in Sect. 3 we describe our approach and discuss possible design criteria. We then give in Sect. 4 a preliminary evaluation of the attainable trade-off between recognition accuracy and processing cost, using the benchmark VIPeR data set and four state-of-the-art descriptors.

2 Related Work

As mentioned in Sect. 1, only a few works have addressed so far the issue of the processing time required to compute the matching scores in person re-identification systems [2,9,15]. In particular, only in [2] the proposed solution is a multi-stage system: the first stage selects a subset of templates using a descriptor with a low processing time for computing matching scores (a bag-of-words feature representation and an indexing scheme based on inverted lists was proposed). The second stage ranks only the selected templates using a different, more complex mean Riemann covariance descriptor. Differently from our approach, only two stages are used in [2], based on *different* and *specific* descriptors; and only a subset of templates is ranked, possibly missing the correct one. We point out that the achieved reduction of processing time, with respect to ranking all templates using the second-stage descriptor, was not reported. In [15] we proposed a dissimilarity-based approach for *generic* descriptors made up of bags of local features, possibly extracted from different body parts, aimed at reducing the processing time for computing the matching scores. It converts any such descriptor into a fixed-size vector of dissimilarity values between the input image and a set of representative bags of local features ("prototypes") extracted from the template gallery; the matching score can then be quickly computed, e.g., as the Euclidean distance. The method of [9] aims at reducing the processing time in the specific multi-shot setting (when several images per individual are available), and for specific descriptors based on local feature matching, e.g., interest points. It first filters irrelevant interest points, then uses a sparse representation for the remaining ones, before computing the matching scores.

Other authors proposed multi-stage systems for improving ranking quality, without considering processing time [5,7,11,13,20]. In the two-stage system of [5] a manually designed descriptor is used in the first stage, which returns the operator the 50 top-ranked templates; if they do not include the probe individual, a classifier is trained to discriminate the latter from other identities, and is then used to re-rank the remaining templates. In [13] person re-identification is addressed as a content-based image retrieval task with relevance feedback, with the aim of increasing recall, assuming that several instances of a probe can be present in the template gallery. In each stage (i.e., iteration of relevance feedback) only the top-ranked templates are shown to the operator, and his feedback is used to adapt the similarity measure to the probe at hand. A similar strategy was proposed in [11]: at each stage only the top-ranked templates are presented to the operator, who is asked to select an individual with a different identity and a very different appearance than the probe. A post-rank function is then learnt, exploiting this feedback and the probe image, and the remaining templates are re-ranked in the next stage. A similar, two-stage approach was proposed in [20]: after presenting to the operator the top-ranked templates from the first stage, the operator is asked to label some pairs of locally similar and dissimilar regions in the probe and template images; this feedback is exploited to re-rank the templates. Another two-stage approach was proposed in [7]: a small subset of the top-ranked templates by of a given first-stage descriptor is re-ranked by the second stage, using a manifold-based method that exploits three specific low-level features.

3 Proposed Approach

Let D denote a given descriptor, $\mathbf{T}$ and $\mathbf{P}$ the descriptors of a template and probe image, respectively, $m(\cdot, \cdot)$ the similarity measure between two descriptors, and $G = \{\mathbf{T}_1, \ldots, \mathbf{T}_n\}$ the template gallery. For a given $\mathbf{P}$, a standard re-identification system computes the matching scores $m(\mathbf{P}, \mathbf{T}_i)$, $i = 1, \ldots, n$, and sorts the template images by decreasing values of the score. Ranking quality is typically evaluated using the cumulative matching characteristic (CMC) curve, i.e., the probability (recognition rate) that the correct identity is within the first r ranks, $r = 1, \ldots, n$. By definition, the CMC curve increases with r, and equals 1 for $r = n$. If t is the processing time for a single matching score, the time for computing all the scores on G is $n \times t$.

To attain a trade-off between recognition accuracy and processing cost, for a *given* descriptor and similarity measure, the solution we investigate is a multi-stage architecture (see Fig. 1) based on the following rationale. Consider first two descriptors D_1 and D_2 with similarity measures m_1 and m_2 and processing cost t_1 and t_2; assume that D_1 is less accurate than D_2, i.e., its CMC curve lies below the one of D_2, as in Fig. 2 (left). The CMC curve of a less accurate descriptor approaches the one of a more accurate one as r increases, and the difference drops below a given threshold Δ after some rank $r_1 < n$ (see Fig. 2, left). This means that D_1 and D_2 exhibit almost the same recognition accuracy for $r > r_1$. If $t_1 < t_2$, one can attain a similar recognition accuracy as D_2, with a lower processing cost, by a two-stage ranking procedure: D_1 is used first to rank all n templates; the top-r_1 ones are then re-ranked using D_2. The corresponding processing time is $T = n \times t_1 + r_1 \times t_2$. If all the n templates are ranked by D_2, instead, the processing time is $T_2 = n \times t_2$. In order for $T < T_2$, r_1 must satisfy:

$$r_1 < n \left(1 - t_1/t_2\right) . \tag{1}$$

If (1) does not hold, one can attain $T < T_2$ by re-ranking in the second stage a lower number of templates than r_1, at the expense of a lower accuracy.

The above approach can be extended to a higher number of stages $N > 2$, using descriptors $D_1, \ldots, D_N$ with increasing accuracy and processing time, $t_1 <$

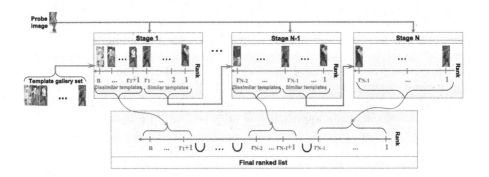

Fig. 1. Overview of the proposed multi-stage ranking approach.

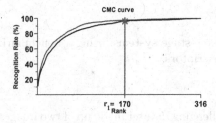 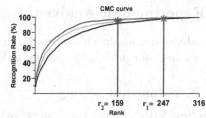

Fig. 2. An example of the criterion used in this work for selecting the number of templates to be ranked in each stage, for a template gallery of size $n = 316$. Left (two-stage system): CMC curve of the first- (black) and second-stage descriptor (blue), and the rank r_1 after which the difference between the two CMC curves is lower than $\Delta = 1\%$. Right: CMC curves of the three stages of a three-stage system, and the corresponding values of r_1 and r_2; note that r_2 has been obtained from the CMC curves of the second and third stages, computed on a template gallery of size $r_1 < n$. (Color figure online)

$t_2 < \ldots < t_N$. Let n_i be the number of matching scores computed by the i-th stage, with $n_1 = n$; since each stage computes a higher number of scores than the next one ($n_1 > n_2 > \ldots > n_N$), the overall processing time is:

$$T = \sum_{i=1}^{N} n_i \times t_i . \tag{2}$$

Let $T_N = n \times t_n$ be the processing time to rank all templates using the most accurate descriptor $\mathbf{D}_N$. For the multi-stage system to attain an accuracy as much similar as possible to the one of $\mathbf{D}_N$, with $T < T_N$, the values n_i ($i > 1$) must be chosen by generalizing the above criterion. More precisely, for $i = 2, \ldots, N$: (i) find the lowest rank r_{i-1} such that the CMC curves of $\mathbf{D}_{i-1}$ and $\mathbf{D}_N$, computed on a template gallery of size n_{i-1}, are closer than a given threshold Δ; (ii) to attain an overall accuracy similar to the one of $\mathbf{D}_N$, choose $n_i = r_{i-1}$ (see Fig. 2, right). If this choice leads to $T \geq T_N$, then T can be decreased by choosing lower values of n_i, $i > 1$, at the expense of a lower recognition accuracy.

To attain a trade-off between accuracy and processing time for a *given* descriptor $\mathbf{D}$ and similarity measure m, the above multi-stage architecture can be implemented by using $\mathbf{D}$ in the last stage, i.e., $\mathbf{D}_N = \mathbf{D}$, and defining $\mathbf{D}_{N-1}$, $\mathbf{D}_{N-2}, \ldots, \mathbf{D}_1$ as increasingly *simpler* versions of the *same* descriptor $\mathbf{D}$, i.e., versions exhibiting a decreasing recognition accuracy and $t_{N-1} > t_{N-2} > \ldots > t_1$. This is the solution we empirically investigate in the rest of this paper. The definition of simpler versions of a given descriptor depends on the specific descriptor at hand. As a simple example, if a descriptor includes color histograms and a distance measure between them, one could reduce the number of bins. In the next section we shall give concrete examples on four different descriptors.

4   Experimental Analysis

We evaluate our approach on two- and three-stage systems, using a benchmark data set and four different appearance descriptors.

4.1   Experimental Setting

Data Set. VIPeR [4] is a benchmark, challenging dataset made up of two images for each of 632 individuals, acquired from two different camera views, with significant pose and illumination changes. As in [3], we repeated our experiments on ten different subsets of 316 individuals each; for each individual we use one image as template and one as probe; we then report the average CMC curve.

Descriptors. SDALF [3][1] subdivides the body into left and right torso and legs. Three kinds of features are extracted from each part: maximally stable color regions, i.e., elliptical regions (blobs) exhibiting distinct color patterns (their number depends on the specific image), with a minimum size of 15 pixels; a $16 \times 16 \times 4$-bins weighted HSV color histogram (wHSV); and recurrent high-structured patches (RHSP) to characterize texture. A specific similarity measure is defined for each feature; the matching score is computed as their linear combination. We did not use RHSP due to its relatively lower performance. We increased the minimum MSCR blob size to 65 and 45 for the first and second stage, respectively (which reduces the number of blobs), and reduced the corresponding number of wHSV histogram bins to $3 \times 3 \times 2$ and to $8 \times 8 \times 3$.

gBiCov is based on biologically-inspired features (BIF) [12],[2] which are obtained by Gabor filters with different scales over the HSV color channels. The resulting images are subdivided into overlapping regions of 16×16 pixels; each region is represented by a covariance descriptor that encodes shape, location and color information. A feature vector is then obtained by concatenating BIF features and covariance descriptors; PCA is finally applied to reduce its dimensionality. We obtained faster versions of gBiCov by increasing the region size to 32×64 and 16×32 pixels for the first and second stage, respectively.

LOMO [10][3] extracts an $8 \times 8 \times 8$-bins HSV color histogram and two scales of the Scale Invariant Local Ternary Pattern histogram (characterizing texture) from overlapping windows of 10×10 pixels; only one histogram is retained from all windows at the same horizontal location, obtained as the maximum value among all the corresponding bins. These histograms are concatenated with the ones computed on a down-sampled image. A metric learning method is used to define the similarity measure. We increased the window size to 20×20 and 15×15 for the first and second stage, respectively, and decreased the corresponding number of bins of the HSV histogram to $3 \times 3 \times 2$ and $4 \times 4 \times 3$.

MCM (Multiple Component Matching) [16][4] subdivides body into torso and legs, and extracts 80 rectangular, randomly positioned image patches from each

---

[1] Source code: http://www.lorisbazzani.info/sdalf.html.

[2] Source code: http://vipl.ict.ac.cn/members/bpma.

[3] Source code: http://www.cbsr.ia.ac.cn/users/scliao/projects/lomo_xqda/.

[4] The source code is available upon request.

part. Each patch is described by a $24 \times 12 \times 4$-bins HSV histogram. The similarity measure is the average k-th Hausdorff distance between the set of patches of each pair of corresponding body parts. In our experiments we reduced the number of patches to 10 and 20 for the first and second stage, respectively, and the corresponding number of bins of the HSV histogram to $3 \times 3 \times 2$ and $12 \times 6 \times 2$.

4.2 Experimental Results

We carried out our experiments using an Intel Core i5 2.6 GHz CPU. For each descriptor we designed a two- and a three-stage system. We used the same version of a given descriptor both in the first stage of two-stage systems and in the second stage of three-stage systems. For each descriptor, the average time for computing a single matching score in each stage is reported in Table 1; note that for MCM the first- and second-stage versions have a much lower processing time than the original one, with respect to the other descriptors: this is due to the use of the Hausdorff distance, which makes the processing time proportional to the *square* of the number of image patches (see above). The number of matching scores n_i computed by the i-th stage ($i > 1$), for each descriptor, is reported in Table 2. These values were computed using the criterion described in Sect. 3, setting a threshold $\Delta = 1\%$ and $\Delta = 0.5\%$ for two- and three-stage systems, respectively. We point out that this criterion aims at keeping recognition accuracy as high as possible (possibly identical to that of the original descriptor), while reducing processing time. We also remind the reader that n_1 always equals the total number of templates, which is $n = 316$ in our experiments. The average CMC curves are reported in Figs. 3 (two-stage systems) and 4 (three-stage systems). Inside each plot we also report a comparison between the CMC curve of multi-stage systems obtained from different values of Δ. The ratio of the corresponding processing time with respect to the one of the original, most accurate descriptor, is reported in Table 2.

Figures 3 and 4 show that the CMC curves of multi-stage systems are nearly identical to the ones of the corresponding original descriptor; some differences are visible in two- and three-stage systems for the SDALF and MCM descriptors, only for ranks higher than 50. The reduction in processing time was however not high: Table 2 shows that it was 25% to 32% for two-stage systems, and 17% to 27% for three-stage systems, depending on the descriptor. Reducing the number of matching scores computed by each stage, through the use of a higher

Table 1. Average processing time (in msec.) for computing one matching score in each stage of two- and three-stage systems, for each of the four descriptors.

		SDALF	gBiCov	LOMO	MCM
Two-stage	1st stage	2.08	0.0057	0.0023	0.060
Three-stage	1st stage	1.60	0.0015	0.0017	0.051
	2nd stage	2.08	0.0057	0.0023	0.060
Original descriptor (last stage)		9.44	0.0400	0.0370	27.400

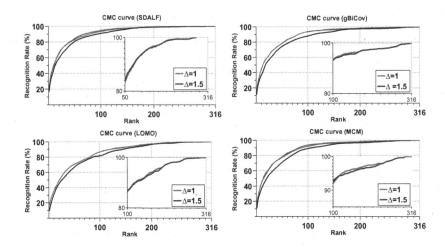

Fig. 3. CMC curves of two-stage systems, for $\Delta = 1\%$ (see text for the details). Blue: original descriptor; black: first-stage; red: two-stage system. The inner plots show a comparison between the CMC curves of two-stage systems obtained for $\Delta = 1\%$ and $\Delta = 1.5\%$: they differ only for the highest ranks shown in these plots. Figure is best viewed in color. (Color figure online)

threshold Δ, affected only slightly the ranking accuracy of multi-stage systems, and only for the highest ranks (see the CMC curves inside the boxes in Figs. 3 and 4). On the other hand, this provided a significant reduction in processing time, especially on three-stage systems, where it ranged from 29 % to 53 % for $\Delta = 1.5\%$.

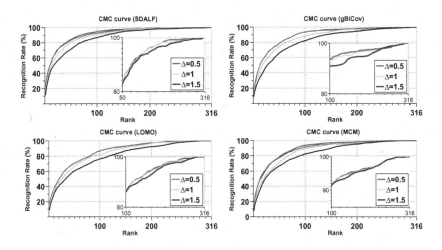

Fig. 4. CMC curves of three-stage systems, for $\Delta = 0.5\%$ (see text for the details). Blue: original descriptor; black: first-stage; green: second stage; red: three-stage system. Inner plots: comparison between the CMC curves of three-stage systems obtained for $\Delta = 0.5\%$, 1 %, and 1.5 %, which differ only in the highest ranks shown in these plots. Figure is best viewed in color. (Color figure online)

The above results provide evidence that the proposed multi-stage ranking approach is capable of improving the trade-off between recognition accuracy and processing time of a given descriptor. The attainable improvement depends on the specific descriptor, i.e., on its similarity measure and on the parameters that can be modified to obtain faster versions of it. This is clearly visible from Table 2; in particular, using the same number of stages and the same criterion to choose the number of matching scores computed by each stage, we consistently attained the lowest reduction in processing time using SDALF. We point out that our experiments were not aimed at finding the best set of parameters, and their best values, to optimize the trade-off between recognition accuracy and processing time of each descriptor. Accordingly, we believe that a more focused choice could provide a higher reduction of processing time than the one attained in our experiments, without affecting recognition accuracy.

Table 2. Number of matching scores computed by each stage but the first one in the two- and three-stage systems, for each of the four descriptors, and for the different values of Δ considered in the experiments. The ratio of the corresponding processing time, with respect to the original descriptor, is also reported.

	Two-stage systems				Three-stage systems								
	$\Delta = 1$		$\Delta = 1.5$		$\Delta = 0.5$			$\Delta = 1$			$\Delta = 1.5$		
	n_2	Time	n_2	Time	n_2	n_3	Time	n_2	n_3	Time	n_2	n_3	Time
SDALF	150	**0.7**	138	**0.66**	271	150	**0.83**	214	151	**0.8**	192	130	**0.71**
gBiCov	170	**0.68**	163	**0.66**	283	178	**0.73**	247	160	**0.66**	209	133	**0.55**
LOMO	196	**0.68**	186	**0.65**	282	198	**0.73**	265	192	**0.71**	254	154	**0.58**
MCM	236	**0.75**	162	**0.51**	248	235	**0.75**	239	170	**0.54**	230	148	**0.47**

5   Conclusion

We proposed a multi-stage ranking approach for person re-identification systems, aimed at attaining a trade-off between ranking quality and processing time of a given appearance descriptor. Our approach focuses on practical application scenarios characterized by a very large template gallery to be ranked in response to a query by a human operator, and/or by a similarity measure exhibiting a high processing time. A first empirical evidence on the benchmark VIPeR data set, using four different descriptors, showed that the proposed approach is capable of reducing processing time with respect to the original descriptor, attaining at the same time almost the same ranking quality. The observed reduction in processing time is not high, although it can be improved by suitably tuning the parameters of the descriptor at hand. In practice, it could also be difficult to accurately estimate the corresponding optimal values of the number of templates to be ranked by each stage (but the first one), as they depend on the size of template gallery. We are currently investigating design criteria focused instead on strict requirements on processing time, for application scenarios where a reduction in ranking quality is acceptable.

References

1. Bedagkar-Gala, A., Shah, S.K.: A survey of approaches and trends in person re-identification. Image Vis. Comput. **32**(4), 270–286 (2014)
2. Dutra, C.R., Schwartz, W.R., Souza, T., Alves, R., Oliveira, L.: Re-identifying people based on indexing structure and manifold appearance modeling. In: SIBGRAPI-Conference on Graphics, Patterns and Images, pp. 218–225 (2013)
3. Farenzena, M., Bazzani, L., Perina, A., Murino, V., Cristani, M.: Person re-identification by symmetry-driven accumulation of local features. In: International Conference on Computer Vision and Pattern Recognition, pp. 2360–2367 (2010)
4. Gray, D., Tao, H.: Viewpoint invariant pedestrian recognition with an ensemble of localized features. In: Forsyth, D., Torr, P., Zisserman, A. (eds.) ECCV 2008. LNCS, vol. 5302, pp. 262–275. Springer, Heidelberg (2008). doi:10.1007/978-3-540-88682-2_21
5. Hirzer, M., Beleznai, C., Roth, P.M., Bischof, H.: Person re-identification by descriptive and discriminative classification. In: Heyden, A., Kahl, F. (eds.) SCIA 2011. LNCS, vol. 6688, pp. 91–102. Springer, Heidelberg (2011). doi:10.1007/978-3-642-21227-7_9-
6. Hirzer, M., Roth, P.M., Bischof, H.: Person re-identification by efficient impostor-based metric learning. In: International Conference on Advanced Video and Signal-Based Surveillance, pp. 203–208 (2012)
7. Huang, S., Gu, Y., Yang, J., Shi, P.: Reranking of person re-identification by manifold-based approach. In: International Conference on Image Processing, pp. 4253–4257 (2015)
8. Kaynak, C., Alpaydin, E.: Multistage cascading of multiple classifiers: one man's noise is another man's data. In: International Conference on Machine Learning, pp. 455–462 (2000)
9. Khedher, M.I., El Yacoubi, M.A.: Two-stage filtering scheme for sparse representation based interest point matching for person re-identification. In: Battiato, S., Blanc-Talon, J., Gallo, G., Philips, W., Popescu, D., Scheunders, P. (eds.) ACIVS 2015. LNCS, vol. 9386, pp. 345–356. Springer, Heidelberg (2015). doi:10.1007/978-3-319-25903-1_30
10. Liao, S., Hu, Y., Zhu, X., Li, S.Z.: Person re-identification by local maximal occurrence representation and metric learning. In: International Conference on Computer Vision and Pattern Recognition, pp. 2197–2206 (2015)
11. Liu, C., Loy, C.C., Gong, S., Wang, G.: POP: person re-identification post-rank optimisation. In: International Conference on Computer Vision, pp. 441–448 (2013)
12. Ma, B., Su, Y., Jurie, F.: Covariance descriptor based on bio-inspired features for person re-identification and face verification. Image Vis. Comput. **32**(6), 379–390 (2014)
13. Metternich, M.J., Worring, M.: Track based relevance feedback for tracing persons in surveillance videos. Comput. Vis. Image Underst. **117**(3), 229–237 (2013)
14. Pudil, P., Novovicova, J., Blaha, S., Kittler, J.: Multistage pattern recognition with reject option. In: International Conference on Pattern Recognition, vol. II, pp. 92–95 (1992)

15. Satta, R., Fumera, G., Roli, F.: Fast person re-identification based on dissimilarity representations. Pattern Recogn. Lett. **33**(14), 1838–1848 (2012)
16. Satta, R., Fumera, G., Roli, F., Cristani, M., Murino, V.: A multiple component matching framework for person re-identification. In: Maino, G., Foresti, G.L. (eds.) ICIAP 2011. LNCS, vol. 6979, pp. 140–149. Springer, Heidelberg (2011). doi:10. 1007/978-3-642-24088-1_15
17. Sperduti, A.: Theoretical and experimental analysis of a two-stage system for classification. IEEE Trans. Pattern Anal. Mach. Intell. **24**(7), 893–904 (2002)
18. Trapeznikov, K., Saligrama, V., Castañón, D.: Multi-stage classifier design. Mach. Learn. **92**(2–3), 479–502 (2013)
19. Viola, P., Jones, M.: Rapid object detection using a boosted cascade of simple features. In: International Conference on Computer Vision and Pattern Recognition, vol. 1, p. 511 (2001)
20. Wang, Z., Hu, R., Liang, C., Leng, Q., Sun, K.: Region-based interactive ranking optimization for person re-identification. In: Ooi, W.T., Snoek, C.G.M., Tan, H.K., Ho, C.-K., Huet, B., Ngo, C.-W. (eds.) PCM 2014. LNCS, vol. 8879, pp. 1–10. Springer, Heidelberg (2014). doi:10.1007/978-3-319-13168-9_1

Unsupervised Parameter Estimation of Non Linear Scaling for Improved Classification in the Dissimilarity Space

Mauricio Orozco-Alzate[1]([⊠]), Robert P.W. Duin[2], and Manuele Bicego[3]

[1] Departamento de Informática y Computación, Universidad Nacional de Colombia, Sede Manizales, km 7 vía al Magdalena, Manizales 170003, Colombia
morozcoa@unal.edu.co
[2] Pattern Recognition Laboratory, Delft University of Technology, Mekelweg 4, 2628 CD Delft, The Netherlands
r.p.w.duin@tudelft.nl
[3] Dipartimento di Informatica, Università degli Studi di Verona, Cá Vignal 2, Strada le Grazie 15, Verona 37134, Italy
manuele.bicego@univr.it

Abstract. The non-linear scaling of given dissimilarities, by raising them to a power in the (0,1) interval, is often useful to improve the classification performance in the corresponding dissimilarity space. The optimal value for the power can be found by a grid search across a leave-one-out cross validation of the classifier: a procedure that might become costly for large dissimilarity matrices, and is based on labels, not permitting to capture the global effect of such a scaling. Herein, we propose an entirely unsupervised criterion that, when optimized, leads to a sub-optimal but often good enough value of the scaling power. The criterion is based on a trade-off between the dispersion of data in the dissimilarity space and the corresponding intrinsic dimensionality, such that the concentrating effects of the power transformation on both the space axes and the spatial distribution of the objects are rationed.

Keywords: Dissimilarity space · Intrinsic dimensionality · Dispersion · Non linear scaling · Nearest neighbor classification · Power transformation

1 Introduction

In statistical pattern recognition, an object is conventionally represented as a vector whose entries correspond to numerical values of its features. Therefore, in such a representation, objects are points in a vector space: the well-known *feature space*. However, this conventional representation is often inconvenient, particularly when the extraction of features from symbolic data (such as graphs and grammars) or from raw sensor measurements (such as signals and images) is difficult or even when it is not clear how to do it in the first place. As an alternative, Pekalska and Duin proposed [13] the option of measuring dissimilarities

© Springer International Publishing AG 2016
A. Robles-Kelly et al. (Eds.): S+SSPR 2016, LNCS 10029, pp. 74–83, 2016.
DOI: 10.1007/978-3-319-49055-7_7

between pairs of objects and organizing them as vectors such that each object is represented as a point in the so-called *dissimilarity space* [7] where any classifier can be trained and applied. The dissimilarity representation is within the field of (dis)similarity pattern recognition that has been actively researched during the last years [14, 15].

In many pattern classification problems it is mandatory to normalize the feature space, i.e. to make comparable the ranges of the different features that can derive from different measures/sensors: the typical approach is to apply a *linear* scaling to the axes of the vector space; this operation guarantees that the classifier decision equally takes into account values in all directions, once the unwanted influences of their original dynamic ranges have been removed. In the dissimilarity space, in contrast, range differences among the directions tend to be less notorious because all the features are of the same nature, i.e. they are all distances to the objects of the reference group, formally called the *representation set*. For this reason linear scaling is less crucial. However, other more complex scaling operations, such as those involving *non linear transformations*, can be very useful and lead to improvements in the classification – this has been suggested also for classical feature spaces [1, 4, 5, 11, 17]. For dissimilarity spaces, Duin et al. [6] found that the non-linear scaling of given dissimilarities by their power transformation appears to be useful for improving the nearest neighbor performance in the dissimilarity space. They studied its behavior in terms of classification error and found that raising dissimilarities to powers less than 1 often contributes to such an improvement. When trying to explain the phenomenon, they suggested that the benefits derive from the following three properties: when applying a power transformation with power less than 1, (i) objects tend to be equally distant from the others, (ii) distances to outliers are shrunk, and (iii) the neighborhood of each object is enlarged by emphasizing distances between close objects.

In their study, as well as in the others related to classical feature spaces cited above [4, 5, 11, 17], the estimation of the proper power parameter represents a crucial open issue; typically such parameter is set by hand, or found by an exhaustive search; in [6] it is estimated via the computationally prohibitive cross validation. In this paper we propose a novel unsupervised criterion which can guide the selection of the parameter of the power transformation: this criterion tries to find a compromise – as the power parameter approaches to zero – between the reduction in the dispersion in the data and the increase in the intrinsic dimensionality of the resulting dissimilarity space (if a too small power is applied all points are converging around 1). This criterion is unsupervised – since it does not require labels – and computationally more feasible than cross validation – since it does not require repeated training of classifiers. A thorough experimental evaluation on several different datasets shows that by applying the power transformation with the best parameter according to the proposed criterion we obtain accuracies which are (i) almost always significantly better than those obtained in the space without the preprocessing and (ii) many times equivalent or better than those obtained by the computationally expensive cross validation procedure.

The rest of the paper is organized as follows: in Sect. 2 we briefly summarize the dissimilarity space and the non linear scaling by power transformation; then, in Sect. 3 we detail the proposed approach; the experimental evaluation is presented in Sect. 4; finally, in Sect. 5, conclusions are drawn and future perspectives are envisaged.

2 Background

2.1 The Dissimilarity Space

The vector arrangement of the dissimilarities computed between a particular object x and other objects from a set $\mathcal{R}$ allows representing x as a point in a vector space. Such a space is called the *dissimilarity space*, having in principle as many dimensions as the cardinality of $\mathcal{R}$, which is known as the *representation set*. For a set of training objects $\mathcal{T}$, the set $\mathcal{R}$ builds a so-called *dissimilarity representation* in the form of a dissimilarity matrix $\mathbf{D}(\mathcal{T}, \mathcal{R})$. The representation set is often the same as the training set, so $\mathbf{D}(\mathcal{T}, \mathcal{R}) = \mathbf{D}(\mathcal{T}, \mathcal{T})$. For notation simplicity, hereafter we simply use $\mathbf{D}$ to refer to the square dissimilarity matrix $\mathbf{D}(\mathcal{T}, \mathcal{T})$.

Several studies [7,13] have shown the possibilities of training classifiers in the dissimilarity space, such that a test object represented in terms of its dissimilarities to $\mathcal{R}$ can be classified by a more sophisticated rule than the nearest neighbor classifier on the given dissimilarities (i.e. template matching, denoted as 1-NN). The classifier in the dissimilarity space can even be the same nearest neighbor rule but now based on distances between points in the dissimilarity space; here we denote that case as 1-NND in order to distinguish it from template matching.

2.2 Non Linear Scaling

Raising all dissimilarities to the same power is a simple and straightforward non linear scaling. For a dissimilarity matrix $\mathbf{D}$, such a transformation can be written as follows:

$$\mathbf{D}^{\star \rho} = (d_{ij}^{\rho}), \qquad \rho > 0 \tag{1}$$

where each entry, $d_{ij} = d(x_i, x_j)$, of the matrix denotes the dissimilarity between two objects x_i and x_j and $\star$ denotes the entrywise (Hadamard) power function [9]. There exists an optimal value for ρ that provides the best 1-NND classification performance. Let's denote it as ρ^*. In most cases, ρ^* is lower than 1. This is reasonable, since with $\rho < 1$ we have a concave function that raises low values and shrinks high values: for dissimilarities, this may have a good impact on the representation in the dissimilarity spaces, since it reduces the impact of outliers (large distances are reduced) and increases the importance of the neighborhood (small distances are increased).

Therefore, we only consider to search for an estimate $\widehat{\rho^*}$ in the interval $(0, 1]$. Below we explain the existing method to estimate ρ^* by cross validation, followed in Sect. 3 by the explanation of our proposed estimation via the optimization of an unsupervised criterion.

Optimization via Cross Validation. A typical procedure to optimize the value of a parameter is by searching over the parameter domain for the lowest cross validation classification error. This strategy was the one used by Duin et al. [6] for finding the best parameter, which we call in this case $\widehat{\rho}^*_{cv}$, as follows:

$$\widehat{\rho}^*_{cv} = \arg\min_{\rho \in (0,1]} \epsilon_{1-NND}(\mathbf{D}^{\star\rho}) \tag{2}$$

where ϵ_{1-NND} denotes the leave-one-out cross validation error of 1-NND. Even though experiments in [6] suggested that this optimization permits a good classification performance, it might become computationally prohibitive for large datasets. Moreover, such criterion does not permit to understand what is happening with the non linear scaling, i.e. it does not provide an explanation of the topological effect of the parameter value in the space.

3   The Proposed Criterion

As introduced before, when applying a power transformation with $\rho < 1$, we obtain a two-fold effect on data in the dissimilarity space. First, the dispersion of the values in each dimension of the space is shrunk (by raising small distances and reducing large distances); second, the neighborhood of each point is highly emphasized (raising small distances). This behaviour is becoming more and more extreme when ρ approaches zero. Clearly, up to some extent these effects are desirable, in order to reduce the impact of outliers (distances to far away points are reduced) and to better characterize the neighborhood of each object (distances to nearby points are raised); however, after a certain point such positive effects are lost, since all points tend to be equally spaced in the space, thus loosing all the information contained in the original dissimilarity matrix. This effect can be monitored by looking at the intrinsic dimensionality of the data, which increases when points tend to be more equally spaced. Therefore, using a criterion that optimizes a trade-off between those two effects (reduction of dispersion and increase of the intrinsic dimensionality) seems a reasonable way to find $\widehat{\rho}^*$.

Among the available dispersion measures, the quartile coefficient of dispersion (qcd) [10, p. 15] is a robust statistical estimator that gives a scale-free measure of data spread. It is given as:

$$qcd = \frac{Q_3 - Q_1}{Q_3 + Q_1}, \tag{3}$$

where Q_3 and Q_1 are the third and first quartiles, respectively. In our case, they are computed as follows: for each column (dimension) of $\mathbf{D}^{\star\rho}$, we find the median of the upper half of the values (which is Q_3, also called the 75th percentile) and the median of the lower half of them (which is Q_1, also called the 25th percentile).

Similarly, there are many methods to estimate the intrinsic dimensionality (id) of a dataset, see for instance the reviews by Camastra [2,3]. We have a adopted the one described in [13, p. 313] which directly computes the estimation from dissimilarity data:

$$\widehat{id}(\mathbf{D}) = \left\lceil 2\frac{(\mathbf{1}^\top \mathbf{D}^{\star 2} \mathbf{1})^2}{n(n-1)\mathbf{1}^\top \mathbf{D}^{\star 4} \mathbf{1} - (\mathbf{1}^\top \mathbf{D}^{\star 2} \mathbf{1})^2} \right\rceil \tag{4}$$

where $\mathbf{D}^{\star 2} = (d_{ij}^2)$, $\mathbf{D}^{\star 4} = (d_{ij}^4)$ and n is the number of columns (and rows) of the square matrix $\mathbf{D}$.

Given these definitions, our criterion tries to determine the best parameter (which we call $\widehat{\rho}^*_{nlm}$) by optimizing the compromise between (i) the average – or, better, its robust estimate, the median – of the dispersion (3) per dimension and (ii) the intrinsic dimension of (4) computed for the pairwise distances in the dissimilarity space, that is, for a matrix of Euclidean distances $\mathbf{D}_{DS}$ between pairs of points in the dissimilarity space. The final criterion can be written as:

$$\widehat{\rho}^*_{nlm} = \arg\min_{\rho \in (0,1]} \left[\underset{1 \leq i \leq n}{\text{median}} \left(qcd_i\right) \times \widehat{id}\left(\mathbf{D}^{\star \rho}_{DS}\right) \right] \tag{5}$$

Notice that, even though there are several alternatives to define a compromise between two variables, we have chosen to minimize the product between them. A multiplicative criterion has also been adopted in other scenarios [8,12] where the two variables of interest are related in a non-trivial way.

3.1 Inductive and Transductive Versions

The criterion introduced in the previous section is completely unsupervised: exploiting this property, we investigate its usefulness in two different flavours, which we called "Version 1" and "Version 2", respectively:

1. Version 1 ($\widehat{\rho}^*_{nlm1}$): the best parameter is the one optimizing the proposed criterion on the training set: this represents the classical learning, also known as inductive inference [16, p. 577], where the criterion is determined by using only the training objects.
2. Version 2 ($\widehat{\rho}^*_{nlm2}$): the best parameter is the one optimizing the proposed criterion on the whole dataset, clearly by ignoring the labels. This represents the so called transductive learning [18] where all the available objects are used: the training objects, for which we can employ the labels, and the testing objects, for which labels are unknown. Since the proposed criterion does not take into account the labels, the transductive learning can be applied.

4   Experimental Results

The proposed approach has been tested using a set of public domain datasets[1] (also employed in [6]) – see Table 1. Most of them are derived from real objects (images, text, protein sequences). The Chickenpieces dataset consists out of 44

---

[1] More information on datasets can be found at http://37steps.com/prdisdata.

Table 1. Datasets employed for empirical evaluation.

Name	Objects	Classes
(1) Catcortex	65	4
(2) Coildelftdiff	288	4
(3) Coildelftsame	288	4
(4) Coilyork	288	4
(5) Delftgestures	1500	20
(6) Flowcytodis1	612	3
(7) Flowcytodis2	612	3
(8) Flowcytodis3	612	3
(9) Flowcytodis4	612	3
(10) Newsgroups	600	4
(11) Prodom	2604	4
(12) Protein	213	4
(13) Woodyplants50	791	14
(14) Zongker	2000	10
(15) Chickenpieces (44 sets)	446	5
(16) Polydish57	4000	2
(17) Polydism57	4000	2

dissimilarity matrices: in the tables, the average characteristics are shown. In our empirical evaluation we compared the errors made by the Nearest Neighbor rule[2] (errors of 1-NND) in four different versions of the dissimilarity space:

1. *Original*: this is unprocessed case (no transformation is applied), i.e. the dissimilarity space is built using the original dissimilarity matrix $\mathbf{D}$.
2. *NL-Cross Val*: in this case the dissimilarity space is built starting from $\mathbf{D}^{\star \widehat{\rho^*_{cv}}}$, i.e. after applying a non linear transformation where the optimal parameter is chosen by optimizing the LOO error on the training set. As said before, this represents the criterion proposed in [6].
3. *NL-Disp (ver. 1)*: in this case the dissimilarity space is built starting from $\mathbf{D}^{\star \widehat{\rho^*_{nlm1}}}$, i.e. after applying non linear transformation with parameter chosen by optimizing the proposed criterion on the training set.
4. *NL-Disp (ver. 2)*: in this case the dissimilarity space is built starting from $\mathbf{D}^{\star \widehat{\rho^*_{nlm2}}}$, i.e. after applying a non linear transformation with parameter chosen by optimizing the proposed criterion on the whole dataset (in a transductive way, see previous section).

---

[2] We restrict ourselves to using a parameterless classifier – the nearest neighbor rule – because we are interested in judging the potential improvement of the data representation after the power transformation, independently from the influence of any classifier parameter.

Table 2. 1NN-D errors for the different datasets. Between brackets we reported the standard errors of the mean.

Dataset	Original	NL-Cross Val	NL-Disp (v1)	NL-Disp (v2)
Catcortex	0.1067(7e-03)	0.1057(7e-03)	0.1012(6e-03)	0.0981(7e-03)
Coildelftdiff	0.4611(2e-03)	0.4498(2e-03)	0.4575(2e-03)	0.4528(2e-03)
Coildelftsame	0.4181(2e-03)	0.4130(2e-03)	0.4158(2e-03)	0.4102(2e-03)
Coilyork	0.3948(2e-03)	0.3265(2e-03)	0.3532(2e-03)	0.3371(2e-03)
Delftgestures	0.0949(2e-04)	0.0526(2e-04)	0.0599(2e-04)	0.0563(2e-04)
Flowcytodis1	0.3857(9e-04)	0.3797(9e-04)	0.3781(9e-04)	0.3770(9e-04)
Flowcytodis2	0.3827(9e-04)	0.3749(1e-03)	0.3754(9e-04)	0.3730(1e-03)
Flowcytodis3	0.4077(9e-04)	0.3911(9e-04)	0.3890(9e-04)	0.3850(9e-04)
Flowcytodis4	0.4251(9e-04)	0.4127(9e-04)	0.4109(8e-04)	0.4083(9e-04)
Newsgroups	0.2960(9e-04)	0.2915(9e-04)	0.2887(9e-04)	0.2887(9e-04)
Prodom	0.0193(9e-05)	0.0072(6e-05)	0.0065(6e-05)	0.0065(6e-05)
Protein	0.0059(6e-04)	0.0063(7e-04)	0.0062(6e-04)	0.0055(6e-04)
Woodyplants50	0.1617(5e-04)	0.1188(5e-04)	0.1379(5e-04)	0.1292(5e-04)
Zongker	0.0529(1e-04)	0.0408(2e-04)	0.0377(2e-04)	0.0377(2e-04)
Chickenpieces	0.1543(1e-04)	0.1252(1e-04)	0.1307(1e-04)	0.1263(1e-04)
Polydish57	0.0306(5e-05)	0.0166(4e-05)	0.0233(4e-05)	0.0233(4e-05)
Polydism57	0.0153(4e-05)	0.0135(3e-05)	0.0226(5e-05)	0.0226(5e-05)

Errors have been computed using averaged hold out cross validation, i.e. by using half of the dataset for training (and representation) and the remaining half for testing. In order to ensure robust estimation of errors, this procedure has been repeated 200 times, and results are averaged. For criteria 2–4, the best value has been chosen in the range $1.25^{-15}, 1.25^{-14.5}, 1.25^{-14}, ..., 1$ for the exponent. Averaged errors, together with standard errors of the mean, are reported in Table 2. In order to get a more direct view on the results, we reported in Table 3 an improvement/degradation table, as resulting from several different pairwise statistical tests. In particular, we compared errors obtained with the proposed criterion (NL-Disp in both versions v1 and v2) with those obtained without transforming the space (Original) and with the parameter chosen via Cross Validation (NL-Cross Val). As statistical test we employed the paired t-test, comparing the 200 errors obtained with the 200 repetitions of the cross validation. In the table, we used five different symbols:

– the symbols "↑" and "↑↑" indicate a statistically significant improvement (results with our criterion are better): the former indicates that the test passed with a p-value less than 0.05 but greater than 0.001, whereas in the latter case the p-value was less than 0.001;

Table 3. Pairwise statistical comparisons: "↑" indicates a statistically significant improvement (results with our criterion are better), "↓" a statistically significant degradation (results with our criterion are worst), whereas "≈" indicates that the two methods are equivalent (i.e. there is no statistically significant difference).

Dataset	NL-Disp (v1) vs Original	NL-Disp (v2) vs Original	NL-Disp (v1) vs NL-Cross Val	NL-Disp (v2) vs NL-Cross Val
Catcortex	↑	↑↑	↑	↑↑
Coildelftdiff	↑↑	↑↑	↓↓	≈
Coildelftsame	↑	↑↑	↓	↑
Coilyork	↑↑	↑↑	↓↓	↓↓
Delftgestures	↑↑	↑↑	↓↓	↓↓
Flowcytodis1	↑↑	↑↑	≈	↑
Flowcytodis2	↑↑	↑↑	≈	≈
Flowcytodis3	↑↑	↑↑	≈	↑↑
Flowcytodis4	↑↑	↑↑	≈	↑↑
Newsgroups	↑↑	↑↑	↑	↑
Prodom	↑↑	↑↑	↑↑	↑↑
Protein	≈	≈	≈	≈
Woodyplants50	↑↑	↑↑	↓↓	↓↓
Zongker	↑↑	↑↑	↑↑	↑↑
Chickenpieces	↑↑	↑↑	↓↓	↓↓
Polydish57	↑↑	↑↑	↓↓	↓↓
Polydism57	↓↓	↓↓	↓↓	↓↓

- "↓" and "↓↓" indicate a statistically significant degradation (results with our criterion are worst); also in this case the former indicates that the test passed with a p-value less than 0.05 but greater than 0.001, whereas in the latter case the p-value was less than 0.001;
- "≈" indicates that the two methods are equivalent (i.e. there is no statistically significant difference).

From the table different observations can be derived. First, as expected, the transductive version (version 2) of our criterion is almost always slightly better than version 1; this interesting result is possible thanks to the unsupervised nature of the proposed criterion. Reasonably, this does not hold if the dataset is large enough (as for Zongker, Polydish57 and Polydism57). Second, non linearly preprocessing the dissimilarity matrix by choosing the parameter with our criterion almost always results in a statistically significant improvement in the classification performances with respect to the original space. The only exceptions are for the protein and the Polydism57 datasets, for which, however, an almost zero error was already achieved in the original space, leaving small room for improvements. This is coherently true for both version 1 and version 2. Finally, the proposed criterion also compares reasonably well with the cross validation approach: if we consider the version 2, in 11 cases out of 17 our results are better or equivalent (in 8 cases they are significantly better), whereas only in 6 cases they are worst. In these latter cases, however, degradations are very small: ≈ 0.01 for CoilYork, WoodyPlants50, Polydism57 and Polydish57, ≈ 0.004 for DelftGestures, and ≈ 0.001 for ChickenPieces. We are convinced that these represent really promising results, also considering that our criterion is completely unsupervised.

5 Conclusions

In this paper a novel unsupervised criterion to tune the parameter of the power transformation (non-linear scaling) of dissimilarities has been proposed. The new tuning criterion is based on a trade-off between the median dispersion per dimension in the dissimilarity space (measured in terms of the quartile coefficient of dispersion) and the intrinsic dimension of the resulting dissimilarity space. The idea behind our approach is that a good performance of the nearest neighbor classifier in the dissimilarity space is associated to such a compromise between how much we shrink the data at the cost of increasing the intrinsic dimensionality – the shrinking is desirable because, by reducing the range, we can potentially reduce the influence of the outliers since we are largely reducing high distances (i.e. the distances to – possible – outliers) more than reducing short distances.

The proposed criterion is unsupervised and, therefore, can be even applied in a transductive learning setting. Empirical results on many different datasets partially support our intuitions. As a future work, we would like to study the properties of the proposed criterion also in classical feature based problems [4,5, 11,17]. Moreover, we aim at providing a more formal – theoretical or numerical – explanation: one possibility is to try to bridge our experimental evidence with the theory on Hadamard powers [9].

Acknowledgments. Discussions for the proposal in this paper started while Mauricio Orozco-Alzate and Manuele Bicego visited the Pattern Recognition Laboratory, Delft University of Technology (Delft, The Netherlands) in September 2015 by a kind invitation from Robert P.W. Duin to attend the "Colors of dissimilarities" workshop.

This material is based upon work supported by Universidad Nacional de Colombia under project No. 32059 (Code Hermes) entitled *"Consolidación de las líneas de investigación del Grupo de Investigación en Ambientes Inteligentes Adaptativos GAIA"* within "Convocatoria interna de investigación de la Facultad de Administración 2015, para la formulación y ejecución de proyectos de consolidación y/o fortalecimiento de los grupos de investigación. Modalidad 1: Formulación y ejecución de proyectos de consolidación".

The fist author also acknowledges travel funding to attend S+SSPR 2016 provided by Universidad Nacional de Colombia through "Convocatoria para la Movilidad Internacional de la Universidad Nacional de Colombia 2016–2018. Modalidad 2: Cofinanciación de docentes investigadores o creadores de la Universidad Nacional de Colombia para la presentación de resultados de investigación o representaciones artísticas en eventos de carácter internacional, o para la participación en megaproyectos y concursos internacionales, o para estancias de investigación o residencias artísticas en el extranjero".

References

1. Bicego, M., Baldo, S.: Properties of the Box-Cox transformation for pattern classification. Neurocomputing (2016, in press)
2. Camastra, F.: Data dimensionality estimation methods: a survey. Pattern Recogn. **36**(12), 2945–2954 (2003)
3. Camastra, F., Staiano, A.: Intrinsic dimension estimation: advances and open problems. Inf. Sci. **328**, 26–41 (2016)
4. Carli, A.C., Bicego, M., Baldo, S., Murino, V.: Non-linear generative embeddings for kernels on latent variable models. In: Proceedings of ICCV 2009 Workshop on Subspace Methods, pp. 154–161 (2009)
5. Carli, A.C., Bicego, M., Baldo, S., Murino, V.: Nonlinear mappings for generative kernels on latent variable models. In: Proceedings of International Conference on Pattern Recognition, pp. 2134–2137 (2010)
6. Duin, R.P.W., Bicego, M., Orozco-Alzate, M., Kim, S.-W., Loog, M.: Metric learning in dissimilarity space for improved nearest neighbor performance. In: Fränti, P., Brown, G., Loog, M., Escolano, F., Pelillo, M. (eds.) S+SSPR 2014. LNCS, vol. 8621, pp. 183–192. Springer, Heidelberg (2014). doi:10.1007/978-3-662-44415-3_19
7. Duin, R.P.W., Pekalska, E.: The dissimilarity space: bridging structural and statistical pattern recognition. Pattern Recogn. Lett. **33**(7), 826–832 (2012)
8. Fahmy, A.A.: Using the Bees Algorithm to select the optimal speed parameters for wind turbine generators. J. King Saud Univ. Comput. Inf. Sci. **24**(1), 17–26 (2012)
9. Guillot, D., Khare, A., Rajaratnam, B.: Complete characterization of Hadamard powers preserving Loewner positivity, monotonicity, and convexity. J. Math. Anal. Appl. **425**(1), 489–507 (2015)
10. Kokoska, S., Zwillinger, D.: CRC Standard Probability and Statistics Tables and Formulae, Student edn. CRC Press, Boca Raton (2000)
11. Liu, C.L., Nakashima, K., Sako, H., Fujisawa, H.: Handwritten digit recognition: investigation of normalization and feature extraction techniques. Pattern Recogn. **37**(2), 265–279 (2004)
12. Mariani, G., Palermo, G., Zaccaria, V., Silvano, C.: OSCAR: an optimization methodology exploiting spatial correlation in multicore design spaces. IEEE Trans. Comput. Aided Des. Integr. Circuits Syst. **31**(5), 740–753 (2012)
13. Pekalska, E., Duin, R.P.W.: The Dissimilarity Representation for Pattern Recognition: Foundations and Applications (Machine Perception and Artificial Intelligence), vol. 64. World Scientific, Singapore (2005)
14. Pelillo, M. (ed.): Similarity-Based Pattern Analysis and Recognition. Advances in Computer Vision and Pattern Recognition. Springer, London (2013)
15. Pelillo, M., Hancock, E.R., Feragen, A., Loog, M. (eds.): SIMBAD. LNCS, vols. 7005, 7953, 9370. Springer, Heidelberg (2011, 2013, 2015)
16. Theodoridis, S., Koutroumbas, K.: Pattern Recognition, 4th edn. Academic Press, London (2009)
17. Van Der Heiden, R., Groen, F.C.A.: The Box-Cox metric for nearest neighbour classification improvement. Pattern Recogn. **30**(2), 273–279 (1997)
18. Vapnik, V.N.: Statistical Learning Theory. Adaptive and Learning Systems for Signal Processing, Communication and Control. Wiley, New York (1998)

The Similarity Between Dissimilarities

David M.J. Tax[1]($\boxtimes$), Veronika Cheplygina[1,2],
Robert P.W. Duin[1], and Jan van de Poll[3]

[1] Pattern Recognition Laboratory,
Delft University of Technology, Delft, The Netherlands
d.m.j.tax@tudelft.nl
[2] Biomedical Imaging Group Rotterdam, Erasmus Medical Center,
Rotterdam, The Netherlands
[3] Transparency Lab, Amsterdam, The Netherlands
transparencylab.com

Abstract. When characterizing teams of people, molecules, or general graphs, it is difficult to encode all information using a single feature vector only. For these objects dissimilarity matrices that do capture the interaction or similarity between the sub-elements (people, atoms, nodes), can be used. This paper compares several representations of dissimilarity matrices, that encode the cluster characteristics, latent dimensionality, or outliers of these matrices. It appears that both the simple eigenvalue spectrum, or histogram of distances are already quite effective, and are able to reach high classification performances in multiple instance learning (MIL) problems. Finally, an analysis on teams of people is given, illustrating the potential use of dissimilarity matrix characterization for business consultancy.

1 Introduction

Consider the problem of evaluating and improving performances of teams in organizations based on the employee responses to questionnaires. The teams differ in size, and the roles of employees may be different for every organization. A key question for an organizations top management is how to support the autonomy of these teams while still keeping an eye on the overall process and the coherency of the teams performance. Assuming a span of control of 10–15 direct reports for an average manager, a middlesize organization may easily comprise of hundreds of teams. So, pattern recognition in organizational development may supply fundamentally important information of how similar – or dissimilar – teams are [1,15,20]. A possible solution is to focus at the diversity within a team – is there a large group of people who are all doing a similar job, or are there some isolated groups of people who are doing very different from the rest? Identifying such groups – clusters of employees – would help to compare the organizational structures on a higher level.

More formally, in this paper we focus on comparing sets (teams) of different samples (employees), residing in different feature spaces (evaluation questions). Comparing the team structures would be equivalent to comparing similarity

© Springer International Publishing AG 2016
A. Robles-Kelly et al. (Eds.): S+SSPR 2016, LNCS 10029, pp. 84–94, 2016.
DOI: 10.1007/978-3-319-49055-7_8

matrices, with each similarity matrix originating from a single team. Comparing similarities alleviates the problem of different feature spaces, yet is still not trivial because the sets can be of different sizes, and there are no natural correspondences between the samples.

Comparing distance matrices has links with comparing graph structures: a distance matrix between N samples can be seen as a fully connected graph with N nodes, where the nodes are unlabeled and the edges are associated with weights. In graph-based pattern recognition, approaches such as graph edit distance [3,21] or graph kernels [10,12] have been used to define distance or similarity measures between graphs. Graph matching approaches search for a best correspondence between the nodes and define the graph distance as a measure of discrepancy between the matched nodes and edges. Graph kernels define similarity by considering all possible correspondences. However, the search space for correspondences becomes very large if the nodes are unlabeled, and the graph is fully connected. In [16] we used a threshold on the distances to reduce the number of edges. However, this threshold had a large influence on the results, suggesting that the larger distances can, too, be informative.

To avoid removing informative edges and to present a computationally efficient solution, in this paper we focus on finding feature representations to represent distance matrices. By representing each distance matrix in the same feature space, they can be compared with each other, for example, using the Euclidean distance. We investigate several representations in this paper, based on spectra [6], histograms of all distances [18], histograms of nearest neighbor distances, and hubness properties [23]. A detailed description of the representations is given in Sect. 2.

In Sect. 3 we investigate how well these features representations can encode the class information for some artificial examples. In Sect. 4 we investigate how good these representations are for multiple instance learning (MIL) datasets, where the goal is to classify sets of feature vectors. In Sect. 4.2 we apply the representation on real-world organisational data, and discuss some of the insights that arise from comparing teams of people.

2   Dissimilarity Matrix Representation

We assume we have a collection of N square dissimilarity matrices $\{D_n \in \mathbb{R}^{m_n \times m_n}; n = 1...N\}$ of size $m_n \times m_n$. One element of matrix D_n is indicated by $D_n(i, j)$. We assume that the matrices have the following characteristics:

- The dissimilarities of objects to themselves is zero (i.e. the D_n have zeros on the diagonal), and the dissimilarity is symmetric ($D_n(i, j) = D_n(j, i)$). In situations the matrices are not symmetric, they are made symmetric by averaging D_n and its transpose: $\tilde{D}_n = (D_n + D_n^\top)/2$.
- The size $m_n \times m_n$ of the matrices can be different for each D_n. It is assumed that the matrices have a minimum size of 3×3.
- The order of the rows and columns is arbitrary and may be permuted without altering the information that is stored in the dissimilarity matrix.

For this data type we investigate a few simple vector representations. Additional similarities between the dissimilarity matrices are also possible (such as embedding each matrix into a low-dimensional space and using the earth-movers distance, or matching the rows and columns of the dissimilarity matrices and computing the Frobenius norm [14]), but these tend to be computationally expensive. Here we focus on vector representations of the dissimilarity matrices.

We consider the following representations:

1. spectrum features `spect`: use the k largest eigenvalues σ of the centered matrix D_n:

$$\mathbf{x}_n = \sigma_{1:k}(C^\top D_n C), \quad \text{where } C = \mathbb{I}_{m_n} - \frac{1}{m_n} \mathbf{1}\mathbf{1}^\top \tag{1}$$

2. histogram of distances `hist`: collect all dissimilarities from all matrices, split the range of dissimilarity values from 0 to the maximum into k equally-sized bins, and count for each D_n the number of occurrences into each bin. Optionally, the count can be converted into a frequency by dividing by $m_n(m_n+1)/2$.

3. equalized histogram of distances `histeq`: split the range of dissimilarity values into k bins with an equal number of counts (instead of using equally wide bins). The bins become wider when dissimilarity values do not occur often, and they become smaller for frequently appearing values.

4. histogram of the k-nearest neighbor distances `distnn`: instead of collecting all dissimilarities, only the dissimilarities up to the k-nearest neighbors are used. Per row of D_n, only $k < m_n$ dissimilarities are used; the total number of dissimilarities is therefore reduced from $m_n(m_n + 1)/2$ to $m_n k$. By this variations in local densities are captured better.

5. histogram of how often samples are the k-th nearest neigbor of other samples `disthub`: a measure used in hub analysis [24]. First the dataset is represented by a k-occurence histogram which stores how often each sample is the k-th nearest neighbor of others. To make this representation comparable across datasets of different sizes, it is summarized by q quantiles of the histogram. For the final representation, we concatenate the quantile-histograms for different values of $k \in \{1, 3, \ldots, |K|\}$, resulting in a $|K| \times q$ dim. feature vector.

In some situations we might want to be invariant to (non-linear) scaling of the dissimilarity values. For example, the expert may only have provided a relative ranking, but not an exact dissimilarity between two elements of a set. In this case, the extracted features should be invariant to the scaling of the dissimilarities. In the above representations, only `disthub` is invariant.

3   Illustrative Examples

To show the characteristics of the different representations, we construct some multi-class artificial datasets. Depending on the experiment we perform, the number of dissimilarity matrices, and the sizes of the matrices are varied.

- The **cluster** dataset is constructed to investigate how well the clustering structure can be characterized. In **cluster** the dissimilarity matrices are computed from 2-dimensional datasets, containing samples belonging to a varying number of clusters (up to four clusters). The class label of a dissimilarity matrix is equal to the number of clusters, and therefore this defines a 4-class classification problem.
- The **subspace** dataset is constructed to investigate how well the subspace structure can be characterized. In **subspace** the dissimilarity matrices are derived from p-dimensional Gaussian distributions, where the dimensionality is one ($p = 1$) for class 1, $p = 2$ for class 2, up to class 4.
- The **outlier** dataset is used to investigate the sensitivity to outliers. In **outlier** the matrices are derived from 2-dimensional Gaussian distributions (zero mean, identity covariance matrix). Class 1 does not contain outliers. Class 2 contains an outlier from a Gaussian distribution with a 10 times larger covariance matrix, and for class 3 contains two such outliers.

Figure 1 shows the five different representations for a sample of 100 dissimilarities drawn from the **cluster** dataset. For the spectrum representation three features are computed, for the (equalized) histograms $k = 10$ and for the **disthub** representation in total 75 features are computed. This **cluster** dataset has a very clear structure, and all the representations are able to distinguish well between the four different classes. In particular, the distinction between 1 cluster and more-than-1 cluster datasets are easy to make. For the **disthub** representation the distinction between the classes is less visible in the figure, due to the large difference in scales between the different features.

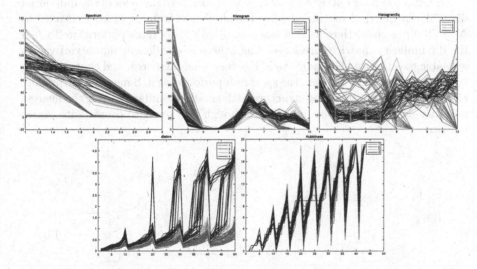

Fig. 1. The different feature representations for dissimilarity matrices derived from the **cluster** dataset with one, two, three or four clusters.

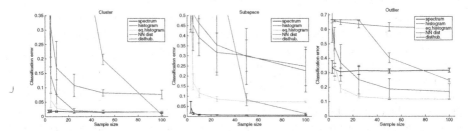

Fig. 2. Learning curves showing the error as a function of the number of training dissimilarity matrices. From left to right: results on `cluster`, `subspace` and `outlier` data.

For each dataset, we compute the dissimilarity matrices using the Euclidean distances between the samples. We then compute the different representations, and use a linear classifier (LDA) to distinguish the different classes per dataset.

Figure 2 shows the classification performance on the three artificial sets as function of the number of training matrices. The size of the individual dissimilarity matrices is fixed to $m_n = 30$. For many situations the (equalized) histogram is able to capture the information needed for good generalization. The histogram estimates start to suffer from noise for very small datasets and for situations where there is no clustering structure, and only the subspace dimensionality is informative. In these situations a spectrum representation is to be preferred. When very large training sizes are available, it is advantageous to use the nearest-neighbor distance histograms. Because `distnn` combines the histograms of the first-, second-, and all higher-nearest neighbors, this representation becomes very high-dimensional, but also very rich.

In Fig. 3 a similar curve to Fig. 2 is shown, only here the sizes of the individual dissimilarity matrices are varied while the number of training matrices is fixed to $N = 100$ per class. Here as well the (equalized) histograms perform well when the dissimilarity matrices are large. Then there is a sufficient number of values available to estimate histograms well. For very small matrices, and characterizing subspace structure or outliers, the spectrum performs well. Somewhat surprising, to characterize the clustering structure with small dissimilarity sizes, the nearest neighbor distances are most effective, although this tends to overfit with larger matrices.

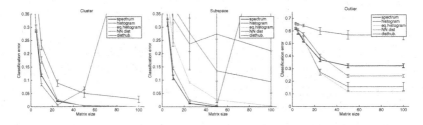

Fig. 3. The classification performance as function of the size m_n of the dissimilarity matrices, for the cluster data (left), the subspace data (middle) and the outlier data (right).

4   Experiments

We distinguish between two sets of experiments, a supervised and an unsupervised set. For the supervised set, we have a collection of labeled dissimilarity matrices. Here we use the bags from MIL data, where the distances between the instances in one bag give one dissimilarity matrix, and each matrix is labeled according to the original bag label (positive or negative). For the unsupervised set we only have a collection of dissimilarities between teams of people, for which we want to investigate how much variability is present in the teams, and what constitutes this variability.

4.1   Supervised Experiments: Multiple Instance Learning

We look at a wide variety of multiple instance learning (MIL) problems. In MIL, the i-th sample is a bag $B_n = \{\mathbf{x}_{n1}, \mathbf{x}_{n2}, \ldots, \mathbf{x}_{nm_n}\}$ of m_n instances. The goal is to classify bags, based on the presence of *concept* feature vectors, or based on the overall distribution of the bag's instances. Consider image classification, where a bag is an image, and an instance is an image patch. When classifying images of tigers, a patch containing a tiger is an example of a concept instance. When classifying images of scenes, it might be more reasonable to examine several patches before deciding what type of environment the image is depicting.

Characteristics of the datasets are listed in Table 1. From our previous experiences with these datasets [5, 25], we expect these datasets to contain a mix of concept-like and distribution-like problems. Note that in our previous work [5, 25] we represented each bag by its dissimilarities relative to a set of prototype bags, whereas here we use an absolute representation where each bag is represented by dissimilarities between its own instances.

Table 1. Characteristics of MIL datasets. Most of the datasets are available for download from http://www.miproblems.org

Dataset	#bags neg/pos	#instances min-mean-max	Dataset	#bags neg/pos	#instances min-mean-max
Musk 1	23/37	3-7-40	alt.atheism	50/50	22-54-76
Musk 2	53/37	4-73-1044	comp.graphics	51/49	12-31-58
Corel African	1410/93	3-5-13	Harddrive	178/190	3-186-299
AjaxOrange	1440/60	31-32-32	Brown Creeper	350/197	3-19-43
Web recomm. 1	55/20	9-46-229	Biocreative comp	2591/396	3-12-53

We removed bags that contained only 1 or 2 instances. We then represented each bag by a $m_n \times m_n$ dissimilarity matrix between its instances, where the dissimilarity is simply the Euclidean distance between the feature vectors. We represented each dissimilarity matrix with the representations described in Sect. 2. We used two classifiers: a linear discriminant classifier and a 1-nearest neighbor classifier. The experiments were performed using 10-fold cross-validation, where the best hyper parameter for each representation type (the optimal value for k), was determined on the training set using a second internal 10-fold cross-validation. We choose $k \in \{5, 10, 25, 50, 100\}$.

We report the AUC performances of both classifiers, using the best parameters for each representation type. For reference, we also list the best performance of traditional MIL classifiers[1]. The classifiers that often perform well are MILES [4], MI-SVM [2], EM-DD [27], a logistic classifier trained on a bag summary representation (based on the mean instance, or the min/max values per feature) [11], and p-posterior classifier [26] (Table 2).

The results are similar to those on artificial data: when the dissimilarity matrices are small, a spectrum representation is preferred. When larger training sets are available, it is often good to choose for an equalized histogram. These histograms tend to become relatively high dimensional, and the classifier can therefore not be too complex, so a linear classifier is a good choice.

What is also surprising is that, although these representations remove the absolute locations of the instances in the feature space, it is still possible to achieve very reasonable classification performance. For some datasets classification performances exceed the best performances achieved up to now (comp.graphics, Biocreative) or are comparable (Corel African, alt.atheism, Harddrive). For datasets that contain a specific concept (Musk1, Musk2, AjaxOrange, Web recomm. 1, Brown Creeper), the classifier that has access to individual feature vectors is better off.

4.2   Unsupervised Experiments: Analysis of Teams of People

Given the required speed in a strategic decision making process, we used an online survey for the unsupervised gathering of a strategic status update from 20,191 employees in 1,378 teams in 277 different client projects on, for example, Human Resource Management, Information Technology and Marketing and Sales. We did not use a Likert scale given the subsequent need for statistical corrections for the structure of the survey [9], for various response styles [7], for a variety of sampling errors [19] and for a wide variety of biases [22]. Instead, we opted to use a Guttman scale with objective verifiable answers [8,13]. The assessment questions were different for different teams. Four different types of assessment can be distinguished: (1) human resource (HR): focusing on team effectiveness, competency assessments, cultural aspects, (2) strategy: how strategy is finally incorporated, innovation assessment, (3) marketing and sales: analysis of client processes, commercial guidance of shops, and (4) IT: project assessment, IT processes, IT governance. From the answers a dissimilarity is derived by computing the pairwise Euclidean distances between the answer scores of all the members in a team.

---

[1] Available from http://homepage.tudelft.nl/n9d04/milweb/.

Table 2. AUC mean (standard deviation) ×100 % of two classifiers on five representations of MIL bags. Bold = best or not significantly worse than best representation per classifier.

Musk 1			Musk 2		
MI-SVM 92.9 (1.3)			MILES 95.3 (1.5)		
repr	LDA	1-NN	repr	LDA	1-NN
spect 25D	**74.2 (18.7)**	55.8 (18.9)	spect 10D	**53.5 (21.1)**	**58.6 (19.5)**
hist 5D	**60.6 (27.6)**	**54.6 (18.7)**	hist 50D	**59.1 (28.6)**	**50.8 (21.5)**
histeq 25D	45.6 (26.1)	52.7 (21.4)	histeq 5D	**59.3 (23.3)**	**51.6 (15.5)**
distnn 10	**68.8 (16.8)**	50.8 (17.2)	distnn 5	**64.3 (26.1)**	**63.0 (23.3)**
disthub 5D	**73.3 (19.4)**	**68.8 (17.1)**	disthub 5D	**55.8 (22.4)**	**63.7 (20.7)**

Corel African			SIVAL AjaxOrange		
EM-DD 91.5 (0.4)			MI-SVM 99.6 (0.1)		
repr	LDA	1-NN	repr	LDA	1-NN
spect 25D	65.3 (9.1)	**74.7 (11.0)**	spect 25D	**87.0 (9.1)**	70.0 (15.4)
hist 100D	81.2 (12.0)	**73.0 (7.7)**	hist 10D	72.3 (11.6)	60.3 (10.8)
histeq 10D	**87.8 (9.1)**	**76.1 (15.7)**	histeq 100D	68.8 (12.5)	61.1 (10.9)
distnn 5	**87.5 (5.8)**	**78.2 (10.4)**	distnn 5	73.6 (10.9)	**66.7 (15.2)**
disthub 5D	59.5 (9.4)	51.6 (12.2)	disthub 20D	64.6 (13.2)	**68.8 (11.1)**

Web recomm. 1			alt.atheism		
MI-SVM 91.9 (0.0)			Logistic on mean 85.2 (2.2)		
repr	LDA	1-NN	repr	LDA	1-NN
spect 5D	48.8 (20.3)	**67.3 (18.6)**	spect 5D	**86.8 (10.7)**	**75.2 (14.7)**
hist 50D	**63.2 (25.0)**	**66.8 (31.1)**	hist 100D	76.0 (10.2)	**75.2 (12.9)**
histeq 5D	58.3 (24.9)	**74.7 (29.2)**	histeq 5D	**76.8 (17.6)**	61.2 (14.0)
distnn 20	**72.8 (22.6)**	**69.5 (21.1)**	distnn 5	**82.8 (10.2)**	**66.4 (16.8)**
disthub 20D	50.8 (21.9)	**53.3 (22.8)**	disthub 10D	56.8 (18.2)	60.0 (19.9)

comp.graphics			Harddrive		
SimpleMIL logistic 73.0 (1.7)			P-posterior 98.5 (0.5)		
repr	LDA	1-NN	repr	LDA	1-NN
spect 5D	**89.0 (11.4)**	61.3 (20.6)	spect 5D	88.5 (7.5)	96.2 (3.8)
hist 50D	73.2 (12.8)	**73.5 (14.8)**	hist 5D	95.0 (6.7)	95.3 (5.0)
histeq 50D	**82.6 (14.0)**	72.4 (19.1)	histeq 25D	**98.7 (2.5)**	**99.1 (1.8)**
distnn 10	**90.9 (13.2)**	**79.4 (10.7)**	distnn 20	76.1 (20.8)	94.1 (3.0)
disthub 10D	55.6 (14.2)	**71.4 (10.2)**	disthub 10D	89.6 (6.7)	80.5 (6.0)

Brown Creeper			Biocreative component		
MILES 95.8 (0.3)			MI-SVM 84.0 (0.0)		
repr	LDA	1-NN	repr	LDA	1-NN
spect 50D	56.4 (14.5)	69.1 (8.7)	spect 50D	78.2 (8.6)	81.0 (9.7)
hist 100D	**87.5 (12.3)**	**76.4 (10.0)**	hist 100D	85.9 (7.4)	78.0 (10.7)
histeq 100D	**81.9 (23.2)**	**82.2 (15.1)**	histeq 25D	**88.6 (7.2)**	**87.1 (8.3)**
distnn 20	64.8 (14.6)	72.0 (10.4)	distnn 10	**89.1 (8.8)**	**86.9 (10.4)**
disthub 10D	76.3 (11.9)	65.7 (9.2)	disthub 20D	76.7 (10.1)	67.7 (8.3)

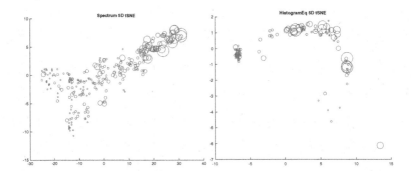

Fig. 4. t-SNE visualisations of 1378 teams of people. Left: spectrum representation of all the teams. The size of the circles indicate the size of the corresponding dissimilarity matrix. Right: equalized histogram representation for teams that got a human resource assessment.

In Fig. 4 the resulting embeddings are shown using the spectrum and equalized histogram representations. Both representations are 5-dimensional, and the 2-dimensional embedding of the 5D data is obtained by using t-SNE [17]. In the left subplot the marker size indices the size m_n of the corresponding matrix D_n. It appears that the first important component is the size team. The plot also shows that there is more variation in the smaller teams, suggesting that in smaller team there is more possibility of specialisation. Larger teams tend to become more similar.

When we normalise for team size, and we focus on one type of questionnaires (Human Resource) we obtain the scatterplot on the right. There is one prominent outlier team. This appears to be a team that got a questionnaire with 160 questions, while normally less than 20 questions are used. Furthermore, there is a large cluster on the left, which contains fairly homogeneous team members, and a long tail up to the right where teams get stronger and stronger clusters of subteams. The teams most far in the tail show a clear clustering, while teams more close to the homogeneous cluster only contain a few outliers in a team.

5   Conclusions

We compared several feature vector representations for characterizing (square) dissimilarity matrices that can vary in size, and for which the rows and columns can be arbitrarily permuted. The spectrum representation is very effective, in particular when the sample sizes are small. It can not only characterize the intrinsic dimensionality, it is also able to characterize cluster structure. When a large sample size is available, it is often advantageous to use the more descriptive histograms of distances. These results can be observed in some artificial, and some real-world MIL problems. For MIL, the representations with a linear or nearest neighbor classifier are sometimes competitive to state-of-the-art classifiers.

We then used the representations in an unsupervised manner in order to characterize real-world organizations. Our analysis revealed some clusters of organizations, that could be interpreted by an expert. Given the current dissimilarity scores we suggest further research into the extent to which organizations are similar with respect to issues that affect a multitude of teams (a top management issue), a single team (a middle management issue) or a single employee (a lower management issue), and whether that similarity is particularly present in specific management topics (for example, in Human Resource Management) and/or in specific industries (e.g. in Professional Services).

References

1. Ahrens, T., Chapman, C.S.: Doing qualitative field research in management accounting: positioning data to contribute to theory. Acc. Organ. Soc. **31**(8), 819–841 (2006)
2. Andrews, S., Tsochantaridis, I., Hofmann, T.: Support vector machines for multiple-instance learning. In: Advances in Neural Information Processing Systems, pp. 561–568 (2002)
3. Bunke, H., Riesen, K.: Recent advances in graph-based pattern recognition with applications in document analysis. Pattern Recogn. **44**(5), 1057–1067 (2011)
4. Chen, Y., Bi, J., Wang, J.: MILES: multiple-instance learning via embedded instance selection. IEEE Trans. Pattern Anal. Mach. Intell. **28**(12), 1931–1947 (2006)
5. Cheplygina, V., Tax, D.M.J., Loog, M.: Multiple instance learning with bag dissimilarities. Pattern Recogn. **48**(1), 264–275 (2015)
6. Cvetkovic, D., Doob, M., Sachs, H.: Spectra of Graphs, 3rd edn. Johann Ambrosius Barth Verlag, Heidelberg (1995)
7. De Jong, M.G., Steenkamp, J.B.E., Fox, J.P., Baumgartner, H.: Using item response theory to measure extreme response style in marketing research: a global investigation. J. Market. Res. **45**(1), 104–115 (2008)
8. Diamond, I.D., McDonald, J., Shah, I.: Proportional hazards models for current status data: application to the study of age at weaning differentials in Pakistan. Demography **23**(4), 607–620 (1986)
9. Edelen, M.O., Reeve, B.B.: Applying item response theory (IRT) modeling to questionnaire development, evaluation, and refinement. Qual. Life Res. **5**, 5–18 (2007)
10. Feragen, A., Kasenburg, N., Petersen, J., de Bruijne, M., Borgwardt, K.: Scalable kernels for graphs with continuous attributes. In: Advances in Neural Information Processing Systems, pp. 216–224 (2013)
11. Gärtner, T., Flach, P.A., Kowalczyk, A., Smola, A.J.: Multi-instance kernels. In: International Conference on Machine Learning, pp. 179–186 (2002)
12. Gärtner, T.: Predictive graph mining with kernel methods. In: Advanced Methods for Knowledge Discovery from Complex Data, pp. 95–121 (2005)
13. Hopkins, L., Ferguson, K.E.: Looking forward: the role of multiple regression in family business research. J. Fam. Bus. Strategy **5**(1), 52–62 (2014)
14. Hubert, L., Arabie, P., Meulman, J.: 9. Anti-Robinson Matrices for Symmetric Proximity Data. ASA-SIAM Series on Statistics and Applied Probability (Book 19), chap. 11, pp. 115–141 (2006)

15. Lau, L., Yang-Turner, F., Karacapilidis, N.: Requirements for big data analytics supporting decision making: a sensemaking perspective. In: Karacapilidis, N. (ed.) Mastering Data-Intensive Collaboration and Decision Making. SBD, vol. 5, pp. 49–70. Springer, Heidelberg (2014). doi:10.1007/978-3-319-02612-1_3
16. Lee, W.J., Cheplygina, V., Tax, D.M.J., Loog, M., Duin, R.P.W.: Bridging structure and feature representations in graph matching. Int. J. Pattern Recogn. Artif. Intell. (IJPRAI) **26**(05), 1260005 (2012)
17. Van der Maaten, L., Hinton, G.: Visualizing data using t-SNE. J. Mach. Learn. Res. **9**(2579–2605), 85 (2008)
18. Papadopoulos, A., Manolopoulos, Y.: Structure-based similarity search with graph histograms. In: Proceedings of the International Workshop on Similarity Search, pp. 174–178 (1999)
19. Piterenko, K.: Business and impact alignment of questionnaire. Master's thesis, Gjovik University College (2013)
20. Plewis, I., Mason, P.: What works and why: combining quantitative and qualitative approaches in large-scale evaluations. Int. J. Soc. Res. Methodol. **8**(3), 185–194 (2007)
21. Riesen, K., Fankhauser, S., Bunke, H., Dickinson, P.J.: Efficient suboptimal graph isomorphism. In: Torsello, A., Escolano, F., Brun, L. (eds.) GbRPR 2009. LNCS, vol. 5534, pp. 124–133. Springer, Heidelberg (2009). doi:10.1007/978-3-642-02124-4_13
22. Roulston, K., Shelton, S.A.: Reconceptualizing bias in teaching qualitative research methods. Qual. Inq. **21**(4), 332–342 (2015)
23. Schnitzer, D., Flexer, A., Schedl, M., Widmer, G.: Local and global scaling reduce hubs in space. J. Mach. Learn. Res. **13**, 2871–2902 (2012)
24. Schnitzer, D., Flexer, A., Tomasev, N.: Choosing the metric in high-dimensional spaces based on hub analysis. In: ESANN (2014)
25. Tax, D.M.J., Loog, M., Duin, R.P.W., Cheplygina, V., Lee, W.-J.: Bag dissimilarities for multiple instance learning. In: Pelillo, M., Hancock, E.R. (eds.) SIMBAD 2011. LNCS, vol. 7005, pp. 222–234. Springer, Heidelberg (2011). doi:10.1007/978-3-642-24471-1_16
26. Wang, H.Y., Yang, Q., Zha, H.: Adaptive p-posterior mixture-model kernels for multiple instance learning. In: Proceedings of 25th International Conference Machine learning, pp. 1136–1143 (2008)
27. Zhang, Q., Goldman, S.A., et al.: EM-DD: an improved multiple-instance learning technique. In: Advances in Neural Information Processing Systems, pp. 1073–1080 (2001)

Bilingual Data Selection Using a Continuous Vector-Space Representation

Mara Chinea-Rios[1]([⊠]), Germán Sanchis-Trilles[2], and Francisco Casacuberta[1]

[1] Pattern Recognition and Human Language Technology Research Center,
Universitat Politècnica de València, 46022 Valencia, Spain
{machirio,fcn}@prhlt.upv.es
[2] Sciling, Universitat Politècnica de València, 46022 Valencia, Spain
gsanchis@sciling.es

Abstract. Data selection aims to select the best data subset from an available pool of sentences with which to train a pattern recognition system. In this article, we present a bilingual data selection method that leverages a continuous vector-space representation of word sequences for selecting the best subset of a bilingual corpus, for the application of training a machine translation system. We compared our proposal with a state-of-the-art data selection technique (cross-entropy) obtaining very promising results, which were coherent across different language pairs.

Keywords: Vector space representation · Data selection · Bilingual corpora

1 Introduction

Entities such as the United Nations and other international organizations need to translate all documentation they generate into different languages, generating very large multilingual corpora which are ideal for training Statistical Machine Translation (SMT) [1] systems. However, such large corpora are difficult to process, increasing the computational requirements needed to train SMT systems robustly. For example, the corpora used in recent SMT evaluations are in the order of 1 billion running words [2].

Different problems that arise when using this huge pool of sentences are, mainly:

– The use of all corpora for training increases the computational requirements.
– Topic mismatch between training and test conditions impacts translation quality.

Nevertheless, standard practice is to train the SMT systems with all the available data, following the premise of "the more data, the better". This assumption is usually correct if all the data belongs to the same domain, but most commercial SMT systems will not have such luck: most SMT systems are designed to translate very specific text, such as user manuals or medical prospects. Hence,

© Springer International Publishing AG 2016
A. Robles-Kelly et al. (Eds.): S+SSPR 2016, LNCS 10029, pp. 95–106, 2016.
DOI: 10.1007/978-3-319-49055-7_9

the challenge is to wisely select a subset of bilingual sentences that performs the best for the topic at hand. This is the specific goal of Data Selection (DS), i.e., to select the best subset of an available pool of sentences. The current paper tackles DS by taking advantage of vector space representations of sentences, feeding on the most recent work on distributed representations [3,4], with the ultimate goal of obtaining corpus subsets that reduce the computational costs entailed, while improving translation quality. The main contributions of this paper are:

- We present a bilingual DS strategy leveraging a continuous space representation of sentences and words. Section 3 is devoted to this task.
- We present empirical results comparing our strategy with state-of-the-art DS method based in cross entropy differences. Sections 5.2 and 5.3 are devoted to this task.

This paper is structured as follows. Section 2 summarises the related work. Section 3 presents our DS method using continuous vector-space representations. Section 4 describes the DS methods we use for comparison. In Sect. 5 the experimental design and results are detailed. Finally, conclusions and future work are discussed in Sect. 6.

2   Related Work

We will refer to the available pool of generic-domain sentences as *out-of-domain* corpus because we assume that it belongs to a different domain than the one to be translated. Similarly, we refer to the corpus belonging to the specific domain of the text to translated as *in-domain* corpus.

The simplest instance of DS can be found in language modelling, where perplexity-based selection methods have been used [5]. Here, out-of-domain sentences are ranked by their perplexity score. Another perplexity-based approach is presented in [6], where cross-entropy difference is used as a ranking function rather than just perplexity, in order to account for normalization. We apply this criterion as comparison with our DS technique. Different works use perplexity-related DS strategies [7,8]. In these papers, the authors report good results when using the strategy presented in [6], and such strategy has become a de-facto standard in the SMT research community. In [7] the authors describe a new bilingual method based on the original proposal by [6], and will be explained in detail in Sect. 4.

Other works have applied information retrieval methods for DS [9], in order to produce different sub-models which are then weighted. The baseline was defined as the result obtained by training only with the corpus that shares the same domain with the test. They claim that they are able to improve the baseline translation quality with their method. However, they do not provide a comparison with a model trained on all the corpora available. More recently, [10] leveraged neural language models to perform DS, reporting substantial gains over conventional n-gram language model-based DS.

3   Data Selection Approaches

Here we describe our bilingual DS for SMT using continuous vector-space representation (CVR) of sentences or documents with the purpose of using each side of the corpus (source and target). This new technique is an extension over previous work describing only monolingual DS using a continuous CVR. Our DS approach requires:

1. A CVR of words (Sect. 3.1) and CVR of sentences or documents (Sect. 3.2).
2. A selection algorithm based on these CVR (Sect. 3.3), and its bilingual extension (Sect. 3.4).

3.1   Continuous Vector-Space Representation of Words

CVR of words have been widely used in a variety of natural language processing applications. These representations have recently demonstrated promising results across a variety of tasks [11–13], such as speech recognition, part-of-speech tagging, sentiment classification and identification, and machine translation.

The idea of representing words in vector space was originally proposed by [14]. The limitation of these proposals were that computational requirements quickly became unpractical for growing vocabulary sizes $|V|$. However, work performed recently in [3,15] made it possible to overcome such drawback, while still representing words as high dimensional real valued vectors: each word w_i in the vocabulary V, $w_i \in V$, is represented as a real-valued vector of some fixed dimension D, i.e., $f(w_i) \in R^D \ \forall i = 1, \ldots, |V|$, capturing the similarity (lexical, semantic and syntactic) between the words.

Two approaches are proposed in [3], namely, the *Continuous Bag of Words Model* (CBOW) and the *Continuous Skip-Gram Model*. CBOW forces the neural net to predict the current word by means of the surrounding words, and Skip-Gram forces the neural net to predict surrounding words using the current word. These two approaches were compared to previously existing approaches, such as the ones proposed in [16], obtaining a considerably better performance in terms of training time. In addition, experimental results also demonstrated that the Skip-Gram model offers better performance on average, excelling especially at the semantic level [3]. These results were confirmed in our own preliminary work, and hence we used the Skip-Gram approach to generate our distributed representations of word.

We used the Word2vec[1] toolkit to obtain the representations of words. Word2vec takes a text corpus as input and produces the word vectors as output. It first constructs a vocabulary V from the training corpus and then learns the CVR of the words.

A problem that arises when using CVR of words is how to represent a whole sentence (or document) with a continuous vector. Following the idiosyncrasy described in the previous paragraph (i.e., semantically close words are also close in their CVR), next section presents the different sentence representations employed in the present work.

---

[1] www.code.google.com/archive/p/word2vec/.

3.2 Continuous Vector-Space Representation of Sentences

Numerous works have attempted to extend the CVR of words to the sentence or phrase level (just to name a few, [4,17,18]). In the present work, we used two different CVRs of sentences, denoted here as $F(\mathbf{x})$ (or, in some cases and to simplify notation, $F_{\mathbf{x}}$):

1. The first one is the most intuitive approach, which relies on using a weighted arithmetic mean of all the words in the document or sentence (as proposed by [17,19]:

$$F_{\mathbf{x}} = F(\mathbf{x}) = \frac{\sum_{w \in \mathbf{x}} N_{\mathbf{x}}(w) f(w)}{\sum_{w \in \mathbf{x}} N_{\mathbf{x}}(w)} \tag{1}$$

   where w is a word in sentence $\mathbf{x}$, $f(w)$ is the CVR of w, obtained as described above, and $N_{\mathbf{x}}(w)$ is the count of w in sentence $\mathbf{x}$. We will refer to this representation by Mean-vec.
2. A more sophisticated approach is presented by [4], where the author adapted the continuous Skip-Gram model [3] to generate representative vectors of sentences or documents by following the same Skip-Gram architecture, generating a special vector $F_{\mathbf{x}}$. We will refer to this representation by Document-vec.

3.3 DS Using Vector Space Representations of Sentences

Since the objective of DS is to increase the informativeness of the in-domain training corpus, it seems important to choose out-of-domain sentences that provide information considered relevant with respect to the in-domain corpus I.

Algorithm 1 shows the procedure. Here, G is the out-domain-corpus, $\mathbf{x}$ is an out-of-domain sentence ($\mathbf{x} \in G$), $F_{\mathbf{x}}$ is the CVR of $\mathbf{x}$, and $|G|$ is the number of sentences in G. Then, our objective is to select data from G such that it is the most suitable for translating data belonging to the in-domain corpus I. For this purpose, we define $F_{\mathbf{S}}$, assuming S as the concatenation of all in-domain corpus I.

Data: $F_{\mathbf{x}}$, $\mathbf{x} \in G$; and $F_{\mathbf{S}}$; threshold τ
Result: Selected-corpus
forall the *sentences* $\mathbf{x}$ *in* G **do**
    if $cos(F_{\mathbf{S}}, F_{\mathbf{x}}) \geq \tau$ **then**
       | add $\mathbf{x}$ to Selected-corpus
    end
end

Algorithm 1. Pseudo-code for our DS technique (Sect. 3.3)

$cos(\cdot, \cdot)$, which implements the cosine similarity between two sentence vectors:

$$cos(F_{\mathbf{S}}, F_{\mathbf{x}}) = \frac{F_{\mathbf{S}} \cdot F_{\mathbf{x}}}{\|F_{\mathbf{S}}\| \cdot \|F_{\mathbf{x}}\|} \tag{2}$$

Note that it would have been possible to use any other similarity metric. Here, the *best* value for $cos(\cdot, \cdot)$ is 1, and the *worst* value for $cos(\cdot, \cdot)$ is 0 and τ is a certain threshold to be established empirically.

3.4 Bilingual DS Using Vector Space Representations of Sentences

In this section, we extend the CVR presented in Sect. 3.3 for making use of bilingual data. Here, the purpose is to tackle directly the bilingual nature of the problem of DS within an SMT setting. By including both sides of the corpus (source and target sentences), Algorithm 1 is modified to obtain Algorithm 2.

Here, G_x and G_y is the out-of-domain corpus (source and target, respectively), and $\mathbf{x_G}$ and $\mathbf{y_G}$ are out-of-domain sentences ($\mathbf{x_G} \in G_x$; $\mathbf{y_G} \in G_y$, respectively). $F_{\mathbf{x_G}}$ is the CVR of $\mathbf{x_G}$, and $F_{\mathbf{y_G}}$ is the CVR of $\mathbf{y_G}$. Similarly as done for $F_{\mathbf{S}}$, we define $F_{\mathbf{S_x}}$ as the CVR of S_x, i.e., the CVR of the concatenation of all source in-domain data and $F_{\mathbf{S_y}}$ as the CVR of S_y, i.e., the CVR of the concatenation of all target in-domain data.

Data: $F_{\mathbf{x_G}}$, $\mathbf{x_G} \in G_x$; $F_{\mathbf{y_G}}$, $\mathbf{y_G} \in G_y$;
and $F_{\mathbf{S_x}}$; $F_{\mathbf{S_y}}$; threshold τ
Result: Selected-corpus
forall the *sentences* $\mathbf{x_G}$ *in* G_x *and* $\mathbf{y_G}$ *in* G_y **do**
    if $[cos(F_{\mathbf{S_x}}, F_{\mathbf{x_G}})] + [cos(F_{\mathbf{S_y}}, F_{\mathbf{y_G}})] \geq \tau$ **then**
    | add $\mathbf{x_G}$ to Selected-corpus
    end
end

Algorithm 2. Pseudo-code for our bilingual DS technique (Sect. 3.4)

---

4 Cross-Entropy Difference Method

As mentioned in Sect. 2, one established DS method consists in scoring the sentences in the out-of-domain corpus by their perplexity. [6] use cross-entropy rather than perplexity, even though they are both monotonically related. The perplexity of a given sentence $\mathbf{x}$ with empirical n-gram distribution p given a language model q is:

$$2^{-\sum_x p(x) \log q(x)} = 2^{H(p,q)} \tag{3}$$

where $H(p,q)$ is the cross-entropy between p and q. The formulation proposed in [6] is: Let I be an in-domain corpus and G be an out-of-domain corpus. Let $H_I(\mathbf{x})$ be the cross-entropy, according to a language model trained on I, of a sentence $\mathbf{x}$ drawn from G. Let $H_G(\mathbf{x})$ be the cross-entropy of $\mathbf{x}$ according to a language model trained on G. The cross-entropy score of $\mathbf{x}$ is then defined as

$$c(\mathbf{x}) = H_I(\mathbf{x}) - H_G(\mathbf{x}) \tag{4}$$

Note that this method is defined in terms of I, as defined by the original authors. Even though it would also be feasible to define this method in terms of S, such re-definition lies beyond the scope of this paper, since our purpose is only to use this method only for comparison purposes.

In [7], the authors propose a extention to their cross entropy method [6] so that it is able to deal with bilingual information. To this end, they sum the cross-entropy difference over each side of the corpus, both source and target. Let (I and G) be a in-domain source corpus and out-of-domain source corpus respectively and (L and J) be a target corpora. Then, the cross-entropy difference is defined as:

$$c(\mathbf{x}) = [H_I(\mathbf{x}) - H_G(\mathbf{x})] + [H_L(\mathbf{y}) - H_J(\mathbf{y})] \tag{5}$$

5   Experiments

In this section, we describe the experimental framework employed to assess the performance of our DS method. Then, we show the comparative with cross-entropy DS.

5.1   Experimental Setup

We evaluated empirically the DS method described in Sect. 3. For the out-of-domain corpus, we used the Europarl corpus [20], which is composed of translations of the proceedings of the European parliament. As in-domain data, we used the EMEA corpus [21], which is available in 22 languages and contains documents from the European Medicines Agency. We conducted experiments with different language pairs (English-French [En-FR]; French-English [Fr-En]; German-English [De-En]; English-German [En-De]) so as to test the robustness of the results achieved. The main figures of the corpora used are shown in Tables 1 and 2.

Table 1. In-domain corpora main figures. (EMEA-Domain) is the in-domain corpus, (Medical-Test) the evaluation data, and (Medical-Mert) development set. $|S|$ stands for number of sentences, $|W|$ for number of words, and $|V|$ for vocabulary size.

Corpus		$|S|$	$|W|$	$|V|$
EMEA-domain	EN	1.0 M	12.1 M	98.1 k
	FR		14.1 M	112 k
Medical-test	EN	1000	21.4 k	1.8 k
	FR		26.9 k	1.9 k
Medical-mert	EN	501	9850	979
	FR		11.6 k	1.0 k
EMEA-domain	DE	1.1 M	10.9 M	141 k
	EN		12.9 M	98.8 k
Medical-test	DE	1000	18.2 k	1.7 k
	EN		19.2 k	1.9 k
Medical-mert	DE	500	8.6 k	874
	EN		9.2 k	979

Table 2. Out-of-domain corpus main figures (abbreviations explained in Table 1).

Corpus		$|S|$	$|W|$	$|V|$
Europarl	EN	2.0 M	50.2 M	157 k
	FR		52.5 M	215 k
Europarl	DE	1.9 M	44.6 M	290 k
	EN		47.8 M	153 k

All experiments were carried out using the open-source phrase-based SMT toolkit Moses [22]. The decoder features a statistical log-linear model including a phrase-based translation model, a language model, a distortion model and word and phrase penalties. The log-lineal combination weights λ were optimized using MERT (minimum error rate training) [23]. Since MERT requires a random initialisation of λ that often leads to different local optima being reached, every point in each plot of this paper constitutes the average of 10 repetitions with the purpose of providing robustness to the results. In the tables reporting translation quality, 95 % confidence intervals of these repetitions are shown, but are omitted from the plots for purpose of clarity. We compared the selection methods with two baseline systems. The first one was obtained by training the SMT system with EMEA-Domain data. We will refer to this setup with the name of baseline-emea. A second baseline experiment has been carried out with the concatenation of the Europarl corpus and EMEA training data (i.e., all the data available). We will refer to this setup as bsln-emea-euro. We also included results for a purely random sentence selection without replacement. In the plots, each point corresponding to random selection represents the average of 5 repetitions.

SMT output will be evaluated by means of BLEU (BiLingual Evaluation Understudy) [24], which measures the precision of uni-grams, bigrams, trigrams, and 4-grams with respect to the reference translation, with a penalty for too short sentences [24].

Word2vec (Sect. 3.1) has different parameters that need to be adjusted. We conducted experiments with different vector dimensions, i.e., $D = \{100, 200, 300, 400, 500\}$. In addition, a given word appears is required to appear n_c times in the corpus so as to be considered for computing its vector. We analysed the effect of considering different values $n_c = \{1, 3, 5, 10\}$. Experiments not reported here for space reasons led to establishing: sentences vector size $v\_s = 200$ and $n_c = 1$ for all further experiments reported.

5.2 Comparative with Cross-Entropy Selection

As a first step, we compare our DS method with the cross-entropy method both in yours monolingual version (Sect. 5.1). Results in Fig. 1 show the effect of adding sentences to the in-domain corpus. We only show cross-entropy results using 2-grams, which was the best result according to previous work. For our DS method, we tested both CVR methods (Document-vec and Mean-vec).

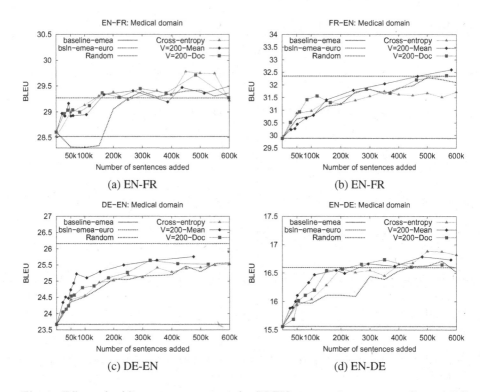

Fig. 1. Effect of adding sentences over the BLEU score using our monolingual DS method, the original cross-entropy method, and random selection. Horizontal lines represent the `baseline-emea` and `bsln-emea-euro` scores.

Several conclusions can be drawn:

- The DS techniques are able to improve translation quality when compared to the `baseline-emea` setting, in all language pairs.
- All DS methods are mostly able to improve over random selection, especially when low amounts of data are added. This is reasonable, since all DS methods including random will eventually converge to the same point: adding all the data available. Even though these results should be expected, previous works (reported in Sect. 2) revealed that beating random was very hard.
- In Fig. 1, the results obtained with our DS method are slightly better (or similar) than the ones obtained with cross-entropy.

5.3   Comparative with Bilingual Cross-Entropy Selection

Results comparing our bilingual DS method with bilingual corss-entropy are shown in Fig. 2. In the case of our DS method, the same approach as in previous section was used. Several conclusions can be drawn:

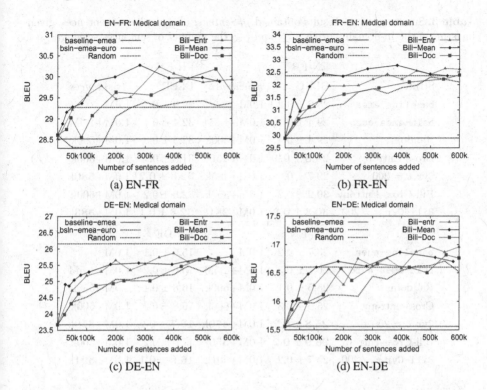

Fig. 2. Effect of adding sentences over the BLEU score using our bilingual DS method, the bilingual cross-entropy method, and random selection. Horizontal lines represent the `baseline-emea` and `bsln-emea-euro` scores.

- Our bilingual DS technique provides better results than including the full out-of-domain corpus (bsln-emea-euro) in language pairs EN-FR, FR-EN, and EN-DE. Specially, the improvements obtained are in the range of $[0, 3 - 0, 9]$ BLEU points using less than $[27\% - 19\%]$ of the out-of-domain corpus. In the DE-EN pair our DS strategy does improve the results over including the full out-of-domain corpus, but results are very similar using less than 33% of the out-of-domain corpus.
- The results achieved by our bilingual DS strategy are consistently better than those achieved by the bilingual cross-entropy method.
- For equal amount of sentences, translation quality is significantly better with the bilingual DS method, as compared to its monolingual form (Fig. 1). Hence, the bilingual DS strategy is able to make good use of the bilingual information, reaching a better subset of the out-of-domain data.

5.4   Summary of the Results

Table 3 shows the best results obtained. As shown, our method is able to yield competitive results for each combination of the language. Note that the bilingual

Table 3. Summary of the results obtained. *#Sentences* for number of sentences, given in terms of the in-domain corpus size, and (+) the number of sentences selected.

	EN-FR		FR-EN	
Strategy	BLEU	*#Sentences*	BLEU	*#Sentences*
baseline-emea	28.6 ± 0.2	1.0M	29.9 ± 0.2	1.0M
bsln-emea-euro	29.4 ± 0.1	1.0M+1.5M	32.4 ± 0.1	1.0M+1.5M
Random	29.4 ± 0.4	1.0M+500k	32.3 ± 0.3	1.0M+500k
Cross-Entropy	$\mathbf{29.8 \pm 0.1}$	1.0M+**450k**	31.8 ± 0.1	1.0M+600k
Cvr $v = 200$	29.7 ± 0.2	1.0M+485k	$\mathbf{32.6 \pm 0.2}$	1.0M+**580k**
Bili-Cross-Entropy	30.2 ± 0.2	1.0M+350k	32.6 ± 0.2	1.0M+600k
Bili-Cvr $v = 200$	$\mathbf{30.3 \pm 0.2}$	1.0M+**281k**	$\mathbf{32.8 \pm 0.1}$	1.0M+**383k**
	DE-EN		EN-DE	
baseline-emea	23.7 ± 0.2	1.0M	15.6 ± 0.1	1.0M
bsln-emea-euro	26.2 ± 0.3	1.0M+1.5M	16.6 ± 0.2	1.0M+1.5M
Random	25.5 ± 0.1	1.0M+600k	16.8 ± 0.1	1.0M+550k
Cross-entropy	25.5 ± 0.3	1.0M+600k	16.8 ± 0.2	1.0M+500k
Cvr $v = 200$	$\mathbf{25.8 \pm 0.2}$	1.0M+**470k**	$\mathbf{16.7 \pm 0.2}$	1.0M+**350k**
Bili-Cross-entropy	$\mathbf{25.9 \pm 0.2}$	1.0M+**350k**	17.0 ± 0.2	1.0M+500k
Bili-Cvr $v = 200$	25.7 ± 0.2	1.0M+501k	$\mathbf{16.9 \pm 0.1}$	1.0M+**391k**

cross entropy methods tends to select more sentences, while translation quality tends to be slightly worse when compared to our method.

6   Conclusion and Future Work

In this work, we presented a bilingual data selection method on CVR of sentence or documents, which intend to yield similar representations for semantically close sentences. In addition, we perform a comparison of our technique with state-of-the-art technique (cross-entropy difference). When comparing our method, an important conclusion stands out: our method is able to yield similar or better quality than the state-of-the-art method and reduce the number of selected sentences. In future work, we will carry out new experiments with bigger and diverse data sets in and different languages for example German language group. In addition, we also intend to combine the both sides of the corpus proposed in more sophisticated ways. Finally, we intend to compare our bilingual data selection method with other data selection techniques.

Acknowledgments. The research leading to these results has received funding from the Generalitat Valenciana under grant PROMETEOII/2014/030 and the FPI (2014) grant by Universitat Politècnica de València.

References

1. Brown, P.F., Cocke, J., Pietra, S.A.D., Pietra, V.J.D., Jelinek, F., Lafferty, J.D., Mercer, R.L., Roossin, P.S.: A statistical approach to machine translation. Comput. Linguist. **16**, 79–85 (1990)
2. Callison-Burch, C., Koehn, P., Monz, C., Peterson, K., Przybocki, M., Zaidan, O.F.: Findings of the 2010 joint workshop on statistical machine translation and metrics for machine translation. In: Proceedings of ACL, pp. 17–53 (2010)
3. Mikolov, T., Chen, K., Corrado, G., Dean, J.: Efficient estimation of word representations in vector space, January 2013. arXiv:1301.3781
4. Le, Q.V., Mikolov, T.: Distributed representations of sentences and documents (2014). arXiv:1405.4053
5. Gao, J., Goodman, J., Li, M., Lee, K.-F.: Toward a unified approach to statistical language modeling for Chinese. ACM TALIP **1**(1), 3–33 (2002)
6. Moore, R.C., Lewis, W.: Intelligent selection of language model training data. In: Proceedings of ACL, pp. 220–224 (2010)
7. Axelrod, A., He, X., Gao, J.: Domain adaptation via pseudo in-domain data selection. In: Proceedings of EMNLP, pp. 355–362 (2011)
8. Schwenk, H., Rousseau, A., Attik, M.: Large, pruned or continuous space language models on a GPU for statistical machine translation. In: Proceedings of the NAACL-HLT, pp. 11–19 (2012)
9. Lü, Y., Huang, J., Liu, Q.: Improving statistical machine translation performance by training data selection and optimization. In: Proceedings of EMNLP, pp. 343–350 (2007)
10. Duh, K., Neubig, G., Sudoh, K., Tsukada, H.: Adaptation data selection using neural language models: experiments in machine translation. In: Proceedings of ACL, pp. 678–683 (2013)
11. Collobert, R., Weston, J.: A unified architecture for natural language processing. In: Proceedings of ICML, pp. 160–167 (2008)
12. Glorot, X., Bordes, A., Bengio, Y.: Domain adaptation for large-scale sentiment classification: a deep learning approach. In: Proceedings of ICML, pp. 513–520 (2011)
13. Cho, K., van Merriënboer, B., Bahdanau, D., Bengio, Y.: On the properties of neural machine translation: encoder-decoder approaches, arxiv:1409.1259 (2014)
14. McClelland, J.L., Rumelhart, D.E., PDP Research Group, et al.: Parallel Distributed Processing, vol. 2. Cambridge University Press, Cambridge (1987)
15. Paulus, R., Socher, R., Manning, C.D.: Global belief recursive neural networks. In: Proceedings of Advances in Neural Information Processing Systems, pp. 2888–2896 (2014)
16. Mikolov, T., Karafit, M., Burget, L., Eernocký, J., Khudanpur, S.: Recurrent neural network based language model. In: Proceedings of INTERSPEECH, pp. 1045–1048 (2010)
17. Mikolov, T., Sutskever, I., Chen, K., Corrado, G.S., Dean, J.: Distributed representations of words and phrases and their compositionality. In: Proceedings of Advances in neural information processing systems, pp. 3111–3119 (2013)
18. Socher, R., Lin, C.C., Manning, C., Ng, A.Y.: Parsing natural scenes and natural language with recursive neural networks. In: Proceedings of ICML, pp. 129–136 (2011)
19. Kågebäck, M., Mogren, O., Tahmasebi, N., Dubhashi, D.: Extractive summarization using continuous vector space models. In: Proceedings of the Workshop on Continuous Vector Space Models and their Compositionality, pp. 31–39 (2014)

20. Koehn, P.: Europarl: a parallel corpus for statistical machine translation. In: Proceedings of MT Summit, pp. 79–86 (2005)
21. Tiedemann, J.: News from OPUS-a collection of multilingual parallel corpora with tools and interfaces. In: Proceedings of RANLP, pp. 237–248 (2009)
22. Koehn, P., et al.: Moses: open source toolkit for statistical machine translation. In: Proceedings of ACL, pp. 177–180 (2007)
23. Och, F.J. : Minimum error rate training in statistical machine translation. In: Proceedings of ACL, pp. 160–167 (2003)
24. Papineni, K., Roukos, S., Ward, T., Zhu, W.-J.: BLEU: a method for automatic evaluation of machine translation. In: Proceedings of ACL, pp. 311–318 (2002)

Improved Prototype Embedding Based Generalized Median Computation by Means of Refined Reconstruction Methods

Andreas Nienkötter and Xiaoyi Jiang[✉]

Department of Mathematics and Computer Science,
University of Münster, Münster, Germany
xjiang@uni-muenster.de

Abstract. Learning a prototype from a set of given objects is a core problem in machine learning and pattern recognition. A popular approach to consensus learning is to formulate it as an optimization problem in terms of generalized median computation. Recently, a prototype-embedding approach has been proposed to transform the objects into a vector space, compute the geometric median, and then inversely transform back into the original space. This approach has been successfully applied in several domains, where the generalized median problem has inherent high computational complexity (typically $\mathcal{NP}$-hard) and thus approximate solutions are required. In this work we introduce three new methods for the inverse transformation. We show that these methods significantly improve the generalized median computation compared to previous methods.

1 Introduction

Learning a prototype from a set of given objects is a core problem in machine learning and pattern recognition, and has numerous applications [4,6,10]. One often needs a representation of several similar objects by a single consensus object. One example is multiple classifier combination for text recognition, where a change in algorithm parameters or the use of different algorithms can lead to distinct results, each with small errors. Consensus methods produce a text which best represents the different results and thus removes errors and outliers.

A popular approach to consensus learning is to formulate it as an optimization problem in terms of generalized median computation. Given a set of objects $O = \{o_1, \ldots, o_n\}$ in domain $\mathcal{O}$ with a distance function $\delta(o_i, o_j)$, the *generalized median* can be expressed as

$$\bar{o} = \arg\min_{o \in \mathcal{O}} SOD(o) \tag{1}$$

where SOD is the sum of distances

$$SOD(o) = \sum_{p \in O} \delta(o, p).$$

© Springer International Publishing AG 2016
A. Robles-Kelly et al. (Eds.): S+SSPR 2016, LNCS 10029, pp. 107–117, 2016.
DOI: 10.1007/978-3-319-49055-7_10

In other words, the generalized median is an object which has the smallest sum of distances to all input objects. Note that the median object is not necessarily part of set O.

The concept of generalized median has been studied for numerous problem domains related to a broad range of applications, for instance strings [10] and graphs [9]. As the generalized median is not necessary part of the set, any algorithm must be of constructive nature and the construction process crucially depends on the structure of the objects under consideration. In addition, the generalized median computation is provably of high computational complexity in many cases. For instance, the computation of generalized median string turns out to be $\mathcal{NP}$-hard [8] for the string edit distance. The same applies to median ranking under the generalized Kendall-τ distance [3] and ensemble clustering for reasonable clustering distance functions, e.g. the Mirkin-metric [11].

Given the high computational complexity, approximate solutions are required to calculate the generalized median in reasonable time. Recently, a prototype embedding approach has been proposed [5], which is applicable to any problem domain and has been successfully applied in several domains [4–6,10]. In this work we investigate the reconstruction issue of this approach towards further improvement of consensus learning quality.

In the prototype embedding approach, the objects are embedded into an Euclidean metric space using the embedding function

$$\varphi(o) \;=\; (\delta(o,p_1), \delta(o,p_2), \ldots, \delta(o,p_d)) \tag{2}$$

where $\delta()$ is a distance between two objects. This embedding function assigns each object o_i to a vector $x_i = \varphi(o_i)$, which consists of its distance to d selected prototype objects $p_1, \ldots, p_d \in O$. Different methods were suggested for the prototype selection, for example k-means clustering, border object selection and others [2]. Using the vectors x_i, the geometric median in vector space is calculated by the Weiszfeld algorithm [16]. Then, the geometric median is transformed back into the original problem space, resulting in the searched generalized median.

In this work we propose new reconstruction methods to use with prototype embedding. A comparison with several established methods is conducted to demonstrate that the proposed new reconstruction methods can significantly improve the quality of generalized median computation (in terms of the optimization function SOD).

The remainder of the paper is organized as follows. In the next section the prototype embedding approach is briefly summarized. In Sect. 3 we then present various reconstruction methods proposed in this work. Experimental results are reported in Sect. 4. Finally, we conclude our findings in Sect. 5.

2 Prototype Embedding Based Generalized Median Computation

The vector space embedding approach consists of three steps (Fig. 1):

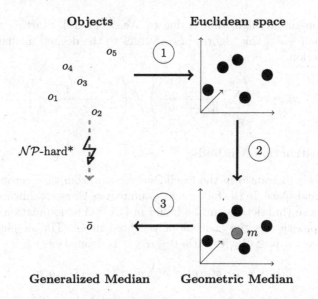

Fig. 1. Overview of prototype embedding based on [5]. *As mentioned in Sect. 1, the generalized median computation is not in all cases $\mathcal{NP}$-hard but in many domains with well-known metrics.

1. Embed the objects into a d-dimensional Euclidean space.
2. Compute the geometric median of the embedded points.
3. Estimate (reconstruct) the generalized median of the objects using an inverse transformation from the geometric median back into the original problem space.

2.1 Embedding Function

Prototype embedding uses (2) to embed objects into vector space. Here two design issues must be considered: the number d of selected prototypes and a selection algorithm. In our work the former issue is considered as a parameter, which will be systematically studied. For prototype selection, the so-called k-medians prototype selector turns out to be a good choice [5,10]. It finds d clusters in the given set of data using the k-means clustering algorithm and declares the object with the smallest sum of distances in each cluster to be a prototype. In this work we thus will use this prototype selection method for the experiments.

2.2 Computation of Geometric Median by Weiszfeld Algorithm

Once the embedded points are determined, the median vector has to be computed. No algorithm is known for exactly computing the Euclidean median in polynomial time, nor has the problem been shown to be $\mathcal{NP}$-hard [7]. The most common approximate algorithm is due to Weiszfeld [16]. Starting with a random vector $m_0 \in \mathbb{R}^d$, this algorithm converges to the desired median using the iterative function

$$m_{i+1} = \left(\sum_{j=1}^{n} \frac{x_j}{||x_j - m_i||} \right) \bigg/ \left(\sum_{j=1}^{n} \frac{1}{||x_j - m_i||} \right)$$

2.3 Reconstruction Methods

The last step is to transform the Euclidean median from the vector space back into the original space. In the following we summarize three established strategies for this purpose (full details can be found in [4,5]). The foundation of all these heuristic approaches is the concept of weighted mean. The weighted mean $\tilde{o}$ between objects $o, p \in \mathcal{O}$ with ratio $0 \leq \alpha \leq 1$ is defined as

$$\delta(o, \tilde{o}) = \alpha \cdot \delta(o, p), \quad \delta(\tilde{o}, p) = (1 - \alpha) \cdot \delta(o, p)$$

In other words, the weighted mean is a linear interpolation between both objects. In many cases, this weighted mean function can be derived from the distance function between objects [4,6].

Linear Reconstruction. Linear reconstruction uses the two objects o_1 and o_2 associated with the two closest points to the geometric median. The geometric median m is projected onto the line between these points using a simple line projection, resulting in a ratio α. The generalized median in the original space $\mathcal{O}$ is then calculated as the weighted mean between o_1 and o_2 using α.

Triangular Reconstruction. The triangulation method first finds the three closest points to the Euclidean median (denoted by x_1, x_2, x_3). Then, the median vector of these points is computed and denoted with x'. Now the method seeks to estimate an object in the original space $\mathcal{O}$ which corresponds to point x' and is thus assumed to be an approximation of the generalized median. Two steps have to be conducted. First, two out of the three points are arbitrarily chosen. Without loss of generality we assume that x_1 and x_2 are selected and we then project the remaining point x_3 onto the line joining x_1 and x_2 by using x'. As a result a point m is received, lying between x_1 and x_2. In this situation the linear interpolation method can be used to recover an object, which corresponds to m. A further application of the interpolation method then yields an object which corresponds to x'.

Recursive Reconstruction. The recursive reconstruction method [4] is more complex, in which the median is recursively projected onto hyperplanes of decreasing dimensionality. The median in n dimensions is calculated as a weighted mean between the median in $n - 1$ dimensions and the n-th object.

This results in a triangular or linear reconstruction if the recursion arrives at dimension three or two. Since this method calculates a generalized median in each step, one can use the intermediate result with the best sum of distances as the final result. This is the best-recursive interpolation.

3   Refined Reconstruction Methods

In this section we present several refined reconstruction methods that can be used in the framework described above towards more accurate generalized median computation.

3.1   Linear Recursive

In linear reconstruction, only the two nearest neighbors of the median vector are considered for the reconstruction. While this is fast and easily done, using only two objects restricts the search for a median object to the weighted mean between these objects. Therefore, we propose to consider the weighted mean between other objects as well. In linear recursive reconstruction, not only the nearest pair of neighbors is considered, but also the next nearest pair and all following ones. The objects are paired as $(o_1, o_2), (o_3, o_4), \ldots, (o_{n-1}, o_n)$ with $\delta_e(m, \varphi(o_i)) \leq \delta_e(m, \varphi(o_{i+1})), \forall 1 \leq i < n$. In this way, the two nearest neighbors are selected as a pair, then the two next neighbors etc. If n is odd, then the last object is not processed in the current iteration, but directly taken over for the next iteration. Each pair is linearly reconstructed (see Sect. 2.3), resulting in $n/2$ median objects. This process is repeated using only the reconstructed objects, resulting in $n/4$ objects. After $\log(n)$ steps, only one object remains. The object with the best SOD from all intermediate results is returned as the generalized median. Naturally, this reconstruction method is at least as good as the linear reconstruction, since it is included as the first pair.

An example using four objects can be found in Fig. 2. The median vector m is first projected on lines connecting each pair of objects. Then, the median is recursively projected on lines connecting pairs of these projections.

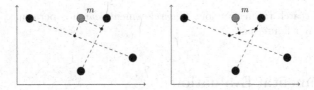

Fig. 2. Linear recursive reconstruction. Result after the first (left) and second iteration (right).

3.2 Triangular Recursive

Triangular recursive reconstruction is very similar to linear recursive reconstruction. The difference is that objects are grouped in triplets instead of pairs, using the triangular reconstruction (Sect. 2.3) to generate $n/3$ objects as a result. This is again repeated until only one object remains. For cases in which only two objects are present, linear reconstruction is used instead.

3.3 Linear Search

The above mentioned reconstruction methods calculate an approximation of the generalized median from the geometric median in vector space. Linear search improves this result by searching for a better approximation. It is reasonable to expect that a true generalized median is very similar to the calculated approximate one. Therefore, the region around the prior result should be searched for an object with lower SOD, see Fig. 3. Using weighted means (black dots) between the previous calculated median object $\bar{o}$ and objects in the set, an object with a lower SOD is searched (red dot). This process is repeated with the new object, until the change in SOD is sufficiently small. Since the generalized median serves as a lower bound of the sum of distance and this method reduces the SOD in each step if possible, it eventually converges to an optimal solution. Using this method only lines between given objects are searched instead of the full neighborhood of $\bar{o}$. Therefore, it is not guaranteed to arrive at the true generalized median and could result in a local minimum.

The linear search is a refinement performed the original domain $\mathcal{O}$. It can be combined with any reconstruction method, which calculates an approximation of the generalized median from the geometric median in vector space.

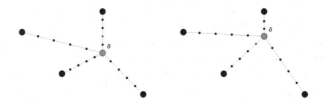

Fig. 3. Linear search reconstruction. This refinement search is performed the original domain $\mathcal{O}$. (Color figure online)

4 Experimental Evaluation

In this section we present experimental results. These results were generated using four datasets of two different types. First, we describe the datasets we used in our evaluation. Then, we show the results using our methods followed by a discussion of these results.

4.1   Datasets

Our methods were evaluated using four datasets in total, each consisting of several sets of objects. Table 1 shows some information about the used datasets.

Table 1. Evaluated datasets

Dataset	Type	Number of sets	Number of objects in each set	Distance function
Darwin	String	36	40	Levenshtein
CCD	String	22	100	Levenshtein
UCI Cluster	Cluster	8	25	Partition distance
Gen. Cluster	Cluster	8	20	Partition distance

The first two datasets consist of strings. The Darwin dataset is artificially generated with lines from Charles Darwins famous work "On the Origin of Species". Each set of 40 strings was created from one line of length between 70 and 140 symbols, which was randomly distorted using probabilities based on real world data [10]. The Copenhagen Chromosome Dataset (CCD) includes 22 sets of genetic strings created by Lundsteen *et al.* [12]. Each string encodes selected parts of a chromosome. Each set consists of 100 strings with lengths from about 20 to 100 symbols. In both cases, the Levenshtein distance [15] is used.

The second two datasets contain clusters, encoded as label vectors. The first sets are clusters generated with data from the UCI Data Repository [1] using k-means clustering with different parameters. The second are artificially generated clusters made with the method of [6]. This method uses a cluster to generate several new ones by the substitution of labels. For the cluster datasets, we use the partition distance [6]. This metric expresses how many labels must be changed to result in identical partitions of the data. The labels themselves are not relevant.

4.2   Quality of Computed Generalized Median

Tables 2 and 3 show the relative sum of distances of the generalized median acquired using the different datasets. Since the maximum number of prototypes is dependent on the number of objects in each set, we used $d = 0.1n, 0.2n, \ldots, 1n$. Table 2 displays the results of the different methods averaged over all tested d. Table 3 only includes the results of the best number of prototypes for each reconstruction method. The sets were embedded two times for each combination of number of prototypes and reconstruction method. The resulting SODs were averaged to compensate for small random influences in the results. The linear search method is combined with linear, best-recursive, and linear recursive methods as base result, respectively.

Table 2. Relative SOD results - average result

		Cluster	Gen. Clusters	CCD	Darwin
previous methods	linear	1.000	1.000	1.000	1.000
	triangular	1.005	0.835	0.972	0.946
	recursive	1.006	0.697	0.856	0.812
	best-recursive	0.904	0.674	0.760	0.712
proposed methods	linear-recursive	0.840	0.239	0.612	0.520
	triangular-recursive	0.865	0.262	0.601	0.518
	linear-search, linear	0.852	0.291	0.682	0.484
	linear-search, best-recursive	0.816	0.279	0.572	0.402
	linear-search, linear-recursive	0.785	0.033	0.533	0.359
	Linear Reconstruction SOD	4857.29	842.82	2799.27	951.65
	Lower Bound SOD	3429.25	586.00	1875.95	641.81

Table 3. Relative SOD results - best result

		Cluster	Gen. Clusters	CCD	Darwin
previous methods	linear	1.000	1.000	1.000	1.000
	triangular	1.002	0.648	0.969	0.902
	recursive	0.964	0.625	0.824	0.675
	best-recursive	0.906	0.625	0.725	0.592
proposed methods	linear-recursive	0.856	0.078	0.646	0.512
	triangular-recursive	0.875	0.126	0.605	0.506
	linear-search, linear	0.845	0.133	0.666	0.443
	linear-search, best-recursive	0.807	0.063	0.574	0.326
	linear-search, linear-recursive	0.805	0.000	0.563	0.327
	Linear Reconstruction SOD	4783.44	773.16	2710.48	899.72
	Lower Bound SOD	3429.25	586.00	1875.95	641.81

As the generalized median minimizes the SOD, a lower result means a more closely approximated result. Since the SOD is highly dependent on the distance function and dataset, a linear transformation $\frac{x-LB}{LR-LB}$ was used to normalize the results x and make them more easily comparable. The result (LR) using linear reconstruction is transformed to 1, the lower bound (LB) of the SOD of the generalized median is transformed to 0. As such, a value of 0.5 means that the calculated generalized median has a SOD which is at half between the result of prototype embedding with linear reconstruction and the lower bound. The lower bound was computed using the method from [9]. Since it may not be a very tight lower bound, it is not guaranteed that the generalized median would really produce a result of zero with this normalization.

The new reconstruction methods described in Sect. 3 show consistent and significant improvements of the SOD compared to previous methods both on average and their best dimension parameter, but the extent of the improvement is dependent on the chosen dataset. Linear-recursive and triangular-recursive reconstruction have similar results, with one resulting in a lower SOD in Cluster datasets, and the other showing better results in string datasets. This indicates influences of the distance and weighted mean functions on the result. As expected, linear search methods improve their respective base results by a large margin. Moreover, the better the base object is, the better the resulting approximated generalized median is. One should therefore use linear recursive as a basis for this method.

4.3   Computational Time

For the experiments we used a computer with Intel Core i5-4590 (4×3.3 GHz) and 16 GB RAM, running Matlab 2015a in Ubuntu 15.10. Figure 4 shows the run time for the UCI Cluster dataset. Since the same methods were used to calculate the embedding and geometric median, only the time of each reconstruction method is included. The results of the other datasets are not shown due to their similarities to these results.

As expected, linear and triangular reconstruction are the fastest methods because only two or three weighted means are used in their computation. The number of dimensions is nearly irrelevant, because only very few distance calculations are needed. Recursive and best-recursive reconstruction use a number of objects equal to the embedding methods, which results in their linear runtime. Interestingly, the need to calculate the SOD for each intermediate result has no negative impact on the speed of the best-recursive approach to the recursive one. Linear recursive and triangular recursive use all objects of the set regardless of the embedding methods, as can be seen by their constant runtime. Linear search methods show the longest runtime, but also one that is nearly independent of the embedding dimension. Although the proposed reconstruction methods are on average slower than the traditional techniques, their significant improvement of the calculated median justifies their use.

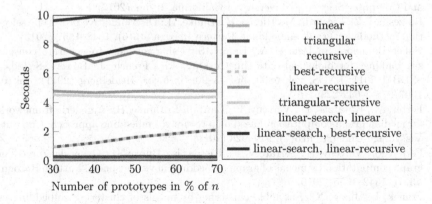

Fig. 4. Run-time of the UCI Cluster dataset.

5   Conclusion

Prototype embedding is a standard method for the embedding of arbitrary objects into Euclidean space with many applications and often used as an integral part of generalized median computation. In this work, we introduce new reconstruction methods for improving the generalized median computation. A study was presented using these methods on several datasets. The proposed linear recursive, triangular recursive, and linear search methods have been shown to consistently improve the results. Especially linear search using linear recursive reconstruction results in a significant improvement for all tested datasets. Although these results are derived from the datasets used in this work, our study provides hint enough to conclude that when dealing with generalized median computation we in general should take these reconstruction methods into account.

This work focused on the influence of reconstruction methods on the generalized median computation. Another point of interest is the use of alternative embedding methods. Prototype embedding has been shown to produce vectors for each object whose distance can be vastly different from their original one [14]. Using distance preserving embedding methods instead shows improvements of the median in several different types of datasets and distance functions [13]. These methods compute vectors x_i, x_j for objects o_i, o_j such that $\delta(o_i, o_j) \approx \delta_e(x_i, x_j)$, resulting in an embedding that more closely reflects the pairwise distances of the objects and thereby lead to a more accurate median computation. Combining the presented reconstruction methods and these distance preserving embedding methods may result in an even more accurate generalized median.

References

1. Bache, K., Lichman, M.: UCI Machine Learning Repository. School of Information and Computer Sciences, University of California, Irvine (2013)
2. Borzeshi, E.Z., Piccardi, M., Riesen, K., Bunke, H.: Discriminative prototype selection methods for graph embedding. Pattern Recogn. **46**(6), 1648–1657 (2013)
3. Cohen-Boulakia, S., Denise, A., Hamel, S.: Using medians to generate consensus rankings for biological data. In: Cushing, J.B., French, J., Bowers, S. (eds.) SSDBM 2011. LNCS, vol. 6809, pp. 73–90. Springer, Heidelberg (2011). doi:10.1007/978-3-642-22351-8_5
4. Ferrer, M., Karatzas, D., Valveny, E., Bardají, I., Bunke, H.: A generic framework for median graph computation based on a recursive embedding approach. Comput. Vis. Image Underst. **115**(7), 919–928 (2011)
5. Ferrer, M., Valveny, E., Serratosa, F., Riesen, K., Bunke, H.: Generalized median graph computation by means of graph embedding in vector spaces. Pattern Recogn. **43**(4), 1642–1655 (2010)
6. Franek, L., Jiang, X.: Ensemble clustering by means of clustering embedding in vector spaces. Pattern Recogn. **47**(2), 833–842 (2014)
7. Hakimi, S.: Location Theory. CRC Press, Boca Raton (2000)

8. de la Higuera, C., Casacuberta, F.: Topology of strings: median string is NP-complete. Theoret. Comput. Sci. **230**(1), 39–48 (2000)

9. Jiang, X., Münger, A., Bunke, H.: On median graphs: properties, algorithms, and applications. IEEE Trans. Pattern Anal. Mach. Intell. **23**(10), 1144–1151 (2001)

10. Jiang, X., Wentker, J., Ferrer, M.: Generalized median string computation by means of string embedding in vector spaces. Pattern Recogn. Lett. **33**(7), 842–852 (2012)

11. Krivánek, M., Morávek, J.: NP-hard problems in hierarchical-tree clustering. Acta Informatica **23**(3), 311–323 (1986)

12. Lundsteen, C., Phillip, J., Granum, E.: Quantitative analysis of 6985 digitized trypsin G-banded human metaphase chromosomes. Clin. Genet. **18**, 355–370 (1980)

13. Nienkötter, A., Jiang, X.: Distance-preserving vector space embedding for generalized median based consensus learning (2016, submitted for publication)

14. Riesen, K., Bunke, H.: Graph classification based on vector space embedding. Int. J. Pattern Recognit Artif Intell. **23**(06), 1053–1081 (2009)

15. Wagner, R.A., Fischer, M.J.: The string-to-string correction problem. J. ACM **21**(1), 168–173 (1974)

16. Weiszfeld, E., Plastria, F.: On the point for which the sum of the distances to n given points is minimum. Ann. Oper. Res. **167**(1), 7–41 (2009)

Graph-Theoretic Methods

On the Relevance of Local Neighbourhoods
for Greedy Graph Edit Distance

Xavier Cortés[1(✉)], Francesc Serratosa[1], and Kaspar Riesen[2]

[1] Universitat Rovira i Virgili, Tarragona, Spain
{xavier.cortes, frances.serratosa}@urv.cat
[2] University of Applied Sciences and Arts Northwestern Switzerland,
Basel, Switzerland
kaspar.riesen@fhnw.ch

Abstract. Approximation of graph edit distance based on bipartite graph matching emerged to an important model for distance based graph classification. However, one of the drawbacks of this particular approximation is its cubic runtime with respect to the number of nodes of the graphs. In fact, this runtime restricts the applicability of bipartite graph matching to graphs of rather small size. Recently, a new approximation for graph edit distance using greedy algorithms (rather than optimal bipartite algorithms) has been proposed. This novel algorithm reduces the computational complexity to quadratic order. In another line of research it has been shown that the definition of local neighbourhoods plays a crucial role in bipartite graph matching. These neighbourhoods define the local substructures of the graphs which are eventually assigned to each other. In the present paper we demonstrate that the type of local neighbourhood and in particular the distance model defined on them is also highly relevant for graph classification using greedy graph edit distance.

Keywords: Graph edit distance · Graph classification · Greedy assignment · Bipartite graph matching

1 Introduction

Graphs offer a convenient way to formally model objects or patterns, which are composed of complex subparts including the relations that might exist between these subparts. In particular, the nodes of a graph can be employed to represent the individual components of a pattern while the edges might represent the structural connections between these components. Attributed graphs, that is, graphs in which nodes and/or edges are labelled with one or more attributes, have been of crucial importance in pattern recognition throughout more than four decades [1–5]. For instance, graphs are successfully used in the field of recognizing biological patterns where one has to formally describe complex protein structures [6] or molecular compounds [7].

This research is supported by projects DPI2013-42458-P and TIN2013-47245-C2-2-R.

A. Robles-Kelly et al. (Eds.): S+SSPR 2016, LNCS 10029, pp. 121–131, 2016.
DOI: 10.1007/978-3-319-49055-7_11

The definition of an adequate dissimilarity model between two patterns is one of the most basic requirements in pattern recognition. The process of evaluating the dissimilarity of two graphs is commonly referred to as *graph matching*. The overall aim of graph matching is to find a correspondence between the nodes and edges of two graphs. There are several graph matching models available. *Spectral methods*, for instance, constitute an important class of error-tolerant graph matching procedures with a quite long tradition [8–12]. *Graph kernels* constitute a second important family of graph matching procedures and various types of graph kernels emerged during the last decade [13–17]. For an extensive review on these and other graph matching methods developed during the last forty years, the reader is referred to [2–4].

In the present paper we focus on the concept of *graph edit distance* [18–20]. This graph matching model is especially interesting because it is able to cope with directed and undirected, as well as with labelled and unlabeled graphs. However, the major drawback of graph edit distance is its computational complexity which is exponential with respect to the size of the graphs. Graph edit distance belongs to the family of *quadratic assignment problems* (QAPs), and thus, graph edit distance is known to be an NP-complete problem. In order to reduce the computational complexity of this particular distance model, an algorithmic framework for the approximation of graph edit distance was introduced in [21]. The basic idea of this approach is to find an optimal assignment of local neighbourhoods in the graphs, reducing the problem of graph edit distance to a *linear sum assignment problem* (LSAP). Recently, a new approximation framework, which extends [21–25] in terms of finding a suboptimal assignment of local neighbourhoods, has been presented in [27]. This new approach further reduces the computational complexity of graph edit distance from cubic to quadratic order. An algorithm has been presented in which the human can interact or guide the algorithm to find a proper correspondence between nodes or interest points [28, 29]. And other graph matching techniques have been presented such as [30, 31].

Both approaches, i.e. the original approximation [21–25] as well as the faster variant [27], use the same definition of local neighbourhoods, viz. single nodes and their adjacent edges. Yet various other definitions for local structures exist [32]. The present paper investigates the impact of these different local neighbourhoods in the context of the recent quadratic approximation algorithm [27].

2   Definitions and Algorithms

2.1   Graph and Neighbourhood

An attributed graph is defined as a triplet $G = (\Sigma_v, \Sigma_e, \gamma_v)$, where $\Sigma_v = \{v_a | a = 1, \ldots, n\}$ is the set of nodes and $\Sigma_e = \{e_{ab} | a, b \in 1, \ldots, n\}$ is the set of undirected and unattributed edges. In the present paper we investigate graphs with labelled nodes only. Yet, all of our concepts can be extended to graphs with labelled edges. Function $\gamma_v : \Sigma_v \to \Delta_v$ assigns attribute values from arbitrary domains to nodes. The order of a graph G is equal to the number of nodes n. The number of edges of node v_a is referred to as $E(v_a)$. Finally, the neighbourhood of a node v_a is termed N_{v_a}.

In this paper we focus on three different types of neighbourhoods N_{v_a} depending on the amount of structural information which is taken into account. The first definition of N_{v_a} is the empty set. That is, no structural information of node v_a is taken into account. Formally, $N_{v_a} = \left(\Sigma_v^{N_{v_a}}, \Sigma_e^{N_{v_a}}, \gamma_v \right)$ with $\Sigma_v^{N_{v_a}} = \Sigma_e^{N_{v_a}} = \{\}$. We refer to this type of neighbourhood as *Node*.

The second definition of N_{v_a} is given by $N_{v_a} = \left(\Sigma_v^{N_{v_a}}, \Sigma_e^{N_{v_a}}, \gamma_v \right)$ with $\Sigma_v^{N_{v_a}} = \{\}$ and $\Sigma_e^{N_{v_a}} = \{e_{ab} \in \Sigma_e\}$. In this case we regard the incident edges of v_a as neighbourhood only (without adjacent nodes). This second definition of N_{v_a} is referred to as *Degree* from now on. The third definition of N_{v_a} used in this paper is given by the set of nodes that are directly connected to v_a including all edges that connect these nodes with v_a. Formally, $N_{v_a} = \left(\Sigma_v^{N_{v_a}}, \Sigma_e^{N_{v_a}}, \gamma_v \right)$ with $\Sigma_v^{N_{v_a}} = \{v_b | e_{ab} \in \Sigma_e\}$ and $\Sigma_e^{N_{v_a}} = \{e_{ab} \in \Sigma_e | v_b \in \Sigma_v^{N_{v_a}}\}$. We refer to this definition of a neighbourhood as *Star*. In Fig. 1 an illustration of the three different neighbourhoods (*Node*, *Star* and *Degree*) is shown.

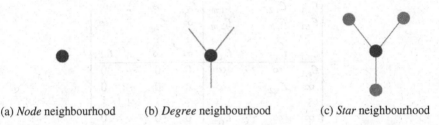

(a) *Node* neighbourhood   (b) *Degree* neighbourhood   (c) *Star* neighbourhood

Fig. 1. *Node*, *Degree* and *Star* neighbourhoods (shown in light-grey) of a single node (shown in dark-grey).

2.2   Graph Edit Distance

A widely used method to evaluate the dissimilarity between two attributed graphs is graph edit distance [18, 19]. The basic idea of this dissimilarity model is to define a distance between two graphs G^p and G^q by means of the minimum amount of distortion required to transform G^p into G^q. To this end, a number of distortions or edit operations, consisting of insertion, deletion, and substitution of both nodes and edges are employed. Edit cost functions are typically introduced to quantitatively evaluate the level of distortion of each individual edit operation. The basic idea of this is to assign a cost to the edit operations proportional to the amount of distortion they introduce in the underlying graphs.

A sequence $(e_1, \ldots, e_k)$ of k edit operations e_i that transform G^p completely into G^q is called an *edit path* $\lambda(G^p, G^q)$ between G^p and G^q . Note that in an edit path $\lambda(G^p, G^q)$ each node of G^p is either deleted or uniquely substituted with a node in G^q, and likewise, each node in G^q is either inserted or matched with a unique node in G^p. The same applies for the edges.

Let $\Upsilon(G^p, G^q)$ denote the set of all edit paths between two graphs G^p and G^q. The edit distance of two graphs is defined as the sum of cost of the minimal cost edit path among all competing paths in $\Upsilon(G^p, G^q)$.

Optimal algorithms for computing the edit distance are typically based on combinatorial search procedures (such as A* based search techniques). These procedures explore the space of all possible mappings of the nodes and edges of G^p to the nodes and edges of G^q (i.e. the search space corresponds to the set of all edit paths $\Upsilon(G^p, G^q)$). Yet, considering m nodes in G^p and n nodes in G^q, the set of possible edit paths $\Upsilon(G^p, G^q)$ contains $O(m^n)$ edit paths. Therefore, exact graph edit distance computation is exponential in the number of nodes of the involved graphs.

The problem of minimizing the graph edit distance can be reformulated as an instance of a Quadratic Assignment Problem (QAP). QAPs belong to the most difficult combinatorial optimization problems for which only exponential run time algorithms are known to date (QAPs are known to be NP-complete). The *Bipartite Graph Matching* algorithm (BP-GED) [21] is an approximation for the graph edit distance that reduces the QAP of graph edit distance computation to an instance of a Linear Sum Assignment Problem (LSAP). This algorithm first generates a cost matrix C which is based on costs of editing local substructures of both graphs. Formally, the cost matrix is defined by:

$$
C = \begin{bmatrix}
C_{1,1} & C_{1,2} & \cdots & C_{1,m} & C_{1,\varepsilon} & \infty & \cdots & \infty \\
C_{2,1} & C_{2,2} & \cdots & C_{2,m} & \infty & C_{2,\varepsilon} & \cdots & \infty \\
\vdots & \vdots & \ddots & \vdots & \vdots & \vdots & \ddots & \vdots \\
C_{n,1} & C_{n,2} & \cdots & C_{n,m} & \infty & \infty & \cdots & C_{n,\varepsilon} \\
C_{\varepsilon,1} & \infty & \cdots & \infty & 0 & 0 & \cdots & 0 \\
\infty & C_{\varepsilon,2} & \cdots & \infty & 0 & 0 & \cdots & 0 \\
\vdots & \vdots & \ddots & \vdots & \vdots & \vdots & \ddots & \vdots \\
\infty & \infty & \cdots & C_{\varepsilon,m} & 0 & 0 & \cdots & 0
\end{bmatrix}
$$

Where $C_{i,j}$ denotes the cost of substituting nodes v_i^p and v_j^q as well as their local neighbourhoods $N_{v_i^p}$ and $N_{v_j^q}$. $C_{i,\varepsilon}$ denotes the cost of deleting node v_i^p and its local neighbourhood $N_{v_i^p}$, and $C_{\varepsilon,j}$ denotes the cost of inserting node v_j^q and its local neighbourhood $N_{v_j^q}$.

A linear assignment algorithm can be applied on C in order to find a (optimal) mapping of nodes and their local neighbourhoods. A large number of solvers for linear sum assignment problems exist [26]. The time complexity of the best performing exact algorithms for LSAPs is cubic in the size of the problem.

Any complete assignment of local substructures derived on C can be reformulated as an admissible edit path from $\Upsilon(G^p, G^q)$. That is, the global edge structure from G^p and G^q can be edited with respect to the node operations captured in the mapping of local substructures (this is due to the fact that edit operations on edges always depend on the edit operations actually applied on their adjacent nodes). Eventually, the total cost of all edit operations (applied on both nodes and edges) can be interpreted as a graph edit distance approximation between graphs G^p and G^q (termed BP-GED from now on).

The edit path found with this particular procedure considers the structural information of the graphs in an isolated way only (singular neighbourhoods). Yet, the derived distance considers the edge neighbourhood of G^p and G^q in a global and consistent way and thus the derived distance is in the best case equal to, or in general larger than the exact graph edit distance.

Recently, a new approach which uses an approximation rather than an exact algorithm to solve the assignment of local substructures has been proposed [27]. This new approach employs a greedy assignment algorithm to suboptimally solve the LSAP stated on cost matrix C. The algorithm iterates through C from top to bottom through all rows, and in every row it assigns the current element to the minimum unused element in a greedy manner. More formally, for each row i in the cost matrix $C = (C_{ij})$ the minimum cost entry $\varphi_i = argmin_{vj} c_{ij}$ is determined and the corresponding node edit operation $(v_i^p \rightarrow v_{\varphi i}^q)$ is added to the mapping of local substructures. By removing column φ_i in C one can ensure that every column of the cost matrix is considered exactly once.

The remaining parts of the approximation algorithm are identical with the original framework. That is, based on the found assignment of local substructures an admissible edit path and its corresponding sum of costs is derived. However, as the complexity of this suboptimal assignment algorithm is only quadratic (rather than cubic), the time complexity of the complete graph edit distance approximation, termed *Greedy-GED* from now on, is further reduced.

3 Distance Models for Neighbourhoods

In this section we review and compare various methods proposed in [32] to obtain the individual cost entries $C_{i,j}, C_{i,\varepsilon}$, and $C_{\varepsilon,j}$ in the cost matrix. These cost entries depend on two weighted disjoint cost values. The first cost is defined with respect to the nodes, while the second cost takes into account the local neighbourhoods of the nodes. Formally, we define $C_{i,j}, C_{i,\varepsilon}$, and $C_{\varepsilon,j}$ according to the following three cases:

(1) If two nodes $v_i^p \in \Sigma_v^p$ and $v_j^q \in \Sigma_v^q$ are mapped to each other, we have

$$C_{ij} = \beta \cdot C(v_i^p \rightarrow v_j^q) + (1 - \beta) \cdot C(N_{v_i^p} \rightarrow N_{v_j^q}) \tag{1}$$

where $\beta \in \,]0,1[$ is a weighting parameter that controls what is more important, the cost of the pure node substitution $(v_i^p \rightarrow v_j^q)$ or the cost of substituting the neighbourhoods $N_{v_i^p}$ and $N_{v_j^q}$ of both nodes.

(2) If one node $v_i^p \in \Sigma_v^p$ in G^p is deleted, we have

$$C_{i,\varepsilon} = \beta \cdot k_v + (1 - \beta) \cdot C(N_{v_i^p} \rightarrow \varepsilon) \tag{2}$$

where $\beta \in \,]0,1[$ (as defined above), k_v refers to a positive constant cost for deleting one node and $C(N_{v_i^p} \rightarrow \varepsilon)$ refers to the cost of deleting the complete neighbourhood of v_i^p

(3) If one node $v_i^q \in \Sigma_v^q$ in G^q is inserted, we finally have (similar to case 2)

$$C_{\varepsilon,j} = \beta \cdot k_v + (1 - \beta) \cdot C(\varepsilon \rightarrow N_{v_j^q}) \tag{3}$$

The cost for node substitutions $C(v_i^p \rightarrow v_j^q)$ is commonly defined with respect to the underlying labelling of the involved nodes. Yet, the definition of an adequate cost model for neighbourhoods is not that straightforward, and this, described in greater detail in the next subsection.

3.1   The Cost of Processing Neighbourhoods

If the structural information of the nodes is not considered, that is we employ the *Node* neighbourhood, we have

$$C(N_{v_i^p} \rightarrow N_{v_j^q}) = C(N_{v_i^p} \rightarrow \varepsilon) = C(\varepsilon \rightarrow N_{v_j^q}) = 0 \qquad (4)$$

For *BP-GED* [21] as well as for *Greedy-GED* [27] the same definition of the neighbourhood of a node v has been used, viz. N_v is defined to be the set of incident edges of node v. That is, the *Degree* neighbourhood, as formally described in Sect. 2, is employed. Remember that in our paper unlabeled edges are considered only, and thus, edge substitution is free of cost. Hence, using this definition of a neighbourhood, the cost of substituting two neighbourhoods with each other is given by the difference of the numbers of edges of the involved nodes. Formally,

$$C(N_{v_i^p} \rightarrow N_{v_j^q}) = k_e \cdot \left| E(v_i^p) - E\left(v_j^q\right) \right| \qquad (5)$$

where k_e refers to a positive constant cost for deleting/inserting edges and $E(.)$ refers to the number of edges of a certain node.

Likewise, the deletion and insertion costs of neighbourhoods depend on the number of incident edges $E(v_i^p)/E\left(v_j^q\right)$ of the deleted or inserted node v_i^p and v_j^q, respectively. Formally,

$$C(N_{v_i^p} \rightarrow \varepsilon) = k_e \cdot E(v_i^p) \qquad (6)$$

$$C(\varepsilon \rightarrow N_{v_j^q}) = k_e \cdot E\left(v_j^q\right) \qquad (7)$$

We name this particular definition of the cost for processing neighbourhoods as *Degree-cost*. We will use this cost model for neighbourhoods as basic reference system.

Next, four other definitions of the cost for processing neighbourhoods are presented. In contrast with the *Degree*-cost model described above, these definitions are based on the *Star* neighbourhood.

The following definition of the insertion as well as the deletion cost assume that the complete neighbourhood has to be inserted or deleted when the corresponding node is inserted or deleted, respectively. That is, insertion and deletion costs consider the cost of processing all edges that connect the central node and the cost of processing all adjacent nodes. Formally, the complete deletion and insertion costs of neighbourhoods depend on the number of adjacent nodes.

$$C(N_{v_i^p} \to \varepsilon) = (k_v + k_e) \cdot E(v_i^p) \tag{8}$$

$$C(\varepsilon \to N_{v_j^q}) = (k_v + k_e) \cdot E\left(v_j^q\right) \tag{9}$$

Where $(k_v + k_e)$ refers to the cost of deleting or inserting one edge and one node. The substitution cost $C(N_{v_i^p} \to N_{v_j^q})$ for star neighbourhoods is based on computing a distance between $N_{v_i^p}$ and $N_{v_j^q}$. For this particular computation every adjacent node and its corresponding edge is interpreted as an indivisible entity. That is, a *Star* neighbourhood can be seen as a set of independent entities (nodes with adjacent edge), and thus, the computation of $C(N_{v_i^p} \to N_{v_j^q})$ refers to an assignment problem between two independent sets. This assignment problem can be solved in the same way as it is done with complete graphs by means of a cost matrix that considers substitutions, insertions and deletions. Formally,

$$C(N_{v_i^p} \to N_{v_j^q}) = AssignmentCost\left(N_{v_i^p}, N_{v_j^q}\right) \tag{10}$$

In order to compute this assignment cost we use four different optimization algorithms.

The first algorithm is given by an optimal algorithm for general LAPs known as *Hungarian* algorithm [33] that runs in cubic time. The second algorithm solves the assignment of $N_{v_i^p}$ and $N_{v_j^q}$ by means of the *Hausdorff* distance for subsets [34]. This assignment algorithm estimates the distance between two sets of entities by removing the restriction of finding a bijective mapping between the individual elements. In contrast with the *Hungarian* algorithm, *Hausdorff* assignments can be computed in quadratic time.

The third algorithm computes the dissimilarity between $N_{v_i^p}$ and $N_{v_j^q}$ by finding a suboptimal assignment of the individual entities by means of the *Greedy* assignment algorithm [27] described in Sect. 2 (also in quadratic time).

Finally, we propose to use a *Planar* distance model. In this case, the relative position of each neighbourhood node is considered. That is, the only allowed assignments of entities from $N_{v_i^p}$ to entities of $N_{v_j^q}$ are the ones that are generated from cyclic combinations of the neighbours. Formally, the sets $N_{v_i^p}$ and $N_{v_j^q}$ are interpreted as strings and the assignment cost is computed through the *Levenshtein* distance on these strings [34–38].

4 Experimental Evaluation

In the present paper three specific definitions for node neighbourhoods are presented, namely *Node*, *Degree* and *Star*. For the computation of both *Node* and *Degree* no additional algorithm is required. Yet, for *Star* neighbourhood a particular assignment solver is needed. We propose four different algorithms for this task. Hence, in total we have six different cost models to compute the individual entries $c_{i,j} \in C$. Major aim of the present paper is to show the relevance of these costs models in terms of classification accuracy and runtime using *Greedy-GED*. We coded the algorithms using MATLAB 2013 and we conducted the experiments on an i5 processor of 1.60 GHz with 4 GB of RAM and Windows 10.

Table 1 shows the recognition ratio of a 1NN classifier and the mean matching runtime achieved with all models on five graph databases, viz. LETTER LOW, LETTER HIGH, GREC, COIL RAG and AIDS from the IAM graph repository [39]. In our experiments parameter β is fixed to 0.5 and we gauge k_v and k_e through a non-exhaustive trial and error process. The best accuracy and the fastest runtime is highlighted in bold face on each database, while the second best is underlined.

Table 1. Recognition ratio (1NN) and Runtime (Øt) of the three neighbourhoods (and for assignment solvers) on five datasets using *greedy graph edit distance*.

Local neighbourhood	Assignment solver	Database									
		LETTER LOW		LETTER HIGH		GREC		COIL RAG		AIDS	
		1NN (%)	Øt (ms)	1NN (%)	Øt (ms)	1NN (%)	Øt (ms)	1NN (%)	Øt (ms)	1NN (%)	Øt (ms)
Degree	–	95.07	0.6	79.07	0.6	97.54	1.3	96.00	0.4	**95.53**	2.1
Node	–	90.67	**0.4**	60.27	**0.4**	93.37	**0.8**	**96.40**	**0.3**	86.80	**1.2**
Star	Hungarian	98.53	12.5	86.53	13.5	96.40	84.7	95.10	5.8	88.00	162.6
	Hausdorff	**98.67**	1.3	88.40	1.4	96.97	6.9	95.10	0.8	89.00	11.7
	Greedy	**98.67**	1.1	86.67	1.3	96.97	5.6	94.50	0.8	88.60	10.2
	Levenshtein	98.53	1.1	**89.33**	1.7	**97.73**	8.7	94.70	1.0	87.53	17.6

We first focus on the mean run time for one matching in ms (Øt). We observe that on all data sets the Node neighbourhood is the fastest model (as it could be expected) followed by the reference model Degree (which is only slightly slower than the Node neighbourhood). The models that are based on the Star neighbourhood suffer from substantial higher runtimes than Node and Degree (especially on the larger graphs of the GREC and AIDS data sets). Comparing the four assignment solvers with each other, we observe that the optimal (cubic time) Hungarian algorithm provides the highest run times among all competing algorithms, while the differences between the other three algorithms are negligible.

Next, we focus on the recognition rates of the different models. We observe that in three out of five cases the Star neighbourhood achieves the best recognition rates (LETTER LOW and HIGH as well as GREC). The Node neighbourhood and the reference model Degree achieve the best recognition rate on COIL RAG and AIDS, respectively. An interesting observation can be made on the AIDS data set where the reference model Degree significantly outperforms all other methods addressed in the present paper. Overall we conclude that it remains difficult to predict an adequate definition of local neighbourhoods for a given data set and application. Yet, there seems to be a weak tendency that increased neighbourhoods might improve the overall matching accuracy.

5 Conclusions

The fast approximation of graph edit distance is still an open and active area of research in the field of structural pattern recognition. A common model for the approximation of graph distances is based on bipartite graph matching. The basic idea of this approach is to reduce the difficult QAP of graph edit distance computation to an LSAP. This particular algorithmic framework consists of three major steps. In a first step the graphs to be matched are subdivided into individual nodes including local neighbourhoods. Next, in step 2, an algorithm solving the LSAP is employed in order to find an assignment of the nodes (plus local neighbourhoods) of both graphs. Finally, in step 3, an approximate graph edit distance is derived from the assignment of step 2. In the present paper we review six different ways to compute the costs of local neighbourhoods and compare them with each other on five data sets. Although no clear winner can be found in the experimental evaluation, the empirical results suggests to use a larger neighbourhood than it is traditionally employed in the context of bipartite graph matching.

References

1. Sanfeliu, A., Alquézar, R., Andrade, J., Climent, J., Serratosa, F., Vergés, J.: Graph-based representations and techniques for image processing and image analysis. Pattern Recogn. **35** (3), 639–650 (2002)
2. Conte, D., Foggia, P., Sansone, C., Vento, M.: Thirty years of graph matching in pattern recognition. Int. J. Pattern Recogn. Artfi. Intell. **18**(3), 265–298 (2004)
3. Vento, M.: A one hour trip in the world of graphs, looking at the papers of the last ten years. In: Artner, N.M., Haxhimusa, Y., Jiang, X., Kropatsch, W.G. (eds.) GbRPR 2013. LNCS, vol. 7877, pp. 1–10. Springer, Heidelberg (2013)
4. Foggia, P., Percannella, G., Vento, M.: Graph matching and learning in pattern recognition in the last 10 years. Int. J. Pattern Recogn. Artif. Intell. **28**(1), 1450001 (2014)
5. Sanroma, G., Penate-Sanchez, A., Alquezar, R., Serratosa, F., Moreno-Noguer, F., Andrade-Cetto, J., Gonzalez, M.A.: MSClique: multiple structure discovery through the maximum weighted clique problem. PLoS ONE **11**(1), e0145846 (2016)
6. Borgwardt, K., Ong, C., Schönauer, S., Vishwanathan, S., Smola, A., Kriegel, H.-P.: Protein function prediction via graph kernels. Bioinformatics **21**(1), 47–56 (2005)
7. Mahé, P., Ueda, N., Akutsu, T., Perret, J., Vert, J.: Graph kernels for molecular structure-activity relationship analysis with support vector machines. J. Chem. Inf. Model. **45**(4), 939–951 (2005)
8. Luo, B., Wilson, R.C., Hancock, E.R.: Spectral feature vectors for graph clustering. In: Caelli, T.M., Amin, A., Duin, R.P., Kamel, M.S., de Ridder, D. (eds.) SPR 2002 and SSPR 2002. LNCS, vol. 2396, pp. 83–93. Springer, Heidelberg (2002)
9. Robles-Kelly, A., Hancock, E.R.: Graph edit distance from spectral seriation. IEEE Trans. Pattern Anal. Mach. Intell. **27**(3), 365–378 (2005)
10. Kosinov, S., Caelli, T.M.: Inexact multisubgraph matching using graph eigenspace and clustering models. In: Caelli, T.M., Amin, A., Duin, R.P., Kamel, M.S., de Ridder, D. (eds.) SPR 2002 and SSPR 2002. LNCS, vol. 2396, pp. 133–142. Springer, Heidelberg (2002)

11. Wilson, R.C., Hancock, E.R., Luo, B.: Pattern vectors from algebraic graph theory. IEEE Trans. Pattern Anal. Mach. Intell. **27**(7), 1112–1124 (2005)
12. Qiu, H., Hancock, E.R.: Graph matching and clustering using spectral partitions. Pattern Recogn. **39**(1), 22–34 (2006)
13. Ramon, J., Gärtner, T.: Expressivity versus efficiency of graph kernels. In: First International Workshop on Mining Graphs, Trees and Sequences, pp. 65–74 (2003)
14. Borgwardt, K., Petri, T., Kriegel, H.-P., Vishwanathan, S.: An efficient sampling scheme for comparison of large graphs. In: International Workshop on Mining and Learning with Graphs (2007)
15. Vishwanathan, S.V.N., Borgwardt, K., Schraudolph, N.N.: Fast computation of graph kernels. In: Annual Conference on Neural Information Processing Systems, pp. 1449–1456. MIT Press (2006)
16. Gaüzère, B., Brun, L., Villemin, D.: Two new graph kernels and applications to chemoinformatics. In: Jiang, X., Ferrer, M., Torsello, A. (eds.) GbRPR 2011. LNCS, vol. 6658, pp. 112–121. Springer, Heidelberg (2011)
17. Gauzere, B., Grenier, P.A., Brun, L., Villemin, D.: Treelet kernel incorporating cyclic, stereo and inter pattern information in chemoinformatics. Pattern Recogn. **48**(2), 356–367 (2015)
18. Bunke, H., Allermann, G.: Inexact graph matching for structural pattern recognition. Pattern Recogn. Lett. **1**, 245–253 (1983)
19. Sanfeliu, A., Fu, K.: A distance measure between attributed relational graphs for pattern recognition. IEEE Trans. Syst. Man Cybern. (Part B) **13**(3), 353–363 (1983)
20. Solé, A., Serratosa, F., Sanfeliu, A.: On the graph edit distance cost: properties and applications. Int. J. Pattern Recogn. Artif. Intell. **26**(5), 1260004 [21 p.] (2012)
21. Riesen, K., Bunke, H.: Approximate graph edit distance computation by means of bipartite graph matching. Image Vis. Comput. **27**(4), 950–959 (2009)
22. Serratosa, F.: Fast computation of bipartite graph matching. Pattern Recogn. Lett. **45**, 244–250 (2014)
23. Serratosa, F.: Computation of graph edit distance: reasoning about optimality and speed-up. Image Vis. Comput. **40**, 38–48 (2015)
24. Serratosa, F.: Speeding up fast bipartite graph matching trough a new cost matrix. Int. J. Pattern Recogn. Artif. Intell. **29**(2),1550010 [17 p.] (2015)
25. Ferrer, M., Serratosa, F., Riesen, K.: Improving bipartite graph matching by assessing the assignment confidence. Pattern Recogn. Lett. **65**, 29–36 (2015)
26. Burkard, R., Dell'Amico, M., Martello, S.: Assignment Problems. Society for Industrial and Applied Mathematics, Philadelphia (2009)
27. Riesen, K., Ferrer, M., Fischer, A., Bunke, H.: Approximation of graph edit distance in quadratic time. In: Liu, C.-L., Luo, B., Kropatsch, W.G., Cheng, J. (eds.) GbRPR 2015. LNCS, vol. 9069, pp. 3–12. Springer, Heidelberg (2015)
28. Serratosa, F., Cortés, X.: Interactive graph-matching using active query strategies. Pattern Recogn. **48**(4), 1364–1373 (2015)
29. Cortés, X., Serratosa, F.: An interactive method for the image alignment problem based on partially supervised correspondence. Expert Syst. Appl. **42**(1), 179–192 (2015)
30. Sanromà, G., Alquézar, R., Serratosa, F., Herrera, B.: Smooth point-set registration using neighbouring constraints. Pattern Recogn. Lett. **33**, 2029–2037 (2012)
31. Sanromà, G., Alquézar, R., Serratosa, F.: A new graph matching method for point-set correspondence using the EM algorithm and softassign. Comput. Vis. Image Underst. **116** (2), 292–304 (2012)
32. Serratosa, F., Cortés, X.: Graph edit distance: moving from global to local structure to solve the graph-matching problem. Pattern Recogn. Lett. **65**, 204–210 (2015)

33. Kuhn, H.W.: The Hungarian method for the assignment problem. Naval Res. Logistics Q. **2**, 83–97 (1955)
34. Huttenlocher, D.P., Klanderman, G.A., Kl, G.A., Rucklidge, W.J.: Comparing images using the Hausdorff distance. IEEE Trans. Pattern Anal. Mach. Intell. **15**, 850–863 (1993)
35. Levenshtein, V.I.: Binary codes capable of correcting deletions, insertions and reversals. Sov. Phys. Dokl. Cybern. Control Theory **10**, 707–710 (1966)
36. Peris, G., Marzal, A.: Fast cyclic edit distance computation with weighted edit costs in classification. Int. Conf. Pattern Recogn. **4**, 184–187 (2002)
37. Serratosa, F., Sanfeliu, A.: Signatures versus histograms: definitions, distances and algorithms. Pattern Recogn. **39**(5), 921–934 (2006)
38. Serratosa, F., Sanromà, G.: A fast approximation of the earth-movers distance between multi-dimensional histograms. Int. J. Pattern Recognit. Artif. Intell. **22**(8), 1539–1558 (2008)
39. Riesen, K., Bunke, H.: IAM graph database repository for graph based pattern recognition and machine learning. In: Vitoria Lobo, N., Kasparis, T., Roli, F., Kwok, J.T., Georgiopoulos, M., Anagnostopoulos, G.C., Loog, M. (eds.) Structural, Syntactic, and Statistical Pattern Recognition. LNCS, vol. 5342, pp. 287–297. Springer, Heidelberg (2008)

GriMa: A Grid Mining Algorithm
for Bag-of-Grid-Based Classification

Romain Deville[1,2], Elisa Fromont[1(✉)], Baptiste Jeudy[1], and Christine Solnon[2]

[1] Université de Lyon, UJM, Laboratoire Hubert Curien (CNRS, UJM, IOGS),
UMR 5516, 42000 Saint-Etienne, France
`elisa.fromont@univ-st-etienne.fr`
[2] Université de Lyon, INSA-Lyon, LIRIS, UMR5205, 69621 Lyon, France

Abstract. General-purpose exhaustive graph mining algorithms have seldom been used in real life contexts due to the high complexity of the process that is mostly based on costly isomorphism tests and countless expansion possibilities. In this paper, we explain how to exploit grid-based representations of problems to efficiently extract frequent grid subgraphs and create Bag-of-Grids which can be used as new features for classification purposes. We provide an efficient grid mining algorithm called GRIMA which is designed to scale to large amount of data. We apply our algorithm on image classification problems where typical Bag-of-Visual-Words-based techniques are used. However, those techniques make use of limited spatial information in the image which could be beneficial to obtain more discriminative features. Experiments on different datasets show that our algorithm is efficient and that adding the structure may greatly help the image classification process.

1 Introduction

General-purpose exhaustive graph mining algorithms are seldom used in real-world applications due to the high complexity of the mining process mostly based on isomorphism tests and countless expansion possibilities during the search [6]. In this paper we tackle the problem of exhaustive graph mining for grid graphs which are graphs with fixed topologies and such that each vertex degree is constant, *e.g.*, 4 in a 2D square grid. This grid structure is naturally present in many boardgames (Checkers, Chess, Go, etc.) or to model ecosystems using cellular automata [10], for example. In addition, this grid structure may be useful to capture low-level topological relationships whenever a high-level graph structure is not obvious to design. In computer vision in particular, it is now widely acknowledged that high-level graph-based image representations (such as region adjacency graphs or interest point triangulations, for example) are sensitive to noise so that slightly different images may result in very different graphs [20,21]. However, at a low-level, images basically are grids: 2D grids of pixels for images, and 3D grids of voxels for tomographic or MRI images modelling 3D objects. When considering videos, we may add a temporal dimension to obtain 2D+t grids. We propose to exploit this grid structure and to characterize images by histograms of frequent subgrid patterns.

© Springer International Publishing AG 2016
A. Robles-Kelly et al. (Eds.): S+SSPR 2016, LNCS 10029, pp. 132–142, 2016.
DOI: 10.1007/978-3-319-49055-7_12

Section 2 describes existing works that use pattern mining approaches for image classification and existing graph mining algorithms related to our proposed approach. Section 3 introduces grid graphs and our sub-grid mining algorithm GRIMA. Section 4 experimentally compares these algorithms and shows the relevance of GRIMA for image classification.

2 Related Work

Mining for Image Classification. Pattern mining techniques have recently been very successfully used in image classification [9,23] as a mean to obtain more discriminative mid-level features. However, those approaches consider the extracted features used to describe images (*e.g.*, Bags-of-Visual-Words/BoWs) as spatially independent from each other. The problem of using bag-of-graphs instead of BoWs has already been mentioned in [2,19,22] for satellite image classification and biological applications. However, none of these papers provide a general graph representation nor a graph mining algorithm to extract the patterns. In [17], authors have already shown that, by combining graph mining and boosting, they can obtain classification rules based on subgraph features that contain more information than sets of features. The GSPAN algorithm [24] is then used to compute the subgraph patterns but a limited number of features per image is used to be able to scale on real-life datasets.

Graph Mining. GSPAN [24] and all similar general exhaustive graph mining algorithms [11] extract frequent subgraphs from a base of graphs. During the mining process, GSPAN does not consider edge angles so that it considers as isomorphic two subgraphs that are different in a grid point of view as shown in Fig. 1(b and c). Because of this, GSPAN may consider as frequent a huge number of patterns and does not scale well. On the other hand, PLAGRAM [20] has been developed to mine plane graphs and thus to scale efficiently on these graphs. However, in PLAGRAM, the extension strategy (which is a necessary step for all exhaustive graph mining algorithms) is based on the faces of the graph which induces an important restriction: All patterns should be composed of faces and the smallest possible subgraph pattern is a single face, i.e., a cycle with 3 nodes. Using PLAGRAM to mine grids is possible but the problem needs to be transformed such that each grid node becomes a face. This transformation is illustrated in Fig. 1: The two isomorphic graphs (b) and (c) (which are both subgraphs of (a)) are transformed into the two non-isomorphic graphs (e) and (f) (so that only (e) is a subgraph of (d)). However, this artificially increases the number of nodes and edges which may again cause scalability issues. Grid graphs have already been introduced in [14], but authors did not take into account the rigid topology of the grid and, in particular, the angle information that we use to define our grid mining problem. Finally, grid graphs are special cases of geometric graphs, for which mining algorithms have been proposed in [1]. These algorithms (FREQGEO and MAXGEO) may be seen as a generalization of our new algorithm GRIMA but, as such, they are not optimized for cases where the graph is known to be a grid

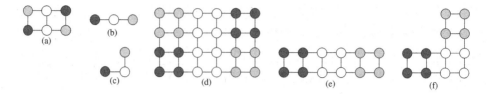

Fig. 1. Examples. (b) and (c) are isomorphic graphs and are both subgraphs of (a). However, (b) and (c) are not grid isomorphic, and (b) is a subgrid of (a) whereas (c) is not a subgrid of (a). (d) (resp. (e) and (f)) is obtained from (a) (resp. (b) and (c)) by replacing each node by a 4-node face (with the same label on the 4 nodes). (e) and (f) are not isomorphic, and (e) is a subgraph of (d) whereas (f) is not a subgraph of (d).

and we show in Sect. 3 that their complexity is higher than the complexity of our algorithm. Besides, the authors do not provide any implementation (thus, also no experiment) of their proposed algorithm which could allow us a comparison with our method.

3   Grid Mining

Grid Graphs. A grid is a particular case of graph such that edges only connect nodes which are neighbors on a grid. More formally, a grid is defined by $G = (N, E, L, \eta)$ such that N is a set of nodes, $E \subseteq N \times N$ is a set of edges, $L : N \cup E \to \mathbb{N}$ is a labelling function, $\eta : N \to \mathbb{Z}^2$ maps nodes to 2D coordinates, and $\forall (u, v) \in E$, the coordinates of u and v are neighbors, *i.e.*, $|x_u - x_v| + |y_u - y_v| = 1$ where $\eta(u) = (x_u, y_u)$ and $\eta(v) = (x_v, y_v)$. Looking for patterns in a grid amounts to searching for subgrid isomorphisms. In an image analysis context, patterns should be invariant to translations and rotations so that grid patterns which are equivalent up to a translation and/or a rotation should be isomorphic. For example, graphs (b) and (c) of Fig. 1 are isomorphic. However, (c) cannot be obtained by translating and/or rotating (b), because the angle between edges of (c) is different from the angle between edges of (b). Therefore, (b) and (c) are not grid-isomorphic. Finally, a grid $G_1 = (N_1, E_1, L_1, \eta_1)$ is subgrid-isomorphic to a grid $G_2 = (N_2, E_2, L_2, \eta_2)$ if there exists $N_2' \subseteq N_2$ and $E_2' \subseteq E_2 \cap N_2' \times N_2'$ such that G_1 is grid isomorphic to $G_2' = (N_2', E_2', L_2, \eta_2)$.

Frequent Grid Mining Problem. Given a database D of grid graphs and a frequency threshold σ, the goal is to output all frequent subgrid patterns in D, *i.e.*, all grids G such that there exist at least σ grids in D to which G is subgrid isomorphic.

Grid Codes. In graph mining algorithms, an important problem is to avoid generating the same pattern multiple times. One successful way to cope with this is to use canonical codes [24] to represent graphs and thus explore a canonical code search space instead of a graph one. In this paragraph, we define the canonical

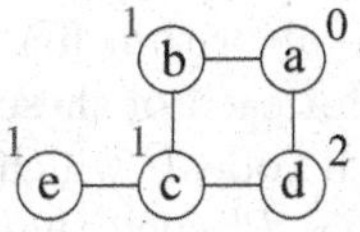

	Code 1		Code 2		Code 3	
edge	$\delta\ i\ j\ a\ L_i\,L_j L_{ij}$	edge	$\delta\ i\ j\ a\ L_i\,L_j L_{ij}$	edge	$\delta\ i\ j\ a\ L_i\,L_j L_{ij}$	
(a,b)	0 0 1 0 0 1 0	(a,b)	0 0 1 0 0 1 0	(d,c)	0 0 1 0 2 1 0	
(b,c)	0 1 2 3 1 1 0	(b,c)	0 1 2 3 1 1 0	(c,e)	0 1 2 2 1 1 0	
(c,e)	0 2 3 1 1 1 0	(c,d)	0 2 3 3 1 2 0	(c,b)	0 1 3 1 1 1 0	
(c,d)	0 2 4 3 1 2 0	(d,a)	1 3 0 3 2 0 0	(b,a)	0 3 4 1 1 0 0	
(d,a)	1 4 0 3 2 0 0	(c,e)	0 2 4 1 1 1 0	(a,d)	1 4 0 1 0 2 0	

Fig. 2. Examples of grid codes (node labels are displayed in blue; edges all have the same label 0). Codes 1 and 2 correspond to traversals started from edge (a,b), and differ on the third edge. Code 3 corresponds to a traversal started from edge (d,c) and it is canonical.

code used in our algorithm. A **code** $C(G)$ of a grid G is a sequence of n edge codes ($C(G) = \langle ec_0, ..., ec_{n-1} \rangle$) which is associated with a depth-first traversal of G starting from a given initial node. During this traversal, each edge is traversed once, and nodes are numbered: The initial node has number 0; Each time a new node is discovered, it is numbered with the smallest integer not already used in the traversal. Each edge code corresponds to a different edge of G and the order of edge codes in $C(G)$ corresponds to the order edges are traversed. Hence, ec_k is the code associated with the kth traversed edge. This edge code ec_k is the tuple $(\delta, i, j, a, L_i, L_j, L_{(i,j)})$ where

- i and j are the numbers associated with the nodes of the kth traversed edge.
- $\delta \in \{0, 1\}$ is the direction of the kth traversed edge:
  - $\delta = 0$ if it is forward, *i.e.*, i already appears in the prefix $\langle ec_0, ..., ec_{k-1} \rangle$ of the code and j is a new node which is reached for the first time;
  - $\delta = 1$ if it is backward, *i.e.*, both i and j already appear in the prefix.
- $a \in \{0, 1, 2, 3\}$ is the angle of the kth traversed edge:
  - $a = 0$ if $k = 0$ (first edge);
  - Otherwise, $a = 2A/\pi$ where $A \in \{\pi/2, \pi, 3\pi/2\}$ is the angle between the edge which has been used to reach i for the first time and (i, j).
- $L_i, L_j, L_{(i,j)}$ are node and edge labels.

For example, let us consider code 1 in Fig. 2. The fourth traversed edge is (c, d). It is a forward edge (because d has not been reached before) so that $\delta = 0$. The angle between (b, c) and (c, d) is $3\pi/2$ so that $a = 3$. The fifth traversed edge is (d, a) which is a backward edge (as a has been reached before) so that $\delta = 1$.

Given a code, we can reconstruct the corresponding grid since edges are listed in the code together with angles and labels. However, there exist different possible codes for a given grid, as illustrated in Fig. 2: Each code corresponds to a different traversal (starting from a different initial node and choosing edges in a different order). We define a total order on the set of all possible codes that may be associated with a given grid by considering a lexicographic order (all edge code components have integer values). Among all the possible codes for a grid, the largest one according to this order is called **the canonical code** of this grid and it is unique.

Description of GRIMA. Our algorithm, GRIMA, follows a standard frequent subgraph mining procedure as described, *e.g.*, in [20]. It explores the search space of all canonical codes in a depth-first recursive way. The algorithm first computes all frequent edges and then calls an `Extend` function for each of these frequent extensions. `Extend` has one input parameter: A pattern code P which is frequent and canonical. It outputs all frequent canonical codes P' such that P is a prefix of P'. To this aim, it first computes the set E of all possible valid extensions of all occurrences of P in the database D of grids: A valid extension is the code e of an edge such that $P.e$ occurs in D. Finally, `Extend` is recursively called for each extension e such that $P.e$ is frequent and canonical. Hence, at each recursive call, the pattern grows.

Node-Induced GRIMA. In our application, nodes are labelled but not edges (all edges have the same label). Thus, we designed a variant of GRIMA, called node-induced-GRIMA, which computes node-induced grids, i.e., grids induced by their node sets. This corresponds to a "node-induced" closure operator on graphs where, given a pattern P, we add all possible edges to P without adding new nodes. In the experiments, we show that this optimization decreases the number of extracted patterns and the extraction time.

Properties of GRIMA *and Complexity.* We can prove (not detailed here for lack of space) that GRIMA is both *correct*, which means that it only outputs frequent subgrids and *complete*, which means that it cannot miss any frequent subgrid. Let k be the number of grids in the set D of input grids, n the number of edges in the largest grid $G_i \in D$ and $|P|$ the number of edges in a pattern P. GRIMA enumerates all frequent patterns in $\mathcal{O}(kn^2.|P|^2) = \mathcal{O}(kn^4)$ time per pattern P. This is a significant improvement over FREQGEO and MAXGEO [1] which have a time complexity of $\mathcal{O}(k^2n^4.\ln n)$ per pattern.

4 Experiments

To assess the relevance of using a grid structure for image classification, we propose to model images by means of grids of visual words and to extract frequent subgrids to obtain Bags-of-Grids (BoGs). We compare these BoGs with a standard classification method which uses simple unstructured Bags-of-Visual-Words (BoWs). Note that neither BoG nor BoW-based image classification give state-of-the-art results for these datasets. In particular, [5] reported that the features discovered using deep learning techniques give much better accuracy results than BoWs and all their extensions on classification problems. However, our aim is to compare an unstructured set of descriptors and a set of descriptors structured by the grid topology. The method presented in this paper is generic and may be used with any low-level features (e.g. deep-learned features) as labels.

Datasets. We consider three datasets. *Flowers* [15] contains 17 flower classes where each class contains 80 images. *15-Scenes* [12] contains 4485 images from 15 scene classes (*e.g.*, kitchen, street, forest, etc.) with 210 to 410 images per class. *Caltech-101* [8] contains pictures of objects belonging to 101 classes. There are 40 to 800 images per class. As it is meaningless to mine frequent patterns in small datasets of images (when there are not enough images in a dataset, there may be a huge number of frequent patterns), we consider a subset of *Caltech-101* which is composed of the 26 classes of *Caltech-101* that contain at least 80 pictures per class. For each class, we randomly select 80 images. This subset will be referred to as *Caltech-26*.

BoW Design. Given a set of patches (small image regions) extracted from an image dataset, visual words are created by quantizing the values of patch descriptors [4,7]. The set of computed visual words is called the *visual vocabulary*. Each image is then encoded as an histogram of this visual vocabulary, called a bag-of-visual-words. A decade ago, patches used to create the visual vocabulary were selected by using interest point detectors or segmentation methods. However, [16] has shown that randomly sampling patches on grids (called dense sampling) gave as good (and often better) results for image classification than when using complex detectors. This made this problem a particularly suited use-case for our algorithm. In our experiments, visual words are based on 16×16 SIFT descriptors [13] which are 128-D vectors describing gradient information in a patch centered around a given pixel. 16×16 SIFT descriptors are extracted regularly on a grid and the center of each SIFT descriptor is separated by $s = 8$ pixels (descriptors are thus overlapping). We use the K-means algorithm to create visual word dictionaries for each dataset. The optimal number of words K is different from one method (BoG) to the other (BoW) so this parameter is studied in the experiments.

BoG Design. The first steps (computation of descriptors densely distributed in each image and creation of the visual vocabulary) are similar for BoW and BoG-based methods. For BoG, we create a square grid based on the grid of visual words by connecting each node to its 4 neighbors (except on the border of the image). In our experiments, grid nodes are labeled by visual words and edges remain unlabeled although the algorithm is generic and could include labels on edges. For efficiency reasons, we preprocess grids before the mining step to remove all nodes which are surrounded by nodes with identical labels (thus creating holes in place of regions with uniform labels). Once grids are created for all images, we mine frequent subgrid patterns with GRIMA class by class. Finally, we consider the union of all frequent patterns and all visual words, and represent each image by an histogram of frequency of patterns and words. These histograms are given as input to the classifier.

Efficiency Results. Figure 3 compares scale-up properties of GRIMA, node-induced-GRIMA, GSPAN and PLAGRAM for the *Flowers* dataset (average results on 10 folds for each class, where each fold contains 72 images). We have set the number K of words to 100, and we have increased the absolute frequency threshold from 45 (62.5%) to 72 (100%). Note that, as explained in Sect. 2, the face-based expansion strategy of PLAGRAM does not allow it to find patterns

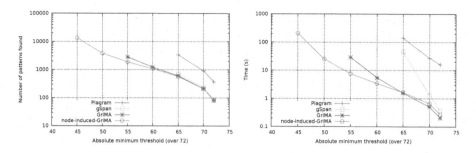

Fig. 3. Number of patterns (left) and time needed (right) with respect to different support thresholds σ to compute all patterns for a grid, on average for all classes of *Oxford-Flowers17* dataset for GSPAN, PLAGRAM, GRIMA and node-induced-GRIMA. Time is limited to 1 h per class.

with no face (such as trees, for example). To allow PLAGRAM to mine the same patterns as its competitors, each node of the original grid graph is replaced by a four-node face (as illustrated in Fig. 1). This way, each frequent subgrid in the original grid is found by PLAGRAM in the expanded grid. However, some patterns found by PLAGRAM do not correspond to patterns in the original grid (e.g., faces in the expanded grid which correspond to edges without their end nodes in the original grid). For this reason, the number of patterns that are mined by PLAGRAM and its computation time are higher than the ones reported for its competitors. GSPAN does not consider edge angles when mining subgraphs, and two isomorphic subgraphs may have different edge angles so that they do not correspond to isomorphic subgrids, as illustrated in Fig. 1. Therefore, GSPAN and GRIMA compute different sets of frequent patterns. However, Fig. 3 shows us that the number of patterns is rather similar (slightly higher for GSPAN than for GRIMA). GRIMA and node-induced-GRIMA scale better than GSPAN, and are able to find all frequent patterns in less than 10 s when $\sigma = 60$ whereas GSPAN has not completed its execution after 3600 s. The number of patterns found by both PLAGRAM and GSPAN for a support lower than 65 are thus not relevant and are not shown in the figure. As explained in Sect. 3, node-induced-GRIMA mines less patterns and is more efficient than GRIMA, and much more efficient than GSPAN and PLAGRAM.

N-ary Classification Results. Let us now evaluate the interest of using grids for image classification on our three datasets. For each dataset we consider different values for the number K of visual words, with $K \in \{100, 500, 1000, 2000, 4000\}$. Note that this corresponds to changing the K of K-means.

GRIMA has one parameter, σ, and Fig. 3 shows us that the number of mined patterns grows exponentially when decreasing σ. CPU time is closely related to the number of mined patterns, and σ should be set so that we have a good tradeoff between the time needed to mine patterns and the gain in accuracy. The number of mined patterns (and the CPU time) depends on σ, but also on the size K of the vocabulary and on the dataset. Therefore, instead of fixing σ,

Table 1. Classification results according to the number of visual words K and the datasets. For BOG, the column #pat displays the number of frequent patterns extracted by node-induced-GRIMA and σ give the frequency freshold (in percentage of the number of images per class) reached to obtain at least 4000 patterns.

K	15 Scenes				Flowers				Caltech-26			
	BoW	BoG	#pat.	σ	BoW	BoG	#pat.	σ	BoW	BoG	#pat.	σ
100	70.7%	70.0%	4,190	60%	48.0%	63.3%	5,284	80%	71.7%	73.7%	6,473	80%
500	73.7%	72.0%	4,027	40%	59.6%	63.9%	4,377	55%	76.3%	75.4%	4,240	60%
1000	73.8%	73.1%	4,877	30%	63.7%	64.6%	4,545	45%	77.3%	76.9%	6,044	45%
2000	73.2%	73.8%	6,128	20%	67.5%	66.9%	4,640	35%	77.0%	77.2%	4,502	40%
4000	74.2%	75.0%	4,218	20%	66.9%	67.0%	5,822	25%	77.2%	75.7%	16,345	35%

we propose to fix the number p of patterns we want to mine: Starting with $\sigma = 100\%$, we iteratively decrease σ and run node-induced-GRIMA for each value of σ until the number of mined patterns reaches p. In our experiments, we have set p to 4,000 as this value corresponds to a good tradeoff between time and gain in accuracy.

For the classification task, we use $N(N-1)/2$ binary Support Vector Machines (SVM), where N is the number of classes. Each SVM is trained to discriminate one class against another. The final classification is obtained using a vote among the classes. SVMs are trained using Libsvm [3] with the intersection kernel presented in [18]. Inputs of SVMs are either BoW histograms (for the unstructured visual word-based method) or BoW $\cup$ BoG histograms (for the grid-based method). The parameter C of SVM has been optimized by cross validation on all datasets. We consider 10-fold cross-validation for *Flowers* and *Caltech-26* since the number of images is the same for each class. For *15-Scenes*, we have created 10 different folds: For each fold, we randomly select 100·images per class for training and 50 images per class for testing. We provide a Student two-tailed t-test to statistically compare both accuracies (BoW and BoW $\cup$ BoG) for each dataset (the results in green are statistically better, in red statistically worse at $p = 0.05$ level).

We report in Table 1 accuracy results obtained for all datasets. As expected, we need to lower the support threshold σ to obtain at least 4000 patterns when the number of visual words increases (since there are more different labels in the grid, the patterns are less frequent). On *15-Scenes*, the use of structural information becomes important only for $K = 4000$ (but this also gives the best accuracy results). On *Flowers* on the contrary, BOG greatly improves the results for low number of visual words ([100,1000]) but it is not statistically better (nor worse) for higher numbers ([2000,4000]) where the accuracy is the best ($K = 2000$). For *Caltech-26*, BOG slightly improves the results when $K = 100$, and it is not significantly different for higher values of K. Overall, those results show that adding structural information does not harm and sometimes improve the representation of images and that our grid mining algorithm can be used in a real-life context.

Table 2. Accuracy of each binary SVM for 7 classes of the *15-scenes* dataset with a vocabulary of $K = 1000$ visual words and a threshold $\sigma \in [75\%, 20\%]$. For each class, we highlight in bold BOG results that are better than BOW, and in green (resp. red) BOG results that are significantly better (resp. worse) than BOW. For the other 8 classes, BOW and BOG are never significantly different.

	BOW	BOG											
		75%	70%	65%	60%	55%	50%	45%	40%	35%	30%	25%	20%
suburb	99.1	99.1	99.1	99.1	99.1	**99.2**	**99.2**	**99.2**	**99.2**	**99.3**	**99.3**	**99.3**	**99.3**
coast	97.4	97.4	**97.5**	**97.5**	**97.5**	**97.5**	**97.5**	**97.5**	**97.5**	**97.5**	97.2	97.1	97.2
forest	98.8	**98.9**	**98.9**	**98.9**	**98.9**	**98.9**	**98.9**	**98.9**	**98.9**	**98.9**	**98.9**	98.8	98.7
highway	97.5	**97.7**	**97.7**	**97.6**	**97.7**	**97.6**	97.5	97.5	97.4	97.3	97.3	97.2	97.1
moutain	97.7	97.6	97.6	97.6	97.7	97.6	97.6	97.5	97.5	97.5	97.5	97.4	97.3
street	97.8	**97.9**	97.8	97.8	97.7	97.7	97.8	97.7	97.7	97.5	97.3	97.1	96.8
industrial	94.6	**94.7**	**94.9**	**94.8**	**95.0**	**95.1**	**95.1**	**95.1**	**95.2**	**95.3**	**95.6**	**95.4**	**95.4**

Binary Classification Results. Finally, we evaluate the interest of using grids for a binary classification task on the *15-scenes* dataset (similar results were observed for the two other datasets). The goal is to provide an insight into each class separately, to see if all classes benefit from the use of grids in a same way. Table 2 reports binary classification results: For each class C, we train a binary SVM to recognize images of C, against images of all other classes. We consider the same parameter settings as for the N-ary classification task described previously, except that we train N binary SVMs (instead of $N(N-1)/2$). Only the 7 classes with statistically significant differences are shown in the table. We can see that some classes really benefit from the use of structured patterns for almost all frequency tresholds (*e.g.*, *industrial*, *coast*, *suburb*, *forest*) whereas for some classes, using unstructured information gives better results (*e.g.*, *street* or *moutain*). This is due to the fact that for some classes, the structure is too similar from this class to another to use it to discriminate classes.

5   Conclusion

We have presented GRIMA, a grid mining algorithm for Bag-of-Grid-based classification. GRIMA takes into account relative angle information between nodes in the grid graph (as well as possible labels on the nodes and the edges) and is designed to scale to large amount of data. We believe that the grid structure can not only be useful for directly modeling some real life problems but can also be used when an underlying high level structure is not obvious to define as it is the case for image classification problems. Experiments show that our algorithm is necessary (compared to existing ones) to efficiently tackle real life problems. They also show on three image datasets that patterns extracted on this grid structure can improve the classification accuracies. However, to further increase the discriminative power of the grid-patterns for image classification we would need to combine at the same time state-of-the-art deep-learned descriptors and design smart post-processing steps as the ones developed in [9] for unstructured models. Besides, we plan to upgrade the GRIMA algorithm to mine 2D+t grids which is necessary to tackle different real-life applications such as the analysis of ecosystems.

Acknowledgement. This work has been supported by the ANR project SoLStiCe (ANR-13-BS02-0002-01).

References

1. Arimura, H., Uno, T., Shimozono, S.: Time and space efficient discovery of maximal geometric graphs. In: Corruble, V., Takeda, M., Suzuki, E. (eds.) DS 2007. LNCS (LNAI), vol. 4755, pp. 42–55. Springer, Heidelberg (2007). doi:10.1007/978-3-540-75488-6_6
2. Brandao Da Silva, F., Goldenstein, S., Tabbone, S., Da Silva Torres, R.: Image classification based on bag of visual graphs. In: IEEE SPS, pp. 4312–4316 (2013)
3. Chang, C.-C., Lin, C.-J.: LIBSVM: a library for support vector machines. ACM Trans. Intell. Syst. Technol. **2**, 27:1–27:27 (2011). http://www.csie.ntu.edu.tw/~cjlin/libsvm
4. Chatfield, K., Lempitsky, V., Vedaldi, A., Zisserman, A.: The devil is in the details: an evaluation of recent feature encoding methods. In: BMVC, pp. 76.1–76.12 (2011)
5. Chatfield, K., Simonyan, K., Vedaldi, A., Zisserman, A.: Return of the devil in the details: delving deep into convolutional nets. In: BMVC (2014)
6. Cook, D., Holder, L.: Mining Graph Data. Wiley, Hoboken (2006)
7. Csurka, G., Dance, C.R., Fan, L., Willamowski, J., Bray, C.: Visual categorization with bags of keypoints. In: ECCV, pp. 1–22 (2004)
8. Fei-Fei, L., Fergus, R., Perona, P.: Learning generative visual models from few training examples: an incremental Bayesian approach tested on 101 object categories. In: IEEE CVPR Workshop of Generative Model Based Vision (2004)
9. Fernando, B., Fromont, É., Tuytelaars, T.: Mining mid-level features for image classification. IJCV **108**(3), 186–203 (2014)
10. Hogeweg, P.: Cellular automata as a paradigm for ecological modelling. Appl. Math. Comput. **27**, 81–100 (1988)
11. Jiang, C., Coenen, F., Zito, M.: A survey of frequent subgraph mining algorithms. KER **28**, 75–105 (2013)
12. Lazebnik, S., Schmid, C., Ponce, J.: Beyond bags of features: spatial pyramid matching for recognizing natural scene categories. In: CVPR, pp. 2169–2178 (2006)
13. Lowe, D.G.: Distinctive image features from scale-invariant keypoints. IJCV **60**(2), 91–110 (2004)
14. Marinescu-Ghemeci, R.: Maximum induced matchings in grids. In: Migdalas, A., Sifaleras, A., Georgiadis, C.K., Papathanasiou, J., Stiakakis, E. (eds.) Optimization Theory, Decision Making, and Operations Research Applications. Springer Proceedings in Mathematics & Statistics, vol. 31, pp. 177–187. Springer, Heidelberg (2013)
15. Nilsback, M.-E., Zisserman, A.: Automated flower classification over a large number of classes. In: ICVGIP, pp. 722–729 (2008)
16. Nowak, E., Jurie, F., Triggs, B.: Sampling strategies for bag-of-features image classification. In: Leonardis, A., Bischof, H., Pinz, A. (eds.) ECCV 2006. LNCS, vol. 3954, pp. 490–503. Springer, Heidelberg (2006). doi:10.1007/11744085_38

17. Nowozin, S., Tsuda, K., Uno, T., Kudo, T., Bakir, G.: Weighted substructure mining for image analysis. In: IEEE CVPR, pp. 1–8 (2007)
18. Odone, F., Barla, A., Verri, A.: Building kernels from binary strings for image matching. IEEE Trans. Image Process. **14**(2), 169–180 (2005)
19. Ozdemir, B., Aksoy, S.: Image classification using subgraph histogram representation. In: ICPR, pp. 1112–1115 (2010)
20. Prado, A., Jeudy, B., Fromont, E., Diot, F.: Mining spatiotemporal patterns in dynamic plane graphs. IDA **17**, 71–92 (2013)
21. Samuel, É., de la Higuera, C., Janodet, J.-C.: Extracting plane graphs from images. In: Hancock, E.R., Wilson, R.C., Windeatt, T., Ulusoy, I., Escolano, F. (eds.) Structural, Syntactic, and Statistical Pattern Recognition. LNCS, vol. 6218, pp. 233–243. Springer, Heidelberg (2010)
22. Brandão Silva, F., Tabbone, S., Torres, R.: BoG: a new approach for graph matching. In: ICPR, pp. 82–87 (2014)
23. Voravuthikunchai, W., Crémilleux, B., Jurie, F.: Histograms of pattern sets for image classification and object recognition. In: CVPR, pp. 1–8 (2014)
24. Yan, X., Han, J.: gSpan: graph-based substructure pattern mining. In: ICDM, pp. 721–724 (2002)

Edge Centrality via the Holevo Quantity

Joshua Lockhart[1], Giorgia Minello[2], Luca Rossi[3(✉)], Simone Severini[1], and Andrea Torsello[2]

[1] Department of Computer Science, University College London, London, UK
[2] DAIS, Università Ca' Foscari Venezia, Venice, Italy
[3] School of Engineering and Applied Science, Aston University, Birmingham, UK
l.rossi@aston.ac.uk

Abstract. In the study of complex networks, vertex centrality measures are used to identify the most important vertices within a graph. A related problem is that of measuring the centrality of an edge. In this paper, we propose a novel edge centrality index rooted in quantum information. More specifically, we measure the importance of an edge in terms of the contribution that it gives to the Von Neumann entropy of the graph. We show that this can be computed in terms of the Holevo quantity, a well known quantum information theoretical measure. While computing the Von Neumann entropy and hence the Holevo quantity requires computing the spectrum of the graph Laplacian, we show how to obtain a simplified measure through a quadratic approximation of the Shannon entropy. This in turns shows that the proposed centrality measure is strongly correlated with the negative degree centrality on the line graph. We evaluate our centrality measure through an extensive set of experiments on real-world as well as synthetic networks, and we compare it against commonly used alternative measures.

Keywords: Edge centrality · Complex networks · Holevo quantity · Quantum information

1 Introduction

The study of complex networks has recently attracted increasing interest in the scientific community, as it allows to model and understand a large number of real-world systems [4]. This is particularly relevant given the growing amount of available data describing the interactions and dynamics of real-world systems. Typical examples of complex networks include metabolic networks [8], protein interactions [7], brain networks [17] and scientific collaboration networks [11].

One of the key problems in network science is that of identifying the most relevant nodes in a network. This importance measure is usually called the *centrality* of a vertex [9]. A number of centrality indices have been introduced in the literature [2,4–6,10,14], each of them capturing different but equally significant aspects of vertex importance. Commonly encountered examples are the degree, closeness and betweenness centrality [5,6,10]. A closely related problem

© Springer International Publishing AG 2016
A. Robles-Kelly et al. (Eds.): S+SSPR 2016, LNCS 10029, pp. 143–152, 2016.
DOI: 10.1007/978-3-319-49055-7_13

is that of measuring the centrality of an edge [3,9]. Most edge centrality indices are developed as a variant of vertex centrality ones. A common way to define an edge centrality index is to apply the corresponding vertex centrality to the line graph of the network being studied. Recall that, given a graph $G = (V, E)$, the line graph $\mathcal{L}(G) = (V', E')$ is a dual representation of G where each node $uv \in V'$ corresponds to an edge $(u, v) \in E$, and there exists and edge between two nodes of $\mathcal{L}(G)$ if and only if the corresponding edges of G share a vertex. By measuring the vertex centrality on $\mathcal{L}(G)$, one can map it back to the edges of G to obtain a measure of edge centrality. However, as observed by Koschützki et al. [9], this approach does not yield the same result as the direct definition of the edge centrality on G. Moreover, the size of the line graph is quadratic in the size of the original graph, thus making it hard to scale to large networks when the chosen centrality measure is computationally demanding.

In this paper, we introduce a novel edge centrality measure rooted in quantum information theory. More specifically, we propose to measure the importance of an edge in terms of its contribution to the Von Neumann entropy of the network [13]. This can be measured in terms of the Holevo quantity, a well known quantum information theoretical measure that has recently been applied to the analysis of graph structure [15,16]. We also show how to approximate this quantity in the case of large networks, where computing the exact value of the Von Neumann entropy is not feasible. This in turns highlights a strong connection between the Holevo edge centrality and the negative degree centrality on the line graph. Finally, we perform a series of experiments to evaluate the proposed edge centrality measure on real-world as well as synthetic graphs, and we compare it against a number of widely used alternative measures.

The remainder of the paper is organised as follows: Sect. 2 reviews the necessary quantum information theoretical background and Sect. 3 introduces the proposed edge centrality measure. The experimental evaluation is presented in Sect. 4 and Sect. 5 concludes the paper.

2   Quantum Information Theoretical Background

2.1   Quantum States and Von Neumann Entropy

In quantum mechanics, a system can be either in a pure state or a mixed state. Using the Dirac notation, a *pure state* is represented as a column vector $|\psi_i\rangle$. A *mixed state*, on the other hand, is an ensemble of pure quantum states $|\psi_i\rangle$, each with probability p_i. The density operator of such a system is a positive unit-trace matrix defined as

$$\rho = \sum_i p_i |\psi_i\rangle \langle\psi_i| . \tag{1}$$

The *Von Neumann entropy* [12] S of a mixed state is defined in terms of the trace and logarithm of the density operator ρ

$$S(\rho) = -\operatorname{tr}(\rho \ln \rho) = -\sum_i \lambda_i \ln(\lambda_i) \tag{2}$$

where $\lambda_1, \ldots, \lambda_{\hat{n}}$ are the eigenvalues of ρ. If $\langle \psi_i | \rho | \psi_i \rangle = 1$, i.e., the quantum system is a pure state $|\psi_i\rangle$ with probability $p_i = 1$, then the Von Neumann entropy $S(\rho) = -\operatorname{tr}(\rho \ln \rho)$ is zero. On other hand, a mixed state always has a non-zero Von Neumann entropy associated with it.

2.2   A Mixed State from the Graph Laplacian

Let $G = (V, E)$ be a simple graph with n vertices and m edges. We assign the vertices of G to the elements of the standard basis of an Hilbert space $\mathcal{H}_G$, $\{|1\rangle, |2\rangle, \ldots, |n\rangle\}$. Here $|i\rangle$ denotes a column vector where 1 is at the i-th position. The *graph Laplacian* of G is the matrix $L = D - A$, where A is the adjacency matrix of G and D is the diagonal matrix with elements $d(u) = \sum_{v=1}^n A(u, v)$. For each edge $e_{i,j}$, we define a pure state

$$|e_{i,j}\rangle := \frac{1}{\sqrt{2}}(|i\rangle - |j\rangle). \tag{3}$$

Then we can define the mixed state $\{\frac{1}{m}, |e_{i,j}\rangle\}$ with density matrix

$$\rho(G) := \frac{1}{m} \sum_{\{i,j\} \in E} |e_{i,j}\rangle \langle e_{i,j}| = \frac{1}{2m} L(G). \tag{4}$$

Let us define the Hilbert spaces $\mathcal{H}_V \cong \mathbb{C}^V$, with orthonormal basis $\mathbf{a}_v$, where $v \in V$, and $\mathcal{H}_E \cong \mathbb{C}^E$, with orthonormal basis $\mathbf{b}_{u,v}$, where $\{u, v\} \in E$. It can be shown that the graph Laplacian corresponds to the partial trace of a rank-1 operator on $\mathcal{H}_V \otimes \mathcal{H}_E$ which is determined by the graph structure [1]. As a consequence, the Von Neumann entropy of $\rho(G)$ can be interpreted as a measure of the amount of entanglement between a system corresponding to the vertices and a system corresponding to the edges of the graph [1].

2.3   Holevo Quantity of a Graph Decomposition

Given a graph G, we can define an ensemble in terms of its subgraphs. Recall that a *decomposition* of a graph G is a set of subgraphs $H_1, H_2, \ldots, H_k$ that partition the edges of G, i.e., for all i, j, $\bigcup_{i=1}^k H_i = G$ and $E(H_i) \cap E(H_j) = \emptyset$, where $E(G)$ denotes the edge set of G. Notice that isolated vertices do not contribute to a decomposition, so each H_i can always be seen a subgraph that contains all the vertices. If we let $\rho(H_1), \rho(H_2), \ldots, \rho(H_k)$ be the mixed states of the subgraphs, the probability of H_i in the mixture $\rho(G)$ is given by $|E(H_i)|/|E(G)|$. Thus, we can generalise Eq. 4 and write

$$\rho(G) = \sum_{i=1}^k \frac{|E(H_i)|}{|E(G)|} \rho(H_i). \tag{5}$$

Consider a graph G and its decomposition $H_1, H_2, \ldots, H_k$ with corresponding states $\rho(H_1), \rho(H_2), \ldots, \rho(H_k)$. Let us assign $\rho(H_1), \rho(H_2), \ldots, \rho(H_k)$ to the elements of an alphabet $\{a_1, a_2, \ldots, a_k\}$. In quantum information theory, the classical concepts of uncertainty and entropy are extended to deal with quantum states, where uncertainty about the state of a quantum system can be expressed using the density matrix formalism. Assume a source emits letters from the alphabet and that the letter a_i is emitted with probability $p_i = |E(H_i)|/|E(G)|$. An upper bound to the accessible information is given by the *Holevo quantity* of the ensemble $\{p_i, \rho(H_i)\}$:

$$\chi(\{p_i, \rho(H_i)\}) = S\left(\sum_{i=1}^{k} p_i \rho(H_i)\right) - \sum_{i=1}^{k} p_i S(\rho(H_i)) \tag{6}$$

3 Holevo Edge Centrality

We propose to measure the centrality of an edge as follows. Let $G = (V, E)$ be a graph with $|E| = m$, and let H_e and $H_{\bar{e}}$ denote the subgraphs over edge sets $\{e\}$ and $E \setminus \{e\}$, respectively. Note that $S(\rho(H_e)) = 0$ and

$$\frac{m-1}{m}\rho(H_{\bar{e}}) + \frac{1}{m}\rho(H_e) = \rho(G). \tag{7}$$

Then the Holevo quantity of the ensemble $\{(m - 1/m, H_{\bar{e}}), (1/m, H_e)\}$ is

$$\chi\left(\left\{\left(\frac{m-1}{m}, H_{\bar{e}}\right), \left(\frac{1}{m}, H_e\right)\right\}\right) = S\left(\rho(G)\right) - \frac{m-1}{m}S\left(\rho(H_{\bar{e}})\right) \tag{8}$$

Definition 1. *For a graph $G = (V, E)$, the* Holevo edge centrality *of $e \in E$ is*

$$HC(e) = \chi\left(\left\{\left(\frac{m-1}{m}, H_{\bar{e}}\right), \left(\frac{1}{m}, H_e\right)\right\}\right) \tag{9}$$

When ranking the edges of a graph G, the scaling factor $(m-1)/m$ is constant for all the edges and thus can be safely ignored. The Holevo edge centrality of an edge e is then a measure of the difference in Von Neumann entropy between the original graph and the graph where e has been removed. In other words, it can be seen as a measure of the contribution of e to the Von Neumann entropy of G. From a physical perspective, this can also be interpreted as the variation of the entanglement between between a system corresponding to the vertices and a system corresponding to the edges of the graph (see the interpretation of the graph Laplacian in Sect. 2).

3.1 Relation with Degree Centrality

In this subsection we investigate the nature of the structural characteristics encapsulated by the Holevo edge centrality. Let $G = (V, E)$ be a graph with n nodes, and let I_n be the identity matrix of size n. We rewrite the Shannon entropy $-\sum_i \lambda_i \ln(\lambda_i)$ using the second order polynomial approximation $k\sum_i \lambda_i(1-\lambda_i)$, where the value of k depends on the dimension of the simplex. We obtain

$$S(\rho(G)) = -\operatorname{tr}\left(\rho(G)\ln\rho(G)\right) \approx \frac{|V|\ln(|V|)}{|V|-1}\operatorname{tr}\left(\rho(G)(I_n - \rho(G))\right) \tag{10}$$

By noting that $\rho(G) = L(G)/(2m)$ and using some simple algebra, we can rewrite Eq. 10 as

$$S(\rho(G)) \approx \frac{|V|\ln(|V|)}{|V|-1}\left(1 - \frac{1}{4m^2}\sum_{v \in V}\left(d^2(v) + d(v)\right)\right) \qquad (11)$$

where $d(v)$ denotes the degree of the vertex v. This in turn allows us to approximate Eq. 9 as

$$HC(e) = S(\rho(G)) - S(\rho(H_{\bar{e}})) \approx -\frac{|V|\ln(|V|)}{|V|-1}\frac{d(u) + d(w)}{2m^2} \qquad (12)$$

where $e = (u, w)$, we omitted the scaling factor $(m-1)/m$ and we made use of the fact that $1/(4m^2) \approx 1/(4(m-1)^2)$.

Equation 12 shows that the quadratic approximation of the Holevo centrality is (almost) linearly correlated with the negative edge degree centrality (see Sect. 4). This in turn gives us an important insight into the nature of the Holevo edge centrality. However, the quadratic approximation captures only part of the structural information encapsulated by the exact centrality measure. In particular, Passerini and Severini [13] suggested that those edges that create longer paths, nontrivial symmetries and connected components result in a larger increase of the Von Neumann entropy. Therefore, such edges should have a high centrality value, higher than what the degree information alone would suggest.

Figure 1 shows an example of such a graph, where the central bridge has a high value of the exact Holevo edge centrality, but a relatively low value of the approximated edge centrality. In Fig. 1(b), the blue edges have all the same degree centrality, i.e., they are all adjacent to four other edges. However, from a structural point of view, the removal of the edges connecting the two cliques at the ends of the barbell graph would have a higher impact, as it would disconnect the graph. As shown in Fig. 1(a), the Holevo centrality captures this structural difference, i.e., the weight assigned to the two bridges (blue) is higher than that assigned to the edges in the cliques (red).

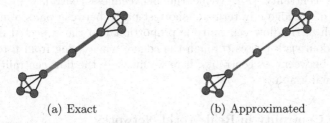

(a) Exact   (b) Approximated

Fig. 1. The Holevo edge centrality and its quadratic approximation on a barbell graph. Here the edge thickness is proportional to the value of the centrality. In (a) the blue edges have a higher centrality than the red edges, but in (b) all these edges (blue) have the same degree centrality. (Color figure online)

4 Experimental Evaluation

In the previous sections we have derived an expression for the Holevo edge centrality, both exact and approximated. Here, we first evaluate this measure on a number of standard networks, and we compare it against other well known edge centralities. We also analyse the behaviour of the proposed centrality measure when graphs endure structural changes.

4.1 Experimental Setup

We perform our experiments on two well known real-world networks, the Florentine families graph and the Karate club network. We then consider the following edge centrality measures:

Degree Centrality: The centrality of an edge e is computed as the degree of the corresponding vertex in the line graph. The idea underpinning the vertex degree centrality is that the importance of a node is proportional to the number of connections it has to other nodes. This is the simplest edge centrality measure, but also the one with the lowest computational complexity.

Betweenness Centrality: The centrality of an edge e is the sum of the fraction of all-pairs shortest paths that pass through e, i.e., $EBC(e) = \sum_{u,v \in V} \frac{\sigma(u,v|e)}{\sigma(u,v)}$ where V is the set of nodes, $\sigma(u,v)$ and $\sigma(u,v|e)$ denote the number of shortest paths between u and v and the number of shortest paths between u and v that pass through e, respectively [3]. An edge with a high betweenness centrality has a large influence on the transfer of information through the network and thus it can be seen as an important bridge-like connector between two parts of a network. Note that the implementation we use does not rely on the line graph, but measure the centrality of an edge directly on the original graph.

Flow Centrality: This centrality measure is also known as random-walk betweenness centrality [10]. While the betweenness centrality measures the importance of an edge e in terms of shortest-paths between pairs of nodes that pass through e, the flow centrality is proportional to the expected number of times a random walk passes through the edge e when going from u to v. Similarly to the betweenness centrality, here we measure the flow centrality directly on the original graph.

4.2 Edge Centrality in Real-World Networks

In order to compare the Holevo edge centrality with the measures described in the previous subsection, we compute, for each network, the correlation between the Holevo quantity and the alternative measures. Figure 2 shows the value of these centralities on the Florentine families graph and the Karate club network.

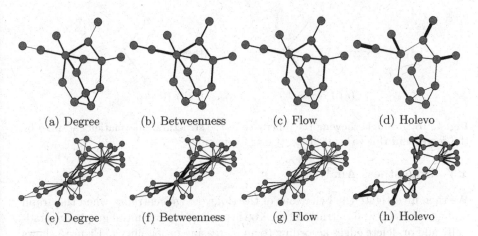

(a) Degree   (b) Betweenness   (c) Flow   (d) Holevo

(e) Degree   (f) Betweenness   (g) Flow   (h) Holevo

Fig. 2. Edge centralities on the Florentine families network (a–d) and the Karate club network (e–h). A thicker edge indicates a higher value of the centrality.

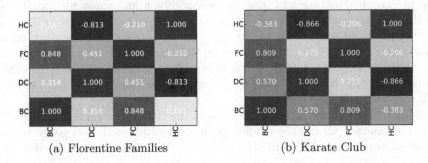

(a) Florentine Families   (b) Karate Club

Fig. 3. Correlation matrices for the centrality measure on the Florentine family network and the karate club network. DC, BC, FC, and HC denote the degree, betweenness, flow and Holevo centralities, respectively.

In these plots, the thickness of an edge is proportional to the magnitude of the centrality index. Figure 3, on the other hand, shows the correlation matrix between the different centralities. Here DC, BC, FC and HC denote the degree, betweenness, flow and Holevo centrality, respectively.

The Holevo centrality is always strongly negatively correlated with the degree centrality. This is in accordance with the properties discussed in Sect. 3. However, there are some significant differences. In general, the Holevo centrality is higher on edges that connect low degree nodes. In this sense, it can be seen as a measure of *peripherality*, rather than centrality. However, when two edges have the same degree centrality, edges that would disconnect the network or break structural symmetries are assigned a higher weight, as Fig. 1 shows. Similarly, in Fig. 4(a) the three edges highlighted in blue have the same degree centrality, but the same edges in Fig. 4(b) have different Holevo centralities. In fact, the removal of the red edge does not result in significant structural changes, while the removal of one of the blue edges increases the length of the tail.

(a) Degree   (b) Holevo

Fig. 4. Toy example showing the difference in the structural information captured by the degree and Holevo centralities. (Color figure online)

4.3 Robustness Analysis

We then investigate the behaviour of the Holevo edge centrality when the graph undergoes structural perturbations. To this end, given an initial graph, we gradually add or delete edges according to an increasing probability p. Figure 5 shows an instance of the noise addition process. Starting from a randomly generated graph, we compute the Holevo edge centrality for all its edges. Then, we perturb the graph structure with a given probability p and again we recompute the Holevo edge centrality for all the graph edges. We compute the correlation between the Holevo centrality of the edges of the original graph and its noisy counterpart. More specifically, we measure the correlation between the centralities of the edges that belong to the intersection of their edge sets. In other words, we analyse how the centrality changes during the perturbation process, with respect to the starting state.

Since we are interested in the variation of the Holevo centrality as the graph structure changes, we use three different random graph models to generate the initial graph: (1) the Erdös-Rényi model, (2) the Watts-Strogatz model and

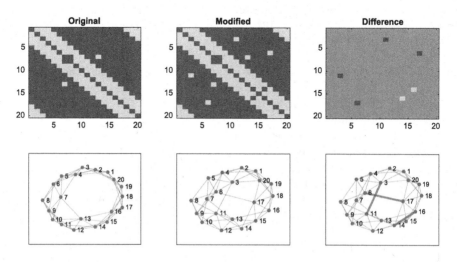

Fig. 5. Perturbation process: on the left, adjacency matrix and plot of the starting graph; in the middle, the edited graph; on the right, the differences between initial and modified graph are highlighted. (Color figure online)

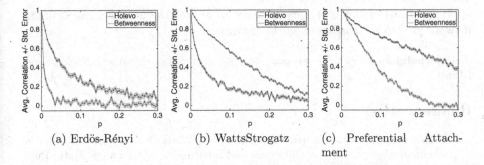

(a) Erdös-Rényi   (b) WattsStrogatz   (c) Preferential Attachment

Fig. 6. Average correlation between the centrality of the edges of the original graph
and those of increasingly noisy version of it. The different columns refer to different
starting graphs: (a) Erdös-Rényi, (b) WattsStrogatz and (c) Preferential Attachment.

(3) the Preferential Attachment model. For each model, we generate a starting
graph with the same number of nodes n and we create 100 noisy instances as p
varies from 0.01 to 0.3. We perform the same experiment for both the Holevo
centrality and the betweenness centrality.

Figure 6 shows the average correlation as we perturb the graph structure,
for both the Holevo and betweenness centrality. As expected, in both cases the
correlation decreases as the similarity between the original graph and the edited
one decreases. However, while the correlation for centrality measures decreases
rapidly in the case of ErdösRényi graphs, on scale-free graphs our centrality
measure decreases linearly with the value of p, while the betweenness centrality
drops significantly more quickly. On the other hand, we observe the opposite
behaviour on small-world graphs. This can be explained by noting that in small-
world graphs there exist multiple alternative paths between every pair of nodes,
and thus the betweenness centrality is less affected by structural modifications.
On the other hand, in scale-free graphs most shortest-paths pass through a
hub, and thus adding a random edge can create shortcuts that greatly affect the
value of the betweenness centrality. The Holevo centrality, however, assigns large
weights to long tails and leaves, which are less affected by the structural noise.

5   Conclusion

In this paper we have introduced a novel edge centrality measure based on
the quantum information theoretical concept of Holevo quantity. We measured
the importance of an edge in terms of the difference in Von Neumann entropy
between the original graph and the graph where that edge has been remove. We
showed that by taking a quadratic approximation of the Von Neumann entropy
we obtain an approximated value of the Holevo centrality that is proportional
to the negative degree centrality. We performed a series of experiments on both
real-world and synthetic networks and we compared the proposed centrality
measure to widely used alternatives. Future work will investigate higher order

approximations of this centrality measure as well as the possibility of defining network growth models based on the Holevo quantity.

Acknowledgments. Simone Severini was supported by the Royal Society and EPSRC.

References

1. de Beaudrap, N., Giovannetti, V., Severini, S., Wilson, R.: Interpreting the Von Neumann entropy of graph laplacians, and coentropic graphs. Panor. Math. Pure Appl. **658**, 227 (2016)
2. Bonacich, P.: Power and centrality: a family of measures. Am. J. Sociol. **92**, 1170–1182 (1987)
3. Brandes, U.: On variants of shortest-path betweenness centrality and their generic computation. Soc. Netw. **30**(2), 136–145 (2008)
4. Estrada, E.: The Structure of Complex Networks. Oxford University Press, Oxford (2011)
5. Freeman, L.C.: A set of measures of centrality based on betweenness. Sociometry **40**, 35–41 (1977)
6. Freeman, L.C.: Centrality in social networks conceptual clarification. Soc. Netw. **1**(3), 215–239 (1979)
7. Ito, T., Chiba, T., Ozawa, R., Yoshida, M., Hattori, M., Sakaki, Y.: A comprehensive two-hybrid analysis to explore the yeast protein interactome. Proc. Natl. Acad. Sci. **98**(8), 4569 (2001)
8. Jeong, H., Tombor, B., Albert, R., Oltvai, Z., Barabási, A.: The large-scale organization of metabolic networks. Nature **407**(6804), 651–654 (2000)
9. Koschützki, D., Lehmann, K.A., Peeters, L., Richter, S., Tenfelde-Podehl, D., Zlotowski, O.: Centrality indices. In: Brandes, U., Erlebach, T. (eds.) Network Analysis. LNCS, vol. 3418, pp. 16–61. Springer, Heidelberg (2005). doi:10.1007/978-3-540-31955-9_3
10. Newman, M.E.: A measure of betweenness centrality based on random walks. Soc. Netw. **27**(1), 39–54 (2005)
11. Newman, M.: Scientific collaboration networks. I. Network construction and fundamental results. Phys. Rev. E **64**(1), 016131 (2001)
12. Nielsen, M.A., Chuang, I.L.: Quantum Computation and Quantum Information. Cambridge University Press, Cambridge (2010)
13. Passerini, F., Severini, S.: Quantifying complexity in networks: the Von Neumann entropy. Int. J. Agent Technol. Syst. (IJATS) **1**(4), 58–67 (2009)
14. Rossi, L., Torsello, A., Hancock, E.R.: Node centrality for continuous-time quantum walks. In: Fränti, P., Brown, G., Loog, M., Escolano, F., Pelillo, M. (eds.) S+SSPR 2014. LNCS, vol. 8621, pp. 103–112. Springer, Heidelberg (2014). doi:10.1007/978-3-662-44415-3_11
15. Rossi, L., Torsello, A., Hancock, E.R.: Measuring graph similarity through continuous-time quantum walks and the quantum Jensen-Shannon divergence. Phys. Rev. E **91**(2), 022815 (2015)
16. Rossi, L., Torsello, A., Hancock, E.R., Wilson, R.C.: Characterizing graph symmetries through quantum Jensen-Shannon divergence. Phys. Rev. E **88**(3), 032806 (2013)
17. Sporns, O.: Network analysis, complexity, and brain function. Complexity **8**(1), 56–60 (2002)

Thermodynamic Network Analysis with Quantum Spin Statistics

Jianjia Wang$^{(\boxtimes)}$, Richard C. Wilson, and Edwin R. Hancock

Department of Computer Science, University of York, York YO10 5DD, UK
jw1157@york.ac.uk

Abstract. In this paper, we explore the thermodynamic analysis of networks using a heat-bath analogy and different choices of quantum spin statistics for the occupation of energy levels defined by the network. We commence from the set of energy states given by the eigenvalues of the normalized Laplacian matrix, which plays the role of the Hamiltonian operator of the network. We explore a heat bath analogy in which the network is in thermodynamic equilibrium with a heat-bath and its energy levels are occupied by either indistinguishable bosons or fermions obeying the Pauli exclusion principle. To compute thermodynamic characterization of this system, i.e. the entropy and energy, we analyse the partition functions relevant to Bose-Einstein and Fermi-Dirac statistics. At high temperatures, the effects of quantum spin statistics are disrupted by thermalisation and correspond to the classical Maxwell-Boltzmann case. However, at low temperatures the Bose-Einstein system condenses into a state where the particles occupy the lowest energy state, while in the Fermi-Dirac system there is only one particle per energy state. These two models produce quite different entropic characterizations of network structure, which are appropriate to different types of structure. We experiment with the two different models on both synthetic and real world imagery, and compare and contrast their performance.

Keywords: Quantum spin statistics · Network entropy

1 Introduction

Physics-based analogies have found widespread use in the analysis and understanding of network structure. Examples include the use of ideas from statistical mechanics [1,2], thermodynamics [3,4] and quantum information [5,6]. For instance, statistical mechanics has been used to characterize the degree distribution of different types of complex networks [1]. Using a thermodynamic heat-bath analogy, the concepts of network communicability and balance have been defined [3]. By using quantum walks the process of preferential attachment has been shown to lead to the intriguing phenomenon of super-cluster condensation in growing networks [7]. Both Bose-Einstein and Fermi-Dirac statistics have been used to describe quantum geometries in networks [5].

© Springer International Publishing AG 2016
A. Robles-Kelly et al. (Eds.): S+SSPR 2016, LNCS 10029, pp. 153–162, 2016.
DOI: 10.1007/978-3-319-49055-7_14

One particularly interesting and widely studied approach is to use thermodynamic analogies as a means of characterizing networks [4]. Here the Laplacian matrix plays the role of network Hamiltonian, and the energy states of a network are the eigenvalues of the Laplacian. By modeling the network as a set of energy states occupied by a system of particles, thermodynamic properties such as total energy and entropy can be computed, and used as network characterizations. The energy states are populated by particles in thermal equilibrium with a heat bath. A key element in this thermalisation approach is to model how the energy states are occupied at a particular temperature. Normally this is assumed to be follow the classical Maxwell-Boltzmann distribution, where the particles are distinguishable and weakly interacting.

In this paper, on the other hand, we explore the case where the particles are quantum mechanical in nature and obey spin statistics. In other words, they are indistinguishable, and are either fermions (half integer spin) or bosons (integer spin). Particles with integer spin are subject to Bose-Einstein statistics and do not obey the Pauli exclusion principle. They can aggregate in the same energy state. At low temperature this leads to the phenomenon of Bose-Einstein condensation. There has been work aimed at extending this model to networks. For instance, by mapping the network model to a Bose gas, phase transitions have been studied in network evolution associated with Bose-Einstein condensation [7]. This model has also been extended to understand processes such as supersymmetry in networks [3]. In the meanwhile, particles with half integer spin are subject to Fermi-Dirac statistics and obey the Pauli exclusion principle. They give rise to models of network structures constrained by the occupancy of the nodes and edges. Examples include traffic flow and also the modelling of certain types of geometric networks such as the Cayley tree [8]. Formally, these physical systems can be described by partition functions with the microscopic energy states, which are represented by a suitable chosen Hamiltonian. In the network theory, the Hamiltonian is computed from the adjacency or Laplacian matrix, but recently, Ye et al. [4], have shown how the partition function can be derived from a characteristic polynomial instead.

Despite this interest in alternative models of the thermalised distribution of energy states under different spin statistics, there has been little systematic study of the resulting thermodynamic characterizations of network structure. Here we consider the effects of occupation statistics on the populations of the energy states where the Hamiltonian operator is the normalized Laplacian matrix and the energy states are given by the network spectrum. We characterize the thermalised system of energy states using partition functions relevant to Bose-Einstein and Fermi-Dirac occupation statistics. From the partition functions we compute average energy and entropy of the system of particles. Because Bose-Einstein particles coalescence in low energy states, and Fermi-Dirac particles have a greater tendency to occupy high energy states because of the Puli exclusion principle, these types of spin statistics lead to very different distributions of energy and entropy for a network with a given structure (i.e. set of normalised Laplacian eigenvalues). Moreover, at low temperature the distributions are also different from the classical Maxwell-Boltzmann case. It is these low-temperature differences in energy and entropy that we wish to investigate as a means of characterizing differences in network structure.

This paper is organised as follows. Section 2 briefly reviews the basic concepts in network representation, especially with density matrix and Hamiltonian operator on graphs. Section 3 reviews the thermodynamic quantities i.e. entropy and energy, and also illustrates Bose-Einstein and Fermi-Dirac statistics. Section 4 provides our experimental evaluation. Finally, Sect. 5 provides the conclusion and direction for future work.

2   Graph Representation

In this section, we provide details of graph representation in quantum theory. We briefly introduce the concept of density matrix for a graph and then give the definition of Hamiltonian operator with the normalized Laplacian matrix.

2.1   Density Matrix

In quantum mechanics the density matrix is used to describe a system whose state is an ensemble of pure quantum states $|\psi_i\rangle$, each with probability p_i. The density matrix is defined as

$$\rho = \sum_{i=1}^{|V|} p_i |\psi_i\rangle\langle\psi_i| \tag{1}$$

In the graph domain, the normalised Laplacian matrix has been used to model the density of states for a network [6,9]. Let $G(V,E)$ be an undirected graph with node set V and edge set $E \subseteq V \times V$, and let $|V|$ represent the total number of nodes on graph $G(V,E)$. The adjacency matrix A of a graph is defined as

$$A = \begin{cases} 0 & \text{if } (u,v) \in E \\ 1 & \text{otherwise.} \end{cases} \tag{2}$$

Then the degree of node u is $d_u = \sum_{v \in V} A_{uv}$.

The normalized Laplacian matrix $\tilde{L}$ of the graph G is defined as $\tilde{L} = D^{-\frac{1}{2}} L D^{\frac{1}{2}}$, where $L = D - A$ is the Laplacian matrix and D denotes the degree diagonal matrix whose elements are given by $D(u,u) = d_u$ and zeros elsewhere. The element-wise expression of $\tilde{L}$ is

$$\tilde{L}_{uv} = \begin{cases} 1 & \text{if } u = v \text{ and } d_u \neq 0 \\ -\frac{1}{\sqrt{d_u d_v}} & \text{if } u \neq v \text{ and } (u,v) \in E \\ 0 & \text{otherwise.} \end{cases} \tag{3}$$

With this notation, Severini et al. [6,9] specify the density matrix to be $\rho = \frac{\tilde{L}}{|V|}$. When defined in this was way the density matrix is Hermitian i.e. $\rho = \rho^\dagger$ and $\rho \geq 0$, $\mathrm{Tr}\rho = 1$. It plays an important role in the quantum observation process, which can be used to calculate the expectation value of measurable quantity.

2.2   Hamiltonian Operator of a Graph

In quantum mechanics, the Hamiltonian operator is the sum of the kinetic energy and potential energy of all the particles in the system. It is the energy operator of the system and the standard formulation on a manifold is $\hat{H} = -\nabla^2 + U(r,t)$.

In our case, we assume the graph to be in contact with a heat reservoir. The eigenvalues of the Laplacian matrix can be viewed as the energy eigenstates, and these determine the Hamiltonian and hence the relevant Schrödinger equation which govern the particles in the system. The particles occupy the energy states of the Hamiltonian subject to thermal agitation by the heat bath. The number of particles in each energy state is determined by the temperature, the assumed model of occupation statistics and the relevant chemical potential.

If we take the kinetic energy operator $-\nabla^2$ to be the negative of the adjacency matrix, i.e. $-A$, and the potential energy $U(r,t)$ to be the degree matrix D, then the Hamiltonian operator is the Laplacian matrix on graph. Similarly, the normalized form of the graph Laplacian can be viewed as the Hamiltonian operator $\hat{H} = \tilde{L}$.

In this case, the energy states of the network $\{\varepsilon_i\}$ are then the eigenvalues of the Hamiltonian $\hat{H}|\psi_i\rangle = \tilde{L}|\psi_i\rangle = E_i|\psi_i\rangle$.

The eigenvalues are all greater than or equal to zero, and the multiplicity of the zero eigenvalue is the number of connected components in the network. Furthermore, the density matrix commutes with the Hamiltonian, i.e. the associated Poisson bracket is zero,

$$[\hat{H}, \rho] = [\tilde{L}, \frac{\tilde{L}}{|V|}] = 0 \tag{4}$$

which means that the network is in equilibrium when there are no changes in the density matrix which describes the system.

3   Quantum Statistics in Networks

Quantum statistics can be combined with network theory to characterize network properties. The network can be viewed as a grand canonical ensemble, and the thermal quantities, such as energy and entropy, depend on the assumptions concerning the Hamiltonian for the system and the corresponding partition function.

3.1   Thermodynamic Quantities

We consider the network as a thermodynamic system specified by N particles with energy states given by the Hamiltonian operator, and it is immersed in a heat bath with temperature T. The ensemble is represented by a partition function $Z(\beta, N)$, where β is inverse of temperature. When specified in this way the various thermodynamic characterizations can be computed for the networks. For instance, the average energy is given by

$$U = \left[-\frac{\partial}{\partial \beta} \log Z(\beta, N) \right]_N = \mathrm{Tr}\,(\rho H) = k_B T^2 \left[\frac{\partial}{\partial T} \log Z \right]_N \tag{5}$$

the thermodynamic entropy by

$$S = k_B \left[\frac{\partial}{\partial T} T \log Z \right]_N \tag{6}$$

and the chemical potential by

$$\mu = -k_B T \left[\frac{\partial}{\partial N} \log Z \right]_\beta \tag{7}$$

For each distribution we capture the statistical mechanical properties of particles in the system using the partition function associated with the different occupation statistics. The network can then be characterized using thermodynamic quantities computed from the partition function, and these include the entropy, energy and temperature.

3.2   Bose-Einstein Statistics

The Bose-Einstein distribution applies to indistinguishable bosons. Each energy state specified by the network Hamiltonian can accommodate an unlimited number of particles. Without obeying the Pauli exclusion principle, Bosons subject to Bose-Einstein statistics can aggregate in the same energy state.

For a system of the network, as the grand-canonical ensemble with a varying number of particles N and a chemical potential μ, the Bose-Einstein partition function is

$$Z_{BE} = \det \left(I - e^{\beta \mu} \exp[-\beta \tilde{L}] \right)^{-1} = \prod_{i=1}^{|V|} \left(\frac{1}{1 - e^{\beta(\mu - \varepsilon_i)}} \right) \tag{8}$$

From Eqs. (5) and (6), the average energy is

$$\langle U \rangle_{BE} = -\frac{\partial \log Z}{\partial \beta} = -\sum_{i=1}^{|V|} \frac{(\mu - \varepsilon_i) e^{\beta(\mu - \varepsilon_i)}}{1 - e^{\beta(\mu - \varepsilon_i)}} \tag{9}$$

while the corresponding entropy is

$$S_{BE} = \log Z + \beta\langle U\rangle = -\sum_{i=1}^{|V|} \log\left(1 - e^{\beta(\mu-\varepsilon_i)}\right) - \beta\sum_{i=1}^{|V|}\frac{(\mu - \varepsilon_i)e^{\beta(\mu-\varepsilon_i)}}{1 - e^{\beta(\mu-\varepsilon_i)}} \quad (10)$$

Both the average energy and entropy depend on the chemical potential for the partition function and hence they are determined by the number of particles in the system. At the temperature β, the corresponding number of particles in the level i with energy ε_i is

$$n_i = \frac{1}{\exp[\beta(\varepsilon_i - \mu)] - 1} \quad (11)$$

As a result, the total number of particles in the system is

$$N = \sum_{i=1}^{|V|} n_i = \sum_{i=1}^{|V|}\frac{1}{\exp[\beta(\varepsilon_i - \mu)] - 1} = \mathrm{Tr}\left[\frac{1}{\exp(-\beta\mu)\exp[\beta\tilde{L}] - I}\right] \quad (12)$$

In order for the number of particles in each energy state to be non-negative, the chemical potential must be less than the minimum energy level, i.e. $\mu < \min \varepsilon_i$.

The equivalent function of density matrix in this case is given by

$$\boldsymbol{\rho}_{BE} = \frac{1}{\mathrm{Tr}(\boldsymbol{\rho}_1) + \mathrm{Tr}(\boldsymbol{\rho}_2)}\begin{pmatrix} \boldsymbol{\rho}_1 & 0 \\ 0 & \boldsymbol{\rho}_2 \end{pmatrix} \quad (13)$$

where $\boldsymbol{\rho}_1 = -\left(\exp[\beta(\tilde{L} - \mu I)] - I\right)^{-1}$ and $\boldsymbol{\rho}_2 = \left(I - \exp[-\beta(\tilde{L} - \mu I)]\right)^{-1}$.

Since Bose-Einstein statistics allow particles to coalesce in the lower energy levels, the corresponding energy and entropy reflect the smaller Laplacian eigenvalues most strongly. As a result the number of connected components (the multiplicity of the zero eigenvalue), and spectral gap (the degree of bi-partiality in a graph) are most strongly reflected.

3.3 Fermi-Dirac Statistics

The Fermi-Dirac distribution applies to indistinguishable fermions with a maximum occupancy of one particle in each energy state. Particles cannot be added to states that are already occupied, and hence obey the Pauli exclusion principle.

These particles behave like a set of free fermions in the complex network with energy states given by the network Hamiltonian. The statistical properties of the networks are thus given by the Fermi-Dirac statistics of the equivalent quantum system, and the corresponding partition function is

$$Z_{FD} = \det\left(I + e^{\beta\mu}\exp[-\beta\tilde{L}]\right) = \prod_{i=1}^{|V|}\left(1 + e^{\beta(\mu-\varepsilon_i)}\right) \quad (14)$$

From Eq.(5) the average energy of the Fermi-Dirac system is

$$\langle U \rangle_{FD} = -\frac{\partial \log Z}{\partial \beta} = -\sum_{i=1}^{|V|} \frac{(\mu - \varepsilon_i)e^{\beta(\mu - \varepsilon_i)}}{1 + e^{\beta(\mu - \varepsilon_i)}} \tag{15}$$

And the associated entropy is given by

$$S_{FD} = \log Z + \beta \langle U \rangle = \sum_{i=1}^{|V|} \log\left(1 + e^{\beta(\mu - \varepsilon_i)}\right) - \beta \sum_{i=1}^{|V|} \frac{(\mu - \varepsilon_i)e^{\beta(\mu - \varepsilon_i)}}{1 + e^{\beta(\mu - \varepsilon_i)}} \tag{16}$$

Under Fermi-Dirac statistics, on the other hand, the number of particles occupying the ith energy state is

$$n_i = \frac{1}{\exp[\beta(\varepsilon_i - \mu)] + 1} \tag{17}$$

and the total number of particles in the network system is

$$N = \sum_{i=1}^{|V|} n_i = \sum_{i=1}^{|V|} \frac{1}{\exp[\beta(\varepsilon_i - \mu)] + 1} = \mathrm{Tr}\left[\frac{1}{\exp(-\beta\mu)\exp[\beta\tilde{L}] + I}\right] \tag{18}$$

With a single particle per energy state, the chemical potential is hence just the nth energy level, and so $\mu = \varepsilon_n$.

Similarly, we find that the equivalent density matrix function

$$\boldsymbol{\rho}_{FD} = \frac{1}{\mathrm{Tr}(\boldsymbol{\rho}_3) + \mathrm{Tr}(\boldsymbol{\rho}_4)}\begin{pmatrix} \boldsymbol{\rho}_3 & 0 \\ 0 & \boldsymbol{\rho}_4 \end{pmatrix} \tag{19}$$

where $\boldsymbol{\rho}_3 = \left(I + e^{-\beta\mu}\exp[\beta\tilde{L}]\right)^{-1}$ and $\boldsymbol{\rho}_4 = \left(I + e^{\beta\mu}\exp[-\beta\tilde{L}]\right)^{-1}$.

Since Fermi-Dirac statistics exclude multiple particles from the same energy level, the corresponding energy and entropy do not just reflect the lower part of the Laplacian spectrum, and are sensitive to a greater portion of the distribution of Laplacian eigenvalues. As a result, we might expect them to be more sensitive to subtle differences in network structure.

4   Experiments and Evaluations

In this section, we provide experiments to evaluate the proposed spin statistical models. We commence by assessing the performance on synthetic data using the entropy for network classification problems. We then apply on the real-world financial networks to distinguish significant structural variance.

4.1   Numerical Results

At first, we investigate how well the different spin statistic models can be used to distinguish synthetic networks generated from the Erdős-Rényi random graphs, Watts-Strogatz small-world networks [10] and Barabási-Albert scale-free network models [11]. We conduct numerical experiments to evaluate whether the thermodynamic quantity, i.e. entropy, can represent differences in the networks.

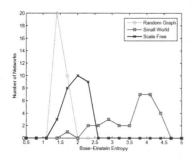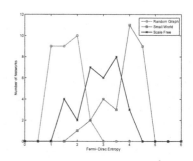

Fig. 1. Histogram of entropy for three classes of synthetic networks in Bose-Einstein statistics and Fermi-Dirac statistics. Temperature $\beta = 10$ and particle number $N = 1$.

These synthetic graphs are created using a variety of model parameters with the number of nodes varying between 100 to 1,000. For small world networks, the rewiring probability is $p = 0.2$ and average node degree is $n = 20$. The scale free networks are derived from Barabási-Albert model [11] with preferential attachment $m = 10$ at each growing step. To simplify the calculation, we set the Boltzmann constant to unity and particle number to one throughout the experiments.

We compare the entropy resulting from the twos spin statistics models. Figure 1 shows the resulting histogram of entropy derived from Bose-Einstein and Fermi-Dirac statistics respectively. In each case the different networks are well separated by the thermodynamic entropy. In the case of Fermi-Dirac statistics, the three clusters of networks are a slightly better clustered than those obtained with Bose-Einstein statistics.

4.2 Experimental Results

The real-world data is extracted from the daily prices of 3,799 stocks traded on the New York Stock Exchange (NYSE). These data provided an empirical investigation in studying the role of communities in the structure of the stock market. We use the correlations of the times-serial stock price to establish networks for the trading days. For each day of trading the correlation is computed between each pair of stock being traded using a time window of 28 days. Edges are created between those pairs of stock whose cross correlation coefficients are in the highest 5%. In this way we obtain a sequence of networks representing the topological structure of the New York stock market from January 1986 to February 2011. This yields a sequence of time-varying networks with a fixed number of 347 nodes and varying edge structure for 5,976 trading days.

We plot the entropy and energy for both Bose-Einstein (blue) and Fermi-Dirac (red) statistics. In order to avoid the thermal disruption in quantum statistics at high temperature, we investigate the spin statistical differences in entropy and energy at low temperature region. Here, to compare the performance, we set the same temperature $\beta = 10$ and particle number $N = 1$ for both two cases.

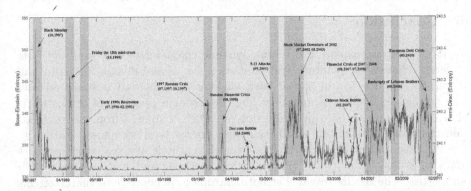

Fig. 2. Entropy in NYSE (1987–2011) derived from Bose-Einstein and Fermi-Dirac statistics. Critical financial events, i.e., Black Monday, Friday the 13th mini-crash, Early 1990s Recession, 1997 Asian Crisis, 9.11 Attacks, Downturn of 2002–2003, 2007 Financial Crisis, the Bankruptcy of Lehman Brothers and the European Debt Crisis, can be represented in thermodynamic entropy with Maxwell-Boltzmann statistic. It is efficient to use partition function associating with entropy to identify events in NYSE. (Color figure online)

Figure 2 shows both entropies with various financial events annotated, including Black Monday, Friday the 13th mini-crash, Early 1990s Recession, 1997 Asian Crisis, 9.11 Attacks, Downturn of 2002–2003, 2007 Financial Crisis, the Bankruptcy of Lehman Brothers and the European Debt Crisis. In each case the entropy undergoes sharp increase corresponding to the financial crises, which are associated with dramatic structural changes in the networks. Similarly in Fig. 3, the energy is also effective in indicating the critical events. The different feature is that energy undergoes a sharp decrease during the financial crises. Moreover, the Bose-Einstein quantities show the greatest variation during the crises, suggesting that changes in cluster-structure (modularity) are important during these episodes.

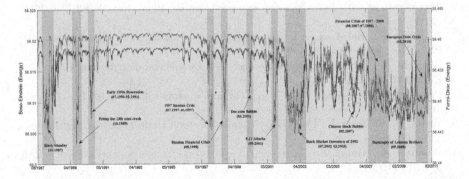

Fig. 3. Thermodynamic energy from Bose-Einstein and Fermi-Dirac statistics occupation statistics for NYSE (1987–2011). Critical financial events, i.e., Black Monday, Friday the 13th mini-crash, Early 1990s Recession, 1997 Asian Crisis, 9.11 Attacks, Downturn of 2002–2003, 2007 Financial Crisis, the Bankruptcy of Lehman Brothers and the European Debt Crisis, all appear as distinct events. (Color figure online)

5   Conclusion

In this paper, we explore the thermodynamic characterizations resulting from different choices of quantum spin statistics, i.e. Bose-Einstein statistics and Fermi-Dirac statistics, in a heat-bath analogy. The method is based on uses the normalized Laplacian matrix as the Hamiltonian operator of the network. The thermodynamic entropy and energy are then computed from the partition functions for Bose-Einstein and Fermi-Dirac energy level occupation statistics.

We have undertaken experiments on both synthetic and real-world network data to evaluate these two spin statistical methods and have analyzed their properties. The results reveal that both entropies are effective in characterizing dynamic network structure, and distinguish different types of network models (random graphs, small world networks, and scale free networks).

Finally, future work will explore the use of the thermodynamic variables in detecting network anomalies and disturbances. Additionally, we will explore the role of the framework for characterizing phase transitions in network structure.

References

1. Albert, R., Barabasi, A.L.: Statistical mechanics of complex networks. Rev. Modern Phys. **74**, 47 (2002)
2. Park, J., Newman, M.: Statistical mechanics of networks. Phys. Rev. E **70**(6), 066117 (2004)
3. Estrada, E., Hatano, N.: Communicability in complex networks. Phys. Rev. E **77**, 036111 (2008)
4. Ye, C., Wilson, R.C., Comin, C.H., Costa, L.F., Hancock, E.R.: Thermodynamic characterization of networks using graph polynomials. Phys. Rev. E **92**, 032810 (2015)
5. Bianconi, G., Rahmede, C., Wu, Z.: Complex Quantum Network Geometries: Evolution and Phase Transitions (2015). arXiv:1503.04739v2
6. Anand, K., Bianconi, G., Severini, S.: Shannon and von Neumann entropy of random networks with heterogeneous expected degree. Phys. Rev. E **83**(3), 036109 (2011)
7. Bianconi, G., Barabasi, A.L.: Bose-Einstein condensation in complex network. Phys. Rev. Lett. **88**, 5632 (2001)
8. Bianconi, G.: Growing Cayley trees described by a Fermi distribution. Phys. Rev. E **036116**, 66 (2002)
9. Passerini, F., Severini, S.: The von Neumann entropy of networks. Int. J. Agent Technol. Syst. **1**, 58–67 (2008)
10. Watts, D., Strogatz, S.: Collective dynamics of small world networks. Nature **393**, 440–442 (1998)
11. Barabasi, A., Albert, R.: Emergence of scaling in random networks. Science **286**, 509–512 (1999)
12. Silva, F.N., Comin, C.H., Peron, T., Rodrigues, F.A., Ye, C., Wilson, R.C., Hancock, E., Costa, L.da F.: Modular Dynamics of Financial Market Networks (2015). arXiv:1501.05040v3

Correlation Network Evolution Using Mean Reversion Autoregression

Cheng Ye$^{(\boxtimes)}$, Richard C. Wilson, and Edwin R. Hancock

Department of Computer Science, University of York, York YO10 5GH, UK
{cy666,richard.wilson,edwin.hancock}@york.ac.uk

Abstract. In this paper, we present a new method for modeling time-evolving correlation networks, using a Mean Reversion Autoregressive Model, and apply this to stock market data. The work is motivated by the assumption that the price and return of a stock eventually regresses back towards their mean or average. This allows us to model the stock correlation time-series as an autoregressive process with a mean reversion term. Traditionally, the mean is computed as the arithmetic average of the stock correlations. However, this approach does not generalize the data well. In our analysis we utilize a recently developed generative probabilistic model for network structure to summarize the underlying structure of the time-varying networks. In this way we obtain a more meaningful mean reversion term. We show experimentally that the dynamic network model can be used to recover detailed statistical properties of the original network data. More importantly, it also suggests that the model is effective in analyzing the predictability of stock correlation networks.

Keywords: Time-evolving correlation network · Mean Reversion Autoregressive Model · Generative probabilistic model

1 Introduction

Generally speaking, a correlation network is the diagrammatic representation of a complex system architecture, where the vertices in the network represent the system components and where the edges contain the connection and correlation information between components. It is for this reason that correlation networks play an increasingly critical role in observing, analyzing and predicting the structure, function and dynamics of realistic large-scale systems. For example, in a stock market there exist a large number of distinct relationships between economic components. By adopting appropriate filtration methods, the most influential correlations can be preserved for constructing the financial market correlation network, which is used for further statistical analyses [1].

Although the bulk of existing correlation network analysis is concerned with static networks, most networks are in reality dynamic in nature [2]. Motivated by the need to understand more deeply the network dynamics, this paper presents a new method for modeling time-evolving correlation networks, and applies the

© Springer International Publishing AG 2016
A. Robles-Kelly et al. (Eds.): S+SSPR 2016, LNCS 10029, pp. 163–173, 2016.
DOI: 10.1007/978-3-319-49055-7_15

resulting method to stock market data. Experimental results show the network model reflects detailed statistical properties of the original network data and more importantly, it can be used to analyze the predictability of stock correlation networks.

1.1 Related Literature

Until recently, one fundamental field of graph theory that has broad applications in network analysis, which has received only marginal attention, is evolutionary graph theory. In fact, many real-world complex systems such as citation networks, communications networks, neural networks and financial networks give rise to structures that change with time. For instance, networks grow and evolve with the addition of new components and connections, or the rewiring of connections from one component to another [3]. In order to analyze such time-evolving graphs, efficient tools for understanding and modeling their time-dependent structure and function are required.

In general, graph evolution can be approached from both macroscopic and microscopic perspectives [4]. On the one hand, the macroscopic approach aims at studying how the global parameters of a dynamic graph evolve from one epoch to another. This can be accomplished by directly employing a number of graph characterizations that have been developed for static graphs to each epoch, and then analyzing the time evolution of these characterizations. For instance, it has been demonstrated that the subgraph centrality can be interpreted as a partition function of a network [5], and as a result the entropy, internal energy and the Helmholtz free energy can be computed and shown to be intimately related to the network dynamics. On the other hand, at the microscopic level, it is the birth-death dynamics of an individual vertex or edge in the graph evolution that are under study. Based on this observation, Grindrod and Higham [4] have introduced a tractable framework for modeling evolving graphs. To do this, they propose a novel range-dependent birth-death mechanism, which allows a variety of evolutionary behaviours to be modeled. The resulting dynamic graph model is set up as a discrete-time Markov chain, and an analogous continuous-time framework can also be developed. This model has proved to be efficient in investigating the evolutionary processes that take place for evolving graphs.

This paper centers on the modeling of time-varying correlation networks. In general, the modeling of the correlation time-series between a pair of components can be achieved using both stochastic and non-stochastic approaches. In particular, stochastic modelling has been widely considered as an essential tool in the analyses of finance, biology and other areas, too. A commonly used approach is to use correlated stochastic processes to map the relationships between components in the financial or biological systems. For example, in the finance literature it is well known that the stochastic process modeling plays a vital role in pricing and evaluation of financial derivatives [6]. On the other hand, the non-stochastic approach also provides a powerful tool for modeling time-series of component correlations. One famous example is furnished by modeling the stock log-returns as autoregressive processes with random disturbances, such as the AR-GARCH model and EGARCH model [7].

1.2 Outline

The remainder of the paper is structured as follows. Section 2 details the development of the time-evolving correlation network modeling using Mean Reversion Autoregression. In this section we also show how the mean reversion term can be obtained from a recently developed generative probabilistic model for graph time-series analysis. In Sect. 3, we show the effectiveness of our method by exploring its experimental performance on realistic stock market data. Finally, Sect. 4 summarizes our contribution present in this paper and points out future research directions.

2   Time Evolving Correlation Network Model

In this section, we provide the detailed development of a model for time-evolving correlation networks. To commence, we introduce an autoregressive model that contains a mean reversion term and use this to simulate the pairwise correlation time-series. Applying such an analysis to each possible pair of vertices in the network, we have to hand a rule that governs the evolution of the edge weight change of the dynamic network. The idea behind the mean reversion model is based on the fact that the log-return of stock price always regresses back to its mean or average. Traditionally, the mean is computed as the arithmetic mean of the stock log-returns. However, this approach clearly does not capture the essential properties of the data. In order to overcome this problem, we show how a generative probabilistic model can be used to determine a more meaningful mean reversion term for the autoregressive model. In short, this generative model provides structure called "supergraph" which can be used to best summarize the structural variations present in the set of network data.

2.1   Mean Reversion Autoregressive Model

In general, a stochastic process is a sequence of measurements representing the evolution of random variables over time. An autoregressive (AR) model represents a type of stochastic processes in which the value is linearly dependent on its previous value and on a stochastic term. Mathematically, let $Y_t = \{y_1, y_2, \cdots, y_t, \cdots\}$ represent a stochastic process of measurements y in time period $[1, 2, \cdots]$, the first-order AR model (AR(1)) implies that

$$y_t = \theta_0 + \theta_1 y_{t-1} + \epsilon_t,$$

where θ_1 is the parameter of the model, θ_0 is a constant and ϵ_t is the white noise. More generally, the p-th order AR model (AR(p)) gives that

$$y_t = \theta_0 + \sum_{i=1}^{p} \theta_i y_{t-i} + \epsilon_t$$

where θ_i represent the parameters of y_{t-i}. In our analysis, we consider the simple case, i.e., AR(1) process in order to reduce the number of parameters in the model.

In economics, the concept of mean reversion has proved to be a widely held belief, i.e., the stock return is likely to regress towards the mean value after a long period. Based on this observation, we add a mean reversion term to the standard AR(1) model in order to obtain the Mean Reversion Autoregressive Model (MRAM)

$$y_t - \bar{y} = \theta_1(y_{t-1} - \bar{y}) + \epsilon_t, \tag{1}$$

where $\bar{y}$ is the mean value of y and $\epsilon_t \sim \mathcal{N}(0, \sigma^2)$ is the white noise. Clearly, the mean reversion term $\bar{y}$ plays a critical role in the MRAM. Broadly speaking, there are a number of different ways to define the mean reversion term. One example would be to simply use the arithmetic mean of y_t, but this approach cannot represent the full, underlying properties of the time-series.

2.2 Generative Model Learning

In the following we present a novel method for constructing a generative model to analyze the structure of labeled data and use this model to determine a more meaningful measure for representing $\bar{y}$. Let $\mathbf{G} = \{\mathcal{G}_1, \mathcal{G}_2, \cdots, \mathcal{G}_t, \cdots, \mathcal{G}_N\}$ represent the time-series graph dataset under study, and $\mathcal{G}_t$ is used to denote the t-th sample graph in the time-series. The generative model, or the supergraph, which we aim to learn from the sample data is denoted by $\tilde{\mathcal{G}} = (\tilde{\mathcal{V}}, \tilde{\mathcal{E}})$, with vertex set $\tilde{\mathcal{V}}$ and edge set $\tilde{\mathcal{E}}$.

We are dealing with labeled graphs. Each vertex in a network has a unique label. In our application involving the New York Stock Exchange data, there are stocks trading in the New York Stock Exchange market. The vertex indices are denoted by lower-case letters including u, v, a, b, α and β, and will interchange these vertex indices with the vertex labels.

We represent the connectivity structure of the sample graph $\mathcal{G}_t$ using a weighted adjacency matrix W^t whose (u, v)-th entry W_{uv}^t indicates the connectivity between vertices u and v in the graph, and clearly, we have $W_{uv}^t \in [0, 1]$. Similarly, we use the matrix $\tilde{W}$ to represent the structure of the supergraph $\tilde{\mathcal{G}}$.

Having introduced the necessary formalism, we now proceed to develop the probabilistic framework for the generative model learning method. To commence, we require the posterior probability of the observed sample graphs given the structure of the generative model $p(\mathbf{G}|\hat{\mathcal{G}})$. Then, the problem of finding the optimal supergraph can be posed in terms of seeking the structure $\tilde{\mathcal{G}}$ that satisfies the condition

$$\tilde{\mathcal{G}} = \operatorname*{argmax}_{\hat{\mathcal{G}}} p(\mathbf{G}|\hat{\mathcal{G}}).$$

We follow the standard approach to constructing the likelihood function, which has been previously used in [8,9]. This involves factorizing the likelihood function over the observed data graphs and making use of the assumption that each individual edge in the sample graphs is conditionally independent of the remainder, given the structure of the supergraph. As a result, we have

$$p(\mathbf{G}|\tilde{\mathcal{G}}) = \prod_t p(\mathcal{G}_t|\tilde{\mathcal{G}}) = \prod_t \prod_u \prod_v p(W_{uv}^t|\tilde{W}_{uv}),\tag{2}$$

where $t = 1, 2, \cdots, N$. Moreover, $p(W_{uv}^t|\tilde{W}_{uv})$ is the probability that the connectivity between u and v in the sample graph $\mathcal{G}_t$ is equal to W_{uv}^t, given that the edge (u, v) in the supergraph $\tilde{\mathcal{G}}$ has connectivity $\tilde{W}_{uv}$. To proceed, we model the distribution $p(W_{uv}^t|\tilde{W}_{uv})$ by adopting a Gaussian distribution $\mathcal{N}(\mu, \sigma^2)$ of the connection weights whose mean is the weight for the edge (u, v) in the supergraph, i.e., $\mu = \tilde{W}_{uv}$. With the observation density model to hand, we write

$$p(W_{uv}^t|\tilde{W}_{uv}) = \frac{1}{\sqrt{2\pi}\sigma} e^{-(W_{uv}^t - \tilde{W}_{uv})^2/2\sigma^2}.$$

To locate the optimal supergraph, we adopt an information theoretic approach and use a two-part minimum description length (MDL) criterion. Underpinning MDL is the principle that the best hypothesis for a given set of data is the one that leads to the shortest code length of the observed data. To formalize this idea, we encode and transmit the data $\mathcal{G}_t$ together with the hypothesis $\tilde{\mathcal{G}}$, leading to a two-part message whose total length is given by

$$\mathcal{L}(\mathbf{G}, \tilde{\mathcal{G}}) = \mathcal{L}(\mathbf{G}|\tilde{\mathcal{G}}) + \mathcal{L}(\tilde{\mathcal{G}}),$$

where $\mathcal{L}(\mathbf{G}|\tilde{\mathcal{G}})$ is the code length of the data graphs given the supergraph and $\mathcal{L}(\tilde{\mathcal{G}})$ is the code length of the estimated supergraph. Determining the most likely supergraph structure can be viewed as seeking the one that minimizes the total code length of the likelihood function. To this end, we take into account the total code length and apply the MDL principle to the model, this allows us to construct a supergraph representation that trades off goodness-of-fit with the sample graphs against the complexity of the model.

To apply the two-part MDL principle, we commence by computing the code length of the data graphs given the supergraph. This can be achieved by simply using the average of the negative logarithm of the likelihood function, with the result that

$$\mathcal{L}(\mathbf{G}|\tilde{\mathcal{G}}) = -\frac{1}{N} \ln p(\mathbf{G}|\tilde{\mathcal{G}})$$

$$= -\frac{1}{N} \sum_t \sum_u \sum_v \left\{ \ln \frac{1}{\sqrt{2\pi}\sigma} - \frac{(W_{uv}^t - \tilde{W}_{uv})^2}{2\sigma^2} \right\},\tag{3}$$

where N is the length of the observed time-series data $\mathbf{G}$.

Next, we compute the code length of the supergraph structure. Traditionally, the complexity of a model is measured by counting the number of parameters in the model. However, this does not generalize well for graphs since the true graph complexity cannot be accurately reflected by information such as the numbers of vertices or edges in the graph. To overcome this problem, we adopt a more meaningful measure of graph complexity, namely the von Neumann entropy, to encode the complexity of the supergraph structure (see [10,11] for detailed information of this entropy). Then, we have the supergraph complexity code length as follows,

$$\mathcal{L}(\tilde{\mathcal{G}}) = 1 - \frac{1}{|\tilde{\mathcal{V}}|} - \frac{1}{|\tilde{\mathcal{V}}|^2} \sum_{(u,v)\in\tilde{\mathcal{E}}} \frac{\tilde{W}_{uv}}{w_u w_v}, \tag{4}$$

where $w_u = \sum_{(u,v)\in\tilde{\mathcal{E}}} \tilde{W}_{uv}$ is the weighted degree of vertex u, which is defined as the sum of the connectivity weights of the edges connected to u and w_v is similarly defined. In effect, the complexity of the supergraph depends on two factors. The first is the order of the supergraph, i.e., the number of the vertices while the second is based on the degree statistics of the vertices in the supergraph.

To recover the supergraph we must optimize the total code length criterion, which can be computed by adding together the two contributions to the total code length, with respect to the weighted adjacency matrix $\tilde{W}$. This can be done in a number of ways. These include gradient descent and soft assign [12]. However here we use a simple fixed-point iteration scheme. We compute the partial derivative of the code length criterion $\mathcal{L}(\mathbf{G}|\tilde{\mathcal{G}})$ given in Eq. (3) with respect to the elements of the weighted adjacency matrix $\tilde{W}_{ab}$. After some analysis the required derivative is

$$\frac{\partial \mathcal{L}(\mathbf{G},\tilde{\mathcal{G}})}{\partial \tilde{W}_{ab}} = \frac{1}{N\sigma^2} \sum_t (\tilde{W}_{ab} - W_{ab}^t) - \frac{1}{|\tilde{\mathcal{V}}|^2} \left\{ \frac{1}{w_a w_b} \right.$$
$$\left. - \frac{1}{w_a^2} \sum_{(a,\beta)\in\tilde{\mathcal{E}}} \frac{\tilde{W}_{a\beta}}{w_\beta} - \frac{1}{w_b^2} \sum_{(\alpha,b)\in\tilde{\mathcal{E}}} \frac{\tilde{W}_{\alpha b}}{w_\alpha} \right\}. \tag{5}$$

where β denote the neighbour vertices of a and α are the neighbours of b.

To set up our fixed-point iteration scheme, we set the above derivative to zero, and re-organize the resulting equation to obtain an update process of the form $\tilde{W}_{ab} \rightarrow g(\tilde{W}_{ab})$, where $g(\cdots)$ is the iteration function. There is of course no unique way of doing this, and for convergence the iteration function $g(\tilde{W}_{ab})$ must have a derivative of magnitude less then unity at the fixed point corresponding to the required solution. One such scheme is

$$\tilde{W}_{ab} \rightarrow \frac{1}{N\sigma^2} \sum_t W_{ab}^t + \frac{1}{|\tilde{\mathcal{V}}|^2} \left\{ \frac{1}{w_a w_b} - \frac{1}{w_a^2} \right.$$
$$\left. \sum_{(a,\beta)\in\tilde{\mathcal{E}}} \frac{\tilde{W}_{a\beta}}{w_\beta^{(n)}} - \frac{1}{w_b^2} \sum_{(\alpha,b)\in\tilde{\mathcal{E}}} \frac{\tilde{W}_{\alpha b}}{w_\alpha} \right\}. \tag{6}$$

The update process is governed by two terms. The first is computed from the local windowed mean of the time-series $\frac{1}{N\sigma^2}\sum_t W_{ab}^t$, while the second term is a step away from the local time-series mean determined by the partial derivative of the von Neumann entropy. This latter update term depends on the local pattern of vertex degrees.

Finally, with the generative structure to hand, we have the MRAM for the time-evolving correlation network. Mathematically, the edge weights of edge (u, v) in the networks follow the process

$$W_{uv}^t - \tilde{W}_{uv} = \theta_1^{uv}(W_{uv}^{t-1} - \tilde{W}_{uv}) + \epsilon_t^{uv} \tag{7}$$

where θ_1^{uv} and ϵ_t^{uv} are the parameter and white noise of edge (u, v).

3   Experiments

In this section, we evaluate the proposed time-evolving correlation network model by applying the model to the stock market data. We confine our attention to two main tasks, the first is to explore whether the MRAM can be used to reflect similar statistical properties of the original correlation time-series; the second is to analyze how the predictability of the stock network changes between different time periods. To commence, we give a brief introduction of the dataset used in the experiments.

NYSE Stock Market Network Dataset. Is extracted from a database consisting of the daily prices of 3799 stocks traded on the New York Stock Exchange (NYSE). This data has been well analyzed in [13], which has provided an empirical investigation studying the role of communities in the structure of the inferred NYSE stock market. The authors have also defined a community-based model to represent the topological variations of the market during financial crises. Here we make use of a similar representation of the financial database. Specifically, we employ the correlation-based network to represent the structure of the stock market since many meaningful economic insights can be extracted from the stock correlation matrices [14]. To construct the dynamic network, 347 stocks that have historical data from January 1986 to February 2011 are selected. Then, we use a time window of 20 days and move this window along time to obtain a sequence (from day 20 to day 6004) in which each temporal window contains a time-series of the daily return stock values over a 20-day period. We represent trades between different stocks as a network. For each time window, we compute the cross-correlation coefficients between the time-series for each pair of stocks. This yields a time-varying stock market network with a fixed number of 347 vertices and varying edge structure for each of 5976 trading days.

In the first experiment, we randomly select two pairs of stocks from the *NYSE Stock Market Network Dataset* and apply the MRAM to their cross-correlation time-series for the entire period, in order to explore whether the model simulation is effective in recovering the statistical properties of the real data. Specifically, we use Eq. (7) to model the correlation time-series of stocks u and v with the mean reversion term determined by the supergraph structure $\tilde{W}_{uv}$. Then, we estimate the model parameters θ_1^{uv} and σ_t^{uv}, which is used to compute the noise term ϵ_t^{uv}. This allows us to obtain a simulation process whose start value is the

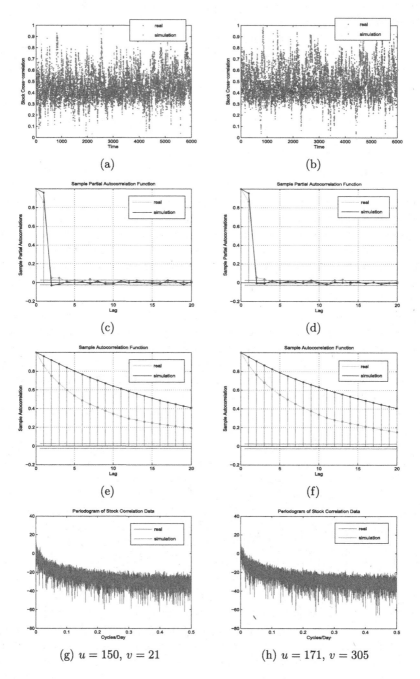

(a)

(b)

(c)

(d)

(e)

(f)

(g) $u = 150$, $v = 21$

(h) $u = 171$, $v = 305$

Fig. 1. Comparison of cross-correlation time-series of different pairs of stocks of real data and simulation data.

same as the real correlation time-series, using the parameters we have estimated from the real time-series in the *NYSE Stock Market Network Dataset*.

Figure 1 shows a comparison of the real data and the model simulation for the stock correlation time-series of two pairs of stocks. The statistical properties under study include the sample partial autocorrelation function, cross-autocorrelation function and the periodicity. Clearly, it is difficult to evaluate the performance of the proposed model from the top plots in Fig. 1, as there is no clear correlation between the real data and the simulation data. However, from the rest of the plots, we observe that the model is able to follow the statistical properties of the real data, especially the partial autocorrelation function and the periodicity. Also from the partial autocorrelation function plots, the values of both real data and simulation data significantly decrease after the first lag, which implies that the choice of using AR(1) process to model correlation time-series is plausible.

The second experimental goal is to analyze the predictability of the financial network, i.e., to explore whether the MRAM can be used to help determine whether the structure of a network is predictable or not. In Fig. 2 we plot the von Neumann entropy for both real data graphs (blue solid line) and model graphs (magenta dot-line) for the time period from year 2000 to 2011. Before year 2007, the von Neumann entropy for realistic graphs is relatively stable, implying that the stock network structure does not experience significant structural changes. After 2007, however, the entropy curve witnesses a number of dramatic

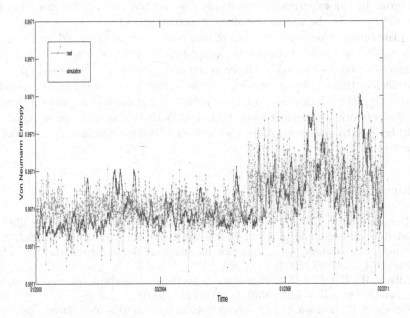

Fig. 2. Comparison of von Neumann entropy for real data and simulation data graphs for time period from 2000 to 2011. (Color figure online)

fluctuations, which means that the network structure is extremely unstable. Turning attention to the simulation data, we note its von Neumann entropy curve clearly exhibits the similar behaviour, following the trend of that of the real data. This again illustrates that the proposed autoregressive model provides an effective method for modeling the time-evolving correlation network in terms of reflecting the structural properties of real data. Moreover, although our model does not offer a way to predict the future network structure, it is indeed useful in understanding and determining the predictability of time-evolving networks.

4 Conclusion

To summarize, in this paper we present a new method for modeling the time-evolving correlation networks, using a Mean Reversion Autoregressive Model. The idea is motivated by the fact that in the finance literature, autoregressive processes are widely employed to model the pricing and evaluation of financial derivatives. Moreover, based on the assumption that stock log-returns always regress back towards the mean or average, we introduce a mean reversion term to the autoregressive model. The mean reversion terms clearly plays a key role in the effectiveness of the model, so it is imperative to have to hand an efficient measure for this term. To this end, we turn our attention to a recently developed generative model, which has been used to analyze the underlying structure of graph time-series, and use this structure to determine a meaningful mean reversion term. In the experiments, we apply the method to realistic stock market data and evaluate the properties of the proposed model.

In the future, the work reported in this paper can be extended in a number of ways. For instance, it would be interesting to explore how the stochastic processes can be used to model the correlation dynamics, which can help develop more efficient time-evolving network models. Moreover, we would be interested in investigating the relationship between the model parameters we have optimized and the network topological characteristics. With this knowledge to hand, we would be able to understand how the network topology influences its structural evolution.

References

1. Tumminello, M., Aste, T., Di Matteo, T., Mantegna, R.N.: A tool for filtering information in complex systems. Proc. Natl. Acad. Sci. **102**, 10421–10426 (2005)
2. Albert, R., Barabási, A.L.: Statistical mechanics of complex networks. Rev. Mod. Phys. **74**(1), 47–97 (2002)
3. Albert, R., Barabási, A.L.: Topology of evolving networks: local events and universality. Phys. Rev. Lett. **85**(24), 5234–5237 (2000)
4. Grindrod, P., Higham, D.J.: Evolving graphs: dynamical models, inverse problems and propagation. In: Proceedings of Royal Society A, vol. 466, pp. 753–770 (2009)
5. Estrada, E., Hatano, N.: Statistical-mechanical approach to subgraph centrality in complex networks. Chem. Phys. Lett. **439**, 247–251 (2007)

6. Christie, A.A.: The stochastic behavior of common stock variances: value, leverage and interest rate effects. J. Financ. Econ. **10**, 407–432 (1982)
7. Ferenstein, E., Gasowski, M.: Modelling stock returns with AR-GARCH processes. Stat. Oper. Res. Trans. **28**(1), 55–68 (2004)
8. Luo, B., Hancock, E.R.: Structural graph matching using the em algorithm and singular value decomposition. IEEE Trans. Pattern Anal. Mach. Intell. **23**, 1120–1136 (2001)
9. Han, L., Wilson, R., Hancock, E.: Generative graph prototypes from information theory. IEEE Trans. Pattern Anal. Mach. Intell. **37**(10), 2013–2027 (2015)
10. Han, L., Escolano, F., Hancock, E., Wilson, R.: Graph characterizations from von neumann entropy. Pattern Recogn. Lett. **33**, 1958–1967 (2012)
11. Ye, C., Wilson, R.C., Comin, C.H., Costa, L.D.F., Hancock, E.R.: Approximate von neumann entropy for directed graphs. Phys. Rev. E **89**, 052804 (2014)
12. Gold, S., Rangarajan, A.: A graduated assignment algorithm for graph matching. IEEE Trans. Pattern Anal. Mach. Intell. **18**(4), 377–388 (1996)
13. Silva, F.N., Comin, C.H., Peron, T.K.D., Rodrigues, F.A., Ye, C., Wilson, R.C., Hancock, E.R., Costa, L.D.F.: On the modular dynamics of financial market networks (2015). arXiv e-prints arXiv:1501.05040
14. Battiston, S., Caldarelli, G.: Systemic risk in financial networks. J. Financ. Manag. Mark. Inst. **1**, 129–154 (2013)

Graph Entropy from Closed Walk and Cycle Functionals

Furqan Aziz[1]($^{(\boxtimes)}$), Edwin R. Hancock[2], and Richard C. Wilson[2]

[1] Department of Computer Science, IM—Sciences, Peshawar, Pakistan
furqan.aziz@imsciences.edu.pk
[2] Department of Computer Science, University of York, York YO10 5GH, UK
{edwin.hancock,richard.wilson}@york.ac.uk

Abstract. This paper presents an informational functional that can be used to characterise the entropy of a graph or network structure, using closed random walks and cycles. The work commences from Dehmer's information functional, that characterises networks at the vertex level, and extends this to structures which capture the correlation of vertices, using walk and cycle structures. The resulting entropies are applied to synthetic networks and to network time series. Here they prove effective in discriminating between different types of network structure, and detecting changes in the structure of networks with time.

Keywords: Graph entropy · Random walks · Ihara coefficients

1 Introduction

The problem of determining the complexity of network structures is an elusive one, which has challenged graph-theorists for over five decades. Broadly speaking the are two approaches to the problem. According to randomness complexity, the aim is to probabilistically characterise the degree of disorganisation of a network, and Shannon entropy provides one convenient way of doing this. One of the earliest attempts at computing network entropy was proposed by Körner [3]. This involves computing the entropy of a vertex packing polytope, and is linked to the chromatic number of the graph. Another simple approach is to use Shannon entropy to compute the homogeneity of the vertex degree distribution. Statistical complexity, on the other hand aims to characterise network complexity beyond randomness. One of the shortcomings of randomness complexity is that it does not capture vertex correlations. To overcome this problem, statistical complexity allows more global structure to be probed. For instance, by using the logical or thermodynamic depth of a network, the details of inhomogeneous degree structure can be problem. One powerful techniques here is to use a variant of the Kologomorov-Chaitin [4,5] complexity to measure how many operations are need to transform a graph into a canonical form (see [9] for a review of network complexity).

© Springer International Publishing AG 2016
A. Robles-Kelly et al. (Eds.): S+SSPR 2016, LNCS 10029, pp. 174–184, 2016.
DOI: 10.1007/978-3-319-49055-7_16

So although entropy based methods provide powerful tools to characterise the properties of a complex network, one of the challenges is to define the entropy in a manner that can capture the correlations or long-range interactions between vertices. One way to do this is to adopt path or cycle-based methods or to use other substructures that allow networks to be decomposed into non-local primitives [7,8]. In this way some of the strengths of both the randomness and statistical approaches to complexity can be combined. One approach that takes an important step in this direction is thermodynamic depth complexity [9]. Here a Birckhoff-vonNeumann polytope is fitted to the heat kernel of a graph. The polytopal expansion uses permutation matrices as a basis, and the Shannon entropy associated with the polytopal expansion coefficients can be used to provide a depth based characterisation of network structure as a function of the diffusion time. However, this approach is time consuming and does link directly to the topological sub-structures which go to form the global network structure.

Here we adopt a different approach, with the aim measuring the entropy associated with closed random walks and cycles in graphs. Our starting point, is the information functional introduced by Dehmer and his co-workers. This allows the entropy of a network or graph to be computed from a functional defined on its vertices. Here, on the other hand we extend this functional to be defined over closed random walks and cycles. The functional for closed random walks builds on Estrada's index [2], while that for cycles uses the coefficients of the Ihara-zeta function, which measure the frequencies of prime cycles in a graph. These two new informational functionals are applied to a variety of synthetic and real world data.

2 Graph Entropy and Information Functionals

In this section we briefly explain the general framework proposed by Dehmer to define graph entropy.

Definition 1 (Dehmer [1]). *Given a graph $G = (V, E)$, its entropy is defined as*

$$I_f(G) := -\sum_{i=1}^{|V|} \frac{f(v_i)}{\sum_{j=1}^{|V|} f(v_j)} log \frac{f(v_i)}{\sum_{j=1}^{|V|} f(v_j)} \tag{1}$$

where $f(v_i)$ is an arbitrary local vertex information functional.   □

A number of information functionals can be defined that capture different local properties of the graph. For example, Dehmer has proposed the following definitions for the information functional that gauges the metrical properties of a graph.

Definition 2. *Given a graph G, the local information functional is defined as*

$$f^V(v_i) := \alpha^{c_1|S_1(v_i,G)|+c_2|S_2(v_i,G)|+....+c_\rho|S_\rho(v_i,G)|}$$

where $|S_k(v_i, G)|$ represents the number of shortest paths of length k reachable from the node v_i and $c_1, c_2, ...c_\rho$ are real valued constants.   □

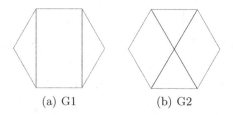

(a) G1   (b) G2

Fig. 1. Two non-isomorphic graphs

The above information functional can easily be obtained by definition of the j-sphere [1].

One of the problem with entropy defined in this way is that it captures only the local properties of a graph, and it is therefore sensitive to the degree distribution of the graph. For example, consider the non-isomorphic graphs in the Fig. 1. The above two graphs are structurally different as G_1 contains two triangles while G_2 does not contain any triangle. However it can be shown that, using Definition 2, the entropy for both the graphs is

$$I_f(G) = - \left[\frac{2\alpha^{16}}{2\alpha^{16} + 4\alpha^{18}} \log \left(\frac{2\alpha^{16}}{2\alpha^{16} + 4\alpha^{18}} \right) + \frac{4\alpha^{18}}{2\alpha^{16} + 4\alpha^{18}} \log \left(\frac{4\alpha^{18}}{2\alpha^{16} + 4\alpha^{18}} \right) \right]$$

Figure 2(a) plots the entropy of the two graphs as a function of α.

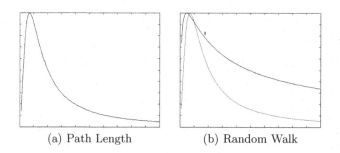

(a) Path Length   (b) Random Walk

Fig. 2. Entropy computed from different information functionals

3   Substructure Based Approach for Graph Entropy

To overcome the problems associated with the methods discussed above, we use random walks on a graph to define graph entropies. Motivated by Esterada's Index (EI), we define the information functional based on closed random walks.

Definition 3. *Let* $|R_k(v_i, G)|$ *denotes the number of random walks in graph* G *of length* k, *starting from and ending at the node* v_i. *We define the information functional as*

$$f^V(v_i) := \alpha^{c_1|R_1(v_i,G)|+c_2|R_2(v_i,G)|+....+c_k|R_k(v_i,G)|}$$

Here $c_1, c_2, ...c_\rho$ *are real valued constants.*   □

Note that if we put $c_n = \frac{1}{n!}$, then the value in the exponent becomes equal to Estrada's Index (EI) [2].

Using Definition 3 the entropy for graph G_1 of Fig. 1 can be shown to be

$$I_f(G_1) = -\left[\frac{2\alpha^{1.5}}{2\alpha^{1.5} + 4\alpha^{3.42}} \log\left(\frac{2\alpha^{1.5}}{2\alpha^{1.5} + 4\alpha^{3.42}} \right) + \frac{4\alpha^{3.42}}{2\alpha^{1.5} + 4\alpha^{3.42}} \log\left(\frac{4\alpha^{3.42}}{2\alpha^{1.5} + 4\alpha^{3.42}} \right) \right]$$

while for G_2 it can be shown to be

$$I_f(G_2) = -\left[\frac{2\alpha^{1.8}}{2\alpha^{1.8} + 4\alpha^{2.68}} \log\left(\frac{2\alpha^{1.8}}{2\alpha^{1.8} + 4\alpha^{2.68}} \right) + \frac{4\alpha^{2.68}}{2\alpha^{1.8} + 4\alpha^{2.68}} \log\left(\frac{4\alpha^{2.68}}{2\alpha^{1.8} + 4\alpha^{2.68}} \right) \right]$$

Figure 2(b) plots the entropies as a function of α for both graphs, which suggests that random walks are a more powerful tool for discriminating between different graph structures.

Our goal in this paper is to define graph entropy using the substructures in the graph. To this end, we decompose the graph into substructures and then use the frequency of a particular substructure to compute the information functional. We now propose a general framework to define entropy.

Definition 4. *Given a graph* G, *we define the graph entropy as*

$$I_f(G) = -\sum_{k=1}^{n} \frac{f(S_k)}{\sum_{i=1}^{n} f(S_i)} \log \frac{f(S_k)}{\sum_{i=1}^{n} f(S_i)} \tag{2}$$

where S_k *represents the information functional computed from* k^{th} *substructure.*   □

There are a number of ways in which a graph can be decomposed and information functionals can be defined. Here we use the coefficients of Ihara zeta function. The reason for using Ihara coefficients is that they are related to the number of simple cycles in the graphs, and hence can be used to gauge the cyclic complexity of a graph.

The Ihara zeta function associated to a finite connected graph G is defined to be a function of $u \in \mathbb{C}$ with $|u|$ sufficiently small by [13]

$$\zeta_G(u) = \prod_{c \in [C]} \left(1 - u^{l(c)} \right)^{-1} \tag{3}$$

The product is over equivalence classes of primitive closed backtrackless, tail-less cycles $c = (v_1, v_2, v_3, ..., v_r = v_1)$ of positive length r in G. Here $l(c)$ represents the length of the cycle.

The reciprocal of Ihara zeta function can also be written in the form of a determinant expression [14]

$$\zeta_G(u) = \frac{1}{\det(I - uT)} \tag{4}$$

where T, the Perron-Frobenius operator, is the adjacency matrix of the oriented line graph of the original graph, and I is the identity matrix of size $2m$. Here m is the number of edges in the original graph. The oriented line graph is constructed by first converting the graph into equivalent digraph, and then replacing each arc of the resulting digraph by a vertex. These vertices are connected if the head of one arc meets the tail of another while preventing backtracking, i.e., arcs corresponding to same edge are not connected.

Since the reciprocal of the Ihara zeta function can be written in the form of a determinant expression, therefore it can also express as a polynomial of degree at most 2. i.e.,

$$\zeta_G(u)^{-1} = \det(I - uT) = c_0 + c_1 u + c_2 u^2 + c_3 u^3 + ... + c_{2m} u^{2m} \tag{5}$$

where $c_0, c_1, ..., c_{2m}$ are called Ihara coefficients and are related to the frequencies of simple cycles in the graph. In particular, it can be shown that if G is a simple graph then $c_0 = 1$, $c_1 = 0$, $c_2 = 0$. Furthermore, the coefficients c_3, c_4 and c_5 are the negative of twice the number of triangles, squares, and pentagons in G respectively. The coefficient c_6 is the negative of the twice the number of hexagons in G plus four times the number of pairs of edge disjoint triangles plus twice the number of pairs of triangles with a common edge, while c_7 is the negative of the twice the number of heptagons in G plus four times the number of edge disjoint pairs of one triangle and one square plus twice the number of pairs of one triangle and one square that share a common edge [6,7]. In [7], we have developed method that can be used to compute Ihara coefficients in a polynomial amount of time.

Definition 5. *Let c_i represents the ith Ihara coefficient. We define the information functionals, $f(c_i)$, as*

$$f(i) := \alpha^{k_i c_i},$$

where k_i are constants.   □

Note that the first three Ihara coefficients are constants [6,7], and therefore we can ignore these coefficients in our computation. Also, since the coefficient beyond the first few coefficients contain redundant information [7], therefore we only retain few coefficients and discard the remainder.

4   Experiments

In this section we explore whether the proposed method can be used to distinguish between graphs that are structurally different. The first dataset comprises synthetically generated networks according to some known network models. Next we apply the proposed methods to three different types of graphs extracted from the COIL [20] dataset. The purpose here is to demonstrate the ability of entropy defined using Ihara coefficients to differentiate between md2 graphs (graphs where the degree of each node is at least 2) with different structures. Finally we investigate the use of the proposed method for the purpose of detecting crises and different stages in an evolving network.

4.1   Random Graphs

We commence by experimenting with the proposed method on synthetically generated graphs according to the following three models.

Erdős-Rényi model(ER) [10]: An ER graph $G(n,p)$ is constructed by connecting n vertices randomly with probability p. i.e., each edge is included in the graph with probability p independent from every other edge. These models are also called *random* networks.

Watts and Strogatz model(WS) [11]: A WS graph $G(n,k,p)$ is constructed in the following way. First construct a regular ring lattice, a graph with n vertices and each vertex is connected to k nearest vertices, k/2 on each side. Then for every vertex take every edge and rewire it with probability p. These models are also called *small*-world networks.

Barabási-Albert model(BA) [12]: A BA graph $G(n,n_0,m)$ is constructed by an initial fully connected graph with n_0 vertices. New vertices are added to the graph one at a time. Each new vertex is connected to m previous vertices with a probability that is proportional to the number of links that the existing nodes already have. These models are also called *scale*-free networks.

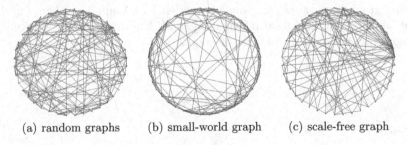

(a) random graphs   (b) small-world graph   (c) scale-free graph

Fig. 3. Graph models

Figure 6 shows an example of each of these models. We have generated 200 graphs for each of these models with n = 21, 22,...,220. The parameters for these models were chosen in such a way that graphs with the same number of vertices have roughly the same number of edges. For ER graphs we choose $p = 15/n$, For WS graphs we choose $k = 16$ and $p = .25$, and for BA graphs we choose $n_0 = 9$ and $k = 8$.

Next we compute the entropy of each graph using both the random walks and Ihara coefficients. We have used the coefficients of Ihara zeta function and closed random walks as information functional to compute entropy. For the Ihara coefficients, we have selected the first six coefficients starting for c_3 to compute the information functional, and used Definition 4 to find entropy. To avoid scaling effects, the last three coefficients were multiplied with $1/|E|$. For random walks, we have selected random walk up to length 6 and used Definition 1 to find entropy. Each coefficient was multiplied with $1/k!$, where k represents the length of the random walk.

For each graph, we have generated a feature vector of length 100. The feature vector is constructed by choosing different values of α and computing information functional and the resulting entropy for each value of α. In our case we have put $\alpha = 0.1, 0.2, ..., 10$. This transforms each graph into a feature vector in a 100 dimensional feature space. To visualise the results, we have performed Principal Component Analysis PCA on the resulting feature vectors and embed the results in a three dimensional vector space. PCA is mathematically defined [17] as an orthogonal linear transformation that transforms the data to a new coordinate system such that the greatest variance by any projection of the data comes to lie on the first coordinate (called the first principal component), the second greatest variance on the second coordinate, and so on. Figure 4(a) shows the resulting embedding on the first three principal components for feature vectors computed using random walks, while Fig. 4(b) shows the resulting embedding on the first three principal components for feature vectors computed using Ihara coefficients. To compare the results, we have also used the local information functional defined by Dehmer [1] that is computed from path lengths. Figure 4(c) shows the resulting embedding. The resulting embedding shows that the entropy computed from random walks gives best results. The Ihara coefficients on the other hand does not provide very good inter-class separation. This is due to the fact that graphs generated using random models have limited number of cycles. Figure also suggest that local information (path length) is not very helpful in distinguishing the different families of graphs.

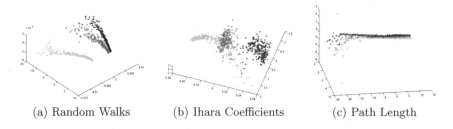

(a) Random Walks   (b) Ihara Coefficients   (c) Path Length

Fig. 4. PCA embedding of different methods

4.2 Graphs Extracted from COIL Dataset

We now perform experiments on the graphs extracted from the images in the Columbia object image library (COIL) dataset [20]. This dataset contains views of 20 different 3D objects under controlled viewer and lighting condition. For each object in the database there are 72 equally spaced views. To establish a graph on the images of objects, we first extract feature points from the image. For this purpose, we use the Harris corner detector [19]. We then construct three different types of graphs using the selected feature points as vertices, i.e., Delaunay triangulation(DT) [15], Gabriel graph(GG) [18], and relative neighbourhood graph(RNG) [16]. Figure 5(a) shows some of the COIL objects, while Fig. 5(b), (c) and (d) show the DT, GG and RNG extracted from the corresponding objects respectively. Next we used Dehmer's definition to compute the entropy of each graph extracted from each of the 72 views of all the 20 objects. We have used the coefficients of Ihara zeta function and closed random walks as information functional to compute entropy. For the Ihara coefficients, we have selected the first six coefficients starting for c_3. To avoid the scaling effect, the last three coefficients were multiplied with $1/|E|$. For each graph, we have generated a signature of length 100 by putting $\alpha = 0.1, 0.2, ..., 10$. For random walk, we have selected random walk upto length 6. Each coefficient was multiplied with $1/k!$, where k is the length of the random walk. Finally, we have also used the local information functional defined by Dehmer to compute entropy. To compare the results, we have performed Principal Component Analysis PCA, on the signatures obtained by choosing different values of α for each of these methods and embed the results in a three dimensional vector space. Figure 4 compares the resulting embedding of the feature vectors on the first three principal components for all the three methods. It is clear from the figure that Ihara coefficients proves to be a powerful tool to distinguish graphs that exhibit a cyclic structure. On the other hand, the entropy computed from random walks and local paths is not very helpful in distinguishing these graphs.

(a) COIL   (b) DT   (c) GG   (d) RNG

Fig. 5. COIL objects and their extracted graphs.

4.3 Time-Varying Networks

In our last experiment we explore whether the proposed methods can be used as a tool for understanding the evolution of a complex network. For this purpose we choose publicly available New York Stock Exchange (NYSE) dataset[25]. This dataset consists of the daily prices of different stocks traded continuously on the New York Stock Exchange for a 25 year span from January 1986 to

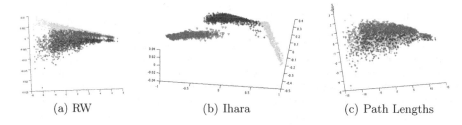

(a) RW   (b) Ihara   (c) Path Lengths

Fig. 6. PCA embedding for COIL dataset

February 2011. A total of 347 stock were selected from this set. We construct
a network over a time-window of 28 days. Here the stocks represent the nodes
of the network. An edge is present, if the correlation value is above a threshold.
We select an empirical value of 0.85 as a threshold value. This was done under
the assumption that, at any given time, a particular stock must interact with
another stock. A new network is generated by sliding the window of 1 day and
repeating the process. In this way a total of 5977 time-varying networks are
generated. Since the networks generated in this way have very limited number
of prime cycles, We have used the entropy defined using closed random walks.
We next applied PCA on the resulting signatures. Figure 7 shows the values of
the eigenvector with the highest variance. The above result suggests that the
proposed method is a very useful tool for detecting changes in a time evolving
network. To compare these results, we have computed the entropy using path
length as defined by Dehmer. Figure 8 shows the values of the first principal
component, after applying PCA on the resulting signatures. This clearly suggests
that entropies defined using local structural properties are not very helpful to
detect changes in a time evolving network.

To compare the results, we have also computed the von Neumann entropy
and the Estrada index of the evolving networks. Figure 9 shows the results.

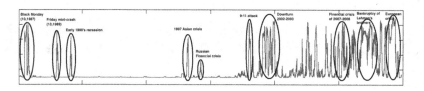

Fig. 7. Entropy computed from random walks on networks

Fig. 8. Entropy computed from path lengths using j-sphere

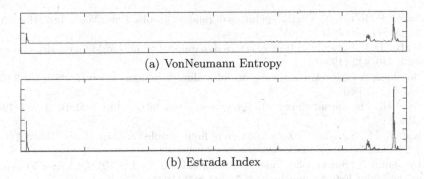

(a) VonNeumann Entropy

(b) Estrada Index

Fig. 9. VonNeumann entropy and Estrada index for NYSE

5   Conclusion

In this paper we have used closed random walks and simple cycles to define information functionals that can be used to define the entropy of a graph (or a network). We have decomposed the graph into substructures and used the frequencies of these substructures to define entropy. We have also presented a generic framework that can be used to define graph entropy by decomposing the graph into substructures. Experiments were performed on numerous datasets including synthetic data, cyclic graphs, and time series data, and the results suggest that the proposed methods can be used to characterise the graphs (and networks) with very higher accuracy as compared to some of the other state-of-the-art methods.

References

1. Dehmer, M.: Information processing in complex networks: graph entropy and information functionals. Appl. Math. Comput. **201**, 82–94 (2008)
2. Estrada, E.: Characterization of 3D molecular structure. Chem. Phys. Lett. **319** (5–6), 713–718 (2000)
3. Körner, J.: Coding of an information source having ambiguous alphabet and the entropy of graphs. In: 6th Prague Conference on Information Theory (1973)
4. Kolmogorov, A.N.: Three approaches to the definition of the concept ? Quantity of information? Probl. Peredachi Informatsii **1**(1), 3–11 (1965)
5. Chaitin, G.J.: On the length of programs for computing finite binary sequences. J. ACM **13**(4), 547–569 (1966)
6. Scott, G., Storm, C.: The coefficients of the Ihara zeta function. Involve - J. Math. **1**(2), 217–233 (2008)
7. Aziz, F., Wilson, R.C., Hancock, E.R.: Backtrackless walks on a graph. IEEE Trans. Neural Netw. Learn. Syst. **24**(6), 977–989 (2013)
8. Ren, P., Wilson, R.C., Hancock, E.R.: Graph characterization via Ihara coefficients. IEEE Trans. Neural Netw. **22**(2), 233–245 (2010)
9. Escolano, F., Hancock, E.R., Lozano, M.A.: Heat diffusion: thermodynamic depth complexity of networks. Phys. Rev. E **85**(3), 036206 (2012)

10. Erdõs, P., Rényi, A.: On the evolution of random graphs. Publ. Math. Inst. Hungar. Acad. Sci. **5**, 17–61 (1960)
11. Watts, D.J., Strogatz, S.H.: Collective dynamics of 'small-world' networks. Nature **393**, 440–442 (1998)
12. Barabási, A., Albert, R.: Emergence of scaling in random networks. Science **286**, 509–512 (1999)
13. Bass, H.: The Ihara-Selberg zeta function of a tree lattice. Int. J. Math. **3**, 717–797 (1992)
14. Kotani, M., Sunada, T.: Zeta function of finite graphs. J. Math. Univ. Tokyo **7**(1), 7–25 (2000)
15. Delaunay, B.: Sur la sphre vide. Izvestia Akademii Nauk SSSR, Otdelenie Matematicheskikh i Estestvennykh Nauk **7**, 793–800 (1934)
16. Toussaint, G.T.: The relative neighbourhood graph of a finite planar set. Pattern Recogn. **12**, 261–268 (1980)
17. Jolliffe, I.T.: Principal Component Analysis. Springer, New York (1986)
18. Gabriel, K.R., Sokal, R.R.: A new statistical approach to geographic variation analysis. Syst. Zool. **12**, 205–222 (1969)
19. Harris, C., Stephens, M.: A combined corner and edge detector. In: Fourth Alvey Vision Conference, Manchester, UK, pp. 147–151 (1988)
20. Nayar, S.K., Nene, S.A., Murase, H.: Columbia object image library (coil 100), Department of Comp. Science, Columbia University, Technical report, CUCS-006-96 (1996)

Dirichlet Graph Densifiers

Francisco Escolano[1]([⊠]), Manuel Curado[1], Miguel A. Lozano[1],
and Edwin R. Hancook[2]

[1] Department of Computer Science and AI,
University of Alicante, 03690 Alicante, Spain
{sco,mcurado,malozano}@dccia.ua.es
[2] Department of Computer Science, University of York, York YO10 5DD, UK
erh@york.ac.uk

Abstract. In this paper, we propose a graph densification method based on minimizing the combinatorial Dirichlet integral for the line graph. This method allows to estimate meaningful commute distances for mid-size graphs. It is fully bottom up and unsupervised, whereas anchor graphs, the most popular alternative, are top-down. Compared with anchor graphs, our method is very competitive (it is only outperformed for some choices of the parameters, namely the number of anchors). In addition, although it is not a spectral technique our method is spectrally well conditioned (spectral gap tends to be minimized). Finally, it does not rely on any pre-computation of cluster representatives.

Keywords: Graph densification · Dirichlet problems · Random walkers

1 Introduction

1.1 Motivation

kNN graphs have been widely used in graph-based learning, since they tend to capture the structure of the manifold where the data lie. However, it has been recently noted [1] that for a standard machine learning setting ($n \to \infty$, $k \approx \log n$ and large d, where n is the number of samples and d is their dimension) we have that kNN graphs result in a sparse, globally uninformative representation. In particular, a kNN-based estimation of the geodesics (for instance through the shortest paths as done in ISOMAP) diverges significantly unless we assign proper weights to the edges of the kNN. Finding such weights is a very difficult task as d increases. As a result, machine learning algorithms for graph-based embedding, clustering and label propagation tend to produce misleading results unless we are able of preserving the distributional information of the data in the graph-based representation. In this regard, recent experimental results with anchor graphs suggest a way to proceed. In [2,7,8], the predictive power of non-parametric regression rooted in the anchors/landmarks ensures a way of constructing very informative weighted kNN graphs from a reduced set of representatives (anchors). Since anchor graphs are bipartite (only *data-to-anchor*

© Springer International Publishing AG 2016
A. Robles-Kelly et al. (Eds.): S+SSPR 2016, LNCS 10029, pp. 185–195, 2016.
DOI: 10.1007/978-3-319-49055-7_17

edges exist), this representation bridges the sparsity of the pattern space because a random walk traveling from node u to node v must reach one or more anchors in advance. In other words, for a sufficient number of anchors it is then possible to find links between distant regions of the space. As a result, the problem of finding suitable weights for the graph is solved through kernel-based regression.

Data-to-anchor kNN graphs are computed from $m \ll n$ representatives (anchors) typically obtained through K-means clustering, in $O(dmnT + dmn)$, where $O(dmnT)$ is due to the T iterations of the K-means process. Since $m \ll m$, the process of constructing the $m \times m$ affinity matrix $W = Z\Lambda Z^T$, where $\Lambda = \mathrm{diag}(Z^T 1)$ and Z is the data-to-anchor mapping matrix, is linear in n. As a byproduct of this construction, we have that the main r eigenvalue-eigenvector pairs associated with $M = \Lambda^{-1/2} Z^T Z \Lambda^{-1/2}$, which has also dimension $m \times m$, lead a compact solution for the spectral hashing problem [14] (see [9] for details). These eigenvectors-eigenvalues may also provide a meaningful estimation of the commute distances between the samples through the spectral expression of this distance [11].

Once considered the benefits of anchor graphs, their use is quite empirical since their foundations are poorly understood. For instance, the choice of the m representatives is quite open and heuristic. The K-means selection process outperforms the uniform selection because it approximates better the underlying distribution. More clever oracles for estimating not only the positions of the representatives but also their number would lead to interesting improvements. However, the developments of these oracles must be compatible with the underlying principle defining an anchor graph, namely *densification*. Densification refers to the process of increasing the number edges (or the weights) of an input graph so that the result preserves and even enforces the structural properties of the input graph. This is exactly what it anchors provide: given a sparse graph associated to a standard machine learning setting, they produce a more compact graph which is locally dense (specially around the anchors) and minimizes inter-class paths.

Graph densification is a principled study of how to significantly increase the number of edges of an input graph G so that the output, H, approximates G with respect to a given test function, for instance whether there exists a given cut. Existing approaches [4] pose the problem in terms of semidefinite programming SDP where a global function is optimized. These approaches have two main problems: (a) the function to optimize is quite simple and does not impose the minimization of inter-class edges while maximizing intra-class edges, and (b) since the number of unknowns is $O(n^2)$, i.e. all the possible edges, and SDP solvers are polynomial in the number of unknowns [10], only small-scale experiments can be performed. However, these approaches have inspired the densification solution proposed in this paper. Herein, instead of proposing an alternative oracle (top-down solution) we contribute with a method for grouping sparse edges so that densification can rely on similarity diffusion (bottom-up solution). Since our long-term scientific strategy is to find a meeting point between bottom-up and top-down densifiers, here we study to what extent we can approximate the performance of anchor graphs from the input sparse graph as a unique source of information.

1.2 Contributions

In this paper, we propose a bottom-up graph densification approach which commences by grouping edges through *return random walks* (Sect. 2). Return random walks (RRW) are designed to enforce intra-class edges while penalizing inter-class weights. Since our strategy is completely unsupervised, return random walks operate under the hypothesis that inter-class edges are rare events. Given input sparse graph G (typically resulting from a thresholded similarity matrix W), RRWs produce a probabilistic similarity matrix W_e. Then, high probability edges are assumed to drive the grouping process. To this end, we exploit the *random walker* [3] but in the edges space (Sect. 3). The random walker minimizes the Dirichlet integral, in this case that associated with the line graph of W_e: $Line_{W_e}$. Given a set of known edges (assumed to be the ones with maximal probability in W_e) we predict the remainder edges. The result is a locally-dense graph H that is suitable for computing commute distances. In our experiments (Sect. 4), we will compare our *Dirichlet densifier* with anchor graph as well as with existing non-spectral alternatives relying exclusivelly on kNN graphs.

2 Return Random Walks

Given a set of points $\chi = \{x_1, ..., x_n\} \subset \mathbb{R}^d$, we map the x_i to the vertices V of an undirected weighted graph $G(V, E, W)$. We have that V is the set of nodes where each v_i represents a data point x_i, $E \subseteq V \times V$ is the set of edges linking adjacent nodes. An edge $e = (i, j)$ with $i, j \in V$, exists if $W_{ij} > 0$ where $W_{ij} = e^{-\sigma \|x_i - x_j\|^2}$, i.e. $W \in \mathbb{R}^{n \times n}$ is a weighted similarity matrix.

Design of W_e. Given W we produce a reweighted similarity matrix W_e by following this rationale: (a) we explore the two-step random walks reaching a node v_j from v_i through any transition node v_k, (b) *on, return* from v_j to v_i we maximize the probability of returning through a different transition node $v_l \neq v_k$. For the first step (going from v_i to v_j through v_k) we have $p_{v_k}(v_j | v_i) = \frac{W_{ik} W_{kj}}{d(v_i) d(v_j)}$ as well as a *standard return* $p_{v_l}(v_i | v_j) = \frac{W_{jl} W_{li}}{d(v_j) d(v_i)}$. Standard return works pretty well if v_i and v_j belong to the same cluster (see Fig. 1-left). However, v_l (the transition node for returning) can be constrained so that $v_l \neq v_k$. In this way, travelling out of a class is penalized since the walker must choose a different path, which in turn is hard to find on average. Therefore, we obtain $W_{e_{ij}}$ from W_{ij} as follows:

$$W_{e_{ij}} = \max_k \max_{\forall l \neq k} \{ p_{v_k}(v_j | v_i) p_{v_l}(v_i | v_j) \}, \qquad (1)$$

i.e. for each possible transition node v_k we compute the probability of go and return (product of independent probabilities) through a different node v_l. We retain the maximum product of probabilities for each v_l referred to a given k and finally we retain the supremum of these maxima. As a result, when inter-class travels are frequent for a given $e = (i, j)$ (Fig. 1-right) its weight $W_{e_{ij}}$ is significantly reduced. Our working hypothesis is that the number of edges

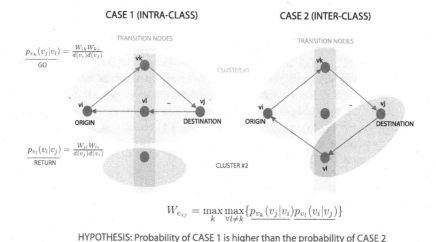

$$W_{e_{ij}} = \max_k \max_{\forall l \neq k} \{p_{v_k}(v_j|v_i) p_{v_l}(v_i|v_j)\}$$

HYPOTHESIS: Probability of CASE 1 is higher than the probability of CASE 2

Fig. 1. Return random walks for reducing inter-class noise.

subject to this condition is small on average, since the number of inter-class edges tends to be small compared with the total number of edges. However, in realistic situations where patterns can be confused due either to their intrinsic similarity or to the use of an unproper similarity measure, this assumption leads to a significant decrease of many weights of W.

3   The Dirichlet Graph Densifier

3.1   The Line Graph

The graph densification problem can be posed as follows: given a graph $G = (V, E, W)$ infer another graph $H = (V, E', W')$ so that $|E'| \geq |E|$ in such a way that the bulk of the increment in the number of edges is constrained to intra-class edges (i.e. the number of inter-class edges is minimized). Therefore, the unknowns of the problem are the new edges to infer, not the vertices. In principle we have a $O(n^2)$ unknowns, where $n = |V|$, but working with all of them is infeasible. This motivates the selection of a small fraction of them (those with the highest values of $W_{e_{ij}}$) according to a given theshold γ_e. The counterintuitive fact that the smaller the fraction the better the accuracy is explained below and showed later in the experimental section. Concerning efficiency, the first impact of this choice is that only $|E''|$ edges, with $|E''| \ll |E|$ are considered for building a graph of edges, i.e. a *line graph* $Line_{W_e}$. Let A the $p \times n$ edge-node incidence matrix defined as follows:

$$A_{e_{ij}v_k} = \begin{cases} +1 \text{ if } i = k, \\ -1 \text{ if } j = k, \\ 0 \quad \text{otherwise,} \end{cases} \tag{2}$$

Then, the $C = AA^T - 2I_p$ is the adjacency matrix of an unweighted line graph, where: I_p is the $p \times p$ identity matrix, the nodes e_a are given by all the possible pairs of $r = |E''|$ edges with a common vertex according to A. The edges of C implement second-order interactions between nodes in the original graph from which A comes from. However, C is still unattributed (although conditioned by W_e). A proper weighting of for this graph is to use standard "go and return" random walks, i.e.

$$Line_{W_e}(e_a, e_b) = \sum_{k=1}^{r} p_{e_k}(e_b|e_a) p_{e_k}(e_a|e_b) , \qquad (3)$$

i.e. return walks are not applied because they become too restrictive. Then, there is an edge in the line graph for every pair (e_a, e_b) with $Line_{W_e}(e_a, e_b) > 0$. We denote the set os edges of the line graph by E_{Line}.

3.2   The Dirichlet Functional for the Line Graph

Given the line graph $Line_{W_e}$ with r nodes (now *edges*) many of them will be highly informative according to W_e and the application of Eq. 3. We retain a fraction of them (again, those with the largest values of W_e) according to a second threshold μ_e. This threshold must be set as smaller as possible since it defines the difference between the "known" and the "unknown". More precisely, W_e acts as a function $W_e : |E''| \to \mathbb{R}$ so that the larger its value the more trustable is a given edge as an stable or known edge in the original graph G. Unknown edges are assumed to have small values of W_e and this is why they are not selected, since the purpose of our method is to infer them.

This is a classical inference problem, now in the space of edges and completely unsupervised, which has been posed in terms of minimizing the disagreements between the weights of existing (*assumed* to be "known") edges and those of the "unknown" or inferred ones. In this regard, since unknown edges are typically neighbors of known ones, the minimization of this disagreement is naturally expressed in terms of finding an harmonic function. Harmonic functions $u(x)$ satisfy $\nabla^2 u = 0$ which in our discrete setting leads to the following property

$$u(e_a) = \frac{1}{d(e_a)} \sum_{(e_a, e_b) \in E_{Line}} Line_{W_e}(e_a, e_b) u(e_b) , \qquad (4)$$

The harmonic function $u(.)$ is not unconstrained, since it is known for some values of the domain (the so called *boundary*). In our case, we set $u(e_a) = W_{e_a}$ for $e_a \in E_B$, referred to as *border nodes* since they are associated with assumed known edges. The harmonic function is unknown for $e_b \in E_I = E'' \sim E_B$ (the *inner nodes*). Then, finding an harmonic function given boundary values is called the *Dirichlet problem* and it is typically formulated in terms of minimizing the following integral

$$D[u] = \frac{1}{2} \int_{\Omega} |\nabla u|^2 d\Omega, \qquad (5)$$

whose discrete version relies on the graph Laplacian [3] (in this case on the Laplacian of the line graph):

$$
\begin{aligned}
D_{Line}[u] &= \frac{1}{2}(Au)^T R(Au) = \frac{1}{2}u^T \mathcal{L}_{Line} u \\
&= \frac{1}{2} \sum_{(e_a, e_b) \in E_{Line}} Line_{W_e}(e_a, e_b)(u(e_a) - u(e_b))^2 ,
\end{aligned}
\tag{6}
$$

where A' is the $|E''| \times |E_{Line}|$ incidence matrix, R is the $|E_{Line}| \times |E_{Line}|$ diagonal constitutive matrix containing all the weights of the edges in the line graph, and $\mathcal{L}_{Line} = D_{Line} - Line_{W_e}$ with $D_{Line} = diag(d(e_a) \ldots d(e_{|E''|}))$ where $d(e_a) = \sum_{e_b \neq e_a} Line_{W_e}(e_a, e_b)$ is the diagonal degree matrix. Then, $\mathcal{L}_{Line}$ is the Laplacian of the line graph.

Given the Laplacian $\mathcal{L}_{Line}$ and the Dirichlet combinatorial integral D_{Line} we have that the nodes in the line graph are partitioned in two classes: "border" and "inner", i.e. $E'' = E_B \cup E_I$. This partition leads to a reordering of the harmonic function $u = [u_B \ u_I]$ as well as the Dirichlet integral:

$$
D[u_I] = \frac{1}{2} \begin{bmatrix} u_B^T & u_I^T \end{bmatrix} \begin{bmatrix} L_B & K \\ K^T & L_I \end{bmatrix} \begin{bmatrix} u_B \\ u_I \end{bmatrix}
\tag{7}
$$

where $D[u_I] = \frac{1}{2}(u_B^T L_B u_B + 2u_I^T K^T U_B + u_I^T L_I u_I)$ and differentiating w.r.t. u_I leads to solve a linear system which relates u_I with u_B:

$$
L_I u_I = -K^T u_B .
\tag{8}
$$

Then, let $s \in [0, 1]$ be a label indicating to what extend a given node of the line graph (an edge in the original graph) is relevant. We define a potential function $Q : E_B \to [0, 1]$ so that for a known node $e_a \in E_B$ we assign a label s, i.e. $Q(e_a) = s$. This leads to declaring the following vector for each label:

$$
m_a^s = \begin{cases} \dfrac{W_{e_a}}{\max_{e_b \in E''}\{W_{e_b}\}} & \text{if } Q(e_a) = s, \\ 0 & \text{if } Q(e_a) \neq s \end{cases} .
\tag{9}
$$

Finally, the linear system is posed in terms of how the known labels do predict the unknown ones, placed in the vector u, as follows:

$$
L_I u^s = -K^T m^s .
\tag{10}
$$

If we consider simultaneously all labels instead of a single one, we have

$$
L_I U = -K^T M \ \Rightarrow \ U = (-K^T M)L_I^{-1} ,
\tag{11}
$$

where U is a vector of $|E_I|$ rows (one per unknown/inner edge, *known solved*) and M has $|E_B|$ rows and columns. Then, let U_b be the b-th row, i.e. the weight of a previously unknown edge e_b. Since there is a bijective correspondence between the nodes in the line graph (some of them are denoted by e_a since they are known, and the remainder are denoted by e_b) and the edges in the original graph $G = (V, E, W)$, then we have that e_k corresponds to edge $(i, j) \in E$. However, since its weight has potentially changed after solving the linear system, we adopt the following densification criterion (labeling) for creating the graph $H = (V, E, W')$:

$$H_{ij} = \begin{cases} \max_{e_k \in U} U_k & \text{if } e_k \in E_I \\ M_{ij} & \text{if } e_k \in E_B, \\ 0 & \text{otherwise} \end{cases} \tag{12}$$

In this way, the edges E' of the dense graphs are given by $H_{ij} > 0$.

4 Experiments and Conclusions

In our experiments we use a reduced version of the NIST digits database: $n = 200$ (20 samples per class) and proceed to estimate commute distances. In all cases, given a similarity matrix we use the $O(n \log n)$ randomized algorithm proposed in [12]. We explore the behavior of the proposed Dirichlet densifier for different values of γ_e, the threshold leading to preserve different fractions of the leading edges (the ones with the highest values in W_e: from 5 % to 50 %. Concerning the theshold μ_e controlling the fraction of leading nodes in the line graph assumed to be "known" (i.e. border data in the terminology of Dirichlet problems), we have explored the same range: from 5 % to 50 % (see Table 1 where we show the accuracies corresponding to each of the 100 experiments performed. A first important conclusion is that the best clustering accuracy (w.r.t. the ground truth) is obtained when the fraction of retained edges for constructing the line graph is minimal. Although the removed edges cannot be reconstructed after solving the Dirichlet equation, i.e. we bound significantly the level of densification, reducing the fraction of retained edges reduces significantly the inter-class

Table 1. Dirichlet densifier: accuracy for the reduced NIST database

Perc. of known labels	Perc. of edges selected									
	5 %	10 %	15 %	20 %	25 %	30 %	35 %	40 %	45 %	50 %
5 %	0.42	0.34	0.31	0.25	0.19	0.18	0.13	0.10	0.10	0.10
10 %	0.49	0.42	0.41	0.36	0.31	0.31	0.23	0.20	0.16	0.16
15 %	0.49	0.46	0.48	0.46	0.40	0.34	0.28	0.24	0.19	0.18
20 %	0.51	0.50	0.53	0.40	0.38	0.34	0.30	0.28	0.24	0.19
25 %	0.53	0.54	0.49	0.47	0.39	0.30	0.32	0.26	0.24	0.17
30 %	0.55	0.53	0.48	0.43	0.34	0.34	0.29	0.20	0.24	0.23
35 %	0.59	0.58	0.51	0.37	0.37	0.26	0.20	0.21	0.18	0.17
40 %	0.56	0.52	0.47	0.36	0.27	0.21	0.23	0.18	0.17	0.14
45 %	0.56	0.54	0.44	0.32	0.25	0.23	0.18	0.18	0.19	0.20
50 %	0.60	0.55	0.39	0.35	0.22	0.18	0.16	0.19	0.16	0.14

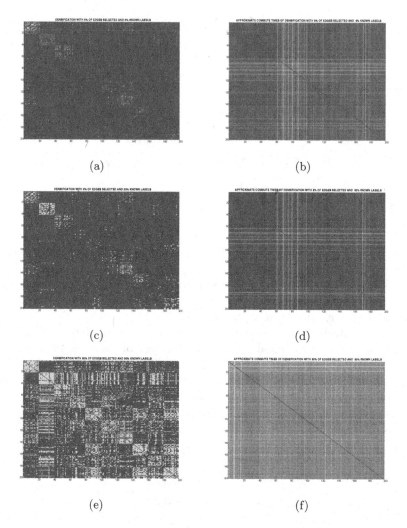

Fig. 2. Densification result and its associate approximate commute times (ACT) matrix for different fractions of known labels $|E_B|$ and leading edges $|E''|$: (a) Densification with $|E_B| = 5\%$, $|E''| = 5\%$, (b) corresponding ACT, (c) Densification with $|E_B| = 50\%$, $|E''| = 5\%$, (d) corresponding ACT, (e) Densification with $|E_B| = 50\%$, $|E''| = 50\%$, (f) corresponding ACT.

noise. We show this effect in Fig. 2 (e)–(f). For instance, a fraction of 5 % (c) produces a better approximation of the commute distance (d) w.r.t. retaining 50 % of the edges to build the line graph. The commute distances after retaining 50 % are meaningless (f) despite the obtained graph is denser. In all cases, the error assumed when approximating the commute times matrix is $\epsilon = 0.25$.

In a second experiment, we compare the commute distances obtained with the optimal Dirichlet densifier (fraction of retained leading edges $|E''| = 5\%|$

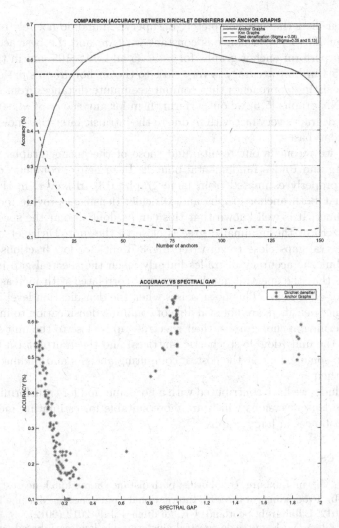

Fig. 3. Top: accuracy of anchors graphs, kNN graphs and dirichlet densifiers. Dirichlet densifiers do not depend on the number of anchors and are completely unsupervised. Bottom: accuracy vs spectral gap.

and fraction known labels $|E_B| = 50\,\%$) with different settings for the anchor graphs. Concerning anchor graphs, in all cases we set $\sigma = 0.08$ for constructing the Gaussian graphs from the raw input data. In our Dirichlet approach we use the same setting. This provides the best result in the range $\sigma \in [0.05, 0.13]$. In Fig. 3-Left we show how the accuracy evolves while increasing the number of (anchors) m: from 5 to 150. The performance of anchor graphs increases with m but degrades after reaching the peak at $m = 70$ (accuracy 0.67). This peak is due to the fact that anchor graphs tend to reduce the amount of inter-class noise. However, this often leads to poor densification. On the other hand,

Dirichlet densifiers they are completely unsupervised and do not rely on anchor computation. Their performance is constant w.r.t. m and the best accuracy is 0.60. We outperform anchor graphs for $m < 35$ and $m > 105$ and in the range $m \in [35, 105]$ our best accuracy is very close to the anchor graph's performance. Regarding existing approaches that compute commute distances from standard weighted kNN graphs [5,6] we outperform them for any choice of m, since their performace degrades very fast with m due to the intrinsic inter-class noise arising in realistic databases.

Finally, we reconcile our results, and those of the anchor graphs with the von Luxburg and Radl's fundamental bounds. In principle, commute distances cannot be properly estimated from large graphs [13]. However, in this paper we show that both anchor graphs and Dirichlet densifiers provide meaningful commute times. It is well known that this can be done insofar the spectral gap is close to zero or the minimal degree is close to the unit. Dirichlet densifiers provide spectral gaps close to zero (see Fig. 3-Right) for low fractions of leading edges, but the accuracy degrades linearly when the spectral gap increases. This means that the spectral gap is negatively correlated with increasing fractions of inter-class noise. This noise arises when the densification level increases since Dirichlet densifiers are not still able of confining densification to intra-class links. Concerning anchor graphs, their spectral gap is close to the unit since the degree also the unit (double-stochastic matrices) and they outperform Dirichlet densifiers to some extent at the cost of computing anchors and finding the best number of them.

To conclude, we have contributed with a novel method for transforming input graphs into denser versions which are more suitable for estimating meaningful commute distances in large graphs.

References

1. Alamgir, M., von Luxburg, U.: Shortest path distance in random k-nearest neighbor graphs. In: Proceedings of the 29th International Conference on Machine Learning, ICML 2012, Edinburgh, Scotland, UK, 26 June–1 July 2012 (2012)
2. Cai, D., Chen, X.: Large scale spectral clustering via landmark-based sparse representation. IEEE Trans. Cybern. **45**(8), 1669–1680 (2015)
3. Grady, L.: Random walks for image segmentation. IEEE Trans. Pattern Anal. Mach. Intell. **28**(11), 1768–1783 (2006)
4. Hardt, M., Srivastava, N., Tulsiani, M.: Graph densification. In: Innovations in Theoretical Computer Science 2012, Cambridge, MA, USA, 8–10 January 2012, pp. 380–392 (2012)
5. Khoa, N.L.D., Chawla, S.: Large scale spectral clustering using approximate commute time embedding. CoRR abs/1111.4541 (2011)
6. Khoa, N.L.D., Chawla, S.: Large scale spectral clustering using resistance distance and spielman-teng solvers. In: Ganascia, J.-G., Lenca, P., Petit, J.-M. (eds.) DS 2012. LNCS (LNAI), vol. 7569, pp. 7–21. Springer, Heidelberg (2012). doi:10.1007/978-3-642-33492-4_4
7. Liu, W., He, J., Chang, S.: Large graph construction for scalable semi-supervised learning. In: Proceedings of the 27th International Conference on Machine Learning (ICML 2010), 21–24 June 2010, Haifa, Israel, pp. 679–686 (2010)

8. Liu, W., Wang, J., Chang, S.: Robust and scalable graph-based semisupervised learning. Proc. IEEE **100**(9), 2624–2638 (2012)
9. Liu, W., Wang, J., Kumar, S., Chang, S.: Hashing with graphs. In: Proceedings of the 28th International Conference on Machine Learning, ICML 2011, Bellevue, Washington, USA, 28 June–2 July 2011, pp. 1–8 (2011)
10. Luo, Z.Q., Ma, W.K., So, A.M.C., Ye, Y., Zhang, S.: Semidefinite relaxation of quadratic optimization problems. IEEE Sig. Process. Mag. **27**(3), 20–34 (2010)
11. Qiu, H., Hancock, E.R.: Clustering and embedding using commute times. IEEE Trans. Pattern Anal. Mach. Intell. **29**(11), 1873–1890 (2007)
12. Spielman, D.A., Srivastava, N.: Graph sparsification by effective resistances. SIAM J. Comput. **40**(6), 1913–1926 (2011)
13. von Luxburg, U., Radl, A., Hein, M.: Hitting and commute times in large random neighborhood graphs. J. Mach. Learn. Res. **15**(1), 1751–1798 (2014)
14. Weiss, Y., Torralba, A., Fergus, R.: Spectral hashing. In: Advances in Neural Information Processing Systems 21, Proceedings of the Twenty-Second Annual Conference on Neural Information Processing Systems, Vancouver, British Columbia, Canada, 8–11 December 2008, pp. 1753–1760 (2008)

A Jensen-Shannon Divergence Kernel for Directed Graphs

Cheng Ye$^{(\boxtimes)}$, Richard C. Wilson, and Edwin R. Hancock

Department of Computer Science, University of York, York YO10 5GH, UK
{cy666,richard.wilson,edwin.hancock}@york.ac.uk

Abstract. Recently, kernel methods have been widely employed to solve machine learning problems such as classification and clustering. Although there are many existing graph kernel methods for comparing patterns represented by undirected graphs, the corresponding methods for directed structures are less developed. In this paper, to fill this gap in the literature we exploit the graph kernels and graph complexity measures, and present an information theoretic kernel method for assessing the similarity between a pair of directed graphs. In particular, we show how the Jensen-Shannon divergence, which is a mutual information measure that gauges the difference between probability distributions, together with the recently developed directed graph von Neumann entropy, can be used to compute the graph kernel. In the experiments, we show that our kernel method provides an efficient tool for classifying directed graphs with different structures and analyzing real-world complex data.

Keywords: Information theoretic kernel · Jensen-Shannon divergence · von Neumann entropy

1 Introduction

Graph kernels have recently evolved into a rapidly developing branch of pattern recognition. Broadly speaking, there are two main advantages of kernel methods, namely (a) they can bypass the need for constructing an explicit high-dimensional feature space when dealing with high-dimensional data and (b) they allow standard machine learning techniques to be applied to complex data, which bridges the gap between structural and statistical pattern recognition [1]. However, although there are a great number of kernel methods aimed at quantifying the similarity between structures underpinned by undirected graphs, there is very little work on solving the corresponding problems for directed graphs. This is unfortunate since many real-world complex systems such as the citation networks, communications networks, neural networks and financial networks give rise to structures that are directed [2].

Motivated by the need to fill this gap in literature, in this paper we propose an information theoretic kernel method for measuring the structural similarity

© Springer International Publishing AG 2016
A. Robles-Kelly et al. (Eds.): S+SSPR 2016, LNCS 10029, pp. 196–206, 2016.
DOI: 10.1007/978-3-319-49055-7_18

between a pair of directed graphs. Specifically, our goal is to develop a symmetric and positive definite function that maps two directed patterns to a real value, by exploiting the information theoretic kernels that are related to the Jensen-Shannon divergence and a recently developed directed graph structural complexity measure, namely the von Neumann entropy.

1.1 Related Literature

Recently, information theoretic kernels on probability distributions have attracted a great deal of attention and have been extensively employed to the domain of classification of structured data [3]. Martins et al. [4] have generalized the recent research advances, and have proposed a new family of nonextensive information theoretic kernels on probability distributions. Such kernels have proved to have a strong link with the Jensen-Shannon divergence, which measures the mutual information between two probability distributions, and which is quantified by gauging the difference between entropies associated with those probability distributions [1].

Graph kernels, on the other hand, particularly aim to assess the similarity between pairs of graphs by computing an inner product on graphs [5]. For example, the classical random walk kernel [6] compares two labeled graphs by counting the number of paths produced by random walks on those graphs. However, random walk kernel has a serious disadvantage, namely "tottering" [7], caused by the possibility that a random walk can visit the same cycle of vertices repeatedly on a graph. An effective technique to avoid tottering is the backtrackless walk kernel developed by Aziz et al. [8], who have explored a set of new graph characterizations based on backtrackless walks and prime cycles.

Turning attention to the directed graph complexity measures, Ye et al. [2] have developed a novel entropy for quantifying the structural complexity of directed networks. Moreover, Berwanger et al. [9] have proposed a new parameter for the complexity of infinite directed graphs by measuring the extent to which cycles in graphs are intertwined. Recently, Escolano et al. [10] have extended the heat diffusion-thermodynamic depth approach from undirected networks to directed networks and have obtained a means of quantifying the complexity of structural patterns encoded by directed graphs.

1.2 Outline

The remainder of the paper is structured as follows. Section 2 details the development of an information theoretic kernel for quantifying the structural similarity between a pair of directed graphs. In particular, we show the graph kernel can be computed in terms of the Jensen-Shannon divergence between the von Neumann entropy of the disjoint union of two graphs, and the entropies of individual graphs. In Sect. 3, we show the effectiveness of our method by exploring its experimental performance on both artificial and realistic directed network data. Finally, Sect. 4 summarizes our contribution present in this paper and points out future research directions.

2 Jensen-Shannon Divergence Kernel for Directed Graphs

In this section, we start from the Jensen-Shannon divergence and explore how this similarity measure can be used to construct a graph kernel method for two directed graphs. In particular, the kernel can be computed in terms of the entropies of two individual graphs and the entropy of their composite graph. We then introduce the recently developed von Neumann entropy which quantifies the structural complexity of directed graphs. This in turn allows us to develop the mathematical expression for the Jensen-Shannon divergence kernel.

2.1 Jensen-Shannon Divergence

Generally speaking, the Jensen-Shannon divergence is a mutual information measure for assessing the similarity between two probability distributions. Mathematically, given a set $M_+^1(X)$ of probability distributions in which X indicates a set provided with some σ-algebra of measurable subsets. Then, the Jensen-Shannon divergence $D_{JS} : M_+^1(X) \times M_+^1(X) \rightarrow [0, +\infty)$ between probability distributions P and Q is computed as:

$$D_{JS}(P,Q) = \frac{1}{2}D_{KL}(P\|M) + \frac{1}{2}D_{KL}(M\|Q), \tag{1}$$

where D_{KL} denotes the classical Kullback-Leibler divergence and $M = \frac{P+Q}{2}$. Specifically, let π_p and π_q ($\pi_p + \pi_q = 1$ and $\pi_p, \pi_q \geq 0$) be the weights of the probability distributions P and Q respectively, then the generalization of the Jensen-Shannon divergence can be defined as:

$$D_{JS}(P,Q) = H_S(\pi_p P + \pi_q Q) - \pi_p H_S(P) - \pi_q H_S(Q), \tag{2}$$

where H_S is used to denote the Shannon entropy of the probability distribution.

To proceed, we introduce the Jensen-Shannon kernel method for undirected graphs developed by Bai and Hancock [1]. Given a pair of undirected graphs $\mathcal{G}_1 = (\mathcal{V}_1, \mathcal{E}_1)$ and $\mathcal{G}_2 = (\mathcal{V}_2, \mathcal{E}_2)$ where $\mathcal{V}$ represents the edge set and $\mathcal{E}$ denotes the vertex set of the graph, the Jensen-Shannon divergence between these two graphs is expressed by:

$$D_{JS}(\mathcal{G}_1, \mathcal{G}_2) = H(\mathcal{G}_1 \oplus \mathcal{G}_2) - \frac{H(\mathcal{G}_1) + H(\mathcal{G}_2)}{2}, \tag{3}$$

where $\mathcal{G}_1 \oplus \mathcal{G}_2$ denotes the disjoint union graph of graphs $\mathcal{G}_1$ and $\mathcal{G}_2$ and $H(\cdots)$ is the graph entropy. With this definition to hand and making use of the von Neumann entropy for undirected graphs [11], the Jensen-Shannon diffusion kernel $k_{JS} : \mathcal{G} \times \mathcal{G} \rightarrow [0, +\infty)$ is then

$$\begin{aligned} k_{JS}(\mathcal{G}_1, \mathcal{G}_2) &= \exp\{-\lambda D_{JS}(\mathcal{G}_1, \mathcal{G}_2)\} \\ &= \exp\left\{\lambda \frac{H_{VN}(\mathcal{G}_1) + H_{VN}(\mathcal{G}_2)}{2} - \lambda H_{VN}(\mathcal{G}_1 \oplus \mathcal{G}_2)\right\}, \end{aligned} \tag{4}$$

where H_{VN} is the undirected graph von Neumann entropy, and λ is a decay factor $0 < \lambda \le 1$. Clearly, the diffusion kernel is exclusively dependent on the individual entropies of the two graphs being compared as the union graph entropy can also be expressed in terms of those entropies.

2.2 von Neumann Entropy for Directed Graphs

Given a directed graph $\mathcal{G} = (\mathcal{V}, \mathcal{E})$ consisting of a vertex set $\mathcal{V}$ together with an edge set $\mathcal{E} \subseteq \mathcal{V} \times \mathcal{V}$, its adjacency matrix A is defined as

$$A_{uv} = \begin{cases} 1 & \text{if } (u,v) \in \mathcal{E} \\ 0 & \text{otherwise.} \end{cases}$$

The in-degree and out-degree at vertex u are respectively given as

$$d_u^{in} = \sum_{v \in \mathcal{V}} A_{vu}, \; d_u^{out} = \sum_{v \in \mathcal{V}} A_{uv}.$$

The transition matrix P of a graph is a matrix describing the transitions of a Markov chain on the graph. On a directed graph, P is given as

$$P_{uv} = \begin{cases} \frac{1}{d_u^{out}} & \text{if } (u,v) \in \mathcal{E} \\ 0 & \text{otherwise.} \end{cases}$$

As stated in [12], the normalized Laplacian matrix of a directed graph can be defined as

$$\tilde{L} = I - \frac{\Phi^{1/2} P \Phi^{-1/2} + \Phi^{-1/2} P^T \Phi^{1/2}}{2}, \tag{5}$$

where I is the identify matrix and $\Phi = diag(\phi(1), \phi(2), \cdots)$ in which ϕ represents the unique left eigenvector of P. Clearly, the normalized Laplacian matrix is Hermitian, i.e., $\tilde{L} = \tilde{L}^T$ where $\tilde{L}^T$ denotes the conjugated transpose of $\tilde{L}$.

Ye et al. [2] have shown that using the normalized Laplacian matrix to interpret the density matrix, the von Neumann entropy of directed graphs is the Shannon entropy associated with the normalized Laplacian eigenvalues, i.e.,

$$H_{VN}^D = -\sum_{i=1}^{|\mathcal{V}|} \frac{\tilde{\lambda}_i}{|\mathcal{V}|} \ln \frac{\tilde{\lambda}_i}{|\mathcal{V}|}, \tag{6}$$

where $\tilde{\lambda}_i, i = 1, \cdots, |\mathcal{V}|$ are the eigenvalues of the normalized Laplacian matrix $\tilde{L}$. Unfortunately, for large graphs this is not a viable proposition since the time required to solve the eigensystem is cubic in the number of vertices. To overcome this problem we extend the analysis of Han et al. [11] from undirected to directed graphs. To do this we again make use of the quadratic approximation to the Shannon entropy in order to obtain a simplified expression for the von Neumann entropy of a directed graph, which can be computed in a time that is quadratic in the number of vertices. Our starting point is the quadratic approximation to the von Neumann entropy in terms of the traces of normalized Laplacian and the squared normalized Laplacian

$$H_{VN}^D = \frac{Tr[\tilde{L}]}{|\mathcal{V}|} - \frac{Tr[\tilde{L}^2]}{|\mathcal{V}|^2}. \tag{7}$$

To simplify this expression a step further, we show the computation of the traces for the case of a directed graph. This is not a straightforward task, and requires that we distinguish between the in-degree and out-degree of vertices. We first consider Chung's expression for the normalized Laplacian of directed graphs and write

$$Tr[\tilde{L}] = Tr[I - \frac{\Phi^{1/2}P\Phi^{-1/2} + \Phi^{-1/2}P^T\Phi^{1/2}}{2}]$$

$$= Tr[I] - \frac{1}{2}Tr[\Phi^{1/2}P\Phi^{-1/2}] - \frac{1}{2}Tr[\Phi^{-1/2}P^T\Phi^{1/2}].$$

Since the matrix trace is invariant under cyclic permutations, we have

$$Tr[\tilde{L}] = Tr[I] - \frac{1}{2}Tr[P\Phi^{-1/2}\Phi^{1/2}] - \frac{1}{2}Tr[P^T\Phi^{1/2}\Phi^{-1/2}]$$

$$= Tr[I] - \frac{1}{2}Tr[P] - \frac{1}{2}Tr[P^T].$$

The diagonal elements of the transition matrix P are all zeros, hence we obtain

$$Tr[\tilde{L}] = Tr[I] = |\mathcal{V}|,$$

which is exactly the same as in the case of undirected graphs.

Next we turn our attention to $Tr[\tilde{L}^2]$:

$$Tr[\tilde{L}^2] = Tr[I^2 - (\Phi^{1/2}P\Phi^{-1/2} + \Phi^{-1/2}P^T\Phi^{1/2}) + \frac{1}{4}(\Phi^{1/2}P\Phi^{-1/2}\Phi^{1/2}P\Phi^{-1/2}$$

$$+\Phi^{1/2}P\Phi^{-1/2}\Phi^{-1/2}P^T\Phi^{1/2} + \Phi^{-1/2}P^T\Phi^{1/2}\Phi^{1/2}P\Phi^{-1/2}$$

$$+\Phi^{-1/2}P^T\Phi^{1/2}\Phi^{-1/2}P^T\Phi^{1/2})]$$

$$= Tr[I^2] - Tr[P] - Tr[P^T] + \frac{1}{4}(Tr[P^2] + Tr[P\Phi^{-1}P^T\Phi]$$

$$+Tr[P^T\Phi P\Phi^{-1}] + Tr[P^{T^2}])$$

$$= |\mathcal{V}| + \frac{1}{2}(Tr[P^2] + Tr[P\Phi^{-1}P^T\Phi]),$$

which is different to the result obtained in the case of undirected graphs.

To continue the development we first partition the edge set $\mathcal{E}$ of the graph $\mathcal{G}$ into two disjoint subsets $\mathcal{E}_1$ and $\mathcal{E}_2$, where $\mathcal{E}_1 = \{(u,v)|(u,v) \in \mathcal{E}$ and $(v,u) \notin \mathcal{E}\}$, $\mathcal{E}_2 = \{(u,v)|(u,v) \in \mathcal{E}$ and $(v,u) \in \mathcal{E}\}$ that satisfy the conditions $\mathcal{E}_1 \bigcup \mathcal{E}_2 = \mathcal{E}$, $\mathcal{E}_1 \bigcap \mathcal{E}_2 = \emptyset$. Then according to the definition of the transition matrix, we find

$$Tr[P^2] = \sum_{u \in \mathcal{V}} \sum_{v \in \mathcal{V}} P_{uv}P_{vu} = \sum_{(u,v) \in \mathcal{E}_2} \frac{1}{d_u^{out}d_v^{out}}.$$

Using the fact that $\Phi = diag(\phi(1), (2), \cdots)$ we have

$$Tr[P\Phi^{-1}P^T\Phi] = \sum_{u \in \mathcal{V}} \sum_{v \in \mathcal{V}} P_{uv}^2 \frac{\phi(u)}{\phi(v)} = \sum_{(u,v) \in \mathcal{E}} \frac{\phi(u)}{\phi(v)d_u^{out^2}}.$$

Then, we can approximate the von Neumann entropy of a directed graph in terms of the in-degree and out-degree of the vertices as follows

$$H_{VN}^D = 1 - \frac{1}{|\mathcal{V}|} - \frac{1}{2|\mathcal{V}|^2}\left\{ \sum_{(u,v) \in \mathcal{E}} \left(\frac{1}{d_u^{out}d_v^{out}} + \frac{d_u^{in}}{d_v^{in}d_u^{out^2}} \right) - \sum_{(u,v) \in \mathcal{E}_1} \frac{1}{d_u^{out}d_v^{out}} \right\} \quad (8)$$

or, equivalently,

$$H_{VN}^D = 1 - \frac{1}{|\mathcal{V}|} - \frac{1}{2|\mathcal{V}|^2}\left\{ \sum_{(u,v) \in \mathcal{E}} \frac{d_u^{in}}{d_v^{in}d_u^{out^2}} + \sum_{(u,v) \in \mathcal{E}_2} \frac{1}{d_u^{out}d_v^{out}} \right\}. \quad (9)$$

Clearly, both entropy approximations contain two terms. The first is the graph size while the second one depends on the in-degree and out-degree statistics of each pair of vertices connected by an edge. Moreover, the computational complexity of these expressions is quadratic in the graph size.

2.3 Jensen-Shannon Divergence Kernel for Directed Graphs

It is interesting to note that the von Neumann entropy is related to the sum of the entropy contribution from each directed edge, and this allows us to compute a local entropy measure for edge $(u, v) \in \mathcal{E}$,

$$I_{uv} = \frac{d_u^{in}}{2|\mathcal{E}||\mathcal{V}|d_v^{in}d_u^{out^2}}.$$

If this edge is bidirectional, i.e., $(u, v) \in \mathcal{E}_2$, then we add an additional quantity to I_{uv},

$$I'_{uv} = \frac{1}{2|\mathcal{E}_2||\mathcal{V}|d_u^{out}d_v^{out}}.$$

Clearly, the local entropy measure represents the entropy associated with each directed edge and more importantly, it avoids the bias caused by graph size, which means that it is the edge entropy contribution determined by the in and out-degree statistics, and neither the vertex number nor edge number of the graph that distinguishes a directed edge.

In our analysis, since we are measuring the similarity between the structures of two graphs, the vertex and edge information plays a significant role in the comparison. As a result, we use the following measure for quantifying the complexity of a directed graph, which is affected by the vertex and edge information,

$$H_E(\mathcal{G}) = \sum_{(u,v) \in \mathcal{E}} \frac{d_u^{in}}{d_v^{in}d_u^{out^2}} + \sum_{(u,v) \in \mathcal{E}_2} \frac{1}{d_u^{out}d_v^{out}}. \quad (10)$$

We let $\mathcal{G}_1 \oplus \mathcal{G}_2 = \{\mathcal{V}_1 \bigcup \mathcal{V}_2, \mathcal{E}_1 \bigcup \mathcal{E}_2\}$, i.e., the disjoint union graph (or composite graph) of $\mathcal{G}_1$ and $\mathcal{G}_2$. Then, we compute the Jensen-Shannon divergence between $\mathcal{G}_1$ and $\mathcal{G}_2$ as

$$D_{JS}(\mathcal{G}_1, \mathcal{G}_2) = H_E(\mathcal{G}_1 \oplus \mathcal{G}_2) - \frac{H_E(\mathcal{G}_1) + H_E(\mathcal{G}_2)}{2}.$$

As a consequence, we finally have the Jensen-Shannon diffusion kernel

$$k_{JS}(\mathcal{G}_1, \mathcal{G}_2) = \exp\left\{\frac{H_E(\mathcal{G}_1) + H_E(\mathcal{G}_2)}{2} - H_E(\mathcal{G}_1 \oplus \mathcal{G}_2)\right\}. \tag{11}$$

The decay factor λ is set to be 1 for simplifying matters. Clearly, since Jensen-Shannon divergence is a dissimilarity measure and is symmetric, the diffusion kernel associated with the divergence is positive definite. Moreover, for a pair of graphs with N vertices, the computational complexity of the kernel is $O(N^2)$.

3   Experiments

In this section, we evaluate the experimental performance of the proposed directed graph Jensen-Shannon divergence kernel. Specifically, we first explore the graph classification performance of our method on a set of random graphs generated from different models. Then we apply our method to the real-world data, namely the NYSE stock market networks, in order to explore whether the kernel method can be used to analyze the complex data effectively.

3.1   Datasets

We commence by giving a brief overview of the datasets used for experiments in this paper. We use two different datasets, the first one is synthetically generated artificial graphs, while the other one is extracted from real-world financial system.

Random Directed Graph Dataset. Contains a large number of directed graphs which are randomly generated according to two different directed random graph models, namely (a) the classical Erdős-Rényi model and (b) the Barabási-Albert model [13]. The different directed graphs in the database are created using a variety of model parameters, e.g., the graph size and the connection probability in the Erdős-Rényi model and the number of added connections at each time step in the Barabási-Albert model.

NYSE Stock Market Network Dataset. Is extracted from a database consisting of the daily prices of 3799 stocks traded on the New York Stock Exchange (NYSE). In our analysis we employ the correlation-based network to represent the structure of the stock market since many meaningful economic insights can be extracted from the stock correlation matrices [14]. To construct the dynamic network, 347 stocks that have historical data from January 1986 to February 2011 are selected. Then, we use a time window of 20 days (i.e., 4 weeks) and move this window along time to obtain a sequence (from day 20 to day 6004) in which each temporal window contains a time-series of the daily return stock values over a 20-day period. We represent trades between different stocks as a network. For each time window, we compute the cross-correlation coefficients between the time-series for each pair of stocks, and create connections between them if the maximum absolute value of the correlation coefficient is among the highest 5 % of the total cross correlation coefficients. This yields a time-varying stock market network with a fixed number of 347 vertices and varying edge structure for each of 5976 trading days.

3.2   Graph Classification

To investigate the classification performance of our proposed kernel method, we first apply it to the *Random Directed Graph Dataset*. Given a dataset consisting of N graphs, we construct a $N \times N$ kernel matrix

$$K = \begin{pmatrix} k_{11} & k_{12} & \cdots & k_{1N} \\ k_{21} & k_{22} & \cdots & k_{2N} \\ \vdots & \vdots & \ddots & \vdots \\ k_{N1} & k_{N2} & \cdots & k_{NN} \end{pmatrix}$$

in which k_{ij}, $i,j = 1, 2, \cdots, N$, denotes the graph kernel value between graphs $\mathcal{G}_i$ and $\mathcal{G}_j$. In our case, we have $k_{ij} = k_{JS}(\mathcal{G}_i, \mathcal{G}_j)$. With the kernel matrix to hand, we then perform the kernel principal component analysis (kernel PCA) [15] in order to extract the most important information contained in the matrix and embed the data to a low-dimensional principal component space.

Figure 1 gives two kernel PCA plots of kernel matrix computed from (a) Jensen-Shannon divergence kernel and (b) shortest-path kernel [7]. In particular, the kernels are computed from 600 graphs that belong to six groups (100 graphs in each group): (a) Erdős-Rényi (ER) graphs with $n = 30$ vertices and connection probability $p = 0.3$; (b) Barabási-Albert (BA) graphs with $n = 30$ vertices and average degree $\bar{k} = 9$; (c) ER graphs with $n = 100$ and $p = 0.3$; (d) BA graphs with $n = 100$ and $\bar{k} = 30$; (e) ER graphs with $n = 30$ and $p = 0.6$; and (f) BA graphs with $n = 30$ and $\bar{k} = 18$. From the left-hand panel, the six groups are clearly separated very well, implying that the Jensen-Shannon divergence kernel is effective in comparing not only the structural properties of directed graphs that belong to different classes, but also the structure difference between graphs generated from the same model. However, the panel on the right-hand side suggests that the shortest-path kernel cannot efficiently distinguish the difference between graphs of the six groups since the graphs cannot be clearly separated.

To better evaluate the properties of the proposed kernel method. We proceed to compare the classification result of the Jensen-Shannon divergence kernels developed for directed and undirected graphs. To this end, we employ the undirected Jensen-Shannon divergence kernel and repeat the above analysis on

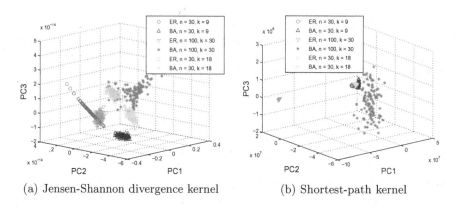

(a) Jensen-Shannon divergence kernel   (b) Shortest-path kernel

Fig. 1. Kernel PCA plots of two graph kernel methods on a number of directed graphs generated from Erdős-Rényi model and Barabási-Albert model.

the undirected version of graphs in *Random Directed Graph Dataset* and report the result in Fig. 2. The undirected kernel method does not separate different groups of graphs well as there are a large number of data points overlapping in the kernel principal component space. Moreover, the BA graphs represented by yellow stars and the ER graphs symbolled by cyan squares ($n = 30$ and $\bar{k} = 18$) are completely mixed. This comparison clearly indicates that for directed and undirected random graphs that are generated using the same model and parameters, the directed graph kernel method is more effective in classifying them than its undirected analogue.

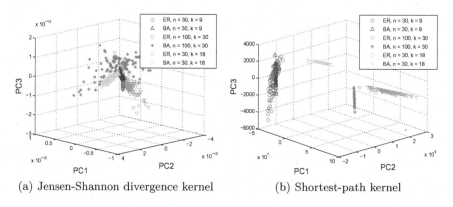

(a) Jensen-Shannon divergence kernel   (b) Shortest-path kernel

Fig. 2. Kernel PCA plots of two graph kernel methods on a number of undirected graphs generated from Erdős-Rényi model and Barabási-Albert model.

3.3   Financial Data Analysis

We turn our attention to the financial data, and explore whether the proposed kernel method can be used as an efficient tool for visualizing and studying the time evolution of the financial networks. To commence, we select two financial crisis time series from the data, namely (a) Black Monday (from day 80 to day 120) and (b) September 11 attacks (from day 3600 to day 3650) and construct their corresponding kernel matrices respectively. Then, in Fig. 3 we show the path of the financial network in the PCA space during the selected financial crises respectively. The number beside each data point represents the day number in the time-series. From the left-hand panel we observe that before Black Monday, the network structure remains relatively stable. However, during Black Monday (day 117), the network experiences a considerable change in structure since the graph Jensen-Shannon divergence kernel changes significantly. After the crisis, the network structure returns to its normal state. A similar observation can also be made concerning the September 11 attacks which is shown in the right-hand panel. The stock network again undergoes a significant crash in which the network structure undergoes a significant change. The crash is also followed by a quick recovery.

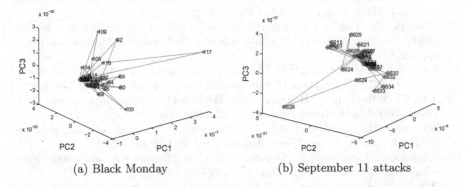

(a) Black Monday   (b) September 11 attacks

Fig. 3. Path of the time-evolving financial network in the PCA space during different financial crises.

4   Conclusion

In this paper, we present a novel kernel method for comparing the structures of a pair of directed graphs. In essence, this kernel is based on the Jensen-Shannon divergence between two probability distributions associated with graphs, which are represented by the recently developed von Neumann entropy for directed graphs. Specifically, the entropy approximation formula allows the entropy to be expressed in terms of a sum over all edge-based entropies of a directed graph, which in turn allows the kernel to be computed from the edge entropy sum of two individual graphs and their disjoint union graph. In the experiments, we have evaluated the effectiveness of our kernel method in terms of classifying structures that belong to various classes and analyzing time-varying realistic networks. In the future, it would be interesting to see what features the Jensen-Shannon divergence kernel reveal in additional domains, such as human functional magnetic resonance imaging data. Another interesting line of investigation would be to explore if the kernel method can be extended to the domains of edge-weighted graphs, labeled graphs and hypergraphs.

References

1. Bai, L., Hancock, E.R.: Graph kernels from the Jensen-Shannon divergence. J. Math. Imaging Vis. **47**(1–2), 60–69 (2013)
2. Ye, C., Wilson, R.C., Comin, C.H., Costa, L.D.F., Hancock, E.R.: Approximate von Neumann entropy for directed graphs. Phys. Rev. E **89**, 052804 (2014)
3. Cuturi, M., Fukumizu, K., Vert, J.P.: Semigroup kernels on measures. J. Mach. Learn. Res. **6**, 1169–1198 (2005)
4. Martins, A.F.T., Smith, N.A., Xing, E.P., Aguiar, P.M.Q., Figueiredo, M.A.T.: Nonextensive information theoretic kernels on measures. J. Mach. Learn. Res. **10**, 935–975 (2009)
5. Vishwanathan, S.V.N., Schraudolph, N.N., Kondor, R., Borgwardt, K.M.: Graph kernels. J. Mach. Learn. Res. **11**, 1201–1242 (2010)
6. Kashima, H., Tsuda, K., Inokuchi, A.: Marginalized kernels between labeled graphs. In: Proceedings in 20th International Conference on Machine Learning, pp. 321–328(2003)
7. Borgwardt, K.M., Kriegel, H.P.: Shortest-path kernels on graphs. In: Proceedings in the 5th IEEE International Conference on Data Mining, pp. 74–81 (2005)
8. Aziz, F., Wilson, R.C., Hancock, E.R.: Backtrackless walks on a graph. IEEE Trans. Neural Netw. Learn. Syst. **24**(6), 977–989 (2013)
9. Berwanger, D., Gradel, E., Kaiser, L., Rabinovich, R.: Entanglement and the complexity of directed graphs. Theor. Comput. Sci. **463**, 2–25 (2012)
10. Escolano, F., Hancock, E.R., Lozano, M.A.: Heat diffusion: thermodynamic depth complexity of networks. Phys. Rev. E **85**, 036206 (2012)
11. Han, L., Escolano, F., Hancock, E.R., Wilson, R.C.: Graph characterizations from von Neumann entropy. Pattern Recogn. Lett. **33**, 1958–1967 (2012)
12. Chung, F.: Laplacians and the Cheeger inequailty for directed graphs. Ann. Comb. **9**, 1–19 (2005)
13. Barabási, A.L., Albert, R.: Emergence of scaling in random networks. Science **286**, 509–512 (1999)
14. Battiston, S., Caldarelli, G.: Systemic risk in financial networks. J. Financ. Manag. Mark. Inst. **1**, 129–154 (2013)
15. Schölkopf, B., Smola, A., Müller, K.B.: Nonlinear component analysis as a kernel eigenvalue problem. Neural Comput. **10**, 1299–1319 (1998)

XNN Graph

Pasi Fränti[1(✉)], Radu Mariescu-Istodor[1], and Caiming Zhong[2]

[1] University of Eastern Finland, Joensuu, Finland
franti@cs.uef.fi
[2] Ningbo University, Ningbo, China

Abstract. K-nearest neighbor graph (KNN) is a widely used tool in several pattern recognition applications but it has drawbacks. Firstly, the choice of k can have significant impact on the result because it has to be fixed beforehand, and it does not adapt to the local density of the neighborhood. Secondly, KNN does not guarantee connectivity of the graph. We introduce an alternative data structure called XNN, which has variable number of neighbors and guarantees connectivity. We demonstrate that the graph provides improvement over KNN in several applications including clustering, classification and data analysis.

Keywords: KNN · Neighborhood graph · Data modeling

1 Introduction

Neighborhood graphs are widely used in data mining, machine learning and computer vision to model the data. Few applications examples are listed below.

- KNN classifier
- Manifold learning
- 3D object matching
- Clustering
- Outlier detection
- Traveling salesman problem
- Word similarity in web mining

Two popular definitions are ε-*neighborhood* and k-*nearest neighbor* (KNN) graph [1]. In the first one, two points are neighbors if they are within a distance ε of each other. KNN neighborhood of a point is defined as its k nearest other data points in the data space. In the corresponding graph, all neighboring points are connected. The ε-neighborhood graph is undirected whereas KNN is directed graph.

The first problem, both in KNN and also ε-neighborhood, is how to select parameters ε and k. The larger the neighborhood, the better the local structures can be captured but the more complex the graph and the higher the processing time. An upper limit is when k equals to the size of the data ($k = N$), which leads to a complete graph. Having fixed value of k, there can be unnecessary long edges in sparse areas that do not capture anything essential about the local structure.

The second problem is that these definitions do not guarantee connectivity of the graph. In Fig. 1, an isolated component is created, which leads to wrong clustering

© Springer International Publishing AG 2016
A. Robles-Kelly et al. (Eds.): S+SSPR 2016, LNCS 10029, pp. 207–217, 2016.
DOI: 10.1007/978-3-319-49055-7_19

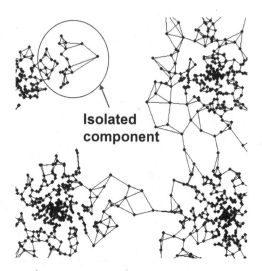

Fig. 1. Example part of 3NN graph where connectivity is broken.

result. With this data the problem can be solved by setting $k = 4$ at the cost of increased complexity. However, with higher dimensional data the value must usually be set to $k = 20$ or even higher. This overwhelms the computation.

In general, there is no good way to set k automatically, and it does not adapt to the local density. According to [2], k has to be chosen rather high, order of N rather than $\log N$. In [3] this neighborhood was reduced by eliminating edges to nodes that were reached via alternative shorter detour via a third node. But it does not solve the connectivity and requires large k to start with. Despite these drawbacks, KNN has become popular – mainly because lack of better alternatives.

In this paper, we introduce a new neighborhood graph called *XNN*. The key idea is to model the local structures in the same way as KNN, but instead of having a fixed value of k, the size of neighborhood is variable and depends on the data locally. In a dense area, there would be more edges than in sparse areas so that both the local structures are captured and the connectivity of the graph is guaranteed.

2   Neighborhood Graphs

Besides KNN, another commonly used structure is *minimum spanning tree* (MST), which guarantees connectivity. However, it is optimized for minimizing the sum of all distances and may lead to complicated chains that do not capture the real structure of the data. MST contains only one edge per data point, on average. This corresponds to 1NN-graph by the size. It is enough to keep the graph connected but not able to capture more complex local properties such as density, or estimate clustering structure.

Good heuristic was proposed in [4] to use $2 \times$ MST. Once the first MST is created, another MST is constructed from the remaining unused edges and the union of the two MSTs makes the neighborhood graph. Intuitively, kind of *second choices* are selected, similarly as the 2nd nearest in 2NN-graph. This doubles the total number of edges in the graph, and can significantly improve the graph. The idea generalizes to any number of k by repeating the MST algorithm k.

The main advantage of k-MST is that it guarantees connectivity. However, it is still possible that important connections are missed leading to wrong analysis of the data because the connectivity may come via long chains that do not capture the local structure of the data. A good neighborhood graph would have the benefit of MST to keep the graph connected, but at the same time, having more edges to capture the structures both in sparse and dense areas. The choice of the value k remains an open problem also with k-MST.

Another well known structure is so-called *Delaunay triangulation*. It is based on *Voronoi* diagram, which is directly related to the partition concept in clustering. Clusters that share the same Voronoi vertex are considered as neighbors. The benefit is that the entire data space will be partitioned and the resulting graph will be naturally connected.

Slightly more practical variant is so-called *Gabriel graph* [5]. It is an old invention but deserves much more attention that it has gained because of two properties shown in [6]: (1) it includes MST as sub graph, and thus guarantees connectivity; (2) it is sub graph of the Delaunay, shares most of its properties but more practical. Elliptical variant was introduced in [7], and an additional density criterion considered in [8].

Both Delaunay and Gabriel graphs suffers the problem that the number of neighbors becomes too high when dimension increases. Another problem is the high time complexity of the graph construction. Delaunay has exponential dependency on the dimensionality, $O(N^{d/2})$, whereas Gabriel graph takes $O(N^3)$. These problems have been studied but general solution is still an open problem [9, 10].

3   X-Nearest Neighborh Graph

The resulting graph should have the following properties:

- Number of neighbors should be decided automatically.
- It should be small, preferably close to a constant.
- The graph will be connected.
- Constructing the graph can be done efficiently.

We define XNN graph as *any sub-graph of Gabriel graph that retains connectivity*, i.e. there are no isolated sub-graphs. We consider three alternatives:

- Full XNN
- Hierarchically-built XNN
- k-limited XNN

We give next their definitions and study their properties with data size, cluster overlap and dimension. Some other design alternatives for this principal idea are also discussed.

3.1   Full XNN

Input is a set of N data points $\{x_1, x_2, ..., x_N\}$ each consisting of d attributes. The points are often in a metric space so that the distance between any two points can be calculated. However, the idea generalizes to similarity measures as well.

Gabriel graph (GG) is defined for points in Euclidean space as follows. Any two points a and b are neighbors if the following holds for all other points c:

$$ab^2 < ac^2 + bc^2 \tag{1}$$

where ab, ac and bc are the distances of the particular points. We use this definition as the basis for the Full XNN as well. It generalizes to other distance measures that are in the range of [0,1] since the only numerical operation is the squared function. In specific, if the data is in Euclidean space then Full XNN equals to GG.

Alternative rule is to calculate the middle point between a and b, which we denote as m. If there exists another point c that is nearer to m than a and b, then this point must be within the circle. Thus, points a and b are neighbors only if they satisfy the following condition:

$$ab < 2mc \tag{2}$$

where m is the middle point (average of a and b) and c is the nearest point to m. The full graph can be found by the brute force algorithm:

```
Algorithm 1: Full XNN(data set) → XNN
FOR i=1 to N-1 DO
 FOR j=i+1 to N DO
 Calculate midpoint m ← (xᵢ+ xⱼ)/2
 x ← Find nearest point for m
 IF x== xᵢ OR x== xⱼ THEN
   Mark xᵢ and xⱼ as neighbors.
```

This formulation is slightly simpler than (1) and no squaring is needed. However, it may not work for non-numerical data because it requires that we can calculate average of any two data points. For instance, taking average of strings "*science*" and "*engineering*" is not trivially defined.

If we have similarities instead of distances, we can convert them to distances by inversion: $d = 1/(s + \varepsilon)$ where $s \in [0,1]$ is a similarity and $\varepsilon = 0.001$ is a small constant to avoid infinite. This leads to the following rule:

$$ab^{-2} < ac^{-2} + bc^{-2} \Leftrightarrow \frac{ab^2 \cdot bc^2 + ab^2 \cdot ac^2}{ac^2 \cdot bc^2} > 1 \tag{3}$$

The main bottleneck of the full graph is that it takes $O(N^3)$ time to compute. Another drawback is that there will be excessive number of neighbors in higher dimensional data.

3.2   Properties

We study next the number of neighbors with the following parameters: data size, dimensionality and overlap. We use the G2 data sets shown in Fig. 2.

Fig. 2. Series of data sets G2-*D*-*O* where *D* is the dimensionality varying from 2 to 1024, and *O* is the overlap parameter varying from 0 to 100. Data size is N = 2048. http://cs.uef.fi/sipu/data/

Figure 3 shows that there is a mild tendency to have more neighbours when data has higher overlap (therefore more dense) but this effect happens only in higher dimensions. For 8-d data there is hardly any visible difference but for 32-d data the number of neighbours almost doubles when the clusters completely overlap compared to when they are separate.

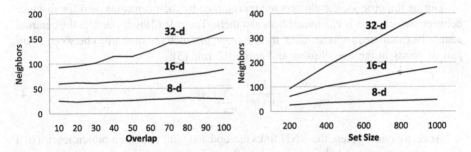

Fig. 3. Average number of neighbors for G2 with increasing overlap (left) and increasing data size (right). Results are for 10 % subsample (left) and overlap is 10 % (right)

The same effect happens when we vary the sub-sample size. For 32-d data, doubling the density (data size) doubles the size of the neighbourhood. The biggest effect, however, comes from the dimensionality, see Fig. 4. The number of neighbours is moderate up to 8-d but then increases quickly and reaches complete graph (1000 neighbours) at latest 256-d. If the clusters are well separated (30 % overlap), it becomes complete graph within the cluster (500 neighbours).

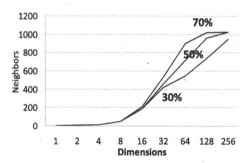

Fig. 4. Average number of neighbors with increasing dimension for G2. Three overlaps of 30 %, 50 % and 70 % were tested for 1 to 256 dimensions. Susample size of 1000 is used.

To sum up, for higher dimensional data we need to limit the size of neighbourhood to keep the graph useful.

3.3 Hierarchical XNN

Hierarchical variant is in Algorithm 2. The idea is to build the graph from clusters instead of data points. We start with only one cluster including all data points. New clusters are then iteratively created by splitting one existing cluster into two. This corresponds to divisive clustering algorithm. However, we do not stop at a certain number of clusters but continue the process until each point belongs to its own cluster. The splitting is done by taking two random points and by applying a few iterations of k–means within the cluster.

During the process, the clusters are represented by their centroids, and the distance between two clusters is calculated between them. The rule (2) is applied. If the centroid cannot be calculated from the data, then average distance (or similarity) between every pair of points in the two clusters and apply the rule (3):

$$\frac{1}{|a| \cdot |b|} \cdot \sum_{i \in a, j \in b} s_{ij} \tag{4}$$

At each dividing step, the XNN links are updated with two main principles. First, a link is always created between a and b. Second, the existing links to the previous cluster ab will be kept. These rules guarantee that if the graph was connected before the split, it will be connected after the split also. The first question is to which one the existing links should be linked to: a or b. We choose the nearest.

The second question is whether the neighbor should be connected to *both*. Consider the situation in Fig. 5. Greedy choice would be to choose only the nearest cluster (c-b, d-b). This would lead to a spanning tree: only one new link is created at every split, and there are N-1 splits in total. Result is a connected graph without redundant links, thus, spanning tree. The other extreme would be to accept both, and link all c-a, c-b, d-a d-b. However, this would lead to a complete graph, which is neither what we want.

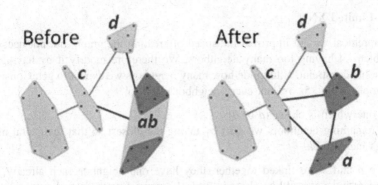

Fig. 5. Example of the hierarhical variant before and after the 4th split. Both *a* and *b* retain the connection to *c*, but only *b* retains the connection to *d*.

Our choice is to accept the second link if the neighbor rule applies. In principle, this can lead to Full XNN, but in practice, the split can have side-effects to other clusters, too. For example, new cluster may break the neighborhood of other clusters. Or vice versa, other clusters may interfere with the neighborhood of *a*, *b*, *c* and *d*. However, taking these side-effects into account would get the time complexity back to $O(N^3)$, which would be too much. We therefore choose to ignore these and settle with approximation. Some links of the Full XNN may therefore be missed, and some extra links may appear. But the key property remains: the graph is connected.

The split is implemented by local k-means. We select two random points in the cluster and iterate k-means 10 times within the cluster. This may not lead to optimal clustering but as it is merely a tool in the process it does not matter that much.

K-means requires $2 \cdot 10 \cdot n = O(n)$ steps. Since the cluster sizes are progressively decreasing this sums up to $O(N \cdot \log^2 N)$ according to [11]. Updating the neighbor links depends on the number of links (X), and on the number of clusters:

$$X \cdot \sum_{i=1}^{N} i = X \cdot N^2 \tag{5}$$

The overall complexity is $O(N \cdot \log^2 N) + O(XN^2)$. If we further limited the neighbor rule by considering neighbor of neighbors, it would lead to $O(N \cdot \log^2 N) + O(X^2 N^2)$.

Algorithm 2: Hierarchical XNN(data) → XNN
Put all in one cluster;
XNN ← ∅;
WHILE |XNN| < N DO
    ab ←SelectLargestCluster(C);
    a, b ← SplitCluster(ab);
    XNN ← XNN ∪ (a,b)
    UpdateXNN;

3.4   k-Limited XNN

The hierarchical variant improves the speed of creating the graph, but not necessarily the problem of having too many neighbors. We therefore modify it by having additional global constraint, k, to guide how many neighbors we expect to get. Considering the example in Fig. 5, we link every neighbor of ab to:

- The one which is closer (a or b)
- The remaining neighbors we start by taking the closest so that their total number won't increase $2k$.

Since a and b are linked together, they have one neighbor each already. Thus, $2(k-1)$ more links can still be chosen. Table 1 summarizes the actual numbers for the two variants. We can see that the Full XNN can have rather high number of neighbors with 16-dimensional data. The k-limited variant, on the other hand, keeps the size of neighbors always moderate.

Table 1. Average number of neighbors in the Full XNN and the k–limited (k = 10) variants. (Numbers in parentheses were run for smaller 10 % sub-sampled data)

Dataset	Dim	Full	Hierarchical	k-limited
Bridge	16	68.8	48	6.5
House	3	14.4	22	7.9
Miss America	16	345	94	6.8
Birch1	2	4.0	(3.4)	(3.4)
Birch2	2	(3.7)	(3.4)	(3.4)
Birch3	2	(3.9)	(3.4)	(3.3)
S1	2	3.8	3.4	3.3
S2	2	3.9	3.4	3.4
S3	2	3.9	3.4	3.4
S4	2	3.9	3.4	3.4

4   Applications

We next demonstrate the graph into three applications:

- Path-based clustering
- KNN classifier
- Traveling salesmen problem.

Path-based clustering [12] measures the distance of points in the dataset by finding the minimum path between two points. The cost of a path is defined as the maximum edge in the path. It can be calculated in Euclidean space or using neighborhood structure such as minimum spanning tree. *Robust Path-based clustering* method [13] uses KNN but it is an open question when to use small or large value of k. With some data the choice can be critical as shown in Fig. 6.

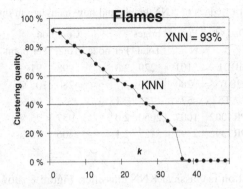

Fig. 6. Clustering quality of *Flames* by Path-based algorithm [13] measured by Pair set index [14].

Second test studied whether XNN has potential for solving traveling salesman problem. We analyzed known optimal solutions of several 2-d problem instances in *TSPLIB*. We counted how many times the same edges were in the optimal solution and also in the neighborhood graph constructed by KNN, k-MST and XNN. Results of KNN and k-MST were obtained by varying k, whereas XNN provides only one result. The aim would be to have as many correct edges in the graph as possible (to allow good optimization), but few other edges (to keep the complexity low). The results in Fig. 7 indicate that XNN has better precision/total ratio.

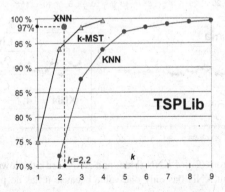

Fig. 7. The number of optimal edges captured by three graphs. The value of XNN is plotted at the location of the average number of neighbors ($k = 2.2$).

We have checked common edges between the optimal TSP and XNN on several datasets from TSPLIB, the result is: for some datasets, the total edges of XNN is kN, where k is a small number (about 2–3), but the kN edges contain the majority of the optimal TSP edges. This observation indicates that if we search the solution of TSP in XNN graph but not the complete graph, the efficiency could be high (Table 2).

Table 2. Number of edges of XNN graph, and how many are in optimal TSP path.

Dataset	Points	Edges		Common	
	N	Total	Per node	Total	Per point
eil101	101	229	2.3	98	97 %
a280	280	750	2.7	280	100 %
RAT575	575	1213	2.1	552	96 %
PR1002	1002	2060	2.0	957	96 %
PR2392	2392	5127	2.1	2306	96 %

The third application is classical KNN classifier. Figure 8 shows that the accuracy depends on k: with *Breast*, better results are obtained simply by increasing k, whereas with *Ionosphere* optimal result is between $k = 20$–25. In both cases, XNN classifier finds the same or better result without parameter tuning. Due to its simplicity, it could be worth further studies for example with distance weighted KNN [15].

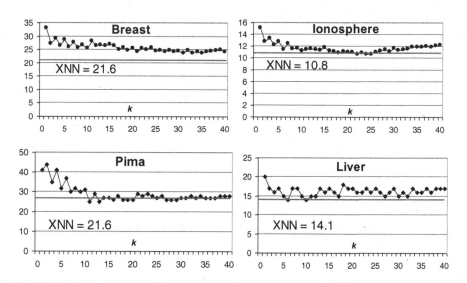

Fig. 8. Classification errors (%) of KNN (blue) and XNN (red); the lower the better. The result of KNN depends on k but XNN has only one value because it has no parameters.

5   Conclusion

New neighborhood graph called XNN is proposed. It is a compromise between fixed size nearest neighborhood graph of KNN, and a much more extensive spatial neighborhood of Gabriel graph. Its main advantages are that it is connected (no isolated components) and that the size of neighborhood is automatically selected for each point separately. Promising results were achieved in three example applications.

References

1. Cover, T., Hart, P.: Nearest neighbor pattern classification. IEEE Trans. Inf. Theory **13**(1), 21–27 (1967)
2. Maier, M., Hein, M., von Luxburg, U.: Optimal construction of kk-nearest-neighbor graphs for identifying noisy clusters. Theoret. Comput. Sci. **410**(19), 1749–1764 (2009)
3. Aoyama, K., Saito, K., Sawada, H., Ueda, N.: Fast approximate similarity search based on degree-reduced neighborhood graphs. In: ACM SIGKDD, San Diego, USA, pp. 1055–1063 (2011)
4. Yang, L.: Building k edge-disjoint spanning trees of minimum total length for isometric data embedding. IEEE Trans. Pattern Anal. Mach. Intell. **27**(10), 1680–1683 (2005)
5. Gabriel, K.R., Sokal, R.R.: New statistical approach to geographic variation analysis. Syst. Zool. **18**, 259–278 (1969)
6. Matula, D.W., Sokal, R.R.: Properties of Gabriel graphs relevant to geographic variation research and the clustering of points in the plane. Geogr. Anal. **12**(3), 205–222 (1980)
7. Park, J.C., Shin, H., Choi, B.K.: Elliptic gabriel graph for finding neighbors in a point set and its application to normal vector estimation. Comput. Aided Des. **38**, 619–626 (2006)
8. Inkaya, T., Kayaligil, S., Özdemirel, N.E.: An adaptive neighborhood construction algorithm based on density and connectivity. Pattern Recogn. Lett. **52**, 17–24 (2015)
9. Cignoni, P., Montani, C., Scopigno, R.: DeWall: a fast divide and conquer Delaunay triangulation algorithm in E^d. Comput. Aided Des. **30**(5), 333–341 (1998)
10. Rezafinddramana, O., Rayat, F., Venturin, G.: Incremental Delaunay triangulation construction for clustering. In: International Conference on Pattern Recognition, ICPR, Stockholm, Sweden, pp. 1354–1359 (2014)
11. Fränti, P., Kaukoranta, T., Nevalainen, O.: On the splitting method for VQ codebook generation. Opt. Eng. **36**(11), 3043–3051 (1997)
12. Fischer, B., Buhmann, J.M.: Path-based clustering for grouping of smooth curves and texture segmentation. IEEE Trans. Pattern Anal. Mach. Intell. **25**, 513–518 (2003)
13. Chang, H., Yeung, D.Y.: Robust path-based spectral clustering. Pattern Recogn. **41**(1), 191–203 (2008)
14. Rezaei, M., Fränti, P.: Set-matching methods for external cluster validity. IEEE Trans. Knowl. Data Eng. **28**(8), 2173–2186 (2016)
15. Gou, J., Du, L., Zhang, Y., Xiong, T.: A new distance-weighted k-nearest neighbor classifier. J. Inf. Comput. Sci. **9**(6), 1429–1436 (2012)

fMRI Activation Network Analysis Using Bose-Einstein Entropy

Jianjia Wang$^{(\boxtimes)}$, Richard C. Wilson, and Edwin R. Hancock

Department of Computer Science, University of York, York YO10 5DD, UK
jw1157@york.ac.uk

Abstract. In this paper, we present a novel method for characterizing networks using the entropy associated with bosonic particles in thermal equilibrium with a heat-bath. According to this analogy, the normalized Laplacian plays the role of Hamiltonian operator, and the associated energy states are populated according to Bose-Einstein statistics. This model is subject to thermal agitation by the heat reservoir. The physics of the system can be captured by using a partition function defined over the normalized Laplacian eigenvalues. Various global thermodynamic characterizations of the network including its entropy and energy then can be computed from the derivative of corresponding partition function with respect to temperature. We explore whether the resulting entropy can be used to construct an effective information theoretic graph-kernel for the purposes of classifying different types of graph or network structure. To this end, we construct a Jensen-Shannon kernel using the Bose-Einstein entropy for a sample of networks, and then apply kernel principle components analysis (kPCA) to map graphs into low dimensional feature space. We apply the resulting method to classify fMRI activation networks from patients with suspected Alzheimer disease.

Keywords: Bose-Einstein statistics · Network entropy · Jensen-Shannon divergence

1 Introduction

Graphs are powerful tools for representing complex patterns of interaction in high dimensional data [1]. For instance, in the application later considered in this paper, graphs provide the activation patterns in fMRI data, which can be indicative of the early onset of Alzheimer's disease. On the other hand, kernel-methods on graphs provide emerging and powerful set of tools to determine the class-structure of different graphs. There are many examples in the literature where graph kernels have successfully exploited topological information, and these include the heat diffusion kernel [2], the random walk kernel [3], and the shortest path kernel [4]. Once a graph kernel is to hand, it provides a convenient starting point from which machine learning techniques can be applied to learn potentially complex class-structure [5]. Despite the success of existing graph

© Springer International Publishing AG 2016
A. Robles-Kelly et al. (Eds.): S+SSPR 2016, LNCS 10029, pp. 218–228, 2016.
DOI: 10.1007/978-3-319-49055-7_20

kernels, one of the main challenges that remains open is to capture the variations present in different classes of graph in a probabilistic manner [6].

Recently, statistical mechanics and information theory have been used to understand more deeply variations in network structure. One of the successes here has been to use quantum spin statistics to describe the geometries of complex networks [7]. For example, using a physical analogy based on a Bose gas, the phenomenon of Bose-Einstein condensation has been applied to study the salient aspects network structure [8]. This has been extended to understand processes such as supersymmetry in networks [9]. Although these types of analogy are useful and provide powerful tools for network analysis, they are not easily accommodated into the kernel-based approach to machine learning.

The aim in this paper is to bridge this gap in the literature. Our aim is to develop a link between statistical mechanics and kernel methods. To do that we define information theoretic kernels in terms of network entropy. The Jensen-Shannon kernel is computed from the Jensen-Shannon divergence, which is an entropic measure of the similarity between two probability distributions. To compute the divergence, the distribution of data must be characterized in terms of a probability distribution and the associated entropy must be to hand. For graph or network structures, both the probability distribution and the associated entropy can be elusive. To solve this problem, in prior work, they have computed the required entropy using von Neumann entropy (essentially the Shannon entropy of the normalized Laplacian eigenvalues) [5]. Here we aim to explore whether the physical heat bath analogy and Bose-Einstein statistics can be used to furnish the required entropy, and implicitly the underlying probability distribution.

We proceed as follows. We commence from a physical analogy in which the normalized Laplacian plays the role as Hamiltonian (energy operator) and the normalized Laplacian eigenvalues are energy states. These states are occupied by bosonic particles (assumed the 0 spin) and the resulting system is in thermodynamic equilibrium with a heat-bath, which is characterized by temperature. The bosons are indistinguishable, and each energy level can accommodate an unlimited number of particles. The effect of the heat bath is to thermalise or randomise the population of energy levels. The occupation of the energy states is therefore governed by Bose-Einstein statistics, and can be characterized using an appropriate partition function. The quantities can be derived from the partition function over the energy states in the network when the system of particles is in thermodynamic equilibrium with the heat bath. From the partition function, we can compute the entropy of the system of particles, and hence compute the Jensen-Shannon kernel. Once the kernel matrix is to hand, we use kernel principle components analysis (kPCA) [10] to embed the graphs into a low dimensional feature space where classification is performed.

The remainder of paper is organised as follows. Section 2 briefly reviews the preliminaries of graph representation in the statistical mechanical domain. Sections 3 and 4 respectively describe the underpinning concepts of how entropy is computed from the Bose-Einstein partition function, and how the Jensen-Shannon kernel is constructed from the resulting entropy. Section 5 presents our experimental evaluation. Finally, Sect. 6 provides conclusions and directions for future work.

2   Graph in Quantum Representation

In this section, we give the preliminaries on the graph representation in quantum domain. We specify the density matrix as the normalized Laplacian matrix. We then introduce the idea of a Hamiltonian operator on a graph and its relationship with the normalized Laplacian, which associated von Neumann entropy.

2.1   Density Matrix

The density matrix, in quantum mechanics, is used to describe a system whose state is an ensemble of pure quantum states $|\psi_i\rangle$, each with probability p_i. It is defined as $\rho = \sum_{i=1}^{V} p_i |\psi_i\rangle\langle\psi_i|$.

With this notation, Passerini and Severini [11] have extended this idea to the graph domain. Specifically, they show that a density matrix for a graph or network can be obtained by scaling the normalized discrete Laplacian matrix by the reciprocal of the number of nodes in the graph.

When defined in this was way the density matrix is Hermitian i.e. $\rho = \rho^\dagger$ and $\rho \geq 0, \mathrm{Tr}\rho = 1$. It plays an important role in the quantum observation process, which can be used to calculate the expectation value of measurable quantity.

2.2   Hamiltonian Operator

In quantum mechanics, the Hamiltonian operator is the sum of the kinetic energy and potential energy of all the particles in the system, and it dictates the Schrödinger equation for the relevant system. The Hamiltonian is given by

$$\hat{H} = -\nabla^2 + U(r,t) \tag{1}$$

Here we specify the kinetic energy operator to be the negative of the normalized adjacency matrix, and the potential energy to be the identity matrix. Then, the normalized form of the graph Laplacian can be viewed as the Hamiltonian operator $\hat{H} = \tilde{L}$.

In this case, the eigenvalues of the Laplacian matrix can be viewed as the energy eigenstates, and these determine the Hamiltonian and hence the relevant Schrödinger equation which govern a system of particles. The graph as a thermodynamic system specified by N particles with energy states given by the network Hamiltonian and immersed in a heat bath at temperature T.

2.3   von Neumann Entropy

The interpretation of the scaled normalized Laplacian as a density matrix opens up the possibility of characterizing a graph using the von Neumann entropy from quantum information theory.

The von Neumann entropy is defined as the entropy of the density matrix associated with the state vector of a system. As noted above, the von Neumann entropy can be computed by scaling the normalized discrete Laplacian matrix for a network [11]. As a result, it is given in terms of the eigenvalues $\lambda_1, \ldots, \lambda_V$ of the density matrix ρ,

$$S = -\text{Tr}(\rho \log \rho) = -\sum_{i=1}^{V} \frac{\hat{\lambda}_i}{V} \log \frac{\hat{\lambda}_i}{V} \tag{2}$$

Since the normalized Laplacian spectrum has been proved to provide an effective characterization for networks or graphs [12], the von Neumann entropy derived from the spectrum may also be anticipated to be an effective tool for network characterization.

In fact, Han et al. [13] have shown how to approximate the calculation of von Neumann entropy in terms of simple degree statistics. Their approximation allows the cubic complexity of computing the von Neumann entropy from the Laplacian spectrum, to be reduced to one of quadratic complexity using simple edge degree statistics, i.e.

$$S = 1 - \frac{1}{V} - \frac{1}{V^2} \sum_{(u,v) \in E} \frac{1}{d_u d_v} \tag{3}$$

This expression for the von Neumann entropy allows the approximate entropy of the network to be efficiently computed and has been shown to be an effective tool for characterizing structural property of networks, with extremal values for cycle and fully connected graphs.

3   Quantum Statistics and Network Entropy

In this section, we describe the statistical mechanics for particles with quantum spin statistics. We then explain how the Laplaican eigenstates are occupied by a system of bosonic particles in equilibrium with a heat bath. This can be characterized using the Bose-Einstein partition function. From the partition function, various thermodynamic quantities, including entropy, can then be computed.

3.1   Thermal Quantities

Based on the heat bath analogy, particles occupy the energy states of the Hamiltonian subject to thermal agitation. The number of particles in each energy state is determined by (a) the temperature, (b) the assumed model of occupation statistics, (c) the relevant chemical potential.

When specified in this way, the various thermodynamic characterizations of the network can be computed from the partition function $Z(\beta, N)$, where $\beta = 1/k_B T$ and k_B is the Boltzmann constant. For instance, the Helmholtz free energy of the system is

$$F(\beta, N) = -\frac{1}{\beta} \log Z(\beta, N) = -k_B T \log Z(\beta, N) \tag{4}$$

and the thermodynamic entropy is given by

$$S = k_B \left[\frac{\partial}{\partial T} T \log Z \right]_N = -\left(\frac{\partial F}{\partial T} \right)_N \tag{5}$$

The incremental change in Helmholtz free energy is related to the incremental change in β and N,

$$dF = \left(\frac{\partial F}{\partial \beta} \right)_N d\beta + \left(\frac{\partial F}{\partial N} \right)_T dN = \frac{S}{k_B} \frac{1}{\beta^2} d\beta + \mu dN \tag{6}$$

where μ is the chemical potential, given by

$$\mu = \left(\frac{\partial F}{\partial N} \right)_T = -k_B T \left[\frac{\partial}{\partial N} \log Z \right]_\beta \tag{7}$$

3.2   Bose-Einstein Statistics

Specified by the network Hamiltonian, each energy state can accommodate an unlimited number of integral spin particles. These particles are indistinguishable and subject to Bose-Einstein statistics. In other words, they do not obey the Pauli exclusion principle, and can aggregate in the same energy state without interacting.

Base on the thermodynamic system specified by N particles with the energy states, the Bose-Einstein partition function is defined as the product of all sets of occupation number in each energy state, and the matrix form is given as

$$Z_{BE} = \det \left(I - e^{\beta \mu} \exp[-\beta \tilde{L}] \right)^{-1} = \prod_{i=1}^{V} \left(\frac{1}{1 - e^{\beta(\mu - \varepsilon_i)}} \right) \tag{8}$$

where the chemical potential μ is defined by Eq. (7), indicating the varying number of particles in the network. From Eq. (5), the corresponding entropy is

$$\begin{aligned}
S_{BE} &= \log Z + \beta \langle U \rangle = -\text{Tr} \left\{ \log[I - e^{\beta \mu} \exp(-\beta \tilde{L})] \right\} \\
&\quad - \text{Tr} \left\{ \beta [I - e^{\beta \mu} \exp(-\beta \tilde{L})]^{-1} (\mu I - \tilde{L}) e^{\beta \mu} \exp(-\beta \tilde{L}) \right\} \\
&= -\sum_{i=1}^{V} \log \left(1 - e^{\beta(\mu - \varepsilon_i)} \right) - \beta \sum_{i=1}^{V} \frac{(\mu - \varepsilon_i) e^{\beta(\mu - \varepsilon_i)}}{1 - e^{\beta(\mu - \varepsilon_i)}}
\end{aligned} \tag{9}$$

Given the temperature $T = 1/\beta$, the average number of particles at the energy level indexed i with energy ε_i is

$$n_i = -\frac{1}{\beta} \left(\frac{\partial \log Z}{\partial \varepsilon_i} \right) = \frac{1}{\exp[\beta(\varepsilon_i - \mu)] - 1} \tag{10}$$

and as a result the total number of particles in the network is

$$N = \sum_{i=1}^{V} n_i = \sum_{i=1}^{V} \frac{1}{\exp[\beta(\varepsilon_i - \mu)] - 1} \tag{11}$$

The matrix form is

$$N = \mathrm{Tr}\left[\frac{1}{\exp(-\beta\mu)\exp[\beta\tilde{L}] - I}\right] \tag{12}$$

where I is the identity matrix.

In order for the number of particles in each energy state to be non-negative, the chemical potential must be less than the minimum energy level, i.e. $\mu < \min \varepsilon_i$. Thus, the entropy derived from Bose-Einstein statistics is related to the temperature, energy states and chemical potential.

4   Graph Kernel Construction

In this section, we show how to compute the Jensen-Shannon divergence between a pair of graphs using the network entropy derived from the heat bath analogy and Bose-Einstein statistics. Once the combined the idea from graph embedding, we establish graph kernel associated with kernel principle component analysis (kPCA) to classify graphs.

4.1   Jensen-Shannon Divergence

The Jensen-Shannon kernel [14] is given by

$$k_{JS}(G_i, G_j) = \log 2 - D_{JS}(G_i, G_j) \tag{13}$$

where $D_{JS}(G_i, G_j)$ is the Jensen-Shannon divergence between the probability distributions defined on graphs G_i and G_j.

We now apply the kernel method of Jensen-Shannon divergence to construct a graph kernel between pairs of graphs. Suppose the graphs are represented by a set $G = G_i, i = 1, \ldots, n$, where $G_i = (V_i, E_i)$. V_i is the set of nodes on graph G_i and $E_i \subset V_i \times V_i$ is the set of edges.

For a pair of graphs G_i and G_j, the union graph is defined as $G_i \oplus G_j$, which has the nodes on both graphs G_i and G_j, that is $G_U = V_i \cup V_j$. It also contains the edge sets between pairs of nodes, such that if $(k, l) \in E_i$, and $(k, l) \in E_j$, then $((k, l), (k, l)) \in E_U$, which means the union graph contains the combined edges of two graphs. With the union graph to hand, the Jensen-Shannon divergence for a couple of graphs G_i and G_j is

$$D_{JS}(G_i, G_j) = H(G_i \oplus G_j) - \frac{H(G_i) + H(G_j)}{2} \tag{14}$$

where $H(G_i)$ is the entropy associated with the graph G_i, and $H(G_i \oplus G_j)$ is the entropy associated with the corresponding union graph G_U.

Using the Bose-Einstein entropy in Eq. (9) for the graphs G_i, G_j and their union $G_i \oplus G_j$, we compute the Jensen-Shannon divergence for the pair of graphs and hence the Jensen-Shannon kernel matrix.

With the graph kernel to hand, we apply kernel-PCA to embed the graphs into a vector space. To compute the embedding, we commence by computing the eigen decomposition of the kernel matrix, which will reproduce the Hilbert space with a non-linear mapping. In such a case, graph features can be mapped to low dimensional feature space with linear separation. So the graph kernel can be decomposed as

$$k_{JS} = U \Lambda U^T \qquad (15)$$

where Λ is the diagonal eigenvalue matrix and U is the matrix with eigenvectors as columns. To recover the matrix X with embedding co-ordinate vectors as columns, we write the kernel matrix in Gram-form, where each element is an inner product of embedding co-ordinate vectors

$$k_{JS} = X X^T \qquad (16)$$

and as a result $X = \sqrt{\Lambda} U^T$.

5   Experiments

In this section, we present experiments on fMRI data. Our aim is to explore whether we can classify the subjects on the basis of similarity of the activation networks from the fMRI scans. To do this we embed the network similarity data into a vector-space by applying kernel-PCA to the Jensen-Shannon kernel. To simplify the calculation, the Boltzmann constant is set to unity through the experiment.

5.1   Dataset

The fMRI data comes from the ADNI initiative [15]. Each patient lies in the MRI scanner with eyes open. BOLD (BOLD: Blood-Oxygenation-Level-Dependent) fMRI image volumes are acquired every two seconds. The fMRI signal at each time point-is measured in volumetric pixels over the whole brain. The voxels here have been aggregated into larger regions of interest (ROIs), and the blood oxygenation time-series in each region has been averaged, yielding a mean time-series for each ROI. The correlation between the average time series in different ROIs represents the functional connectivity between regions.

We construct the graph using the cross-correlation coefficients for the average time serial pairs of ROIs. To do this we create an undirected edge between two ROI's if the cross-correlation co-efficient between the time series is in the top 40 % of values. The threshold is fixed for all the available data, which provides an optimistic bias for constructing graphs. Those ROIs which have missing time series data are discarded. Subjects fall into four categories according to their degree of disease severity, namely full Alzheimer's (AD), Late Mild Cognitive

Impairment (LMCI), Early Mild Cognitive Impairment (EMCI) and Normal. The LMCI subjects are more severely affected and close to full Alzheimer's, while the EMCI subjects are closer to the healthy control group (Normal). There are 30 subjects in the AD group, 34 in LMCI, 47 in EMCI and 38 in the Normal group.

5.2 Experimental Results

Now we describe the application of the above methods to investigate the structural dissimilarity of the fMRI activation networks, which is used to distinguish different groups of patients. We compute the Jensen-Shannon kernel matrix using the Bose-Einstein entropy and compare the performance with that obtained from von Neumann entropy. Given the spectra of a graph and the total number of particles, the chemical potential can be derived from Eq. (11), which is used to calculate the entropy. Fig. 1 shows the results of mapping the graphs into a 3-dimensional feature space obtained by kernel principal components analysis (kPCA). We use first three eigenvectors to show the cluster of each group. The common feature is that both the Bose-Einstein and von Neumann entropies separate the four groups of subjects. In the case of Bose-Einstein statistics, the clusters are better separated than those obtained with von Neumann entropy.

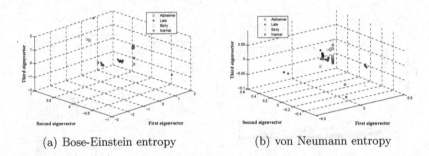

(a) Bose-Einstein entropy   (b) von Neumann entropy

Fig. 1. Kernel PCA performance of Jensen-Shannon divergence in Bose-Einstein entropy (Fig. 1(a)) and von Neumann entropy (Fig. 1(b)). Temperature $\beta = 10$ and particle number $N = 1$.

To place our analysis on a more quantitative footing, we apply Fisher's linear discriminant analysis to classify graphs with the kernel features, and compute the classification accuracy. Since the number of sampling in the datatset is small, we apply the leave-one-out cross-validation and use all the graphs as the testing data. Table 1 summaries the results of classification accuracy obtained by Jensen-Shannon kernels computed from the two entropies. Compared to the accuracy with von Neumann entropy, that obtained with Bose-Einstein entropy exhibits a higher classification accuracy. The Maxwell-Boltzmann entropy outperforms the von Neumann entropy on three classes of data presented by a margin of

Table 1. Classification accuracy for entropy from Bose-Einstein statistics and von Neumann entropy

Classification accuracy	Alzheimer	LMCI	EMCI	Normal
Bose-Einstein statistics	93.33 % (28/30)	100 % (34/34)	89.36 % (42/47)	92.11 % (35/38)
von-Neumann entropy	93.33 % (28/30)	88.24 % (30/34)	82.98 % (39/47)	86.84 % (33/38)

about 10 %. This reveals that the proposed graph kernel computed with Jensen-Shannon Divergence and Bose-Einstein entropy improve the performance for the fMRI data classification.

5.3   Discussion

The main parameters of the Bose-Einstein entropy are the temperature and number of particles in the system. In this section, we discuss the effects of the temperature on the energy level occupation statistics and hence upon the entropic kernel performance at low and high temperatures. We first focus on the average number of particles given the temperature β at each energy level ε_i from Eq. (10). In Fig. 2(a), we plot the occupation number for the different normalised Laplacian energy states with different values of temperature.

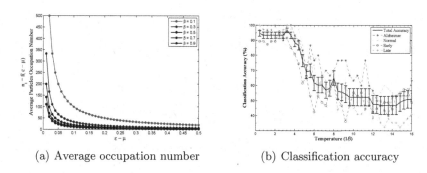

(a) Average occupation number   (b) Classification accuracy

Fig. 2. Average occupation number for energy state set different temperature for Bose-Einstein statistics (Fig. 2(a)). Classification accuracy changes with temperature in Jensen-Shannon divergence with entropy from Bose-Einstein statistics (Fig. 2(b)).

As shown in this figure, with fixed temperature and increasing energy, the number of particles in each energy level decreases. As a result the lower energy levels are occupied with the largest number of particles. Furthermore, as the temperature decreasing, the number of particles in each energy state decreases. From Eq. (10), it should be noted that the number of particles in each state is determined by two factors, namely (a) the Bose-Einstein occupation statistics, and (b) the number of particles as determined by the chemical potential.

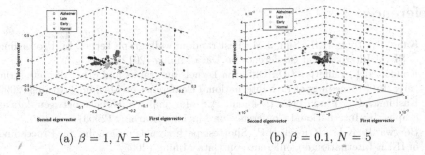

(a) $\beta = 1$, $N = 5$ (b) $\beta = 0.1$, $N = 5$

Fig. 3. Kernel PCA performance of Jensen-Shannon divergence with entropy from Bose-Einstein statistics at different values of temperature.

In order to evaluate how temperature effects the performance of the Jensen-Shannon kernel, we compare its behaviour at low and high temperature. For the fMRI brain activation data, we set $\beta = 1$ and $\beta = 0.1$, leaving the total particle number $N = 5$ unchanged. Compared to the low temperature case ($\beta = 10$) in Fig. 1, increasing temperature makes the four classes of graphs more densely clustered in feature space, shown in Fig. 3(a) and (b). This is term which reduces the performance of kernel PCA. Figure 2(b) shows the performance of classification changes with temperature. As the temperature increasing, the occupation number at each energy level increases and particles become to propagate at the high energy states. The probabilities of energy states in the system become identical to each other, reaching the uniform distribution as the temperature approaching to infinite. So all the groups with the same number of states, in statistical perspective, are rather similar to each other at high temperature. This reduces the performance of classification accuracy.

6   Conclusion

In this paper, we show how to compute an information theoretic graph-kernel using Bose-Einstein entropy and the Jensen-Shannon divergence. This method is based on quantum statistics associated with bosonic population of the normalized Laplacian eigenstates.

By applying kernel PCA to the Jensen-Shannon kernel matrix, we embed sets of graphs into a low dimensional space. In order to evaluate the performance of thermal entropies, we use discriminant classifier analysis to assign the graphs to different groups. Experimental results reveal that the method improves the classification performance for graphs extracted from fMRI data. The kernel method combined Bose-Einstein entropy and the Jensen-Shannon divergence provides an effective and efficient method in fMRI network analysis. Further work maybe focus on investigating the confusion matrix to evaluate the performance of classification.

References

1. Kang, U., Tong, H.H., Sun, J.M.: Fast random walk graph kernel. In: Proceedings of SIAM International Conference on Data Mining (2012)
2. Kondor, R.I., Lafferty, J.D.: Diffusion kernels on graphs and other discrete structures. In: Proceedings of 19th International Conference on Machine Learning (2002)
3. Kashima, H., Tsuda, K., Inokuchi, A.: Marginalized kernels between labeled graphs. In: International Conference on Machine Learning (2003)
4. Borgwardt, K.M., Kriegel, H.P.: Shortest-path kernels on graphs. In: Proceedings of IEEE International Conference on Data Mining (2005)
5. Bai, L., Hancock, E.R.: Graph kernels from the Jensen-Shannon divergence. J. Math. Imaging Vis. **47**, 60–69 (2012)
6. Torsello, A., Rossi, L.: Supervised learning of graph structure. In: Pelillo, M., Hancock, E.R. (eds.) SIMBAD 2011. LNCS, vol. 7005, pp. 117–132. Springer, Heidelberg (2011). doi:10.1007/978-3-642-24471-1_9
7. Bianconi, G., Rahmede, C., Wu, Z.: Complex Quantum Network Geometries: Evolution and Phase Transitions (2015). arxiv:1503.04739v2
8. Bianconi, G., Barabasi, A.L.: Bose-Einstein condensation in complex network. Phys. Rev. Lett. **86**, 5632–5635 (2001)
9. Bianconi, G.: Supersymmetric multiplex networks described by coupled Bose and Fermi statistics. Phys. Rev. E **91**, 012810 (2015)
10. Schölkopf, B., Smola, A., Müller, K.-R.: Kernel principal component analysis. In: Gerstner, W., Germond, A., Hasler, M., Nicoud, J.-D. (eds.) ICANN 1997. LNCS, vol. 1327, pp. 583–588. Springer, Heidelberg (1997). doi:10.1007/BFb0020217
11. Passerini, F., Severini, S.: The von Neumann entropy of networks. Int. J. Agent Technol. Syst. **1**, 58–67 (2008)
12. Chung, F.: Spectral graph theory. In: CBMS Regional Conference Series in Mathematics, no. 92 (1997)
13. Han, L., Hancock, E., Wilson, R.: Characterizing graphs using approximate von Neumann entropy. Pattern Recogn. Lett. **33**, 1958–1967 (2012)
14. Martins, A.F.T., Smith, N.A., Xing, E.P., Aguiar, P.M.Q., Figueiredo, M.A.T.: Nonextensive information theoretic kernels on measures. J. Mach. Learn. Res. **10**, 935–975 (2009)
15. fMRI Image Data Alzheimer's Disease Neuroimaging Institute (ADNI). http://adni.loni.usc.edu/

Model Selection, Classification and Clustering

Classification of Volumetric Data Using Multiway Data Analysis

Hayato Itoh[1]([✉]), Atsushi Imiya[2], and Tomoya Sakai[3]

[1] Graduate School of Advanced Integration Science, Chiba University,
Yayoi-cho 1-33, Inage-ku, Chiba 263-8522, Japan
hayato-itoh@graduate.chiba-u.jp
[2] Institute of Management and Information Technologies,
Chiba University, Yayoi-cho 1-33, Inage-ku, Chiba 263-8522, Japan
[3] Graduate School of Engineering, Nagasaki University,
Bunkyo-cho 1-14, Nagasaki 852-8521, Japan

Abstract. We introduce a method to extract compressed outline shapes of objects from global textures of volumetric data and to classify them by multiway tensor analysis. For the extraction of outline shapes, we applied three-way tensor principal component analysis to voxel images. A small number of major principal components represent the shape of objects in a voxel image. For the classification of objects, we use tensor subspace method. Using extracted outline shapes and tensor-based classification method, we achieve pattern recognition for volumetric data.

1 Introduction

For shape analysis in medicine and biology, outline shapes are fundamental feature for recognition and retrieval in information filtering from large amount of data. Therefore, for fast recognition and retrieval, we are required to extract outline shapes from volumetric medical data. Since the nature of medical and biological data is volumetric, these data are expressed in multiway data array. Using multiway data analysis [1], we extract compressed outline shapes of objects from global texture of volumetric data. For this extraction, we applied three-way tensor principal component analysis [2] to voxel images, although in traditional data analysis, three-way data are embedded in vector space. A small number of major principal components represent the shape of the objects in a voxel image. Applying tensor subspace method (TSM) [3] to these major principal components, we construct tensor-based classification. In the classification, the TSM measures the similarity between a query and a tensor subspace spanned by principal axes of a category.

For numerical computation, we deal with sampled patterns. In traditional pattern recognition, these sampled patterns are embedded in an appropriate-dimensional Euclidean space as vectors. The other way is to deal with sampled patterns as higher-dimensional array data. These array data are expressed by tensor to preserve multilinearity of function in the original pattern space. Tensors allow expressing multidimensional array data in multilinear forms.

© Springer International Publishing AG 2016
A. Robles-Kelly et al. (Eds.): S+SSPR 2016, LNCS 10029, pp. 231–240, 2016.
DOI: 10.1007/978-3-319-49055-7_21

For the analysis of three-way array, three-mode factor analysis has been proposed [1] in statistics. This method called Tucker decomposition. The Tucker decomposition with orthogonal constraint is equivalent to the third-order tensor principal component analysis (TPCA) [2]. Applying this three-way TPCA to voxel images, we can directly extract the compressed outline shape of objects from volumetric data. In the decomposition procedure of the TPCA, we can extract multilinear structure of each category from data. Using these structures of each category, we can construct subspace learning method.

For the extraction of the outline of shapes in two-dimensional images, active contour model [4] and active shape model [5] has been proposed. The active contour model, which is known as snakes, extracts the outline of a shape by an energy-minimising spline without specific models for categories. The active shape model extracts the shape of objects using the specific models, which are obtained from learning data of categories. As the extensions of these shape extraction models to points cloud in three-dimensional space, statistical models are proposed [6,7]. In two- and three-dimensional images, these models rely on the set of points, which represent each boundary of objects on images. These points are manually and semi-automatically extracted in advance as preprocessing, while our method extracts an outline shape without the extraction of points.

Furthermore, we show that the compression of volumetric data by the three-way TPCA can be approximated by the reduction based on the three-dimensional discrete cosine transform (3D-DCT).

2   Extraction Based on Tensor Form

2.1   Tensor Representation for N-way Arrays

A Nth-order tensor $\mathcal{X}$ defined in $\mathbb{R}^{I_1 \times I_2 \times \ldots I_N}$ is expressed as

$$\mathcal{X} = (x_{i_1, i_2, \ldots i_N}) \tag{1}$$

for $x_{i_1, i_2, \ldots i_N} \in \mathbb{R}$ by N indices i_n. Each subscript n denotes the n-mode of $\mathcal{X}$. For the outer products of N vectors, if the tensor $\mathcal{X}$ satisfies the condition

$$\mathcal{X} = \boldsymbol{u}^{(1)} \circ \boldsymbol{u}^{(2)} \circ \cdots \circ \boldsymbol{u}^{(N)}, \tag{2}$$

where $\circ$ denotes the outer product, we call this tensor $\mathcal{X}$ a rank-one tensor. For $\mathcal{X}$, the n-mode vectors, $n = 1, 2, \ldots, N$, are defined as the I_n-dimensional vectors obtained from $\mathcal{X}$ by varying this index i_n while fixing all the other indices. The unfolding of $\mathcal{X}$ along the n-mode vectors of $\mathcal{X}$ is defined by

$$\mathcal{X}_{(n)} \in \mathbb{R}^{I_n \times I_{n'}}, \tag{3}$$

where $I_{n'} = I_1 \times I_2 \times \ldots I_{n-1} \times I_{n+1} \times \cdots \times I_N$, and the column vectors of $\mathcal{X}_{(n)}$ are the n-mode vectors of $\mathcal{X}$. The n-mode product $\mathcal{X} \times_n \boldsymbol{U}$ of a matrix $\boldsymbol{U} \in \mathbb{R}^{J_n \times I_n}$ and a tensor $\mathcal{X}$ is a tensor $\mathcal{G} \in \mathbb{R}^{I_1 \times I_2 \times \cdots \times I_{n-1} \times J_n \times I_{n+1} \times \cdots \times I_N}$ with elements

$$g_{i_1, i_2, \ldots, i_{n-1}, j_n, i_{n+1}, \ldots, i_N} = \sum_{i_n=1}^{I_n} x_{i_1, i_2, \ldots, I_N} u_{j_n, i_n}, \tag{4}$$

by the manner in ref. [8]. For the m- and n-mode product by matrices U and V, we have

$$\mathcal{X} \times_n U \times_m V = \mathcal{X} \times_m V \times_n U \tag{5}$$

since n-mode projections are commutative [8]. We define the inner product of two tensors $\mathcal{X} = (x_{i_1,i_2,\dots,i_N})$ and $\mathcal{Y} = (y_{i_1,i_2,\dots,i_N})$ in $\mathbb{R}^{I_1 \times I_2 \times \cdots \times I_N}$ by

$$\langle \mathcal{X}, \mathcal{Y} \rangle = \sum_{i_1} \sum_{i_2} \cdots \sum_{i_N} x_{i_1,i_2,\dots,i_N} y_{i_1,i_2,\dots,i_N}. \tag{6}$$

Using this inner product, the Frobenius norm of a tensor $\mathcal{X}$ is

$$\|\mathcal{X}\|_{\mathrm{F}} = \sqrt{\langle \mathcal{X}, \mathcal{X} \rangle} = \|\mathcal{X}\|_{\mathrm{F}} = \|\mathrm{vec}\ \mathcal{X}\|_2, \tag{7}$$

where vec and $\|\cdot\|_2$ are the vectorisation operator and Euclidean norm of a tensor, respectively. For the two tensors $\mathcal{X}_1$ and $\mathcal{X}_2$, we define the distance between them as

$$d(\mathcal{X}_1, \mathcal{X}_2) = \|\mathcal{X}_1 - \mathcal{X}_2\|_{\mathrm{F}}. \tag{8}$$

Although this definition is a tensor-based measure, this distance is equivalent to the Euclidean distance between the vectorised tensors $\mathcal{X}_1$ and $\mathcal{X}_2$ from Eq. (7).

2.2 Projection to Tensor Subspace

As the tensor $\mathcal{X}$ is in the tensor space $\mathbb{R}^{I_1} \otimes \mathbb{R}^{I_2} \otimes \cdots \otimes \mathbb{R}^{I_N}$, the tensor space can be interpreted as the Kronecker product of N vector spaces $\mathbb{R}^{I_1}, \mathbb{R}^{I_2}, \dots, \mathbb{R}^{I_N}$. To project $\mathcal{X} \in \mathbb{R}^{I_1} \otimes \mathbb{R}^{I_2} \otimes \cdots \otimes \mathbb{R}^{I_N}$ to another tensor $\mathcal{Y}$ in a lower-dimensional tensor space $\mathbb{R}^{P_1} \otimes \mathbb{R}^{P_2} \otimes \cdots \otimes \mathbb{R}^{P_N}$, where $P_n \leq I_n$ for $n = 1, 2, \dots, N$, we need N orthogonal matrices $\{U^{(n)} \in \mathbb{R}^{I_n \times P_n}\}_{n=1}^N$. Using the N projection matrices, the tensor-to-tensor projection (TTP) is given by

$$\hat{\mathcal{X}} = \mathcal{X} \times_1 U^{(1)\top} \times_2 U^{(2)\top} \cdots \times_N U^{(N)\top}. \tag{9}$$

This projection is established in N steps, where at the nth step, each n-mode vector is projected to a P_n-dimensional space by $U^{(n)}$. The reconstruction from a projected tensor $\hat{\mathcal{X}}$ is achieved by

$$\mathcal{X} = \hat{\mathcal{X}} \times_1 U^{(1)} \times_2 U^{(2)} \cdots \times_N U^{(N)}. \tag{10}$$

2.3 Principal Component Analysis for Third-Order Tensors

A third-order tensor $\mathcal{X} \in \mathbb{R}^{I_1 \times I_2 \times I_3}$, which is the array $X \in \mathbb{R}^{I_1 \times I_2 \times I_3}$, is denoted as a triplet of indices (i_1, i_2, i_3). We set the identity matrices I_j, $j = 1, 2, 3$ in $\mathbb{R}^{I_j \times I_j}$. Here we summarise higher-order singular value decomposition (HOSVD) [9] for third-order tensors. For a collection of tensors $\{\mathcal{X}_i\}_{i=1}^N \in \mathbb{R}^{I_1 \times I_2 \times I_3}$ satisfying the zero expectation condition $\mathrm{E}(\mathcal{X}_i) = 0$, we compute

$$\hat{\mathcal{X}}_i = \mathcal{X}_i \times_1 U^{(1)\top} \times_2 U^{(2)\top} \times_3 U^{(3)\top}, \tag{11}$$

where $U^{(j)} = [u_1^{(j)}, \ldots, u_{I_j}^{(j)}]$, that minimises the criterion

$$J_- = \mathrm{E}\left(\|\mathcal{X}_i - \hat{\mathcal{X}}_i \times_1 U^{(1)} \times_2 U^{(2)} \times_3 U^{(3)}\|_\mathrm{F}^2 \right) \tag{12}$$

and maximises the criterion

$$J_+ = E(\|\hat{X}_i\|_\mathrm{F}^2), \tag{13}$$

with respect to the conditions $U^{(j)\top}U^{(j)} = I_j$. By fixing $\{U^j\}_{j=1}^N$ except $U^{(j')}$, $j' = \{1, 2, 3\}$, we have

$$J_j = \mathrm{E}\left(\|U^{(j)\top}\mathcal{X}_{i,(j)}\mathcal{X}_{i,(j)}^\top U^{(j)}\|_\mathrm{F}^2 \right). \tag{14}$$

Eigendecomposition problems are derived by computing the extremes of

$$E_j = J_j + tr((I_j - U^{(j)\top}U^{(j)})\Sigma^{(j)}), \; j = 1, 2, 3. \tag{15}$$

For matrices $M^{(j)} = \frac{1}{N}\sum_{i=1}^N \mathcal{X}_{i,(j)}\mathcal{X}_{i,(j)}^\top$, $j = 1, 2, 3$ of rank $M^{(j)} = K$, the optimisation of J_- derives the eigenvalue decomposition

$$M^{(j)}U^{(j)} = U^{(j)}\Sigma^{(j)}, \tag{16}$$

where $\Sigma^{(j)} \in \mathbb{R}^{I_j \times I_j}$, $j = 1, 2, 3$, are diagonal matrices satisfying the relationships $\lambda_k^{(j)} = \lambda_k^{(j')}$, $k \in \{1, 2, \ldots, K\}$ for

$$\Sigma^{(j)} = \mathrm{diag}(\lambda_1^{(j)}, \lambda_2^{(j)} \cdots, \lambda_K^{(j)}, 0 \cdots, 0). \tag{17}$$

For the optimisation of $\{J_j\}_{j=1}^3$, there is no closed-form solution to this maximisation problem [9]. Algorithm 1 is the iterative procedure of multilinear TPCA [2]. We adopt Algorithm 1 [2] to optimise $\{J_j\}_{j=1}^3$. For $p_k \in \{e_k\}_{k=1}^K$, we set orthogonal projection matrices $P^{(j)} = \sum_{k=1}^{k_j} p_k p_k^\top$ for $j = 1, 2, 3$. Using these $\{P^{(j)}\}_{j=1}^3$, the low-rank tensor approximation [9] is given by

$$\mathcal{Y} = \mathcal{X} \times_1 (P^{(1)}U^{(1)})^\top \times_2 (P^{(2)}U^{(2)})^\top \times_3 (P^{(3)}U^{(3)})^\top, \tag{18}$$

where $P^{(j)}$ selects k_j bases of projection matrices $U^{(j)}$. The low-rank approximation using Eq. (18) is used for compression in TPCA.

For HOSVD for third-order tensors, we have the following theorems.

Theorem 1. *The compression computed by HOSVD is equivalent to the compression computed by TPCA.*

(*Proof*) The projection that selects $K = k_1 k_2 k_3$ bases of the tensor space spanned by $u_{i_1}^{(1)} \circ u_{i_2}^{(2)} \circ u_{i_3}^{(3)}$, $i_j = 1, 2, \ldots, k_j$ for $j = 1, 2, 3$, is

$$\begin{aligned} (P^{(3)}U^{(3)} &\otimes P^{(2)}U^{(2)} \otimes P^{(1)}U^{(1)}) \\ &= (P^{(3)} \otimes P^{(2)} \otimes P^{(1)})(U^{(3)} \otimes U^{(2)} \otimes U^{(1)}) = PW, \end{aligned} \tag{19}$$

where W and P are the projection matrix and a unitary matrix, respectively. Therefore, HOSVD is equivalent to TPCA for third-order tensors.   (Q.E.D.)

Algorithm 1. Iterative method for volumetric data

Input: A set of tensors $\{\mathcal{X}_i\}_{i=1}^{N}$. Dimension of projected tensors $\{k_j\}_{j=1}^{3}$.
  A maximum number of iterations K. A sufficiently small number η.

Output: A set of orthogonal matrices $\{U^{(j)}\}_{j=1}^{3}$.

1: Compute the eigendecomposition of a matrix $M^{(j)} = \frac{1}{N}\sum_{i=1}^{N}\mathcal{X}_{i,(j)}\mathcal{X}_{i,(j)}^{\top}$,
  where $\mathcal{X}_{i,(j)}$ is a j-mode unfolded $\mathcal{X}_i$, for $j = 1, 2, 3$.

2: Construct orthogonal matrices by selecting eigenvectors
  corresponding to the k_j largest eigenvalues for $j = 1, 2, 3$.

3: Compute $\Psi_0 = \sum_{i=1}^{N}\|\mathcal{X}_i \times_1 U^{(1)\top} \times_2 U^{(2)\top} \times_3 U^{(3)\top}\|_{\mathrm{F}}$.

4: Iteratively compute the following procedure.
  for $k = 1, 2, \ldots, K$
    for $j = 1, 2, 3$
      Update $U^{(j)}$ by decomposing the matrix
      $\sum_{i=1}^{N}(\mathcal{X} \times_{\xi_\alpha} U^{(\xi_\alpha)\top} \times_{\xi_\beta} U^{(\xi_\beta)\top})(\mathcal{X} \times_{\xi_\alpha} U^{(\xi_\alpha)\top} \times_{\xi_\beta} U^{(\xi_\beta)\top})^{\top}$,
      where $\xi_\alpha, \xi_\beta \in \{1, 2, 3\} \setminus \{j\}$, $\xi_\alpha < \xi_\beta$
    end
    Compute $\Psi_k = \sum_{i=1}^{N}\|\mathcal{X}_i \times_1 U^{(1)\top} \times_2 U^{(2)\top} \times_3 U^{(3)\top}\|_{\mathrm{F}}$
    if $|\Psi_k - \Psi_{k-1}| < \eta$
      break
  end

Furthermore, we have the following theorem.

Theorem 2. *The HOSVD method is equivalent to the vector PCA method.*

(*Proof*) The equation

$$\mathcal{Y} = \mathcal{X} \times_1 (P^{(1)}U^{(1)})^{\top} \times_2 (P^{(2)}U^{(2)})^{\top} \times_3 (P^{(3)}U^{(3)})^{\top} \tag{20}$$

is equivalent to

$$\mathrm{vec}\mathcal{Y} = (P^{(3)}U^{(3)} \otimes P^{(2)}U^{(2)} \otimes P^{(1)}U^{(1)})^{\top}\mathrm{vec}\mathcal{X} = (PW)^{\top}\mathrm{vec}\mathcal{X}. \tag{21}$$

(Q.E.D.)

This theorem implies that the 3D-DCT-based reduction is an acceptable approximation of HOSVD for third-order tensors since this is an analogy of the approximation of PCA for two-dimensional images by the reduction based on the two-dimensional discrete cosine transform [10].

2.4 Reduction by Three-Dimensional Discrete Cosine Transform

For sampled one-dimensional signal $x = (x_1, x_2, \ldots, x_n)^{\top}$, we have transformed signal $\tilde{x} = (\tilde{x}_1, \tilde{x}_2, \ldots, \tilde{x}_n)^{\top}$ by using discrete cosine transform (DCT)-II [11,12]. Matrix representation of the DCT transform is given by

$$\tilde{x} = Dx, \quad D = ((d_{ij})), \quad d_{ij} = \cos\left(\frac{\pi}{n}\left((j-1) + \frac{1}{2}\right)(i-1)\right), \tag{22}$$

where $i, j = 1, 2, \ldots n$.

For a set of a third-order tensor $\{\boldsymbol{X}_i\}_{i=1}^N$ such that $\boldsymbol{X}_i \in \mathbb{R}^{n \times n \times n}$, setting a DCT matrix $\boldsymbol{D} \in \mathbb{R}^{n \times n}$ and projection matrices $\boldsymbol{P}^{(1)} \in \mathbb{R}^{k_1 \times n}$, $\boldsymbol{P}^{(2)} \in \mathbb{R}^{k_2 \times n}$ and $\boldsymbol{P}^{(3)} \in \mathbb{R}^{k_3 \times n}$, we define the 3D-DCT-based reduction by

$$\hat{\boldsymbol{X}}_i = \boldsymbol{X}_i \times_1 (\boldsymbol{P}^{(1)} \boldsymbol{D}) \times_2 (\boldsymbol{P}^{(2)} \boldsymbol{D}) \times_3 (\boldsymbol{P}^{(3)} \boldsymbol{D}), \tag{23}$$

where $k_1, k_2, k_3 < n$. The 3D-DCT-based reduction is an acceptable approximation of the compression by the PCA, TPCA and HOSVD. If we apply the fast Fourier transform to the computation of the 3D-DCT for tensor projections of each mode, the computational complexity is $\mathcal{O}(n \log n)$.

3   Classification Based on Tensor Form

We adopt multilinear tensor subspace method for third-order tensors. This method is a three-dimensional version of the 2DTSM [3]. For a third-order tensor $\mathcal{X}$, setting $\boldsymbol{U}^{(j)}$, $j = 1, 2, 3$, to be orthogonal matrices, we call the operation

$$\mathcal{Y} = \mathcal{X} \times_1 \boldsymbol{U}^{(1)\top} \times_2 \boldsymbol{U}^{(2)\top} \times_3 \boldsymbol{U}^{(3)\top} \tag{24}$$

the orthogonal projection of $\mathcal{X}$ to $\mathcal{Y}$. Therefore, using this expression for a collection of matrices $\{\mathcal{X}_i\}_{i=1}^M$, such that $\mathcal{X}_i \in \mathbb{R}^{I_1 \times I_2 \times I_3}$ and $\mathrm{E}(\mathcal{X}_i) = 0$, the solutions of

$$\{\boldsymbol{U}^{(j)}\}_{j=1}^3 = \arg \max \mathrm{E} \left(\frac{\|\mathcal{X} \times_1 \boldsymbol{U}^{(1)\top} \times_2 \boldsymbol{U}^{(2)\top} \times_3 \boldsymbol{U}^{(3)\top}\|_{\mathrm{F}}}{\|\mathcal{X}_i\|_{\mathrm{F}}} \right) \tag{25}$$

with respect to $\boldsymbol{U}^{(j)\top} \boldsymbol{U}^{(j)} = \boldsymbol{I}$ for $j = 1, 2, 3$ define a trilinear subspace that approximates $\{\mathcal{X}_i\}_{i=1}^M$. Therefore, using orthogonal matrices $\{\boldsymbol{U}_k^{(j)}\}_{j=1}^3$ obtained as the solutions of Eq. (25) for kth category, if a query tensor $\mathcal{G}$ satisfies the condition

$$\arg \left(\max_l \frac{\|\mathcal{G} \times_1 \boldsymbol{U}_l^{(1)\top} \times_2 \boldsymbol{U}_l^{(2)\top} \times_3 \boldsymbol{U}_l^{(3)\top}\|_{\mathrm{F}}}{\|\mathcal{G}\|_{\mathrm{F}}} \right) = \{\boldsymbol{U}_k^{(j)}\}_{j=1}^3, \tag{26}$$

we conclude that $\mathcal{G} \in \mathcal{C}_k$, $k, l = 1, 2, \ldots, N_{\mathcal{C}}$, where $\mathcal{C}_k$ and $N_{\mathcal{C}}$ are the tensor subspace of kth category and the number of categories, respectively. For the practical computation of projection matrices $\{\boldsymbol{U}_k^{(j)}\}_{j=1}^3$, we adopt the iterative method described in Algorithm 1.

4   Numerical Examples

We present two examples for extraction of outline shapes of volume data, and abilities of our method for classification of volumetric data. For experiments, we use the voxel images of human livers obtained as CT images. This image set contains 25 male-livers and seven female-livers. Note that these voxel images are aligned to their centre of gravity. In the experiments, we project these voxel images to small size tensors. For the projections, we adopt TPCA and the 3D-DCT. In the iterative method of TPCA, setting the number of bases to the size of the original tensors in Algorithm 1, we call the method full projection (FP). If we set the number of bases to smaller than the size of the original tensors in Algorithm 1, we call the method full projection truncation (FPT). Table 1 summarises the sizes and numbers of original and dimension-reduced voxel images.

Firstly, we show the approximation of a voxel image of a liver by three methods. The FP, FPT and 3D-DCT reduce the size of the data from $89 \times 97 \times 76$ voxels to $32 \times 32 \times 32$ voxels. Figure 1 illustrates volume rendering of original data and reconstructed data by these compressed tensors. Compared to Figs. 1(a) and 1(e), in Figs. 1(b)–(c) and (f)–(h), the FP, FPT and 3D-DCT preserve outline shapes of liver. In Fig. 1, the reconstructed data by the 3D-DCT gives a closer outline shape and more similar interior texture to those of the original than the FP and FPT. In Fig. 1, these results show that projections to small-size tensors extract outline shapes.

For the analysis of projected data by the FP, FPT and 3D-DCT, we decompose these projected tensors by Algorithm 1. Here, we set the size of bases in Algorithm 1 to $32 \times 32 \times 32$ and use 35 projected tensors of livers for each reduction methods. In decompositions, we reordered eigenvalues $\lambda_i^{(j)}$, $j = 1, 2, 3$, $i = 1, 2, \ldots, 32$ of the three modes to λ_i, $i = 1, 2, \ldots, 96$ in the descending order. Figure 2 shows the cumulative contribution ratios of reordered eigenvalues for the projected tensors obtained by the FP, FPT and 3D-DCT. Figure 3 illustrates reconstructed data obtained by using the 20 major principal components.

In Fig. 2, profiles of curves for three methods are almost coincident while the CCR of the 3D-DCT is a little bit higher than the others. In three methods, the CCRs become higher than 0.8 if we select more than 19 major principal components. In Fig. 3, shapes and interior texture for three methods are almost the same. In Figs. 3(d)–(f), the interior texture of a liver is not preserved and the outer shape is burred. In these results for three methods, major principal components represent outline shapes.

Secondly, we show results of the classification of voxel images of livers by the TSM. For the classification, we use 25 male-livers and seven female-livers since the sizes and shapes of livers between male and female are statistically different. Figures 4(a) and 4(b) illustrates the examples of livers of male and female, respectively. We use the voxel images of livers of 13 males and 4 females as training data. The residual voxel images are used as test data. In the recognition, we estimate the gender of livers. The recognition rate is defined as the successful estimation ratio for 1000 gender estimations. In each estimation of a gender for a query, queries are randomly chosen from the test dataset. For the 1-, 2- and 3-modes, we evaluate the results for multilinear subspaces with sizes from one to the dimension of the rejected tensors. Figure 4(c) shows the results of the classification. The TSM give 90 % recognition rate at the best with tensor subspace spanned by every two major principal axis of the three modes.

Table 1. Sizes and number of volumetric data of livers. ♯data represents the number of livers obtained from different patients. The data size is the original size of the volumetric data. The reduced data size is the size of the volume data after tensor-based reduction.

	♯data	Data size [voxel]	Reduced data size [voxel]
Volumetric liver data	32	$89 \times 97 \times 76$	$32 \times 32 \times 32$

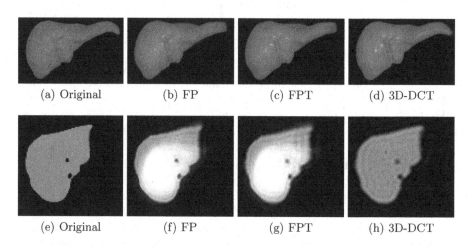

(a) Original   (b) FP   (c) FPT   (d) 3D-DCT

(e) Original   (f) FP   (g) FPT   (h) 3D-DCT

Fig. 1. Original and reconstructed volumetric data of liver data. (a) shows the rendering of original data. (b)–(d) show the rendering of reconstructed data after the FP, FPT and 3DDCT, respectively. (e)–(f) illustrate axial slice images of these volumetric data in (a)–(d), respectively. The sizes of reduced tensors are shown in Table 1.

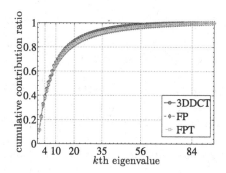

Fig. 2. Cumulative contribution ratios for three compressed tensors. For compression, we adopt FP, FPT and 3D-DCT. For the computation of the cumulative contribution ratio of eigenvalues obtained by the FP, we used all eigenvalues of modes 1, 2 and 3 after sorting them into descending order.

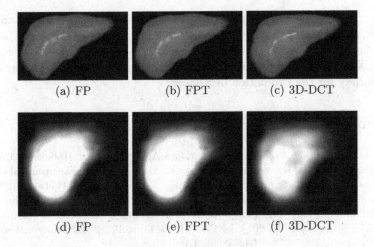

(a) FP   (b) FPT   (c) 3D-DCT

(d) FP   (e) FPT   (f) 3D-DCT

Fig. 3. Reconstruction by using only major principal components of the decomposition by the FP. Top and bottom rows illustrate volume rendering and axial slice of reconstructed data, respectively. For reconstruction, we use the 20 major principal components. Left, middle and right columns illustrate the results for the tensors projected by the FP, FPT and 3D-DCT.

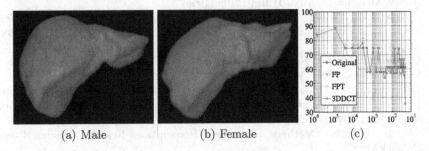

(a) Male   (b) Female   (c)

Fig. 4. Recognition rates of gait patterns and liver data for original and compressed tensors. (a) and (b) illustrate the examples of livers of male and female, respectively. (c) shows the recognition rates for three compression methods. For compression, we use the HOSVD, FP, FPT and 3D-DCT. In (c), the horizontal and vertical axes represent the compression ratio and recognition ratio [%], respectively. For the original size $D = 89 \times 97 \times 76$ and the reduced size $K = k \times k' \times k''$, the compression ratio is given as D/K for reduced size k.

5   Conclusions

We applied the three-way TPCA to the extraction of outline shapes of volumetric data and classification of them. In the numerical examples, we demonstrated that the three-way TPCA extracts the outline shape of human livers. Furthermore, the TSM accurately classified the extracted outline shapes. Moreover, we showed that the 3D-DCT-based reduction approximated both the outline shape and texture of livers.

This research was supported by the "Multidisciplinary Computational Anatomy and Its Application to Highly Intelligent Diagnosis and Therapy" project funded by a Grant-in-Aid for Scientific Research on Innovative Areas from MEXT, Japan, and by Grants-in-Aid for Scientific Research funded by the Japan Society for the Promotion of Science.

References

1. Kroonenberg, P.M.: Applied Multiway Data Analysis. Wiley, Hoboken (2008)
2. Lu, H., Plataniotis, K., Venetsanopoulos, A.: MPCA: multilinear principal component analysis of tensor objects. IEEE Trans. Neural Netw. **19**(1), 18–39 (2008)
3. Itoh, H., Sakai, T., Kawamoto, K., Imiya, A.: Topology-preserving dimension-reduction methods for image pattern recognition. In: Kämäräinen, J.-K., Koskela, M. (eds.) SCIA 2013. LNCS, vol. 7944, pp. 195–204. Springer, Heidelberg (2013). doi:10.1007/978-3-642-38886-6_19
4. Kass, M., Witkin, A., Trzopoulos, D.: Snakes: active contour models. IJCV **1**(4), 321–331 (1988)
5. Cootes, T.F., Cooper, D., Taylor, C.J., Graham, J.: Active shape models - their training and application. CVIU **61**, 38–59 (1995)
6. McInerney, T., Terzopoulos, D.: Deformable models in medical image analysis: a survey. Med. Image Anal. **1**(2), 91–108 (1996)
7. Davies, R.H., Twining, C.J., Cootes, T.F., Taylor, C.J.: Building 3-D statistical shape models by direct optimization. IEEE Trans. Med. Imaging **29**(4), 961–981 (2010)
8. Cichoki, A., Zdunek, R., Phan, A.H., Amari, S.: Nonnegative Matrix and Tensor Factorizations. Wiley, Hoboken (2009)
9. Lathauwer, L.D., Moor, B.D., Vandewalle, J.: On the best rank-1 and rank-(r_1, r_2, r_n) approximation of higher-order tensors. SIAM J. Matrix Anal. Appl. **21**(4), 1324–1342 (2000)
10. Oja, E.: Subspace Methods of Pattern Recognition. Research Studies Press, Baldock (1983)
11. Ahmed, N., Natarajan, T., Rao, K.R.: Discrete cosine transform. IEEE Trans. Comput. **C−23**, 90–93 (1974)
12. Hamidi, M., Pearl, J.: Comparison of the cosine and Fourier transforms of Markov-1 signals. IEEE Trans. Acoust. Speech Sig. Process. **24**(5), 428–429 (1976)

Commute Times in Dense Graphs

Francisco Escolano[1,2]([✉]), Manuel Curado[1,2], and Edwin R. Hancock[1,2]

[1] Department of Computer Science and AI,
University of Alicante, 03690 Alicante, Spain
{sco,mcurado}@dccia.ua.es
[2] Department of Computer Science, University of York, York YO10 5DD, UK
erh@cs.york.ac.uk

Abstract. In this paper, we introduce the approach of graph densification as a means of preconditioning spectral clustering. After motivating the need of densification, we review the fundamentals of graph densifiers based on cut similarity and then analyze their associated optimization problems. In our experiments we analyze the implications of densification in the estimation of commute times.

Keywords: Graph densification · Cut similarity · Spectral clustering

1 Introduction

1.1 Motivation

Machine learning methods involving large graphs face a common problem, namely the natural sparsification of data as the number of dimensions d increases. In this regard, obtaining the proximity structure of the data is a key step for the subsequent analysis. This problem has been considered from two complementary perspectives: *efficiency* and *utility*. On the one hand, an efficient, i.e. scalable, proximity structure typically emerges from reducing the $O(dn^2)$ time complexity of kNN graphs, where n is the number of samples. The classical approach for dealing with large graphs is the Nyström method. It consists of sampling either the feature space or the affinity space so that the eigenproblems associated with clustering relaxations become tractable. For instance, in [10] there is a variational version of this method. In [6] an approximated kNN is obtained in $O(dn^t)$ with $t \in (1, 2)$ by recursively dividing and glueing the samples. More recently, anchor graphs [13,15] provide *data-to-anchor* kNN graphs, where $m \ll n$ is a set of representatives (anchors) typically obtained through K-means clustering, in $O(dmnT + dmn)$ where $O(dmnT)$ is due to the T iterations of the K-means process. These graphs tend to make out-of-the-sample predictions compatible with those of Nyström approximations, and in turn their approximated adjacency/affinity matrices are ensured to be positive semidefinite.

On the other hand, the utility of the kNN representation refers to its suitability to predict or infer some properties of the data. These properties include

© Springer International Publishing AG 2016
A. Robles-Kelly et al. (Eds.): S+SSPR 2016, LNCS 10029, pp. 241–251, 2016.
DOI: 10.1007/978-3-319-49055-7_22

(a) their underlying density and (b) the geometry induced by both the shortest path distances and the commute time distances. Concerning the density, it is well known that it can be estimated from the degrees of the kNN graph if its edges contain the local similarity information between the data, i.e. when the graph is weighted. However, when the kNN graph is unweighted the estimation is only acceptable for *reasonably dense* graphs, for instance when $k^{d+2}/(n^2 \log^d n) \to \infty$ as proposed in [20]. However, these densities are unrealistic, since the typical regime, the one adopted in practice, is $k \approx \log n$. A similar conclusion is reached when shortest path distances are analyzed both in weighted and unweighted kNN graphs. The shortest path distance computed from an unweighted kNN graph typically diverges from the geodesic distance. However this is not the case of the one computed from a weighed kNN graph. The solution proposed in [1] consists of assigning proper weights to the edges of the unweighted kNN graphs. Since these weights depend heavily on the ratio $r = (k/(n\mu_d))^{1/d}$, where μ_d is the volume of a d−dimensional unit ball, one expects $r \to 0$ for even moderate values of d, meaning that for high dimensional data both unweighted and weighted graphs yield similar, i.e. diverging, estimations. Finally, it is well know that for large k−NN (unweighted) graphs the commute time distance can be misleading since it only relies on the local densities (degrees) of the nodes [21,22].

Therefore, for a standard machine learning setting ($n \to \infty$, $k \approx \log n$ and large d) we have that kNN graphs result in a sparse, globally uninformative representation. This can be extended to ϵ−graphs and Gaussian graphs as well. As a result, machine learning algorithms for graph-based embedding, clustering and label propagation tend to produce misleading results unless we are able of preserving the distributional information of the data in the graph-based representation. In this regard, recent experimental results with anchor graphs suggest a way to proceed. In [5], the predictive power of non-parametric regression rooted in the anchors/landmarks ensures a way of constructing very informative weighted kNN graphs. Since anchor graphs are bipartite (only *data-to-anchor* edges exist), this representation bridges the sparsity of the pattern space because a random walk traveling from node u to node v must reach one or more anchors in advance. In other words, for a sufficient number of anchors it is then possible to find links between distant regions of the space. This opens a new perspective for computing meaningful commute distances in large graphs. It is straightforward to check that the spectral properties of the approximate weight matrix $W = Z\Lambda Z^T$, where $\Lambda = \mathrm{diag}(Z^T 1)$ and Z is the data-to-anchor mapping matrix, rely on its low-rank. Then, it is possible to compute a reduced number of eigenvalue-eigenvector pairs associated with a small $m \times m$ matrix, where m is the number of anchors (see [16] for details). In this way, the spectral expression of the commute distance [18] can accomodate these pairs for producing meaningful distances. Our interpretation is that the goodness of the eigenvalue-eigenvector pairs is a consequence of performing kernel PCA process over ZZ^T where the columns of Z act as kernel functions. This interpretation is consistent with the good hashing results obtained with anchor graphs [14,16] where the kernel encoded in the columns of Z is extensively exploited.

Although anchor graphs provide meaningful commute distances with low-complexity spectral representations, some authors have proposed more efficient methods where anchor graphs are bypassed for computing these distances. For instance, Chawla and coworkers [9,11] exploit the fact that commute distances can be approximated by a randomized algorithm in $O(n \log n)$ [19]. Then, using standard kNN graphs with low k for avoiding intra-class noise, their method beats anchor graphs, in terms of clustering accuracy, in several databases. These results are highly contradictory with respect to the von Luxburg and Radl's fundamental bounds (in principle commute distances cannot be properly estimated from large kNN graphs [22]). The authors argue that this can only be explained by the fact that their graphs are quite different from those explored for defining the fundamental bounds (particularly the ϵ−geometric graphs). Their estimator works better than anchor graphs in *dense datasets*, i.e. in settings with a low number of classes and many samples. Our preliminary experiments with the NIST database, with ten classes, confirm that their technique does not improve anchor graphs when data is sparse enough as it happens in a standard machine learning setting.

1.2 Contributions

We claim that one way of providing meaningful estimations of commute distances is to transform the input sparse graph into a *densified graph*. This implies the inference of novel links between data from existing ones. This is exactly what anchor graphs do when incorporate data-to-anchor edges. In this paper, we show that the inference of novel edges can be done by applying recent results in theoretical computer science, namely *cut densification* which in turn is an instance of *graph densification*. Graph densification consists in populating an input graph G with new edges (or weights if G is weighted) so that the output graph H preserves or enforces some structural properties of G. Graph densification offers a principled way of dealing with sparse graphs arising in machine learning so that commute distances can be properly estimated. In this paper we will introduce the main principles of densification and will explore their implications in Pattern Recognition (PR). In our experiments (see the Discussion section) we will show how the associated optimization problems (primal and dual) lead to a reasonable densification (in terms of PR). To the best of our knowledge this is the first application of densification principles to estimate the commute distance.

2 Graph Densification

2.1 Combinatorial Formulation

Graph densification [8] is a principled study of how to significantly increase the number of edges of an input graph G so that the output, H, approximates G with respect to a given test function, for instance whether there exists a given cut. This study is motivated by the fact that certain NP-hard problems have a PTAS

(Polynomial Time Approximation Scheme) when their associated graphs are dense. This is the case of the MAX-CUT problem [2]. Frieze and Kannan [7] raise the question whether this "easyness" is explained by the Szemerédi Regularity Lemma, which states that large dense graphs have many properties of random graphs [12].

For a standard machine learning setting, we have that G is typically sparse either when a kNN representation is used or when a Gaussian graph, usually constructed with a bandwidth parameter t satisfying $t \to 0$, is chosen. Then, the densification of G so that the value of any cut is at most C times the value of the same cut in G is called a *one-sided C-multiplicative cut approximation*. This (normalized) cut approximation must satisfy:

$$\frac{cut_H(S)}{m(H)} \leq C \cdot \frac{cut_G(S)}{m(G)}, \tag{1}$$

for any subset $S \subset V$ of the set of vertices V, where $cut_G(S) = \sum_{u \in S, v \in V \sim S} x_{uv}$ considers edge weights $\{x_{uv}\}_{u,v \in V}$ and $x_{uv} \in [0,1]$. For H we have $cut_G(S) = \sum_{u \in S, v \in V \sim S} x'_{uv}$ for edge weights $\{x'_{uv}\}_{u,v \in V}$ also satisfying $x'_{uv} \in [0,1]$. Cuts are normalized by the total edge weight $m(.)$ of each graph, i.e. $m(G) = \sum_{u,v} x_{uv}$ and $m(H) = \sum_{u,v} x'_{uv}$.

Cut Similarity and Optimization Problem. The cut approximation embodies a notion of similarity referred to as $C-cut$ *similarity*. Two graphs G and H are C-cut similar if $cut_H(S) \leq C \cdot cut_G(S)$ for all $S \subset V$, i.e. if the sum of the weights in the edges cut is approximately the same in every division. Considering the *normalized version* in Eq. 1, finding the optimal *one-sided C*–multiplicative cut densifier can be posed in terms of the following linear program:

$$\textbf{P1} \quad Max \quad \sum_{u,v} x'_{uv}$$

$$s.t. \quad \forall\, u,v: \; x'_{uv} \leq 1$$

$$\forall\, S \subseteq V: \sum_{u \in S, v \in V \sim S} x'_{uv} \leq C \cdot cut_G(S) \sum_{u,v} x'_{uv}$$

$$x'_{uv} \geq 0. \tag{2}$$

Herein, the term *one-sided* refers only to satisfy the upper bound in Eq. 1. The program **P1** has 2^n constraints, where $n = |V|$, since for every possible cut induced by S, the sum of corresponding edge weights $\sum_{u \in S, v \in V \sim S} x'_{uv}$ is bounded by C times the sum of the weights for the same cut in G. The solution is the set of edge weights x'_{uv} with maximal sum so that the resulting graph H is C–cut similar to G. The NP-hardness of this problem can be better understood if we formulate the dual LP. To this end we must consider a *cut metric* $\delta_S(.,.)$ where [4]

$$\delta_S(u,v) = \begin{cases} 1 \text{ if } |\{u,v\} \cap S| = 1 \\ 0 \text{ otherwise} \end{cases} \tag{3}$$

i.e. δ_S accounts for *pairs of nodes* (not necessarily defining an edge) with an end-point in S. As there are 2^n subsets S of V we can define the following metric ρ on $V \times V$, so that $\rho = \sum_S \lambda_S \delta_S$, with $\lambda_S \geq 0$, is a non-negative combination of a exponential number of cut metrics. For a particular pair $\{u, v\}$ we have that $\rho(u, v) = \sum_S \lambda_S \delta_S(u, v)$ accounts for the number subsets of V where either u or v (but not both) is an end-point. If a graph G has many cuts where $\frac{cut_G(S)/m(G)}{\sum_{u,v} \delta_S(u,v)} \to 0$ then we have that $\rho(u, v) \geq \mathbb{E}_{(u',v') \in E} \, \rho(u', v')$ since

$$\mathbb{E}_{(u',v') \in E} \, \rho(u', v') = \sum_S \lambda_S \mathbb{E}_{(u',v') \in E} \, \delta_S(u', v') = \sum_S \lambda_S \frac{cut_G(S)}{m(G)}. \qquad (4)$$

These cuts all called *sparse cuts* since the number of pairs $\{u, v\}$ involved in edges is a small fraction of the overall number of pairs associated with a given subset S, i.e. the graph stretches at a sparse cut. The existence of sparse cuts, more precisely *non-ovelapping sparse cuts* allows the separation of a significant number of vertices $\{u, v\}$ where their distance, for instance $\rho(u, v)$, is larger (to same extent) than the average distance taken over edges. This rationale is posed in [8] as satisfying the condition

$$\sum_{u,v \in V} \min \left\{ \rho(u, v) - C \cdot \mathbb{E}_{(u',v') \in E} \, \rho(u', v'), 1 \right\} \geq (1 - \alpha) n^2, \qquad (5)$$

where C is a constant as in the cut approximation, and $\alpha \in (0, 1)$. This means that a quadratic number of non-edge pairs are bounded away from the average length of an edge. In other words, it is then possible to embed the nodes involved in these pairs in such a way that their distances in the embedding do not completely collapse. This defines a so called (C, α) *humble embedding*.

Finding the metric, $\rho(u, v)$ that best defines a humble embedding is the dual problem of **P1**:

$$\textbf{P2} \quad Min_{\rho = \sum_S \lambda_S \delta_S} \sum_{u,v} \sigma_{uv}$$

$$s.t. \qquad \forall \, u, v: \qquad \rho(u, v) - C \cdot \mathbb{E}_{(u',v') \in E} \, \rho(u', v') \geq 1 - \sigma_{uv}$$

$$\sigma_{uv}, \lambda_S \geq 0, \qquad (6)$$

where the search space is explicitly the power set of V. Since the optimal solution of **P2** must satisfy

$$\sigma_{uv} = \max \left\{ 0, C \cdot \mathbb{E}_{(u',v') \in E} \, \rho(u', v') + 1 - \sigma_{uv} \right\}, \qquad (7)$$

we have that **P2** can be written in a more compact form:

$$\min_\rho \sum_{u,v} \max \left\{ 0, C \cdot \mathbb{E}_{(u',v') \in E} \, \rho(u', v') + 1 - \sigma_{uv} \right\}, \qquad (8)$$

which is equivalent to $n^2 - \max_\rho \sum_{u,v} \min \left\{ 1, \rho(u, v) - C \cdot \mathbb{E}_{(u',v') \in E} \right\}$. Therefore, a solution satisfying $\sum_{u,v} \sigma_{uv} = \alpha n^2$ implies that the graph has a humble embedding since

$$\max_\rho \sum_{u,v} \min \left\{ 1, \rho(u, v) - C \cdot \mathbb{E}_{(u',v') \in E} \right\} = (1 - \alpha) n^2. \qquad (9)$$

Since the σ_{uv} variables in the constraints of **P2** are the dual variables of x_{uv} in **P1**, the existence of a (C, α) humble embedding rules out a C-densifier with an edge weight greater than αn^2 and vice versa.

2.2   Spectral Formulation

Since $Q_G(z) = z^T L_G z = \sum_{e_{uv} \in E} x_{uv}(z_u - z_v)^2$, if z is the characteristic vector of S (1 inside and 0 outside) then Eq. 1 is equivalent to

$$\frac{z^T L_H z}{m(G)} \leq C \cdot \frac{z^T L_G z}{m(G)}, \tag{10}$$

for $0 - 1$ valued vectors z, where L_G and L_H are the respective Laplacians. However, if H satisfies Eq. 10 for any real-valued vector z, then we have a *one-sided C-multiplicative spectral approximation* of G, where L_G and L_H are the Laplacians. This spectral approximation embodies a notion of similarity between the Laplacians L_G and L_H. We say that G and H are $C-$*spectrally similar* if $z^T L_H z \leq C \cdot z^T L_G z$ and it is denoted by $L_H \preceq C \cdot L_G$. Spectrally similar graphs share many algebraic properties [3]. For instance, their effective resistances (rescaled commute times) are similar. This similarity is bounded by C and it leads to nice interlacing properties. We have that the eigenvalues of $\lambda_1, \ldots, \lambda_n$ of L_G and the eigenvalues $\lambda'_1, \ldots, \lambda'_n$ of H satisfy: $\lambda'_i \leq C \cdot \lambda_i$. This implies that H does not necessarily increases the spectral gap of G and the eigenvalues of L_G are not necessarily shifted (i.e. increased).

Whereas the spectral similarity of two graphs can be estimated to precission ϵ in time polynomial in n and $\log(1/\epsilon)$, it is NP-hard to approximately compute the cut similarity of two graphs. This is why existing theoretical advances in the interplay of these two concepts are restricted to *existence theorems* as a means of characterizing graphs. However, the semi-definite programs associated with finding both optimal cut densifiers and, more realistically, optimal spectral densifiers are quite inspirational since they suggest scalable computational methods for graph densification.

Spectral Similarity and Optimization Problem. When posing **P1** and **P2** so that they are *tractable* (i.e. polynomial in n) the cut metric ρ, which has a combinatorial nature, is replaced by a norm in $\mathbb{R}^n$. In this way, the link between the existence of humble embeddings and that of densifiers is more explicit. Then, let $z_1, \ldots, z_n \in \mathbb{R}^n$ the vectors associated with a given embedding. The concept (C, α) humble embedding can be redefined in terms of satisfying:

$$\sum_{u,v \in V} \min \left\{ ||z_u - z_v||^2 - C \cdot \mathbb{E}_{(u',v') \in E} \, ||z'_u - z'_v||^2, 1 \right\} \geq (1 - \alpha)n^2, \tag{11}$$

where distances *between pairs* should not globally collapse when compared with those *between pairs associated with edges*. Then the constraint in **P2** which is associated with the pair $\{u, v\}$ should be rewritten as:

$$||z_u - z_v||^2 - C \cdot \mathbb{E}_{(u',v') \in E} \, ||z'_u - z'_v||^2 \geq 1 - \sigma_{uv}. \tag{12}$$

Therefore, **P2** is a linear problem with quadratic constraints. For $Z = [z_1, \ldots, z_n]$ we have that $||z_u - z_v||^2 = b_{uv}^T Z^T Z b_{uv}$ where $b_{uv} = e_u - e_v$. Then, a Semipositive Definite (SPD) relaxation leads to express the first term of the left part of each inequality in terms of $b_{uv}^T Z b_{uv}$ provided that $Z \succeq 0$. Similarly, for the SPD relaxation corresponding to the expectation part of each inequality, we consider the fact that the Laplacian of the graph can be expressed in terms of $L_G = \sum_{u,v} w_{uv} b_{uv} b_{uv}^T$. Since $z^T L_G z = \sum_{(u',v') \in E} w_{uv} ||z(u') - z(v')||^2$, if $z \sim \mathcal{N}(0, Z)$, i.e. z is assumed to be a zero mean vector in $\mathbb{R}^n$ with covariance $Z \succeq 0$, we have that $\mathbb{E}_{(u',v') \in E} ||\tilde{z}_u' - \tilde{z}_v'||^2$ can be expressed in terms of $\mathrm{tr}(L_G Z)$ (see [17] for details). Therefore the SDP formulation of **P2** is as follows

$$\textbf{P2}_{\text{SDP}} \quad Min \sum_{u,v} \sigma_{uv}$$

$$s.t. \quad b_{uv}^T Z b_{uv} - C \cdot \mathrm{tr}(L_G Z) \geq 1 - \sigma_{uv}$$

$$Z \succeq 0, \ \sigma_{uv} \geq 0. \tag{13}$$

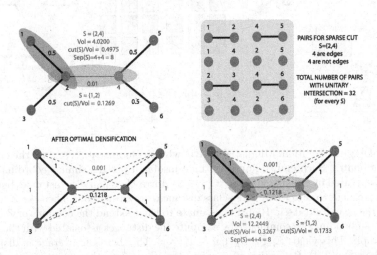

Fig. 1. Densification example. Graph G with $n = 6$ nodes. Top-left: Cuts associated with a couple of sets $S = \{2, 4\}$ and $S = \{1, 2\}$. We define $Sep(S) = \sum_{u,v} \delta_S(u, v)$. For the cut $S = \{2, 4\}$ there are 4 pairs associated with edges and 4 pairs not associated with edges (top-right). This means that this cut is sparse since $\frac{cut(S)}{Vol(G)Sep(S)} = 0.0622$. In bottom-left we show the densification H result solving the spectral version of problem **P1** (Eq. 2) through the dual problem **P2** (Eq. 6) for $C = 2$. Red-dotted lines have weight 0.001. Some cuts have lower values, for instance the one for $S = \{2, 4\}$, whereas others such as the cut for $S = \{1, 2\}$ increase (bottom-right). This is important since the new volume has also increased. All cuts satisfy $\frac{cut_H(S)}{m(H)} \leq C \cdot \frac{cut_G(S)}{m(G)}$. (Color figure online)

Then, the dual problem of $\mathbf{P2_{SDP}}$, i.e. the SDP relaxation of $\mathbf{P1}$ is

$$\mathbf{P1_{SDP}} \quad Max \sum_{u,v} x'_{uv}$$

$$s.t. \quad \forall\, u,v : x'_{uv} \leq 1$$

$$\sum_{u,v} x'_{uv} b_{uv} b_{uv}^T \preceq \left(C \cdot \sum_{u,v} x'_{uv} \right) L_G$$

$$x'_{uv} \geq 0. \tag{14}$$

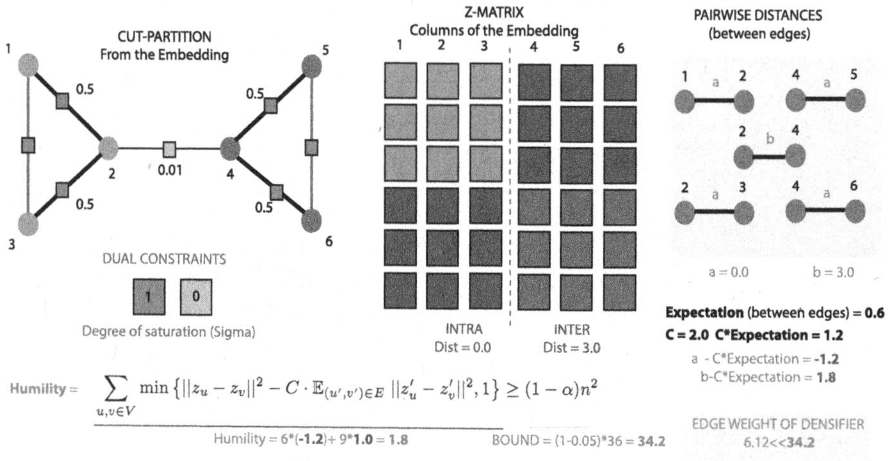

Fig. 2. SDP Dual Solution. Middle: Z-matrix whose columns $z_1, \ldots, z_n$ are the embedding coordinates. Such embedding is optimal insofar it assigns similar coordinates to vertices separated by the sparse cut $S = \{1, 2, 3\}$. Intra-class pairwise distances between columns are close to zero where inter-class distances are close to 3.0. Then the Z matrix encodes the sparse cut itself. Right: to estimate to what extend the columns of Z define a humble embedding, we commence by compute the distances associated with the edges of the graph. This yields $\mathbb{E}_{(u',v')\in E} \|z'_u - z'_v\|^2 = 0.6$ where the average is distorted due to the edge $(2, 4)$. Regarding edge pairs, deviations from the expectation are -1.2 for inter-class edges and $+1.8$ for the only inter-class edge. When considering non-edge pairs, for inter-class pairs we have a deviation of $3.0 - 0.6 = 2.4$, whereas for inter-class non-edge pairs, basically $(1, 3)$ and $(5, 6)$ the deviation is negative: -0.6. Therefore, for computing the *humility* of the embedding (see text) we have only 6 deviation smaller than the unit: 4 of these deviations correspond to inter-class edges and 2 of them to intra-class edges. The remainder correspond to 9 non-edge pairs. The resulting humility is 1.8 meaning that $(1 - \alpha)n^2 = 1.8$, i.e. $\alpha = 0.95$. Therefore, the graph has not a one-sided C-multiplicative spectral densifier with edge weight more than $\alpha n^2 = 34.2$. Actually, the weight of the obtained spectral densifier is 6.12. Left: summary of the process in the graph. The colors of the vertices define the grouping given by Z. The colors of the squares indicate whether σ_{uv} are close to 0 (unsaturated constraint) or close to 1 (saturated constraint). Only σ_{24} is unsaturated since $(2, 4)$ distorts the expectation. Variables corresponding two non-edges but linking intra-class vertices are also saturated, namely σ_{13} and σ_{56} (both have a negative deviation). The remaining pairs are unsaturated and they are not plotted for the sake of simplicity.

As in the combinatorial version of densification, first we solve the dual and then the primal. The solution of $\mathbf{P2}_{SDP}$ provides σ_{uv} as well as the coordinates of the optimal embedding (in terms of avoiding the collapse of distances) in the columns of Z. In Fig. 2 we explain how the dual solution is obtained for the graph in Fig. 1. We denote the right hand of Eq. 11 as *humility*. The higher the humility the lower the maximum weight of the spectral densifier (as in the combinatorial case).

3   Discussion and Conclusions

With the primal SDP problem $\mathbf{P1}_{SDP}$ at hand we have that $\lambda_i' \leq \left(C \cdot \sum_{u,v} x_{uv}' \right) \lambda_i$ where λ_i' are the eigenvalues of the Laplacian $L_H = \sum_{u,v} x_{uv}' b_{uv} b_{uv}^T$ associated with the densified graph H. For $C > 1$ we have that densification tends to produce a quasi complete graph $\mathcal{K}_n$. When we add to the cost of the dual problem $\mathbf{P2}_{SDP}$ the term $-K \log \det(Z)$ (a log-barrier) enforces choices for $Z \succeq 0$ (i.e. ellipsoids) with maximal volume which also avoids $\mathcal{K}_n$. In this way, given a fixed $K = 1000$, the structure of the pattern space emerges[1]

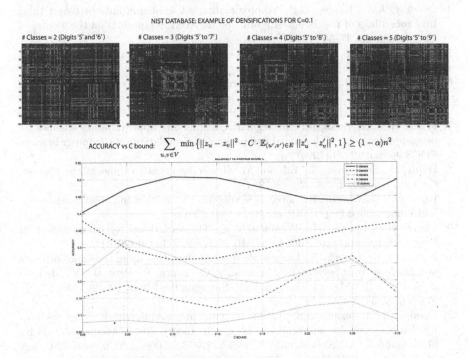

Fig. 3. Densification results for the NIST database.

---

[1] All examples/experiments were obtained with the SDPT3 solver [23] version 4.0. In our experiments, the number of variables is $|E| \approx 4500$ and the SDP solver is polynomial with $|E|$.

as we modify the $C < 1$ bound so that the spectral gap is minimized in such a way that reasonable estimations of the commute distance emerge. In Fig. 3 we summarize some experiments done by subsampling the NIST digit database. Given the densifications (more dense in red) the commute time matrix is estimated and the accuracy w.r.t. the ground truth is plotted. Accuracy decreases with the number of classes and in many cases the optimal value is associated with low values of C. The quality of the results is conditioned by the simplicity of the optimization problem (guided only by a *blind* cut similarity, which does not necessarily impose to reduce inter-class noise) but it offers a nice path to explore.

References

1. Alamgir, M., von Luxburg, U.: Shortest path distance in random k-nearest neighbor graphs. In: Proceedings of ICML 2012 (2012)
2. Arora, S., Karger, D., Karpinski, M.: Polynomial time approximation schemes for dense instances of NP-hard problems. J. Comput. Syst. Sci. **58**(1), 193–210 (1999)
3. Batson, J.D., Spielman, D.A., Srivastava, N., Teng, S.: Spectral sparsification of graphs: theory and algorithms. Commun. ACM **56**(8), 87–94 (2013)
4. Benczúr, A.A., Karger, D.R.: Approximating s-t minimum cuts in $O(n^2)$ time. In: Proceedings of the Twenty-Eighth Annual ACM Symposium on the Theory of Computing, pp. 47–55 (1996)
5. Cai, D., Chen, X.: Large scale spectral clustering via landmark-based sparse representation. IEEE Trans. Cybern. **45**(8), 1669–1680 (2015)
6. Chen, J., Fang, H., Saad, Y.: Fast approximate kNN graph construction for high dimensional data via recursive lanczos bisection. J. Mach. Learn. Res. **10**, 1989–2012 (2012)
7. Frieze, A.M., Kannan, R.: The regularity lemma and approximation schemes for dense problems. In: 37th Annual Symposium on Foundations of Computer Science, FOCS 96, pp. 12–20 (1996)
8. Hardt, M., Srivastava, N., Tulsiani, M.: Graph densification. Innovations Theoret. Comput. Sci. **2012**, 380–392 (2012)
9. Khoa, N.L.D., Chawla, S.: Large scale spectral clustering using approximate commute time embedding. CoRR abs/1111.4541 (2011)
10. Vladymyrov, M., Carreira-Perpinan, M.A.: The Variational Nystrom method for large-scale spectral problems. In: ICML 2016, pp. 211–220 (2016)
11. Khoa, N.L.D., Chawla, S.: Large scale spectral clustering using resistance distance and spielman-teng solvers. In: Ganascia, J.-G., Lenca, P., Petit, J.-M. (eds.) DS 2012. LNCS (LNAI), vol. 7569, pp. 7–21. Springer, Heidelberg (2012). doi:10.1007/978-3-642-33492-4_4
12. Komlós, J., Shokoufandeh, A., Simonovits, M., Szemerédi, E.: The regularity lemma and its applications in graph theory. In: Khosrovshahi, G.B., Shokoufandeh, A., Shokrollahi, A. (eds.) TACSci 2000. LNCS, vol. 2292, pp. 84–112. Springer, Heidelberg (2002). doi:10.1007/3-540-45878-6_3
13. Liu, W., He, J., Chang, S.: Large graph construction for scalable semi-supervised learning. In: Proceedings of ICML 2010, pp. 679–686 (2010)
14. Liu, W., Mu, C., Kumar, S., Chang, S.: Discrete graph hashing. In: NIPS 2014, pp. 3419–3427 (2014)

15. Liu, W., Wang, J., Chang, S.: Robust and scalable graph-based semisupervised learning. Proc. IEEE **100**(9), 2624–2638 (2012)
16. Liu, W., Wang, J., Kumar, S., Chang, S.: Hashing with graphs. In: Proceedings of ICML 2011, pp. 1–8 (2011)
17. Luo, Z., Ma, W., So, A.M., Ye, Y., Zhang, S.: Semidefinite relaxation of quadratic optimization problems. IEEE Sig. Process. Mag. **27**(3), 20–34 (2010)
18. Qiu, H., Hancock, E.R.: Clustering and embedding using commute times. IEEE TPAMI **29**(11), 1873–1890 (2007)
19. Spielman, D.A., Srivastava, N.: Graph sparsification by effective resistances. SIAM J. Comput. **40**(6), 1913–1926 (2011)
20. von Luxburg, U., Alamgir, M.: Density estimation from unweighted k-nearest neighbor graphs: a roadmap. In: NIPS 2013, pp. 225–233 (2013)
21. von Luxburg, U., Radl, A., Hein, M.: Getting lost in space: large sample analysis of the resistance distance. In: NIPS 2010, pp. 2622–2630 (2010)
22. von Luxburg, U., Radl, A., Hein, M.: Hitting and commute times in large random neighborhood graphs. J. Mach. Learn. Res. **15**(1), 1751–1798 (2014)
23. Toh, K.C., Todd, M., Tutuncu, R.: SDPT3 - A MATLAB software package for semidefinite programming. Optim. Methods Softw. **11**, 545–581 (1998)

Outlier Robust Geodesic K-means Algorithm for High Dimensional Data

Aidin Hassanzadeh[1(✉)], Arto Kaarna[1], and Tuomo Kauranne[2]

[1] Machine Vision and Pattern Recognition Laboratory,
School of Engineering Science, Lappeenranta University of Technology,
Lappeenranta, Finland
{aidin.hassanzadeh,arto.kaarna}@lut.fi
[2] Mathematics Laboratory, School of Engineering Science,
Lappeenranta University of Technology, Lappeenranta, Finland
tuomo.kauranne@lut.fi

Abstract. This paper proposes an outlier robust geodesic K-mean algorithm for high dimensional data. The proposed algorithm features three novel contributions. First, it employs a shared nearest neighbour (SNN) based distance metric to construct the nearest neighbour data model. Second, it combines the notion of geodesic distance to the well-known local outlier factor (LOF) model to distinguish outliers from inlier data. Third, it introduces a new ad-hoc strategy to integrate outlier scores into geodesic distances. Numerical experiments with synthetic and real world remote sensing spectral data show the efficiency of the proposed algorithm in clustering of high-dimensional data in terms of the overall clustering accuracy and the average precision.

Keywords: Clustering · K-means · High-dimensional data · Geodesic distance · Shared nearest neighbour · Local outlier factor

1 Introduction

The K-means algorithm is one of the widely used clustering algorithms in the area of cluster analysis and it has been integrated into various image processing, data mining and pattern recognition applications. K-means is basically an objective function based optimization scheme that iteratively assigns data to K clusters while attempting to minimize intra-cluster variation. The K-means algorithm is simple and scalable to a wide range of data types. K-means assumes Euclidean distance as the dissimilarity metric and thus tends to produce clusters of spherical shape. Although this assumption has been shown to be reasonable for many applications [10,11], it is not universally true with data clusters of non-spherical and complex shapes, such as spatial data and hyperspectral remote sensing imaging. Moreover, the classic K-means algorithm can adversely be affected by outliers and thus is not able to achieve realistic results if the clusters are contaminated by outlying data.

© Springer International Publishing AG 2016
A. Robles-Kelly et al. (Eds.): S+SSPR 2016, LNCS 10029, pp. 252–262, 2016.
DOI: 10.1007/978-3-319-49055-7_23

Several variants of K-means algorithm have been introduced to address these shortcomings [4,7,11]. The density sensitive geodesic K-means algorithm, henceforth DGK-means, proposed by [1] is an approach that tries to address the issues of non-spherical clusters and outlying data. The DGK-means algorithm replaces the Euclidean distance with a manifold based geodesic distance metric which is resistant to outliers. This algorithm suffers from two main difficulties: first, it can easily be affected by the curse of dimensionality, and second it may fail if the data clusters come from different density patterns.

This paper investigates the DGK-means and proposes an outlier robust geodesic distance based K-means algorithm, called ORGK-means. The proposed algorithm is similar to the DGK-means but utilizes a new geodesic distance metric. By this means, the ORGK-means algorithm attempts to address the issues of high-dimensionality, data of varying cluster densities and data with outliers.

The proposed ORGK-means algorithm includes three main contributions. First, an alternative distance measure based on the notion of shared nearest neigbor (SNN) is proposed for pairwise distance estimation. SNN, originally introduced as a similarity measure based on nearest neighbours [6], is considered an efficient method for problems involving high-dimensional data, clustering of data of varying size and distribution and data contaminated with outliers [5]. Its utilization has been reported in several applications with high-dimensional data [3,8,9,12] and outlier-scoring algorithms [13]. Second, the well-known local outlier factor (LOF) based on the notion of geodesic distance is used to rank outlierness of data. By using geodesic distance based LOF, the ORGK-means algorithm is expected to be more robust to density fluctuations. Third, to provide more flexibility in modelling and improved usability, a double sigmoid function with adaptive parameter estimation is proposed to integrate outlier scores into the distances.

The remainder of the paper is organized as follows. Section 2 briefly reviews the DGK-means algorithm, presenting the main steps involved. Section 3 introduces the new elements proposed to address the shortcomings in the DGK-means algorithm. Section 4 presents experimental results and evaluations. Section 5 concludes the paper.

2 Density Sensitive Geodesic K-means Algorithm

There are three main features in the DGK-means algorithm: general distance K-means, density sensitive geodesic distance, and geodesic K-means.

2.1 General Distance K-means

The DGK-means reformulates the whole update process in the classic K-means to a generative procedure, called general distance K-means, that can utilize any distance metric. Let $X = \{\mathbf{x}_i\}_{i=1}^{n}$ be the set of n real-valued data points of dimension p to be clustered onto m data clusters $C = \{C_l\}_{l=1}^{m}$. Given the distance metric $d : X \times X \to \mathbb{R}^+$, the general K-means algorithm aims to minimize the objective loss function:

$$W_{GD}(X) = \sum_{l=1}^{m} \sum_{\mathbf{x}_i \in C_l} \sum_{\mathbf{x}_j \in C_l} d^2(\mathbf{x}_i, \mathbf{x}_j). \tag{1}$$

Let (X, d) be the metric space. Provided that X can be mapped onto Euclidean space, the minimization of the objective function in Eq. 1 can be performed iteratively without explicitly calculating the cluster centroids. Denoting $\gamma_{t+1}(\mathbf{x}_i) : X \to \{l\}_{l=1}^{m}$ as the cluster membership function at iteration $t+1$, the update cluster assignment for every data instance $\mathbf{x}_i$ is given as follows:

$$\gamma_{t+1}(\mathbf{x}_i) = \arg \min_{1 \le l \le m} \left(\frac{2}{n_l(t)} \sum_{\mathbf{x}_r \in C_l(t)} d^2(\mathbf{x}_i, \mathbf{x}_r) - \frac{1}{n_l^2(t)} \sum_{\mathbf{x}_r, \mathbf{x}_{r'} \in C_l(t)} d^2(\mathbf{x}_r, \mathbf{x}_{r'}) \right). \tag{2}$$

2.2   Density Sensitive Geodesic Distance

The geodesic distance is the fundamental element in the DGK-means algorithm imposing the global structure of the data. Through the sparse k-neighbourhood graph representation of the data, denoting the data points as the graph nodes and the corresponding pairwise distances as the edge-weights, the geodesic distance between two data points is given by the sum of the edge weights of the shortest path connecting them. By this means, the geodesic distance of points residing on different geometrical structures is of higher values compared to those from similar geometrical structures.

Geodesic distance based on pure Euclidean distance may be inaccurate in the presence of outliers. In order to reduce the effects of outliers, the DGK-means algorithm incorporates the outlierness profile of data into geodesic distance estimation. It achieves this by defining the graph edge weights through combining the pairwise distances of the data points with their local densities. In particular, DGK-means uses an exponential transfer model to compute graph edge weights ω_{ij} that adjusts the pairwise Euclidean distances based on the local densities of their end points:

$$\omega_{ij} = \exp \left(\frac{1}{\sigma^2} \max \left(\hat{f}(\mathbf{x}_i), \hat{f}(\mathbf{x}_j) \right) \right) \|\mathbf{x}_i - \mathbf{x}_j\|, \tag{3}$$

Here $\hat{f}(\mathbf{x}_i)$ is the local density estimate of point $\mathbf{x}_i$ with respect to its local neigborhood and is computed using the multivariate k-NN density estimator:

$$\hat{f}(\mathbf{x}_i) = \frac{k-1}{n \ vol(\mathbf{x}_i)} \tag{4}$$

This formulation produces robust geodesic distances in low dimensions, but it does not perform well in high dimensions. First, the geodesic distances on the

k-nearest neighbour graph based on Euclidean distance suffer from concentration problem in high dimensions and can not capture thoroughly the similarity of data points [4]. Second, the k-NN density estimator in the DGK-means algorithm can easily be affected by the curse of dimensionality. k-NN density estimator relies on the computation of the volume of the sphere to represent the far-end local neighbourhood of data which is numerically intractable with a low number of data samples in high dimensions.

2.3   Geodesic K-means

The DGK-means as per the general distance approach requires all intra-cluster distances to be computed and is computationally expensive. The DGK-means algorithm may be reformulated within a randomized baseline that eliminates the multiple invocation of intra-cluster pairwise distance computations. The DGK-means achieves this through a randomized process in which virtual cluster centroids are estimated over a random sample set of each data cluster at each iteration.

3   Outlier Robust Geodesic K-means Algorithm

The ORGK-means follows the same outline as that of DGK-means algorithm but introduces three particular improvements in the formulations of the pairwise distances, the density estimation and the weighting transform model.

3.1   Distance Metric Based on SNN

The proposed ORGK-means algorithm adopts an alternative strategy based on the concept of SNN similarity to compute pairwise distances. Distance measure based on SNN, also referred to as the secondary SNN-distance measure, have been shown efficient in high dimensional data space [5,9] that can perform well with data of different size, shape and varying distributions [3]. SNN-based similarity of two data points is the degree by which their underlying patterns overlap with one another. In terms of the sparse k-neighbourhood graph as described in Sect. 2.2, the SNN-similarity of two data points is seen as the number of points shared by the k-nearest neighbour lists of those points.

Given the set of k-nearest neighbours $\mathcal{N}_k(\mathbf{x}_i)$ and $\mathcal{N}_k(\mathbf{x}_j)$ of the points $\mathbf{x}_i$ and $\mathbf{x}_j$, the SNN similarity is given by the number of their common neighbours:

$$sim_{ssN_k}(\mathbf{x}_i, \mathbf{x}_j) = |\mathcal{N}_k(\mathbf{x}_i) \cap \mathcal{N}_k(\mathbf{x}_j)|. \tag{5}$$

The normalized SNN similarity measure $simcos_k$ is defined as follows:

$$simcos_k(\mathbf{x}_i, \mathbf{x}_j) = \frac{sim_{ssN_k}(\mathbf{x}_i, \mathbf{x}_j)}{k}. \tag{6}$$

Several SSN-based distance measures can be constructed based on the $simcos_k$ metric [5]. In this work, the SNN based inverse distance $dinv_k$ is utilized.

$$dinv_k(\mathbf{x}_i, \mathbf{x}_j) = 1 - simcos_k(\mathbf{x}_i, \mathbf{x}_j) \tag{7}$$

3.2   Geodesic Based Local Outlier Factor

The ORGK-means algorithm utilizes a geodesic based local outlier factor ($gLOF$) to rank the outlierness of data. LOF, originally introduced by Breunig et al. [2], is an outlier scoring algorithm that relies on the k-nearest neighbour model and the notion of local reachability density. $gLOF$ enhances LOF with the geodesic distance.

As the same way in LOF, $gLOF$ benefits from several advantages. It is a non-parametric and model-free approach that does not make any assumption regarding the distribution of data. It is resistant to changes in density of data distribution patterns. In addition, $gLOF$ compared to LOF incorporates both local and global structure of the data and it provides richer outlier scoring scheme.

To compute the outlier score of an individual point, $gLOF$ compares its local density with the points in its neighbourhood. It takes the ratio of the local reachability density of the data point to the median local reachability density of its surrounding neighbours. This is different from the original LOF where the arithmetic mean is utilized to approximate the local neighbourhood reachability density. The idea in utilizing the median is to make the density estimator more robust to the outlying points. When an outlying point or a point belonging to other clusters is located in the neighbourhood, the mean is likely to misrepresent (underestimate) the dispersion of neighbouring local densities as they are dominated by the local density of that outlying point. In such cases, the median is considered a reasonable choice that is more robust to outliers.

Formally, the $gLOF$ score of a point $\mathbf{x}_i$ is given by:

$$gLOF_k(\mathbf{x}_i) = \frac{median \ \{lrd_k(\mathbf{x}_j)\}_{\mathbf{x}_j \in \mathcal{N}_k(\mathbf{x}_i)}}{lrd_k(\mathbf{x}_i)}, \tag{8}$$

where $lrd(\mathbf{x}_i)$ describes the local reachability density of the point $\mathbf{x}_i$ over its local neighbourhood.

The local reachability density is loosely estimated as the inverse of the median of the reachability geodeisc distances to the point $\mathbf{x}_i$ from its neighbours. The local reachability density at point $\mathbf{x}_i$ is defined as follows:

$$lrd_k(\mathbf{x}_i) = \left(median \ \left\{ rd_k(\mathbf{x}_i, \mathbf{x}_j) \right\}_{\mathbf{x}_j \in \mathcal{N}_k(\mathbf{x}_i)} \right)^{-1}, \tag{9}$$

where $rd_k(\mathbf{x}_i, \mathbf{x}_j)$ refers to the reachability geodesic distance from $\mathbf{x}_i$ to $\mathbf{x}_j$ given by:

$$rd_k(\mathbf{x}_i, \mathbf{x}_j) = max(R(\mathcal{N}_k(\mathbf{x}_i)), d_{G,W}(\mathbf{x}_i, \mathbf{x}_j)). \tag{10}$$

Here, $R(\mathcal{N}_k(\mathbf{x}_i))$ is the geodesic distance from $\mathbf{x}_i$ to its k-th nearest neighbour and $d_{G,W}(\mathbf{x}_i, \mathbf{x}_j)$ is the geodesic distance from $\mathbf{x}_i$ to $\mathbf{x}_j$. The geodesic distance is defined by the shortest path over the k shared nearest neighbourhood graph representation with the SNN based inverse distance metric $dinv_k$. The reachability distance $rd_k(\mathbf{x}_i, \mathbf{x}_j)$ is asymmetric and its role is to enhance the stability of results. It is defined to smooth out the statistical fluctuations in $d_{G,W}(.,.)$ when it is small compared to the distance of k-th neighbouring point. The larger the value of k, the higher smoothing is applied.

The $gLOF$ value of a point $\mathbf{x}_i$ located in a region of homogeneous density (inlier) is approximately 1, but it is of higher value if the density of the local neighbourhood of its neighbours is higher that the density of the local neighbourhood of the point itself (outlier).

3.3 Weighting Transform Model

The first-order exponential weighting model used in the DGK-means algorithm is designed to map density values within $[0\ 1)$ interval onto $[1\ \infty)$. Such weighting model does not suit to the ORGK-mean algorithm where the outlierness of the data is ranked by $gLOF$ scores not limited to the range $[0\ 1]$, and either the normalization is not straightforward. In addition, the model used in DGK-means strongly depends on the scaling parameter σ whose selection is not well defined [1].

The ORGK-means algorithm relies on a sigmoid function model to transform the outlier scores to the geodesic distances where the extreme outlier scores are eliminated. Specifically, the weighting transform model is built upon the double sigmoid function whose parameters can be adaptively tuned by the statistics of outlier score distribution in an ad hoc manner.

Denote the outlier scores of the points $\mathbf{x}_i$ and $\mathbf{x}_j$, obtained from $gLOF$ model, by s_i and s_j respectively. The proposed weighting function is given by:

$$
\omega_{ij} = \begin{cases} \left(1 + \beta\ exp\left[-2\ \dfrac{max(s_i, s_j) - \tau}{\alpha_1}\right]\right)^{-1} & \text{if } s < \tau, \\[2em] \left(1 + \beta\ exp\left[-2\ \dfrac{max(s_i, s_j) - \tau}{\alpha_2}\right]\right)^{-1} & otherwise \end{cases} \tag{11}
$$

where β is a scaling parameter and τ is the threshold parameter beyond which a data point is considered as an outlier. α_1 and α_2 are edge parameters that define the region $[\tau - \alpha_1\ \tau + \alpha_2]$ at which the weighing function is approximately linear.

Choosing a proper value for the threshold parameter τ is highly dependent on the data as $gLOF$ produces a varying range of scores relative to the underlying local and global structures of data. However, since the scores obtained by $gLOF$ are typically positively skewed, the value of parameter τ can be set to the mean of the score distribution. In positively skewed distribution, the mean pulls toward the direction of skew (the direction of the outliers) and therefore can provide an approximate basis for the decision about the outlierness of data.

The mean of the score distribution can reasonably approximate the maximal inlier score value but it can be of too large values if there are erratic deviations in score values or the distribution is extremely skewed. To address this issue, the mean is estimated over truncated data such that a certain percentage of data, o_p, corresponding to the largest scores is discarded. The mean obtained in this manner resembles the truncated mean estimator that is less sensitive to extreme outliers.

Inspired by the definition of skewness as the measure of asymmetry about the mean, the value of α_1 can be set to $truncmean_{o_p}(s) - mode(s)$ and the value of α_2, not as critical as τ and α_1, can be set to any value lower than $max(s) - truncmean_{o_p}(s)$.

4   Experimental Results

Evaluation of the proposed ORGK-means algorithm for clustering in high dimensions was carried using synthetic and real remote sensing hyperspectral data.

4.1   Test Data

Synthetic Data: To investigate the ability of the proposed ORGK-means algorithm several datasets of different dimensionality were generated. The datasets were constructed to generate unbiased data sampled from the Gaussian distribution over an increasing range of dimensions that share certain attributes, [2, 4, 8, 16, 32, 64, 128, 256, 512]. Outliers were uniformly scattered to the space with the ratio of 5% and 10% of the total number of samples. Inspired by [5], data series were generated from three classes of *8-Relevant*, *Half-Relevant* and *All-Relevant* each differing in the portion of informative variables that are relevant to clusters. Each dataset contained 500 samples that are uniformly divided into 7 clusters whose mean and variance were uniformly randomized ensuring data comes from well separated clusters of various distributions and that every pair of cluster has 10% overlap at most.

Real Data: Two different benchmark hyperspectral image datasets were used for experiments: *Botswana* and *SalinasA*. The *Botswana* dataset was acquired by NASA EO-1 satellite over the Okavango Delta, and the corrected version of the data includes 145 bands. 7 out of the 14 classes with an equal number of samples were chosen in the experiment. 1575 data samples of dimension 145 are distributed to 7 classes. The *SalinasA* dataset was collected by AVIRIS over Salinas Valley in southern California, USA. The test hypercube consists of 7138 samples of dimension 204 comprising of 6 different classes[1].

4.2   Results and Discussion

The performance of the proposed ORGK-means was compared to a number of methods including classic K-means, k-NN based geodesic K-means, SNN based geodesic K-means and density based geodesic K-means, respectively denoted as K-mean, GK-means/kNN, GK-means/SNN and DGK-means. To ensure fair evaluation, all the algorithms were experimented with the pre-known number of clusters and identical randomly initialized data-to-cluster assignments. Given the number of clusters as well as the initial data-to-cluster assignments, the proposed ORGK-means algorithm requires the parameters the top percentage of outliers o_p and the number of nearest neighbours k to be specified.

---

[1] These datasets are available at http://www.ehu.eus/ccwintco/index.php?title=Hyperspectral_Remote_Sensing_Scenes.

The top outlier percentage o_p defines the percentage of data points that are considered as the extreme outliers to be excluded. It can be set empirically by a domain expert or can be approximated by the analysis of the outlier score distribution. Here, in the case of the synthetic datasets, the value for o_p was chosen based on the original percentage of added outliers and in the case of the real datasets, it was chosen empirically by searching the range from 1 % to 10 %.

The number of nearest neighbours k defines the neighbourhood size in the NN-model and significantly affects the performance of computing geodesic distances and outlier scores. The number of nearest neighbour versus overall accuracy was searched within the range form 5 to 150 and the best cases were only reported for the methods used.

Performance comparisons of the proposed ORGK-means algorithm and the other methods applied to synthetic data are shown in Fig. 1, giving the

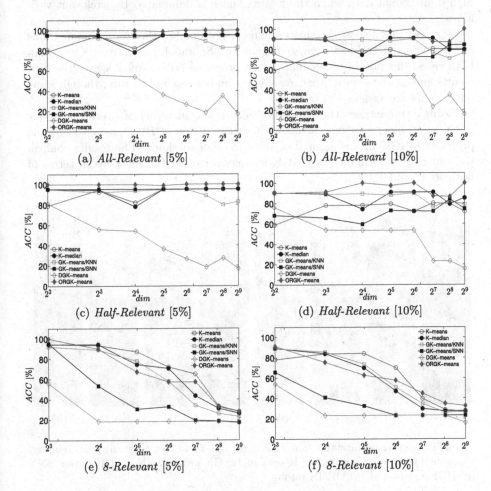

(a) *All-Relevant* [5%]   (b) *All-Relevant* [10%]

(c) *Half-Relevant* [5%]   (d) *Half-Relevant* [10%]

(e) *8-Relevant* [5%]   (f) *8-Relevant* [10%]

Fig. 1. Clustering performance comparison on the synthetic dataset over increasing dimensionality; (a–b) *All-Relevant (%)*, (c–d) *Half-Relevant*, (e–f) *8-Relevant*.

overall accuracy versus the scaling dimension. Simulation results on *All-Relevant* and *Half-Relevant* data series show the improvement introduced by ORGK-means over the other competing methods, specifically GK-means/kNN, GK-means/SNN and DGK-means. However, there are a few exceptions in which the K-median stands as the out-performer. This observation can be explained by the synthetic data having been generated from an isotropic Gaussian mixture model residing on linear space where K-means equipped with geodesic distance does not necessarily yield higher separability power.

The results on *8-Relevant* as expected show that the performance of all compared clustering algorithms affected as dimension grows. This is to confirm the hampering effects of irrelevant attributes on the distinguishability of data clusters present in high dimensions. This observation also indicates that the ORGK-means algorithm, similarly to the other compared methods, is not able to handle high dimensional data when the feature space is dominated by irrelevant variables.

Figures 2 and 3 show the clustering maps obtained by the competing methods on real hyperspectral test data, *Botswana* and *SalinasA*, respectively. Overall, in both cases, but notably in *SalinasA*, the proposed ORGK-means gave the best results among the competitors, reaching a higher separation rate, though some data points are assigned as outliers.

Table 1 summarises the clustering accuracies, in terms of overall accuracy (ACC) and macro average Positive Predictive Value (PPV_m), achieved by the proposed ORGK-means algorithm and the other methods. The results confirm the superior performance of ORGK-means over other the methods in terms of both ACC and PPV_m.

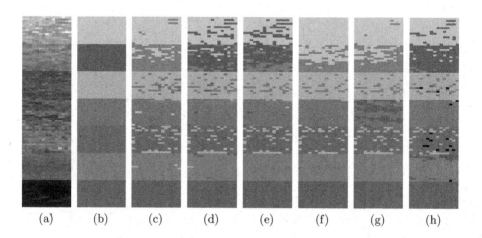

(a)   (b)   (c)   (d)   (e)   (f)   (g)   (h)

Fig. 2. Clustering performance comparison on *Botswana* test data; (a) RGB rendering, (b) ground truth (c) K-means, (d) K-median, (e) GK-means/kNN, (f) GK-means/SNN (g) DGK-means and (h) ORGK-means.

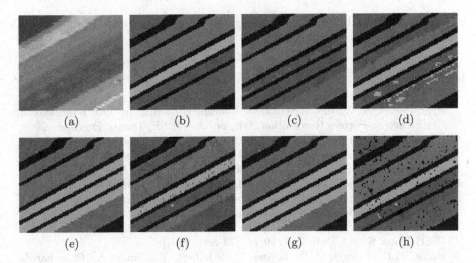

Fig. 3. Clustering performance comparison on *SalinasA* test data; (a) RGB rendering, (b) ground truth (c) K-means, (d) K-median, (e) GK-means/kNN, (f) GK-means/SNN (g) DGK-means and (h) ORGK-means.

Table 1. Highest overall clustering ACC and macro averaged PPV (in percentage) obtained by the considered clustering algorithms.

Methods	Botswana			SalinasA		
	k	ACC [%]	PPV_m [%]	k	ACC [%]	PPV_m [%]
K-means	-	77.8	71.5	-	61.8	46.0
K-median	-	86.2	87.7	-	52.7	58.7
GK-means/kNN	75	84.2	87.1	10	59.4	45.5
GK-means/SNN	125	81.5	73.2	10	59.0	57.5
DGK-means	15	74.2	73.1	10	32.4	45.5
ORGK-means	125	89.3	90.9	50	83.8	91.3

5   Conclusion

In this paper, an outlier robust geodesic K-means (ORGK-means) algorithm is proposed for clustering of high dimensional data. The proposed ORGK-means extends the standard K-means algorithm by using an outlier-adjusted geodesic distance. In the proposed ORGK-means algorithm, SNN based distance metric is utilized as the pairwise dissimilarity measure. Geodesic based LOF, exploiting both local and global structural information, is introduced to rank the degree of outlierness, and an adaptive weighting transform model based on the double sigmoid function is proposed to adjust geodesic distances. The efficiency of the proposed ORGK-means algorithm in clustering high-dimensional data was evaluated using synthetic and real world remote sensing spectral data. The numerical

A. Hassanzadeh et al.

results on the overall clustering accuracy and the average precision showed the utility of the proposed algorithm.

References

1. Asgharbeygi, N., Maleki, A.: Geodesic k-means clustering. In: 19th International Conference on Pattern Recognition 2008, pp. 1–4. IEEE, Tampa, December 2008
2. Breunig, M.M., Kriegel, H.P., Ng, R.T., Sander, J.: LOF: identifying density-based local outliers. In: Proceedings of the 2000 ACM SIGMOD International Conference on Management of Data, SIGMOD 2000, pp. 93–104. ACM, New York (2000)
3. Ertöz, L., Steinbach, M., Kumar, V.: Finding clusters of different sizes, shapes, and densities in noisy, high dimensional data. In: Proceedings of Second SIAM International Conference on Data Mining. SIAM (2003)
4. Francois, D., Wertz, V., Verleysen, M.: The concentration of fractional distances. IEEE Trans. Knowl. Data Eng. **19**(7), 873–886 (2007)
5. Houle, M.E., Kriegel, H.-P., Kröger, P., Schubert, E., Zimek, A.: Can shared-neighbor distances defeat the curse of dimensionality? In: Gertz, M., Ludäscher, B. (eds.) SSDBM 2010. LNCS, vol. 6187, pp. 482–500. Springer, Heidelberg (2010). doi:10.1007/978-3-642-13818-8_34
6. Jarvis, R., Patrick, E.A.: Clustering using a similarity measure based on shared near neighbors. IEEE Trans. Comput. **C–22**(11), 1025–1034 (1973)
7. Kaufman, L., Rousseeuw, P.J.: Finding Groups in Data: An Introduction to Cluster Analysis, chap. 2, pp. 68-125. Wiley (2008)
8. Moëllic, P.A., Haugeard, J.E., Pitel, G.: Image clustering based on a shared nearest neighbors approach for tagged collections. In: Proceedings of the 2008 International Conference on Content-Based Image and Video Retrieval, CIVR 2008, pp. 269–278. ACM, New York (2008)
9. Tomasev, N., Mladeni, D.: Hubness-aware shared neighbor distances for high-dimensional k-nearest neighbor classification. Knowl. Inf. Syst. **39**(1), 89–122 (2014)
10. Wang, D., Ding, C., Li, T.: K-subspace clustering. In: Buntine, W., Grobelnik, M., Mladenić, D., Shawe-Taylor, J. (eds.) ECML PKDD 2009. LNCS (LNAI), vol. 5782, pp. 506–521. Springer, Heidelberg (2009). doi:10.1007/978-3-642-04174-7_33
11. Wu, J.: Cluster analysis and k-means clustering: an introduction. In: Wu, J. (ed.) Advances in K-means Clustering. Springer Theses, pp. 1–16. Springer, Heidelberg (2012)
12. Yin, J., Fan, X., Chen, Y., Ren, J.: High-dimensional shared nearest neighbor clustering algorithm. In: Wang, L., Jin, Y. (eds.) FSKD 2005. LNCS (LNAI), vol. 3614, pp. 494–502. Springer, Heidelberg (2005). doi:10.1007/11540007_60
13. Zimek, A., Schubert, E., Kriegel, H.P.: A survey on unsupervised outlier detection in high-dimensional numerical data. Stat. Anal. Data Min. **5**(5), 363–387 (2012)

Adaptive Sparse Bayesian Regression with Variational Inference for Parameter Estimation

Satoru Koda[✉]

Graduate School of Mathematics, Kyushu University, Fukuoka, Japan
s-kouda@math.kyushu-u.ac.jp

Abstract. A relevance vector machine (RVM) is a sparse Bayesian modeling tool for regression analysis. Since it can estimate complex relationships among variables and provide sparse models, it has been known as an efficient tool. On the other hand, the accuracy of RVM models strongly depends on the selection of their kernel parameters. This article presents a kernel parameter estimation method based on variational inference theories. This approach is quite adaptive, which enables RVM models to capture nonlinearity and local structure automatically. We applied the proposed method to artificial and real datasets. The results showed that the proposed method can achieve more accurate regression than other RVMs.

1 Introduction

A relevance vector machine (RVM) is a sparse Bayesian modeling tool for regression analysis. The RVM has several advantages over other well-known regression models such as Support Vector Regression (SVR) and k-Nearest Neighbor. Specifically, since the RVM holds statistical backgrounds, it can make probabilistic outputs and utilize general theories regarding model selection in statistical fields. In addition, the RVM can provide sparser models and is easier to implement than the SVR. These properties have made the RVM a valuable application in various fields such as economics, engineering, and medical science.

One drawback of the RVM, however, is that it is not robust to outliers because the typical RVM assumes homogeneous variance for all samples. Hence, some methods assume *heteroscedastic variance* to deliver robust outputs, which can reduce the effect of outliers and even improve accuracy in predictions [4,5]. The expanded implementation for robust estimation is so simple that we use the algorithm in our proposal.

Another challenge is the selection of kernel parameters in basis functions. Since the accuracy of the RVM depends on them, they should be selected very carefully. Although the parameters are basically decided by cross-validation (CV) or the RVM evidence procedure, they suffer from computation time and over-fitting problems [8]. Model selection criteria are more practical approaches based on statistical viewpoints [6,11]. They can avoid the risk of over-fitting simultaneously with the parameter optimization, but they still have difficulty handling

© Springer International Publishing AG 2016
A. Robles-Kelly et al. (Eds.): S+SSPR 2016, LNCS 10029, pp. 263–273, 2016.
DOI: 10.1007/978-3-319-49055-7_24

multiple parameters. Other approaches [7,12] maximize the likelihood with a penalty term for model complexity, but these methods are effective only for data without outliers.

Taking these downsides into account, we developed a kernel parameter estimation approach on the basis of variational inference principles. The process approximates a parameter posterior distribution by minimizing the distance to its true posterior distribution [2]. Specifically, we applied the variational inference theories for non-conjugate models [13] to the kernel parameter estimation in the RVM so that the RVM optimizes these parameters during its normal training process. We also used the heteroscedastic variance and individual kernel parameters for each basis function to obtain robust and adaptive regression models. Consequently, the RVM using the proposed approach achieved more accurate and robust regression analysis.

The remainder of this paper is organized as follows. In Sect. 2, we briefly explain the basic ideas of the RVM. Our new approach is described in Sect. 3. Section 4 shows the results of our numerical experiments for benchmark datasets. Section 5 summarizes the paper.

2   Review of Relevance Vector Machine

2.1   Models

Let us suppose that training data $(\boldsymbol{x}_n, t_n) \in \mathbb{R}^d \times \mathbb{R}$ are given for $n = 1, \ldots, N$, where $\boldsymbol{x}_n$ denotes an explanatory variable and t_n denotes a scalar output. In the normal RVM, the conditional distribution for t_n is assumed to be

$$p(t_n|\boldsymbol{x}_n, \sigma^2) = N\left(t_n|y(\boldsymbol{x}_n), \sigma^2\right), \quad \sigma^2 \in \mathbb{R}_+. \tag{1}$$

The mean value of the distribution takes the following form of a linear combination with $\boldsymbol{w} = (w_0, w_1, \ldots, w_N)^T \in \mathbb{R}^{N+1}$:

$$y(\boldsymbol{x}_n) = \sum_{i=1}^{N} w_i K(\boldsymbol{x}_n, \boldsymbol{x}_i) + w_0 = \boldsymbol{\phi}(\boldsymbol{x}_n)^T \boldsymbol{w}. \tag{2}$$

Here, $\boldsymbol{\phi}(\boldsymbol{x}_n) = (1, K(\boldsymbol{x}_n, \boldsymbol{x}_1), \ldots, K(\boldsymbol{x}_n, \boldsymbol{x}_N))^T$ and

$$K(\boldsymbol{x}_n, \boldsymbol{x}_i) = \exp\left(-\gamma \|\boldsymbol{x}_n - \boldsymbol{x}_i\|^2\right), \quad \gamma \in \mathbb{R}_+. \tag{3}$$

In this paper, we use the Gaussian kernel. Note that the index $n \in \{1, \ldots, N\}$ is used for samples, and the index $i \in \{0, \ldots, N\}$ is used to describe the weight w_i and the corresponding i-th basis function $\boldsymbol{\psi}_i = (K(\boldsymbol{x}_1, \boldsymbol{x}_i), \ldots, K(\boldsymbol{x}_N, \boldsymbol{x}_i))^T$. The 0-th basis function represents a bias term.

The prior distribution of the weight $\boldsymbol{w}$ is given as the *automatic relevance determination* (ARD) prior as follows:

$$p(\boldsymbol{w}|\boldsymbol{\alpha}) = \prod_{i=0}^{N} N(w_i|0, \alpha_i^{-1}) = N(\boldsymbol{w}|\boldsymbol{0}, A^{-1}), \tag{4}$$

where $\boldsymbol{\alpha} = (\alpha_0, \ldots, \alpha_N)^T \in \mathbb{R}_+^{N+1}$ and $A = \text{diag}(\boldsymbol{\alpha})$. Each α_i is the precision parameter assigned to every weight.

Now, suppose that $\boldsymbol{t} = (t_1, \ldots, t_N)^T$, $X = (\boldsymbol{x}_1, \ldots, \boldsymbol{x}_N)^T$ and $\Phi = (\boldsymbol{\phi}(\boldsymbol{x}_1), \ldots, \boldsymbol{\phi}(\boldsymbol{x}_N))^T$. Following the Bayes' rule, the posterior distribution of the weight $\boldsymbol{w}$ can be derived as

$$p(\boldsymbol{w}|\boldsymbol{t}, X, \boldsymbol{\alpha}, \sigma^2) = \frac{p(\boldsymbol{t}|X, \boldsymbol{w}, \sigma^2)p(\boldsymbol{w}|\boldsymbol{\alpha})}{p(\boldsymbol{t}|X, \boldsymbol{\alpha}, \sigma^2)} = N(\boldsymbol{w}|\boldsymbol{m}, \Sigma), \tag{5}$$

where $\boldsymbol{m} = \sigma^{-2}\Sigma\Phi^T\boldsymbol{t}$ and $\Sigma = (\sigma^{-2}\Phi^T\Phi + A)^{-1}$. The maximum a posterior (MAP) estimate depends on the hyperparameters $\boldsymbol{\alpha}$ and σ^2, which are determined by maximizing the following marginal likelihood

$$p(\boldsymbol{t}|X, \boldsymbol{\alpha}, \sigma^2) = \int p(\boldsymbol{t}|X, \boldsymbol{w}, \sigma^2)p(\boldsymbol{w}|\boldsymbol{\alpha})d\boldsymbol{w} = N(\boldsymbol{t}|\boldsymbol{0}, C), \tag{6}$$

where the covariance matrix $C = \sigma^2 I + \Phi A^{-1}\Phi^T$.

Before we derive the analytical update procedures for the hyperparameters, let us review some advanced RVM algorithms in previous studies.

Tipping [8] and Tzikas et al. [12] both proposed using multiple kernel parameters for more adaptive regression analysis. For example, Tzikas et al. [12] assumed that each basis function has different parameters like

$$K(\boldsymbol{x}_n, \boldsymbol{x}_i) = \exp\left(-\gamma_i\|\boldsymbol{x}_n - \boldsymbol{x}_i\|^2\right), \quad \gamma_i \in \mathbb{R}_+. \tag{7}$$

In this case, suitable kernel parameters become more difficult to select. Tzikas et al. optimized the parameters by maximizing the marginal likelihood in (6) together with a penalty term for model complexity developed by Schmolck and Everson [7]. The penalty term takes

$$p(\boldsymbol{\alpha}|\sigma^2) \propto \exp\{-c\,\text{Tr}(S)\}, \tag{8}$$

where $S = \sigma^{-2}\Phi\Sigma\Phi^T$ is the hat matrix, and the constant c varies on the basis of information criteria. We follow the style of the basis functions in (7) but estimate their parameters with the variational inference described in the following section.

Another expansion is for robust estimation. Han and Zhao [4] assumed the heteroscedastic variance,

$$p(t_n|\boldsymbol{x}_n, \beta_n, \sigma^2) = N\left(t_n|y(\boldsymbol{x}_n), \sigma^2/\beta_n\right), \tag{9}$$

in the same way as Tipping and Lawrence [9]. Each precision parameter β_n associated with every sample follows the gamma distribution $\text{Gamma}(\beta_n|c_n, d_n)$. Han and Zhao estimated the parameters $\boldsymbol{\alpha}$, σ^2 and $\boldsymbol{\beta} = (\beta_1, \ldots, \beta_N)^T$ by maximizing the marginal likelihood jointly with the prior information of $\boldsymbol{\beta}$. Also, we use the heteroscedastic variance in our proposed method. In this case, the likelihood is given by

$$p(\boldsymbol{t}|X, \boldsymbol{w}, \boldsymbol{\beta}, \sigma^2) = N(\boldsymbol{t}|\Phi\boldsymbol{w}, B^{-1}), \tag{10}$$

where $B = \text{diag}\left(\boldsymbol{\beta}/\sigma^2\right)$. The posterior mean and covariance are given by $\boldsymbol{m} = \Sigma\Phi^T B\boldsymbol{t}$ and $\Sigma = (\Phi^T B\Phi + A)^{-1}$ respectively, and the covariance of the marginal likelihood turns into $C = B^{-1} + \Phi A^{-1}\Phi^T$.

2.2 Parameter Optimization

We review the sequential optimization algorithm of the RVM in the same way as Faul and Tipping [3,10] and Han and Zhao [4]. First, we review the optimization for the precision α using [10]. The basic idea is that we maximize the log marginal likelihood,

$$l(\alpha) = \log p(t|X, \alpha, \beta, \sigma^2) = -\frac{1}{2}\left(N\log 2\pi + |C| + t^T C^{-1} t\right), \qquad (11)$$

w.r.t a single α_i individually and repeat the process in turn. From the decomposition of the covariance $C = C_{-i} + \alpha_i \psi_i \psi_i^T$, we can separate the likelihood (11) into the term $l(\alpha_i)$ depending on α_i and the other term $l(\alpha_{-i})$. The $l(\alpha_i)$ is given by

$$l(\alpha_i) = \frac{1}{2}\left\{\log \alpha_i - \log(\alpha_i + s_i) + \frac{q_i^2}{\alpha_i + s_i}\right\}, \qquad (12)$$

where $s_i = \psi_i^T C_{-i}^{-1}\psi_i$ and $q_i = \psi_i^T C_{-i}^{-1}t$. The stationary point of $l(\alpha_i)$ is solved uniquely as

$$\alpha_i = \frac{s_i^2}{q_i^2 - s_i}, \qquad \text{if} \quad q_i^2 > s_i, \qquad (13)$$

$$\alpha_i = \infty, \qquad \text{otherwise.} \qquad (14)$$

When α_i goes to infinity, the posterior distribution of the corresponding weight w_i sharply peaks around zero, which means the basis function ψ_i can be pruned out from the model. The samples whose corresponding weights are nonzero are called *relevance vectors* (RVs). We also define a set of basis functions related to RVs as the active set.

The iterative formulas to update β and σ^2 are as follows:

$$\beta_n = \frac{2c_n + 1}{2d_n + \sigma^{-2}\left\{(t_n - \phi(x_n)^T m)^2 + \phi(x_n)^T \Sigma\phi(x_n)\right\}}, \qquad (15)$$

$$\sigma^2 = \frac{(t - \Phi m)^T B_0(t - \Phi m)}{N - \text{Tr}\{I - \Sigma A\}}, \qquad (16)$$

where $B_0 = \text{diag}(\beta)$.

The overall algorithm is described in Algorithm 1. We initialize $\alpha_0 = 1$ and $\alpha_i = \infty$ for all $i \neq 0$. More detailed descriptions are in the literature [4,10].

3 Variational Inference for Model Selection

The variational inference techniques based on the mean field theory have been applied to various kinds of statistical models as well as the RVM [1]. In contrast, we adopt the so-called local variational inference, which constructs the lower bound of the likelihood directly.

Algorithm 1. Sequential optimization algorithm for RVM

1: Initialize the parameters $\alpha, \beta, \sigma^2, c_n, d_n$, and $V = \{\psi_0\}$.
2: Calculate m, Σ, s_i and q_i.
3: **while** not converged **do**
4:     Select a candidate basis function ψ_i randomly.
5:     **if** $q_i^2 > s_i$ **then**
6:         Set $\alpha_i = \frac{s_i^2}{q_i^2 - s_i}$ and $V = V \cup \{\psi_i\}$.
7:         Optimize the kernel parameter γ_i in ψ_i.
8:     **else**
9:         Set $\alpha_i = \infty$ and $V = V \setminus \{\psi_i\}$.
10:    **end if**
11:    Remove all basis functions with $q_i^2 < s_i$ from V.
12:    Update β, σ^2.
13:    Update m, Σ, s_i and q_i.
14: **end while**

3.1   Variational Inference and Proposed Optimization

The proposed approach estimates a kernel parameter γ_i when the corresponding basis function ψ_i is taken into the model in Step 6 of Algorithm 1. Hence, we explain the estimation procedure for a single γ_i. Suppose that each kernel parameter γ_i has the prior distribution $p(\gamma_i) = \text{Gamma}(\gamma_i | a_i, b_i)$. Under this assumption, the marginal likelihood function can be reformulated as

$$p(t) = \int \int p(t|w, \gamma_i) p(w) p(\gamma_i) dw d\gamma_i = \int p(t|\gamma_i) p(\gamma_i) d\gamma_i, \qquad (17)$$

where $p(t|\gamma_i) = N(t|0, C)$. The covariance matrix C depends on γ_i. For simplicity, we omit the notation for X, α, β, and σ^2. The log likelihood function can be divided into the following two terms:

$$\log p(t) = L(q) + KL(q\|p), \qquad (18)$$

where

$$L(q) = \int q(\gamma_i) \log \left\{ \frac{p(t|\gamma_i) p(\gamma_i)}{q(\gamma_i)} \right\} d\gamma_i, \qquad (19)$$

$$KL(q\|p) = -\int q(\gamma_i) \log \left\{ \frac{p(\gamma_i|t)}{q(\gamma_i)} \right\} d\gamma_i. \qquad (20)$$

The $KL(q\|p)$ denotes the Kullback-Leibler (KL) divergence between distributions p and q and satisfies the inequality $KL(q\|p) \geq 0$. The lower bound $L(q)$ is maximized when the variational posterior distribution $q(\gamma_i)$ accords exactly with its true posterior distribution $p(\gamma_i|t)$. In non-conjugate priors, however, we cannot acquire any closed-form posterior distributions. Thus, we assume the posterior distribution of γ_i also has the parametric gamma distribution $q(\gamma_i) = \text{Gamma}(\gamma_i | a, b)$. Although a and b are the specific parameters of the posterior distribution, but we omit the index i for simplicity. We approximate the distribution by using the variational inference techniques with a Taylor series.

Let us describe the lower bound to be maximized as

$$L(q) = E_q[f(\gamma_i)] + H[q], \tag{21}$$

where $f(\gamma_i) = \log p(t|\gamma_i)p(\gamma_i)$. The $H[q]$ is the entropy given by

$$H[q] = -\int q(\gamma_i) \log q(\gamma_i) d\gamma_i = \log \Gamma(a) - (a-1)\psi(a) - \log b + a, \tag{22}$$

where $\psi(a) = \frac{d}{da} \log \Gamma(a)$ is the digamma function. Note that the notation ψ differs from that for the basis function ψ_i. The $f(\gamma_i)$ is given by

$$f(\gamma_i) = l(\alpha_i) + a_i \log b_i - \log \Gamma(a_i) + (a_i - 1) \log \gamma_i - b_i \gamma_i + \text{Const}, \tag{23}$$

where the term Const is independent of γ_i. Note that $l(\alpha_i)$ given as (12) depends on γ_i through the s_i and q_i. Let us define $\hat{\gamma}_i$ as a stationary point of $f(\gamma_i)$ searched for by gradient methods and expand $f(\gamma_i)$ around $\hat{\gamma}_i$, then

$$E_q[f(\gamma_i)] = f(\hat{\gamma}_i) + \frac{1}{2}f''(\hat{\gamma}_i)\left(\frac{a}{b} - \hat{\gamma}_i\right)^2 + \frac{1}{2}f''(\hat{\gamma}_i)\frac{a}{b^2}. \tag{24}$$

The lower bound $L(q)$ is the sum of (22) and (24). Once we obtain the explicit form of the lower bound, we have to maximize it w.r.t its parameters a and b. The iterative formulas are obtained by using the derivatives of (21),

$$a = \left\{\frac{f''(\hat{\gamma}_i)}{b^2} - \psi'(a)\right\}^{-1} \times \left\{\frac{\hat{\gamma}_i f''(\hat{\gamma}_i)}{b} - \frac{f''(\hat{\gamma}_i)}{2b^2} - \psi'(a) - 1\right\}, \tag{25}$$

$$b = \frac{a\hat{\gamma}_i f''(\hat{\gamma}_i)}{2} + \frac{1}{2}\sqrt{af''(\hat{\gamma}_i)\left(a\hat{\gamma}_i^2 f''(\hat{\gamma}_i) - 4a - 4\right)}. \tag{26}$$

We can obtain the best approximated posterior distribution by repeating these updates as long as the lower bound increases. Then the MAP estimate of the approximated gamma posterior distribution gives us the optimal γ_i. After the updates of a and b are converged once, they can be reused as the initial values in the next optimization phase of γ_i.

3.2   Discussion

We note some desirable properties of the proposed approach. Since it estimates kernel parameters on the basis of a Bayesian framework, we can incorporate prior information in the parameter estimation. This prior information plays an important role in avoiding over-fitting. The robust RVM modeling also helps to reduce the risk of taking outliers into the model. The combination of the Bayesian parameter estimation and the robust modeling enables us to carry out more precise and adaptive learning regardless of the presence or absence of outliers.

However, there still remains the arbitrariness about how to initialize the hyperparameters of the prior a_i and b_i. Although every γ_i as well as a_i and b_i can be estimated during the inference process, the initial values affect results as shown in the following section. To solve the problem, we recommend using the Bayesian information criterion (BIC) given by

$$BIC(\gamma) = -2 \cdot \text{log-likelihood} + \log N \cdot \text{Tr}\{S\}. \tag{27}$$

Here, the hat matrix $S = \Phi \Sigma \Phi^T B$ measures the effective number of parameters. Tripathi and Govindaraju [11] have already reported that the BIC is a suitable criterion for kernel parameter selection in the normal RVM. Therefore, the following steps would be reasonable in the proposed modeling. (1) Training normal RVM models with several values of γ. (2) Estimating their prediction risks on the basis of the BIC values and choosing the best $\gamma_0 = \text{argmin}_\gamma BIC(\gamma)$. (3) Setting $a_i = p\gamma_0 + 1$ and $b_i = p$, where $p \in \mathbb{R}_+$ coordinates the trade-off between the likelihood and the prior. (4) Implementing the proposed approach using the initial values. Here, we recommend that $p = 0.1$ judging from our careful experiments. In practice, it can yield better results to take a few initial γ_0 within the reasonable range of BIC values, which causes extra computation but is still more effective than CV.

4 Numerical Experiments

We applied our approach to some datasets together with two previously proposed methods. One is Adaptive RVM (ARVM) [12], which uses the adaptive kernel function in (7) and maximizes the marginal likelihood with the model complexity term in (8) w.r.t the hyperparameters and kernel parameters directly. The other is Robust RVM (RRVM) [4], which adopts the heteroscedastic variance in (9). The proposed method can be viewed as combining the advantages of these RVMs.

All the parameters are initialized in accordance with previous work [4,10]. The iterative RVM algorithm terminates after 1000 iterations in all experiments.

4.1 Artificial Data

We used the artificial dataset of 200 samples generated from the function

$$f(x) = \begin{cases} \sin(x) & 0 \le x < 2\pi \\ \sin(4x) & 2\pi \le x < 4\pi \\ \frac{1}{2\pi}(x - 4\pi) & 4\pi \le x \le 6\pi, \end{cases} \tag{28}$$

for $x \in (0, 6\pi)$. We added Gaussian noise $\epsilon \sim N(0, 0.2^2)$ to the targets. We also computed the mean squared error (MSE) with 800 test points generated from the true function to assess efficiency.

At first, we briefly conducted 50 simulations without outliers.

Table 1. Simulation results of artificial data without outliers

ARVM		RRVM		Proposal	
MSE	# RV	MSE	# RV	MSE	# RV
0.0340	11.6	0.0120	24.8	**0.0093**	19.2

Table 1 shows the results. The results of ARVM were not stable; it sometimes estimated models that were too smooth, which suggests that outcomes strongly depend on the randomness of ψ_i selection in Step 4 of Algorithm 1. The proposed method achieved better MSE and sparser models than RRVM thanks to the adjustment of the kernel parameters.

Next, we added 20 outliers to the same model and simulated more closely using some initial γ values. Table 2 shows the result summaries, and Fig. 1 shows examples of the results.

Table 2. Simulation results of artificial data with outliers

γ	ARVM		RRVM		Proposal		BIC
	MSE	# RV	MSE	# RV	MSE	# RV	
1	0.3306	2.48	0.1756	14.86	0.1030	11.18	253.48
2	0.3268	2.72	0.0127	24.96	0.0101	16.52	62.20
4	0.2970	4.04	0.0130	30.06	**0.0094**	16.90	63.82
8	0.3003	4.18	0.0158	37.14	0.0098	19.12	80.99
16	0.3244	3.86	0.0213	47.66	0.0114	23.08	107.60
32	0.3264	4.48	0.0283	62.80	0.0147	31.62	153.97

The results revealed that ARVM is completely inappropriate for data containing outliers. In contrast, RRVM was able to ignore the effect of outliers, but it estimated models that were too smooth or over-fit depending on the values of kernel parameters as shown in Fig. 1(b), (h). Even in the best model of RRVM, the estimated curve fits excessively in the linear interval like Fig. 1(e). The proposed method with $\gamma = 2$ was the best model of all. Figure 1(f) shows that it can estimate a smooth curve in every region accurately. When the initial $\gamma = 1$, estimated curves were under-fit because basis functions having broad variance cannot capture high frequency region like Fig. 1(c). However, if such a basis function is included in the model accidentally, the model can adjust its kernel parameter so that the curve fits well. Therefore, we can expect that far more iterative steps can improve predictive performance. Although models with a relatively large initial gamma provided smoother curves than RRVM, they slightly over-fit like in Fig. 1(i).

As mentioned in Sect. 3, the results can be affected by initial γ values. However, an initial guess for the γ using BICs of normal RVM models can give us

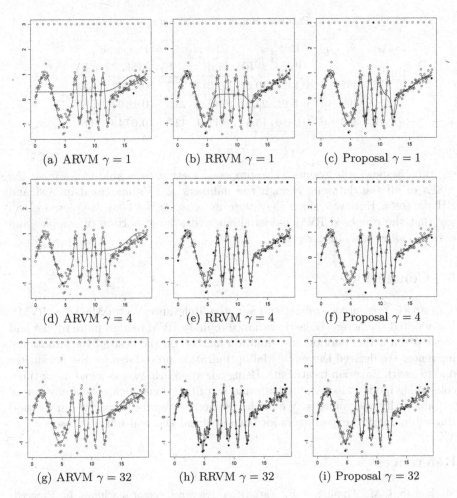

Fig. 1. Simulation results of artificial data with outliers. The black line shows the true curve, the blue line shows the estimated curve, and the shaded circles show RVs. (Color figure online)

accurate indication as shown in the last column of Table 2. Using BICs, we can avoid poor results like those in Fig. 1(c), (i).

4.2 Real Data

We also conducted experiments for three real-world benchmark datasets obtainable from the UCI website [14]. The Auto MPG (AM) dataset concerns city-cycle fuel consumption, the Boston Housing (BH) dataset concerns house values in Boston, and the Yacht Hydrodynamics (YH) dataset concerns resistance estimation of ships. Let each variable be normalized to have zero mean and unit variance. We evaluated the performances through 10-fold CV.

Table 3. Results for real data

Data	N	d	ARVM		RRVM		Proposal	
			MSE	# RV	MSE	# RV	MSE	# RV
AM	392	8	0.1328	10.6	0.1217	27.7	**0.1207**	59.1
BH	506	13	0.1701	22.8	0.1842	23.9	**0.1404**	80.8
YH	308	6	0.2410	13.3	0.1101	42.3	**0.0748**	13.8

Table 3 shows the results. The proposed method was able to improve the MSEs of all the datasets, though the numbers of RVs increased in AM and BH datasets. However, since they were at most 18 % of all samples, we can say that the proposed RVM model eliminated a large portion of samples and corresponding basis functions.

5 Conclusion

This article presented a robust and adaptive Relevance Vector Machine (RVM). We adopted the heteroscedastic variance to make RVM models more robust and developed a kernel parameter estimation method on the basis of the variational inference. We derived the sequential optimization procedures for the parameter updates with Bayesian treatments. Using our approach, we can avoid using time-consuming model selection methods and learn functional localities of any curves automatically and robustly. We further demonstrated the ability of our method through numerical experiments for both artificial and real-world datasets.

References

1. Bishop, C.M., Tipping, M.E.: Variational relevance vector machines. In: Proceedings of 16th Conference on Uncertainty in Artificial Intelligence, pp. 46–53 (2000)
2. Bishop, C.M.: Pattern Recognition and Machine Learning. Springer, New York (2006)
3. Faul, A.C., Tipping, M.E.: Analysis of sparse Bayesian learning. In: Advances in Neural Information Processing, vol. 14, pp. 383–389. MIP Press (2002)
4. Han, M., Zhao, Y.: Robust relevance vector machine with noise variance coefficient. In: Proceedings of the 2010 International Joint Conference on Neural Networks, pp. 1–6 (2010)
5. Hwang, M., Jeong, M.K., Yum, B.J.: Robust relevance vector machine with variational inference for improving virtual metrology accuracy. IEEE Trans. Semicond. Manuf. **27**(1), 83–94 (2014)
6. Matsuda, D.: Predictive model selection criteria for relevance vector regression models. Josai Math. Monogr. **8**, 97–113 (2015)
7. Schmolck, A., Everson, R.: Smooth relevance vector machine: a smoothness prior extension of the RVM. Mach. Learn. **68**(2), 107–135 (2007)
8. Tipping, M.E.: Sparse Bayesian learning and the relevance vector machine. J. Mach. Learn. Res. **1**, 211–244 (2001)

9. Tipping, M.E., Lawrence, K.D.: A variational approach to robust Bayesian interpolation. In: Neural Networks for Signal Processing, pp. 229–238 (2003)
10. Tipping, M.E., Faul, A.: Fast marginal likelihood maximization for sparse Bayesian models. In: Proceedings of 9th International Workshop Artificial Intelligence and Statistics (2003)
11. Tripathi, S., Govindaraju, R.C.: On selection of kernel parameters in relevance vector machines for hydrologic applications. Stoch. Environ. Res. Risk Assess. **21**(6), 747–764 (2007)
12. Tzikas, D.G., Likas, A.C., Galatsanos, N.P.: Sparse Bayesian modeling with adaptive kernel learning. IEEE Trans. Neural Netw. **20**(6), 926–937 (2009)
13. Wang, C., Blei, D.M.: Variational inference in nonconjugate models. J. Mach. Learn. Res. **14**(1), 1005–1031 (2013)
14. Lichman, M.: UCI Machine Learning Repository. University of California, School of Information and Computer Science, Irvine, CA. http://archive.ics.uci.edu/ml

Multiple Structure Recovery via Probabilistic Biclustering

M. Denitto[1(✉)], L. Magri[1], A. Farinelli[1], A. Fusiello[2], and M. Bicego[1]

[1] Department of Computer Science, University of Verona, Verona, Italy
matteo.denitto@univr.it
[2] DPIA, University of Udine, Udine, Italy

Abstract. Multiple Structure Recovery (MSR) represents an important and challenging problem in the field of Computer Vision and Pattern Recognition. Recent approaches to MSR advocate the use of clustering techniques. In this paper we propose an alternative method which investigates the usage of biclustering in MSR scenario. The main idea behind the use of biclustering approaches to MSR is to isolate subsets of points that behave "coherently" in a subset of models/structures. Specifically, we adopt a recent generative biclustering algorithm and we test the approach on a widely accepted MSR benchmark. The results show that biclustering techniques favorably compares with state-of-the-art clustering methods.

1 Introduction

The extraction of multiple models from noisy or outlier-contaminated data – a.k.a. Multiple Structure Recovery (MSR) – is an important and challenging problem that emerges in many Computer Vision applications [7,10,31]. With respect to single-model estimation in presence of noise and outliers, MRS aims at facing the so called *pseudo*-outliers (i.e. "outliers to the structure of interest but inliers to a different structure" [27]), which push robust estimation to its limit. If, in addition, the number of structures is not known in advance, MSR turns into a thorny model-selection problem, as one have to pick, among all the possible interpretations of the data, the most appropriate one.

In the literature, the problem of MSR has been successfully tackled by leveraging on clustering techniques [13,18,19,25]. Generally, the data matrix to analyze reports the points to cluster on one dimension and the features/descriptors on the other dimension [1]. Clustering approaches group the rows (or the columns) of a given data matrix on the basis of a similarity criterion. For example in these recent approaches *J-linkage*[30], *T-linkage* [18] and *RPA* [19]. The feature vector used to represent data is derived form the preferences expressed by the data points for a pool of tentative structures obtained by random sampling. Hence cluster analysis is performed via either agglomerative or partitional methods where distances measure the (dis)agreement between preferences.

Although it has been shown that clustering provides good solution to the MSR problem, there are situations where the performances of clustering can

A. Robles-Kelly et al. (Eds.): S+SSPR 2016, LNCS 10029, pp. 274–284, 2016.
DOI: 10.1007/978-3-319-49055-7_25

be highly compromised by data matrix structure (e.g. noisy data matrices; or rows behaving similarly only in a small portion of the data matrix). Retrieving information in scenarios where clustering struggles can be done through a recent class of approaches called biclustering.

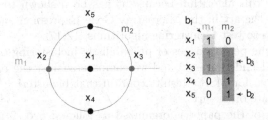

Fig. 1. Example of overlapping biclusters on Multiple Structure Recovery. In this case the possible biclusters are the: b_1) the points lying on m_1, b_2) the points lying on m_2 and b_3) the points lying on the intersections.

With the term *bi-clustering* we refer to a specific category of algorithms performing clustering on both rows and columns of a given data matrix [17]. The goal of biclustering is to isolate sub-matrices where the rows present a "coherent behavior" in a restricted columns subset, and vice-versa. Compared with clustering, biclustering exploits *local* information to retrieve structures that cannot be found performing analysis on whole rows or columns. The problem of biclustering (also known as co-clustering, and strongly related to subspace clustering) is gaining increasing attention in the Pattern Recognition community, with many papers being published in recent years (e.g. [4, 8, 11, 22]). Even if it was originally proposed in biological scenarios (i.e. analysis of gene expression microarray datasets [2, 17, 20]), biclustering has been widely adopted in many other contexts ranging from market segmentation to data mining [6, 14, 21].

This paper is positioned exactly in this context, investigating the performance of biclustering on MSR. The main advantage of using biclustering on MSR is that biclustering can isolate portions of the data matrix where a subset of points share similar attitudes in a subset of models. Compared to clustering, biclustering retrieves additional information on which models better describes a particular subset of points. Moreover, biclustering provides better understanding of the data since different biclusters can overlap (on rows, columns or both). This means that a certain point can belong to different biclusters and hence it can be characterized by distinct subsets of models (see Fig. 1 for an example). Thus biclustering approaches can easily deal with intersections, which is a known problematic situation when using clustering algorithms [28].

To the best of our knowledge there exists only a preliminar work applying biclustering techniques to MSR [28]. In [28] authors show that the application of biclustering techniques to MSR is promising and provides superior solution when compared with clustering. While this provides a significant contribution to the state of the art, there is large room for improvements since the method adopted by the authors has some limitations (i.e. it works with sparse binary matrices and it needs some pre-processing/post-processing operations to retrieve the final solutions).

In this paper we investigate the use of a recent probabilistic biclustering approach, namely Factor Analysis for BIcluster Acquisition (FABIA) [12] for MSR. We choose this algorithm because it has been shown to perform better than the state of the art in the Microarray Gene Expression analysis, a widely exploited scenario to test biclustering algorithms [12,17].

We evaluate the performances of probabilistic biclustering on a real benchmark dataset (Adelaide dataset) and we compare against the recent clustering algorithms adopted for MSR. Results confirm that biclustering favorably compares with the state-of-the-art.

The reminder of the paper is organized as follows: Sect. 2 provides a brief review of MSR and the clustering methods we compare to. Section 3 formalizes the biclustering problem and the algorithm we adopt. Section 4 presents and discusses the experimental evaluation. Finally, some concluding remarks are presented in Sect. 5

2   Background and Related Work

This section provides background knowledge about the proposed framework. It formalizes the MSR problem and the clustering approaches currently adopted.

2.1   Multiple Structure Recovery

MSR aims at retrieving parametric models from unstructured data in order to organize and aggregate visual content in significant higher-level geometric structures. This task is commonly found in many Computer Vision applications, a typical example being 3D reconstruction, where MSR is employed either to estimate multiple rigid moving objects (to initialize multibody Structure from Motion [7,23]), or to produce intermediate geometric interpretations of reconstructed 3D point cloud [10,31]. Other instances include face clustering, body-pose estimation and video motion segmentation. In all these scenarios the information of interest can be extracted from the observed data and aggregated in suitable structures by estimating some underlying geometric models, e.g. planar patches, homographies, fundamental matrices or linear subspaces.

More formally, to set a general context, let μ be a model e.g. lines, subspace, homography, fundamental matrices or other geometric primitives and $X = \{x_1, \ldots, x_n\}$ be a finite set of n points, possibly corrupted by noise and outlier. The problem of MSR consists in extracting k instances of μ termed structures from the data, defining, at the same time, subsets $C_i \subset X, i = 1, \ldots,$ such that all points described by the i-th structures are aggregated in C_i. Often the models considered are parametric, i.e. the structures can be represented as vectors in a proper parameter space.

2.2 Clustering

The extensive landscape of approaches aimed at MSR can be broadly categorized along two mutually orthogonal strategies, namely *consensus analysis* and *preference analysis*. Focusing on preference analysis, these methods reverse the role of data and models: rather than considering models and examining which points match them, the residuals of individual data points are taken into account [3, 29, 32]. This information is exploited to shift the MSR problem from the ambient space where data lives to a *conceptual* [24] one where it is addressed via cluster analysis techniques.

T-Linkage [18] and RPA [19] can be ascribed to these clustering-based methods as they share the same *first-represent-then-clusterize* approach. Both the algorithms represent data points in a m-dimensional unitary cube as vectors whose components collect the preferences granted to a set of m hypotheses structures instantiated by drawing at random m minimal sample sets – the minimum-sized set of data points necessary to estimate a structure. Preferences are expressed with a soft vote in $[0, 1]$ according to the continuum of residuals in two different fashions. As regards T-linkage, a voting function characterized by an hard cutoff is employed. RPA, instead, exploits the Cauchy weighting function (of the type employed in M-estimators) that has an infinite rejection point mitigating the sensitivity of the inlier threshold.

The rationale beyond both these representations is that the agreement between the preferences of two points in this conceptual space reveals the multiple structures hidden in the data: points sharing the same preferences are likely to belong to the same structures.

T-Linkage captures this notion through the Tanimoto distance, which in turn is used to segment the data via a tailored version of average linkage that succeeds in detecting automatically the number of models. If rogue points contaminate the data, outlier structures need to be pruned via ad hoc-post processing techniques.

RPA, on the contrary, requires the number of desired structures as input but inherently caters for gross contamination. At first a kernelized version of the Tanimoto distances is feed to Robust Principal Analysis to remove outlying preferences. Then Symmetric Non Negative Factorization [15] is performed on the low rank part of the kernel to segment the data. Hence, the attained partition is refined in a MSAC framework. More precisely, the consensus of the sampled hypotheses are scrutinized and the structures that, within each segment, support more points are retained as solutions.

While T-linkage can be considered as a pure preference method, RPA attempts to combine also the consensus-side of the MSR problem. However it does not fully reap the benefit of working with both the dimensions of the problem, as biclustering does, for preference and consensus are considered only sequentially.

All these methods can be regarded as processing the *Preference Matrix*, where each entry (i, j) represents the vote granted by the i-th point to the j-th tentative structures. Rows of that matrix provides the representation of points that are used to derive the affinity matrices for clustering. Column of that matrix are the consensus set of a model hypothesis.

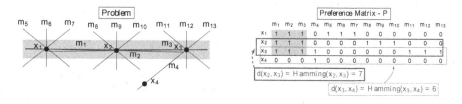

Fig. 2. Shortfalls of clustering on MRS.

An example where clustering struggles is provided in Fig. 2, which describes a simple MSR problem: to group similar points on the basis of their behavior with respect to the proposed models we should perform clustering on the Preference Matrix P which describes the relationship between the points $\{x_1, x_2, x_3, x_4\}$ and the models $\{m_1, \cdots, m_{13}\}$. Assume we perform clustering adopting the Hamming distance (i.e. number of different bits): since the distance between the x_3 and the x_4 is smaller than the distance between x_3 and x_2, clustering would assign the third and the fourth point to the same group. However looking at the problem diagram it is clear that points x_1, x_2 and x_3 should belong to the same cluster. This information can be retrieved performing a simultaneous clustering of both rows and columns of the Preference Matrix, isolating a subset of models (m_1, m_2 and m_3) where the points x_1, x_2, x_3 share a similar behavior (shaded area in Fig. 2). This is exactly what biclustering techniques do.

Next section provides a more formal definition of the biclustering problem, focusing on the approach adopted to analyze the Preference Matrix (FABIA [12]).

3   Biclustering

As mentioned in Sects. 1 and 2.2 the goal of biclustering applied to MSR is the simultaneous clustering of points and structures/models of a given Preference Matrix, merging the well known concepts of consensus analysis and preference analysis. Due to the similarity of RPA (where the kernelized matrix is factorized in order to obtain point clusters) we present biclustering from a *sparse low-rank matrix factorization* perspective, also the most suitable to understand the insights behind the FABIA algorithm [12].

We denote as $D \in \mathbb{R}^{n \times m}$ the given data matrix, and let $R = \{1, \ldots, n\}$ and $C = \{1, \ldots, m\}$ be the set of row and column indices. We adopt D_{TK}, where $T \subseteq R$ and $K \subseteq C$, to represent the submatrix with the subset of rows in T and the subset of columns in K. Given this notation, we can define a *bicluster* as a submatrix D_{TK}, such that the subset of rows of D with indices in T exhibits a "coherent behavior" (in some sense) across the set of columns with indices in K, and vice versa. The choice of coherence criterion defines the type of biclusters to be retrieved (for a comprehensive survey of biclustering criteria, see [8,17,22]).

A possible coherence criterion for a bicluster (sub-matrix) is for the corresponding entries to have a similar value, significantly different from the other entries of the matrix. For example, a data matrix containing one bicluster with rows $T = \{1, 2, 3, 4\}$ and columns $K = \{1, 2\}$ may look like

$$D = \begin{bmatrix} 10 & 10 & 0 & 0 \\ 10 & 10 & 0 & 0 \\ 10 & 10 & 0 & 0 \\ 10 & 10 & 0 & 0 \\ 0 & 0 & 0 & 0 \\ 0 & 0 & 0 & 0 \end{bmatrix}, \quad v = \begin{bmatrix} 5 \\ 5 \\ 5 \\ 5 \\ 0 \\ 0 \end{bmatrix}, \quad z = \begin{bmatrix} 2 \\ 2 \\ 0 \\ 0 \end{bmatrix}.$$

From an algebraic point of view, this matrix can be represented by the outer product $D = vz^T$ of the sparse vectors v and z. We call these vectors *prototypes* (for v) and *factors* (for z). Generalizing to k biclusters, we can formulate the biclustering problem as the decomposition of the given data matrix D as the sum of k outer products,

$$D = \sum_{i=1}^{k} v_i z_i^T = VZ, \tag{1}$$

where $V = [v_1, \ldots, v_k] \in \mathbb{R}^{n \times k}$ and $Z = [z_1, \ldots, z_k]^T \in \mathbb{R}^{k \times m}$.

The connection between biclustering and sparse low-rank matrix factorization can be highlighted by observing that the factorization of the original data matrix shows that it has rank no larger than the number of biclusters (usually much lower than the number of rows or columns). Moreover, if the size of the matrix D is much bigger than the bicluster size (as it is typically the case in many applications), the resulting prototype and factor vectors should be composed mostly by zeros (*i.e.*, the prototypes and factors should be sparse).

3.1   FABIA

In the biclustering literature, there are several proposals of biclustering methods through matrix factorization (*e.g.*, [16,33]); however, to the best of our knowledge, the only probabilistic approach is FABIA.

FABIA is a generative model for biclustering based on factor analysis [12]. The model proposes to decompose the data matrix by adding noise to the strict low rank decomposition in (1),

$$D = \sum_{i=1}^{k} v_i z_i^T + Y = VZ + Y, \tag{2}$$

where matrix $Y \in \mathbb{R}^{n \times m}$ accounts for random noise or perturbations, assumed to be zero-mean Gaussian with a diagonal covariance matrix. As explained above (Sect. 3) the prototypes in V and factors in Z should be sparse. To induce sparsity, FABIA uses two types of priors: (i) an independent Laplacian prior, and (ii) a prior distribution that is non-zero only in region where prototypes are sparse (for further details, see [12]). The model parameters are estimated using a variational EM algorithm [9,12], for all the details about FABIA derivation and implementation[1], please refer to [12].

4   Experimental Evaluation

This section provides the performances comparison between some clustering methods recently applied to MSR [18,19,25] and the probabilistic biclustering approach presented in Sect. 3.1. The comparison with [28] was not possible since the code is not available. The workflow of the overall procedure can be sketched as follows: starting from an image (i) we generate the hypothesis and compute the Preference Matrix following the guidelines in [19]; (ii) then the probabilistic biclustering technique have been applied.

To assess the quality of the approaches we used the widely adopted *Adelaide* real benchmark dataset[2]. Moreover we conduct a reproducibility analysis, since it is known that RPA algorithm can produce very different solutions due to the random initialization required by the Symmetric NNMF step.

4.1   Adelaide Dataset

We explored the performances of probabilistic biclustering on two type of experiments, namely motion and plane estimation. In motion segmentation experiments, we were provided with two different images of the same scene composed by several objects moving independently; the aim was to recover fundamental matrices to subsets of point matches that undergo the same motion. With respect to the plane segmentation scenario, given two uncalibrated views of a scene, the goal was to retrieve the multi-planar structures by fitting homographies to point correspondences. The AdelaideRMF dataset is composed of 38 image pairs (19 for motion segmentation and 19 for plane segmentation) with matching points contaminated by gross outliers. The ground-truth segmentations are also available. In order to assess the quality of the results, we adopted the misclassification errors, that counts the number of wrong point assignment according to the map between ground-truth labels and estimated ones that minimize the overall number of misclassified points (as in [26]). For fair comparison, the Preference Matrix fed to FABIA was generated relying on the guided sampling scheme presented in [19].

FABIA parameters which regulate the factors/prototypes sparsity and the threshold to retrieve biclusters memberships have been varied in the range suggested by the authors in [12]. The best results on the whole Adelaide dataset (motion and plane estimation) are reported in Table 1. The performances of other methods are taken from [19]; results show that FABIA provides higher quality solutions on the motion segmentation dataset, and on average it performs better on the planar segmentation. Focusing on the motion segmentation dataset, there

---

[1] Code available from http://www.bioinf.jku.at/software/fabia/fabia.html.

[2] The dataset can be downloaded from https://cs.adelaide.edu.au/~hwong/doku.php?id=data.

Table 1. Misclassification error (ME %) for motion segmentation (left) and planar segmentation (right). k is the number of models and % out is the percentage of outliers.

	k	%out	T-lnkg	RCMSA	RPA	FABIA
biscuitbookbox	3	37.21	**3.10**	16.92	3.88	3.86
breadcartoychips	4	35.20	14.29	25.69	7.50	**4.22**
breadcubechips	3	35.22	3.48	8.12	5.07	**0.87**
breadtoycar	3	34.15	9.15	18.29	7.52	**0.60**
carchipscube	3	36.59	4.27	18.90	6.50	**1.52**
cubebreadtoychips	4	28.03	9.24	13.27	4.99	**1.07**
dinobooks	3	44.54	20.94	23.50	15.14	**9.72**
toycubecar	3	36.36	15.66	13.81	9.43	**9.50**
biscuit	1	57.68	16.93	14.00	1.15	**0**
biscuitbook	2	47.51	3.23	8.41	3.23	**1.32**
boardgame	1	42.48	21.43	19.80	11.65	**8.96**
book	1	44.32	3.24	4.32	2.88	**0**
breadcube	2	32.19	19.31	9.87	**4.58**	19.42
breadtoy	2	37.41	5.40	3.96	**2.76**	19.62
cube	1	69.49	7.80	8.14	3.28	**1.66**
cubetoy	2	41.42	3.77	5.86	4.04	**2.21**
game	1	73.48	1.30	5.07	3.62	**0**
gamebiscuit	2	51.54	9.26	9.37	2.57	**2.44**
cubechips	2	51.62	6.14	7.70	4.57	**0.53**
mean			9.36	12.37	5.49	**4.61**
median			7.80	9.87	4.57	**1.66**

	k	%out	T-lnkg	RCMSA	RPA	FABIA
unionhouse	5	18.78	48.99	2.64	**10.87**	21.54
bonython	1	75.13	11.92	17.79	15.89	**6.82**
physics	1	46.60	29.13	48.87	**0.00**	0.00
elderhalla	2	60.75	10.75	29.28	**0.93**	3.04
ladysymon	2	33.48	24.67	39.50	24.67	**11.81**
library	2	56.13	24.53	40.72	31.29	**20.47**
nese	2	30.29	7.05	46.34	**0.83**	4.92
sene	2	44.49	7.63	20.20	**0.42**	2.20
napiera	2	64.73	28.08	31.16	**9.25**	21.85
hartley	2	62.22	21.90	37.78	**17.78**	23.59
oldclassicswing	2	32.23	20.66	21.30	25.25	**7.92**
barrsmith	2	69.79	49.79	**20.14**	36.31	29.88
neem	3	37.83	25.65	41.45	19.86	**11.20**
elderhallb	3	49.80	31.02	35.78	**17.82**	18.63
napierb	3	37.13	**13.50**	29.40	31.22	36.68
johnsona	4	21.25	34.28	36.73	**10.76**	17.96
johnsonb	7	12.02	24.04	16.46	26.76	**24.50**
unihouse	5	18.78	33.13	**2.56**	5.21	15.76
bonhall	6	6.43	21.84	19.69	41.67	24.02
mean			24.66	28.30	**17.20**	15.94
median			23.38	29.40	**17.53**	17.96

are only three situations where FABIA works worse than clustering approaches. A possible explanation on why FABIA struggles could be because general biclustering approaches are tested in scenarios where the number of biclusters is much higher than in MSR (i.e. ~100 in Gene Expression analysis versus 3–7 in this dataset). To overcome this behavior we run FABIA increasing the number of biclusters to retrieve and aggregating the results on the basis of column overlap as done in [5], this leads to an improvement of the solution quality; results are reported in Table 2.

Table 2. Increasing the number of biclusters improve the results obtained by FABIA on the motion segmentation dataset.

biscuitbookbox	k = 3	3.86
	k = 4	**1.35**
breadcube	k = 2	19.42
	k = 4	11.36
breadtoy	k = 2	19.62
	k = 4	**1.22**

4.2 Reproducibility

In this section we assess the reproducibility of the two methods that better perform on the Adelaide dataset: RPA and FABIA. The goal is to demonstrate that probabilistic approaches can overcome the problem of reproducibility present in RPA. For a fair comparison we adopted the *2R3RTCRT* video sequence from the Hopkins dataset[3]: a dataset where both the approaches retrieve good and similar solutions.

Hopkins dataset is a motion segmentation benchmark where the input data consists in a set of features trajectories across a video taken by a moving camera, and the problem consist in recovering the different rigid-body motions contained in the dynamic scene. Motion segmentation can be seen as a subspace clustering problem under the modeling assumption of affine cameras. In fact, under the assumption of affine projection, it is simple to demonstrate that all feature trajectories associated with a single moving object lie in a linear subspace of dimension at most 4 in $\mathbb{R}^{2F}$ (where F is the number of video frames). Feature trajectories of a dynamic scene containing k rigid motion lie in the union of k low dimensional subspace of $\mathbb{R}^{2F}$ and segmentation can be reduced to clustering data points in a union of subspaces.

To test the reproducibility of RPA and FABIA we run the algorithms on the same Preference Matrix hundred times, and for each trial we assess the misclassification error; the results are reported in Fig. 3. The figure shows the result obtained by the approaches in each iteration and its distance from the average results. Results clearly show that while the two approaches are compatible on average, FABIA retrieves the same solution in each iteration while RPA is much less stable.

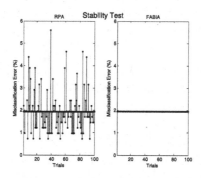

Fig. 3. Reproducibility. The methods have been run 100 times on the same Preference Matrix. Plots show the misclassification error in of each trials along with the distance from the mean (RPA mean = 1.93 %, FABIA mean = 1.95 %).

5   Conclusion and Discussion

In this paper we present an alternative to clustering for the problem of Multiple Structure Recovery (MSR), namely biclustering. In general, biclustering techniques allow to retrieve superior and more accurate information than clustering

---

[3] The dataset can be downloaded at http://www.vision.jhu.edu/data/hopkins155/.

approaches, characterizing each cluster of points with the subset of features that better describes them. The goal of biclustering approaches applied to MSR is isolate submatrices inside the Preference Matrix where a subset of points behave "coherently" in a subset of models/structures. We tested the recent probabilistic biclustering approach FABIA on the Adelaide benchmark dataset, proving that it favorably compares with the state of the art. Moreover we tested the reproducibility of the analyzed methods showing that FABIA is much more stable than the second competitor RPA.

References

1. Bishop, C.M.: Pattern recognition and Machine Learning. Springer, Heidelberg (2006)
2. Cheng, Y., Church, G.: Biclustering of expression data. In: Proceedings of 8th International Conference on Intelligent Systems for Molecular Biology (ISMB 2000), pp. 93–103 (2000)
3. Chin, T., Wang, H., Suter, D.: Robust fitting of multiple structures: the statistical learning approach. In: International Conference on Computer Vision, pp. 413–420 (2009)
4. Denitto, M., Farinelli, A., Bicego, M.: Biclustering gene expressions using factor graphs and the max-sum algorithm. In: Proceedings of 24th International Conference on Artificial Intelligence, pp. 925–931. AAAI Press (2015)
5. Denitto, M., Farinelli, A., Franco, G., Bicego, M.: A binary factor graph model for biclustering. In: Fränti, P., Brown, G., Loog, M., Escolano, F., Pelillo, M. (eds.) S+SSPR 2014. LNCS, vol. 8621, pp. 394–403. Springer, Heidelberg (2014). doi:10.1007/978-3-662-44415-3_40
6. Dolnicar, S., Kaiser, S., Lazarevski, K., Leisch, F.: Biclustering overcoming data dimensionality problems in market segmentation. J. Travel Res. **51**(1), 41–49 (2012)
7. Fitzgibbon, A.W., Zisserman, A.: Multibody structure and motion: 3-D reconstruction of independently moving objects. In: Vernon, D. (ed.) ECCV 2000. LNCS, vol. 1842, pp. 891–906. Springer, Heidelberg (2000). doi:10.1007/3-540-45054-8_58
8. Flores, J.L., Inza, I., Larrañaga, P., Calvo, B.: A new measure for gene expression biclustering based on non-parametric correlation. Comput. Methods Prog. Biomed. **112**(3), 367–397 (2013)
9. Girolami, M.: A variational method for learning sparse and overcomplete representations. Neural Comput. **13**(11), 2517–2532 (2001)
10. Häne, C., Zach, C., Zeisl, B., Pollefeys, M.: A patch prior for dense 3D reconstruction in man-made environments. In: 2012 2nd International Conference on 3D Imaging, Modeling, Processing, Visualization and Transmission (3DIMPVT), pp. 563–570. IEEE (2012)
11. Henriques, R., Antunes, C., Madeira, S.C.: A structured view on pattern mining-based biclustering. Pattern Recogn. **48**(12), 3941–3958 (2015)
12. Hochreiter, S., Bodenhofer, U., Heusel, M., Mayr, A., Mitterecker, A., Kasim, A., Khamiakova, T., Van Sanden, S., Lin, D., Talloen, W., et al.: Fabia: factor analysis for bicluster acquisition. Bioinformatics **26**(12), 1520–1527 (2010)
13. Jain, S., Govindu, V.M.: Efficient higher-order clustering on the Grassmann manifold (2013)

14. Kaytoue, M., Codocedo, V., Buzmakov, A., Baixeries, J., Kuznetsov, S.O., Napoli, A.: Pattern structures and concept lattices for data mining and knowledge processing. In: Bifet, A., May, M., Zadrozny, B., Gavalda, R., Pedreschi, D., Bonchi, F., Cardoso, J., Spiliopoulou, M. (eds.) ECML PKDD 2015. LNCS (LNAI), vol. 9286, pp. 227–231. Springer, Heidelberg (2015). doi:10.1007/978-3-319-23461-8_19

15. Kuang, D., Yun, S., Park, H.: SymNMF: nonnegative low-rank approximation of a similarity matrix for graph clustering. J. Glob. Optim. 1–30 (2014)

16. Lee, M., Shen, H., Huang, J.Z., Marron, J.: Biclustering via sparse singular value decomposition. Biometrics 66(4), 1087–1095 (2010)

17. Madeira, S., Oliveira, A.: Biclustering algorithms for biological data analysis: a survey. IEEE Trans. Comput. Biol. Bioinform. 1, 24–44 (2004)

18. Magri, L., Fusiello, A.: T-linkage: a continuous relaxation of J-linkage for multi-model fitting. In: Proceedings of IEEE Conference on Computer Vision and Pattern Recognition, pp. 3954–3961 (2014)

19. Magri, L., Fusiello, A.: Robust multiple model fitting with preference analysis and low-rank approximation. In: Xie, X., Tam, Jones, M.W., Tqam, G.K.L. (eds.) Proceedings of British Machine Vision Conference (BMVC), pp. 20.1–20.12. BMVA Press, September 2015

20. Mitra, S., Banka, H., Pal, S.K.: A MOE framework for biclustering of microarray data. In: 18th International Conference on Pattern Recognition, 2006, ICPR 2006, vol. 1, pp. 1154–1157. IEEE (2006)

21. Mukhopadhyay, A., Maulik, U., Bandyopadhyay, S., Coello, C.A.C.: Survey of multiobjective evolutionary algorithms for data mining: Part II. IEEE Trans. Evol. Comput. 18(1), 20–35 (2014)

22. Oghabian, A., Kilpinen, S., Hautaniemi, S., Czeizler, E.: Biclustering methods: biological relevance and application in gene expression analysis. PloS ONE 9(3), e90801 (2014)

23. Ozden, K.E., Schindler, K., Van Gool, L.: Multibody structure-from-motion in practice. IEEE Trans. Pattern Anal. Mach. Intell. 32(6), 1134–1141 (2010)

24. Pekalska, E., Duin, R.P.: The Dissimilarity Representation for Pattern Recognition: Foundations And Applications (Machine Perception and Artificial Intelligence). World Scientific Publishing, Singapore (2005)

25. Pham, T.T., Chin, T.J., Yu, J., Suter, D.: The random cluster model for robust geometric fitting. Pattern Anal. Mach. Intell. 36(8), 1658–1671 (2014)

26. Soltanolkotabi, M., Elhamifar, E., Candès, E.J.: Robust subspace clustering. Ann. Stat. 42(2), 669–699 (2014)

27. Stewart, C.V.: Bias in robust estimation caused by discontinuities and multiple structures. IEEE Trans. Pattern Anal. Mach. Intell. 19(8), 818–833 (1997)

28. Tepper, M., Sapiro, G.: A biclustering framework for consensus problems. SIAM J. Imaging Sci. 7(4), 2488–2525 (2014)

29. Toldo, R., Fusiello, A.: Robust multiple structures estimation with J-linkage. In: European Conference on Computer Vision (2008)

30. Toldo, R., Fusiello, A.: Robust multiple structures estimation with J-linkage. In: Forsyth, D., Torr, P., Zisserman, A. (eds.) ECCV 2008. LNCS, vol. 5302, pp. 537–547. Springer, Heidelberg (2008). doi:10.1007/978-3-540-88682-2_41

31. Toldo, R., Fusiello, A.: Image-consistent patches from unstructured points with J-linkage. Image Vis. Comput. 31(10), 756–770 (2013)

32. Zhang, W., Kosecká, J.: Nonparametric estimation of multiple structures with outliers. In: Vidal, R., Heyden, A., Ma, Y. (eds.) WDV 2005-2006. LNCS, vol. 4358, pp. 60–74. Springer, Heidelberg (2007). doi:10.1007/978-3-540-70932-9_5

33. Zhang, Z.Y., Li, T., Ding, C., Ren, X.W., Zhang, X.S.: Binary matrix factorization for analyzing gene expression data. Data Min. Knowl. Disc. 20(1), 28–52 (2010)

Generalizing Centroid Index to Different Clustering Models

Pasi Fränti$^{(\boxtimes)}$ and Mohammad Rezaei

University of Eastern Finland, Joensuu, Finland
franti@cs.uef.fi

Abstract. Centroid index is the only measure that evaluates cluster level differences between two clustering results. It outputs an integer value of how many clusters are differently allocated. In this paper, we apply this index to other clustering models that do not use centroid as prototype. We apply it to centroid model, Gaussian mixture model, and arbitrary-shape clusters.

Keywords: Clustering · Validity index · External index · Centroid index

1 Introduction

Clustering aims at partitioning a data set of n points into k clusters. *External index* measures how similar a clustering solution is to a given ground truth (if available), or how similar two clustering solutions are. This kind of measure is needed in clustering ensemble, measuring stability and evaluating performance of clustering algorithms.

In clustering algorithms, one of the main challenges is to solve the global allocation of the clusters instead of just tuning the partition borders locally. Despite of this, most external cluster validity indexes calculate only point-level differences without any direct information about how similar the cluster-level structures are.

Rand index (RI) [1] and *Adjusted Rand index* (ARI) [2] count the number of pairs of data points that are partitioned consistently in both clustering solutions (*A* and *B*); if a pair of points is allocated in the same cluster in *A*, they should be allocated into the same cluster also in *B*. This provides estimation of point-level similarity but does not give much information about the similarity at cluster level.

More sophisticated methods operate at the cluster level. *Mutual information* (MI) [3] and *Normalized mutual information* (NMI) [4] measure the amount of information (conditional entropy) that can be obtained from a cluster in *A* using the clusters in *B*. Set-matching based measures include *Normalized van Dongen* (NVD) [5] and *Criterion-H* (CH) [6]. They match the clusters between *A* and *B*, and measure the amount of overlap between the clusters, see Fig. 1. However, all of them measure point-level differences. What is missing is a simple structural cluster-level measure.

Figure 2 demonstrates the situation with three clustering results: *k-means* (KM) [7], *random swap* (RS) [8] and *agglomerative clustering* (AC) [9]. The clustering structure of RS and AC is roughly the same, and their differences come mostly from the minor inaccuracies of the centroids in the agglomerative clustering. The result of the k-means

© Springer International Publishing AG 2016
A. Robles-Kelly et al. (Eds.): S+SSPR 2016, LNCS 10029, pp. 285–296, 2016.
DOI: 10.1007/978-3-319-49055-7_26

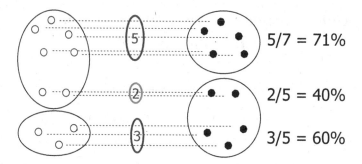

Fig. 1. Principle of set-matching based external validity indexes. The values are the number of overlap between the clusters. Blue indicate clusters that would be selected for matching. Sample index values for this example would be NMI = 0.42, NVD = 0.20, CH = 0.20, CI = 0. (Color figure online)

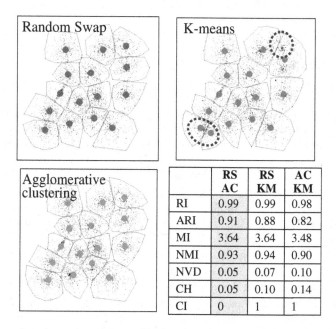

Fig. 2. Comparing three clustering results: Random Swap (RS), Agglomerative clustering (AC) and K-means (KM) with eight measures.

has a more significant, structural difference: one centroid is missing at the top, and there are too many centroids at the bottom.

Some indexes (ARI, NVD, CH) indicate that RS-vs-AC in Fig. 2 are more similar (ARI = 0.91; NVD = 0.05; CH = 0.05) than RS-vs-KM (ARI = 0.88; NVD = 0.07; CH = 0.10), or AC-vs-KM (ARI = 0.82; NVD = 0.10; CH = 0.14) but the numbers do not tell that their global structure is the same.

In a recent work [10], we introduced a cluster level index called *centroid index* (CI) to cope with this problem by measuring the number of clusters allocated differently. For each centroid in A, the method finds its nearest centroid in B. Then it calculates the *indegree*-values for each centroid in B; how many times it was selected as the nearest. An orphan centroid (indegree = 0) indicates that a cluster is differently located in the clustering structure. The index value is the count of these orphans so that CI-value indicates the number of cluster-level differences. Value CI = 0 indicates that the clustering structures are identical, CI = 1 that one cluster mismatch, and so on.

The measure is somewhat rough as it ignores the point-level differences. A simple point-level extension called *centroid similarity index* (CSI) was therefore also considered. However, the main idea to measure the cluster level differences is best captured by the raw CI-value, or relative to the number of clusters: CI/k. In Fig. 2, CI-value clearly tells that RS and AC have similar clustering structure (CI = 0), and that k-means has one difference (CI = 1).

A limitation of CI is that we must have the centroids. It is possible to use other clustering models like *k-medoids* [11] and *k-modes* [12] by finding the nearest medoid/mode using the distances in the feature space. However, in more complex clustering models like Gaussian mixture model (GMM), density-based or arbitrary-shape clustering, it is not as simple.

In this paper, we show how to apply centroid index by using partitions instead of the centroids. For every cluster in A, we first find its most similar cluster in B by calculating the amount of overlap by Jaccard coefficient, which is the number of shared points divided by the total number of distinctive data points in the two clusters:

$$J = \frac{|A_i \cap B_j|}{|A_i \cup B_j|} \tag{1}$$

where A_i and B_j are the matching (most similar) clusters.

However, we do not sum up the overlaps but we further analyze the nearest neighbor mapping. The cluster-to-cluster similarity is now determined at point level, but the overall value is still measured at the cluster level. The measure is calculated as the number of orphan clusters (indegree = 0) as in the original definition of centroid index. Thus, the method now generalizes from the centroid-model to other clustering models independent on how the cluster is represented: centroid, medoid, mode, or even no prototype used at all.

For example, in Fig. 1, the topmost clusters are mapped to each other, and the bottom-most clusters to each other. This results in mapping where all indegree = 1 for both clusters, and CI-value becomes 0. Therefore, these two solutions have the same clustering structure and they differ only at the point level.

2 External Cluster Validity

Existing cluster validity indexes can be divided into three categories: *pair-counting*, *information theoretic* and *set-matching based* measures:

Pair-counting measures
- RI = Rand index [1]
- ARI = Adjusted Rand index [2]

Information theoretic measures
- MI = Mutual information [3]
- NMI = Normalized Mutual information [4]

Set-matching based measures
- NVD = Normalized van Dongen [5]
- CH = Criterion H [6]
- CI = Centroid Index [10]
- CSI = Centroid Similarity Index [10]

We next briefly recall the idea of set-matching based methods [13]. The clusters are first either *paired* (NVD) or *matched* (CH and CI). In pairing, best pair of the clusters in A and B are found by minimizing the sum of the similarities of the paired clusters. Hungarian [13], or greedy algorithm [6, 14] has been used to solve it.

In matching, nearest neighbour is searched. This does not always result in bijective mapping where exactly two clusters are paired, but several clusters in A can be mapped to the same cluster in B (or vice versa), see Fig. 3. The mapping is not symmetric, and is usually done in both ways: $A{\to}B$ and $B{\to}A$.

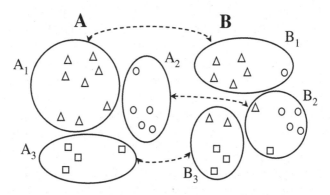

Fig. 3. Matching of clusters using pointwise similarities.

The similarities of the matched/paired clusters are then calculated by summing up the total overlap. In [13], the values are first normalized by the maximum cluster size. In NVD, CH and CSI, the normalization is performed after the summation by dividing the total number of shared points by the total number of points:

$$S = \frac{1}{N} \cdot \sum_{i=1}^{k} |A_i \cap NN(A_i)| \tag{2}$$

where $NN(A_i)$ is the cluster in B that A_i is matched/paired with. Centroid index (CI) finds the nearest centroid without any use of the partitions:

$$NN(A_i) = \arg\min_{1 \leq j \leq k} \left\| c[A_i] - c[B_j] \right\| \qquad (3)$$

where $c[A_i]$ and $c[B_j]$ are the centroids of A and B. The other difference is that CI does not use the point-level measure (2). Instead, it calculates the number of mappings (*indegree*) made for each cluster B_j, and then sums up the number of clusters in B that has not been mapped at all (*indegree* = 0). These are called orphans:

$$CI_1(A \rightarrow B) = \sum_{j=1}^{k} Orphan(B_j) \qquad (4)$$

where *Oprhan(B)* has value 1 if no cluster in A is mapped to it:

$$Orphan(B) = \begin{cases} 1 & InDegree(A) = 0 \\ 0 & InDegree(A) > 0 \end{cases} \qquad (5)$$

In order to have symmetric index, we perform mappings in both ways: $A \rightarrow B$ and $B \rightarrow A$. The CI-value is then defined as the maximum of these two:

$$CI(A, B) = \max\{CI_1(A \rightarrow B), CI_1(B \rightarrow A)\} \qquad (6)$$

To sum up, the index is easy to calculate, and the result has clear interpretation: how many clusters are differently allocated. An example is shown in Fig. 4.

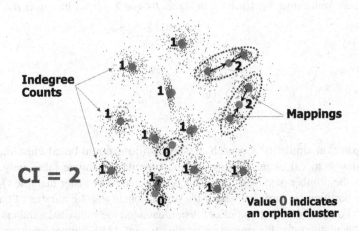

Fig. 4. Example of Centroid index (CI). Only mapping RED→PURPLE is shown. Some purple circles are under the red and not visible. (Color figure online)

3 Centroid Index Using Different Clustering Models

The main limitation of the centroid index is that it requires centroids, which may not exist in other clustering models such as *k-medoids* [11], *k-modes* [12], *Gaussian mixture model* (*GMM*), *density-based* and *arbitrary-shape* clustering. To apply the index to other models, we consider the following alternative approaches:

- Prototype similarity
- Partition similarity
- Model similarity

3.1 Prototype vs. Partition Similarity

K-medoids and k-modes clustering operate also in a feature space (usually Euclidean), so generalization to them is trivial using the prototype similarity approach. We just find the nearest prototype in the other solution using the distance in the feature space. It does not matter whether the prototype is centroid, medoid or mode.

The second approach uses the partitions of the two clustering solutions. The matching of each cluster is done by finding the most similar cluster in the other solution using (1). This applies to any partition-based clustering and it can be calculated from the contingency table in $O(N \cdot k^2)$ time [2]. Contingency table tells how many shared points two clusters A_i and B_j have. See Table 1 for an example.

Table 1. Contingency table for the clusterings presented in Fig. 3. For instance, clusters A_1 shares 5 points with cluster B_1, 1 point with cluster B_2, and 2 points with cluster B_3.

	B_1	B_2	B_3	$\sum$
A_1	5	1	2	8
A_2	1	4	0	5
A_3	0	1	3	4
$\sum$	6	6	5	17

This partition similarity approach applies also to centroid-based clustering but it may be slower to calculate. Finding the most similar prototype takes only $O(dk^2)$, where d is the number of dimensions. This is expected to be faster than the $O(N) + O(k^2)$ of the partition similarity; unless if the dimensionality or the number of clusters is very high. For S1–S4 data (http://cs.uef.fi/sipu/datasets) the estimated numbers are 450 distance calculations for the prototype similarity, and 4352 counter additions for the partition similarity. However, observed values in Table 2 show that the calculation of the squared distance takes so much longer that the speed benefit is practically lost. Thus, the partition-based variant is virtually as fast already when $d = 2$.

Table 2. Processing times for CI using prototype and partition similarity. The third row shows how many times they provided different CI-values. Data sets S1-S4 are from [15].

	S1	S2	S3	S4	Birch2
Prototype	15 ms	15 ms	15 ms	15 ms	120 ms
Partition	14 ms	14 ms	14 ms	14 ms	250 ms
Different	5.8 %	7.1 %	5.2 %	7.9 %	30 %

It is also possible that the two approaches provide different mappings. Even then, the resulting CI-value is expected to be mostly the same. Table 2 also reports the number of times the two approaches give different CI-value.

The third approach is to derive the similarity of clusters directly from the probability density function of the model. Next we study this for Gaussian mixture model.

3.2 Model Similarity

Gaussian mixture model (GMM) represents every cluster by its centroid and covariance matrix. This increases the size of the model from $O(1)$ to $O(d^2)$ per cluster. However, there is often not enough data to estimate the covariances reliably. A simplified variant therefore considers the diagonal of the covariance matrix, thus, reducing the model size to $O(d)$. We also use this simplified variant here.

Expectation maximization (EM) [16] algorithm optimizes GMM analogous to k-means. It iterates *Expectation* and *Maximization* steps in turn to optimize loglikelihood. It also suffers the same problem as k-means: stucks into a local optimum. Better variants include split-and-merge [17], genetic algorithm [18] and random swap [19].

For comparing two clusters (mixtures), we can use any of the three approaches. Prototype similarity approach ignores the covariance and just finds the nearest centroid in the other solution. Partition similarity performs hard partition by assigning each point into the cluster with maximum likelihood, after which the partition similarity can be calculated using the contingency table.

For the model similarity, we use here *Bhattacharyya coefficient*. It measures the similarity between two probability distributions p and q, and is calculated as follows:

$$S_{BC} = \sum \sqrt{p_i \cdot q_j} \tag{7}$$

For two multivariate normal distributions, it can be written as:

$$S_{BC} = \frac{1}{8} \left(c[A_i] - c[B_j] \right)^T \Sigma^{-1} \left(c[A_i] - c[B_j] \right) + \frac{1}{2} \ln \left(\frac{|\Sigma|}{\sqrt{|\Sigma_1||\Sigma_2|}} \right) \tag{8}$$

where $c[A_i]$ and $c[B_j]$ are the means (centroids), Σ_1 and Σ_2 are the covariance matrices of the two clusters A_i and B_j. and Σ is the average of Σ_1 and Σ_2. The first term in (8) represents the *Mahalonobis distance*, which is a special case of Bhattacharyya when the covariance matrixes of the two distributions are the same as is the case in GMM.

Figure 5 shows an example of two GMM models by SMEM [17] and RSEM [19] algorithms. We compare the results by calculating the various indexes from the resulting partitions. The result of CI was calculated by all the three approaches; all resulted into the same mapping, and gave exactly the same value CI = 2.

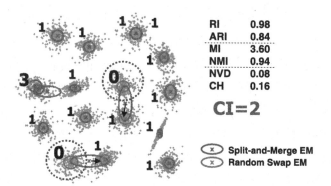

Fig. 5. Centroid index for Gaussian mixture model. Only mapping SMEM→RSEM is shown.

3.3 Arbitrary Partitions

Arbitrary-shape clustering differs from the model-based clustering in that there is no model or prototype for the clusters. This kind of clustering problem is often solved based on connectivity. For example, single link algorithm [20] results in a minimum spanning forest where each cluster is represented by the minimum spanning tree of the points in the cluster.

To compare such clustering results by CI, we use the partition similarity because it is independent on the chosen clustering model. All what is required is that we can access the partitions. However, in on-line clustering where huge amount of data is processed, the original data points might not be stored but deleted immediately when their contribution to the model is calculated. In this case, the partition-based similarity cannot be applied but in most offline applications, we do have access to partitions.

Examples of data with arbitrary-shape clusters is shown in Fig. 6 when clustered by k-means (left) and single link (right). K-means misses the two smaller clusters, and

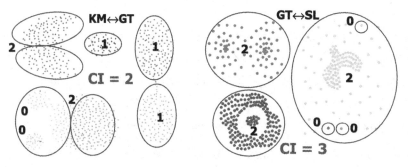

Fig. 6. K-means result for Aggregate (left) and Single Link result for Compound (right) when compared to the ground truth (GT). Mapping of the max value is shown (KM→GT; GT→SL).

divides the bigger ones in the middle and on the top. Single link makes three mistakes in total, by merging three real clusters and by creating three false clusters of size one.

4 Experiments

Here we provide numerical comparison of the centroid index (CI) against the selected existing indexes. We use the following data sets:

- S1–S4 [15]: 5000 points in 15 clusters.
- Unbalanced [13]: 6500 points in 8 clusters
- Birch2 [23]: 100,000 points in 100 clusters
- Aggregate [24]: 788 points in 7 clusters
- Compound [25]: 399 points in 6 clusters

We apply the following algorithms: K-means (KM), K-means++ (KM++) [26], Random Swap (RS) [8], Genetic Algorithm (GA) [27] for the data with spherical clusters; Single Link (SL) [21], DBSCAN [21], Split-and-Merge (SAM) [22] for data with arbitrary-shape clusters. K-means is applied for both.

Same clustering results with the corresponding validity index values are given in Tables 3. Unbalanced is rather easy to cluster by good algorithm but k-means fails because random initialization tends to select all centroids from the bigger clusters to the left, and only one centroid will move to cover the five small clusters, see Fig. 7. Thus, leaving four other clusters empty, which results in CI = 4. Most other indexes react to this but their exact values tell very little about how severe the error is, whereas the CI-value tells that half (4/8) of the clusters are wrongly allocated.

Table 3. Sample clustering results with validity values. CI = 0 indicates correct structure.

	RI	ARI	MI	NMI	NVD	CH	CSI	CI
Birch2								
KM	1.00	0.81	6.26	0.96	0.12	0.24	0.88	18
KM ++	1.00	0.95	6.54	0.99	0.03	0.06	0.97	4
RS	1.00	1.00	6.64	1.00	0.00	0.00	1.00	0
GA	1.00	1.00	6.64	1.00	0.00	0.00	1.00	0
S1								
KM	0.98	0.82	3.57	0.93	0.09	0.17	0.83	2
KM ++	1.00	1.00	3.90	0.98	0.00	0.00	1.00	0
RS	1.00	1.00	3.90	0.98	0.00	0.00	1.00	0
GA	1.00	1.00	3.90	0.98	0.00	0.00	1.00	0
S2								
KM	0.97	0.80	3.46	0.90	0.11	0.18	0.82	2
KM ++	1.00	0.99	3.87	0.99	0.00	0.00	1.00	0
RS	1.00	0.99	3.87	0.99	0.00	0.00	1.00	0
GA	1.00	0.99	3.87	0.99	0.00	0.00	1.00	0

(continued)

Table 3. (*continued*)

	RI	ARI	MI	NMI	NVD	CH	CSI	CI
Unbalanced								
KM	0.92	0.79	1.85	0.81	0.14	0.29	0.86	4
KM ++	1.00	1.00	2.03	1.00	0.00	0.00	1.00	0
RS	1.00	1.00	2.03	1.00	0.00	0.00	1.00	0
GA	1.00	1.00	2.03	1.00	0.00	0.00	1.00	0
SL	1.00	0.99	1.91	0.97	0.02	0.05	0.98	3
DBSCAN	1.00	1.00	2.02	0.99	0.00	0.00	1.00	0
SAM	0.93	0.81	1.85	0.82	0.12	0.25	0.88	4
Aggregate								
KM	0.91	0.71	2.16	0.84	0.14	0.24	0.86	2
SL	0.93	0.80	1.96	0.88	0.09	0.18	0.91	2
DBSCAN	0.99	0.98	2.41	0.98	0.01	0.01	0.99	0
SAM	1.00	1.00	2.45	1.00	0.00	0.00	1.00	0
Compound								
KM	0.84	0.54	1.71	0.72	0.25	0.34	0.75	2
SL	0.89	0.74	1.54	0.80	0.13	0.26	0.87	3
DBSCAN	0.95	0.88	1.90	0.87	0.10	0.12	0.90	2
SAM	0.83	0.53	1.78	0.76	0.19	0.34	0.81	2

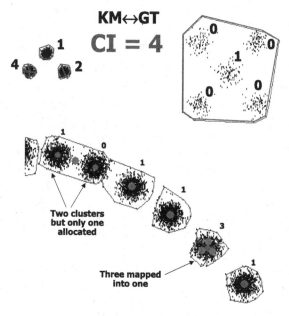

Fig. 7. Datasets unbalance (above) and Birch2 (below) clustered by k-means.

The results for the data with arbitrary-shaped clusters are similar. DBSCAN and SAM work well for the Aggregate providing perfect clustering structure (CI = 0) although DBSCAN leaves out few points as outliers. Compound is more challenging and all the methods make 2 or 3 errors in the clustering structure, usually merging the leftmost clusters and creating too many on the right.

Finally, we study how the indexes react when randomness is added increasingly to artificially created partitions (for details see [13]). Figure 8 shows that the centroid index does not react at all for these point-level changes as long as most points keeps in the original cluster. The values of the set-based measures (NVD, CH, Purity) decrease linearly, which shows that they are most appropriate to measure point-level changes.

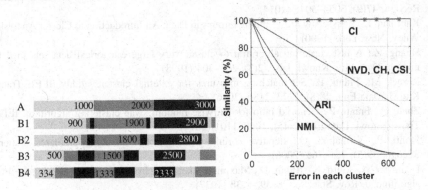

Fig. 8. Effect of increasing randomness in partitions to the clustering result.

5   Conclusions

Centroid Index (CI) is the only validity index that provides cluster level measure. It tells exactly how many clusters are differently allocated, which is more useful information than counting point-level differences. In this paper, we applied it to other clustering models such as Gaussian mixture and data with arbitrary-shaped clusters. Its main advantage is that the significance of the index value can be trivially concluded: value CI > 0 indicate that there is a significant difference in the clustering structure.

References

1. Rand, W.M.: Objective criteria for the evaluation of clustering methods. J. Am. Stat. Assoc. **66**(336), 846–850 (1971)
2. Hubert, L., Arabie, P.: Comparing partitions. J. Classif. **2**, 193–218 (1985)
3. Vinh, N.X., Epps, J., Bailey, J.: Information theoretic measures for clusterings comparison: variants, properties, normalization and correction for chance. J. Mach. Learn. Res. **11**, 2837–2854 (2010)
4. Kvalseth, T.O.: Entropy and correlation: some comments. IEEE Trans. Syst. Man Cybern. **17**(3), 517–519 (1987)

5. Dongen, S.V.: Performance criteria for graph clustering and Markov cluster experiments. Technical report INSR0012, Centrum voor Wiskunde en Informatica (2000)
6. Meila, M., Heckerman, D.: An experimental comparison of model based clustering methods. Mach. Learn. **41**(1–2), 9–29 (2001)
7. MacKay, D.: An example inference task: clustering. In: MacKay, D. (ed.) Information Theory, Inference and Learning Algorithms, pp. 284–292. Cambridge University Press, Cambridge (2003)
8. Fränti, P., Kivijärvi, J.: Randomised local search algorithm for the clustering problem. Pattern Anal. Appl. **3**(4), 358–369 (2000)
9. Fränti, P., Virmajoki, O., Hautamäki, V.: Fast agglomerative clustering using a k-nearest neighbor graph. IEEE TPAMI **28**(11), 1875–1881 (2006)
10. Fränti, P., Rezaei, M., Zhao, Q.: Centroid index: cluster level similarity measure. Pattern Recogn. **47**(9), 3034–3045 (2014)
11. Kaufman, L., Rousseeuw, P.J.: Finding Groups in Data: An Introduction to Cluster Analysis. Wiley, New York (1990)
12. Huang, Z.: A fast clustering algorithm to cluster very large categorical data sets in data mining. Data Min. Knowl. Disc. **2**(3), 283–304 (1998)
13. Rezaei, M., Fränti, P.: Set matching measures for external cluster validity. IEEE Trans. Knowl. Data Eng. **28**(8), 2173–2186 (2016)
14. Zhao, Q., Fränti, P.: Centroid ratio for pairwise random swap clustering algorithm. IEEE Trans. Knowl. Data Eng. **26**(5), 1090–1101 (2014)
15. Fränti, P., Virmajoki, O.: Iterative shrinking method for clustering problems. Pattern Recogn. **39**(5), 761–765 (2006)
16. Dempster, A., Laird, N., Rubin, D.: Maximum likelihood from incomplete data via the EM algorithm. J. Roy. Stat. Soc. B **39**, 1–38 (1977)
17. Udea, N., Nakano, R., Gharhamani, Z., Hinton, G.: SMEM algorithm for mixture models. Neural Comput. **12**, 2109–2128 (2000)
18. Pernkopf, F., Bouchaffra, D.: Genetic-based em algorithm for learning Gaussian mixture models. IEEE Trans. Pattern Anal. Mach. Intell. **27**(8), 1344–1348 (2005)
19. Zhao, Q., Hautamäki, V., Kärkkäinen, I., Fränti, P.: Random swap EM algorithm for Gaussian mixture models. Pattern Recogn. Lett. **33**, 2120–2126 (2012)
20. Jain, A.K., Dubes, R.C.: Algorithms for Clustering Data. Prentice-Hall Inc., Upper Saddle River (1988)
21. Ester, M., Kriegel, H.P., Sander, J., Xu, X.: A density-based algorithm for discovering clusters in large spatial databases with noise. In: KDDM, pp. 226–231 (1996)
22. Zhong, C., Miao, D., Fränti, P.: Minimum spanning tree based split-and-merge: a hierarchical clustering method. Inf. Sci. **181**, 3397–3410 (2011)
23. Zhang, T., et al.: BIRCH: a new data clustering algorithm and its applications. Data Min. Knowl. Disc. **1**(2), 141–182 (1997)
24. Gionis, A., Mannila, H., Tsaparas, P.: Clustering aggregation. ACM Trans. Knowl. Disc. Data (TKDD) **1**(1), 1–30 (2007)
25. Zahn, C.T.: Graph-theoretical methods for detecting and describing gestalt clusters. IEEE Trans. Comput. **100**(1), 68–86 (1971)
26. Arthur, D., Vassilvitskii, S.: K-means ++: the advantages of careful seeding. In: ACM-SIAM Symposium on Discrete Algorithms (SODA 2007), pp. 1027–1035, January 2007
27. Fränti, P.: Genetic algorithm with deterministic crossover for vector quantization. Pattern Recogn. Lett. **21**(1), 61–68 (2000)

Semi and Fully Supervised Learning Methods

The Peaking Phenomenon in Semi-supervised Learning

Jesse H. Krijthe[1,2(✉)] and Marco Loog[1,3]

[1] Pattern Recognition Laboratory, Delft University of Technology,
Delft, Netherlands
jkrijthe@gmail.com
[2] Department of Molecular Epidemiology,
Leiden University Medical Center, Leiden, Netherlands
[3] The Image Section, University of Copenhagen, Copenhagen, Denmark

Abstract. For the supervised least squares classifier, when the number of training objects is smaller than the dimensionality of the data, adding more data to the training set may first increase the error rate before decreasing it. This, possibly counterintuitive, phenomenon is known as peaking. In this work, we observe that a similar but more pronounced version of this phenomenon also occurs in the semi-supervised setting, where instead of labeled objects, unlabeled objects are added to the training set. We explain why the learning curve has a more steep incline and a more gradual decline in this setting through simulation studies and by applying an approximation of the learning curve based on the work by Raudys and Duin.

Keywords: Semi-supervised learning · Peaking · Least squares classifier · Pseudo-inverse

1   Introduction

In general, for most classifiers, classification performance is expected to improve as more labeled training examples become available. The dipping phenomenon is one exception to this rule, showing for specific combinations of datasets and classifiers that error rates can actually increase with increasing numbers of labeled data [9]. For the least squares classifier and some other classifiers, the *peaking phenomenon* is another known exception. In this setting, the classification error may first increase, after which the error rate starts to decrease again as we add more labeled training examples. The term peaking comes from the form of the learning curve: an example of which is displayed in Fig. 1.

The term 'peaking' is inspired by a different peaking phenomenon described by [5] (see also [6]), who studies the phenomenon that the performance of many classifiers peaks for a certain number of features and then decreases as more features are added. In this work we consider a different peaking phenomenon that occurs when the number of training objects is increased, and the peak does

© Springer International Publishing AG 2016
A. Robles-Kelly et al. (Eds.): S+SSPR 2016, LNCS 10029, pp. 299–309, 2016.
DOI: 10.1007/978-3-319-49055-7_27

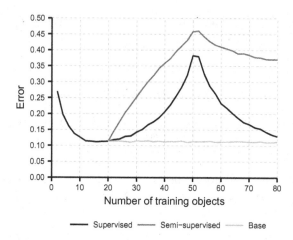

Fig. 1. Empirical learning curves for the supervised least squares classifier (Eq. (4)) where labeled data is added and the semi-supervised least squares classifier (Eq. (5)) which uses 10 labeled objects per class and the remaining objects as unlabeled objects. "Base" corresponds to the performance of the classifier that uses the first 10 labeled objects for each class, without using any additional objects. Data are generated from two Gaussians in 50 dimensions, with identity covariance matrices and a distance of 4 between the class means.

not refer to a peak in performance, but a peak in terms of the classification error, after which performance starts increasing again. While this type of peaking also shows up in feature curves, where we increase the number of features, we focus on learning curves in terms of the number of training objects because it relates more closely to the question whether unlabeled data should be used at all in the semi-supervised setting.

The peaking phenomenon considered here is observed in [2,3,10,11,14] for various classifiers in the supervised learning setting and [2,3,10,15] additionally describe different ways to get rid of this unwanted behaviour, notably, by only considering a subset of relevant objects, by adding regularization to the parameter estimation, adding noise to objects, doing feature selection, or by injecting random features.

While this peaking phenomenon has been observed for the least squares classifier when the amount of labeled data is increased, we find similar but worse behaviour in the semi-supervised setting. Following the work in [2,4], we study a particular semi-supervised adaptation of the least squares classifier in greater depth. An example of the actual behaviour is shown in Fig. 1. When the amount of labeled objects remains fixed (20 in the figure) while we increase the amount of unlabeled data used by this semi-supervised learner, the peaking phenomenon changes in two ways: the error increases more rapidly when unlabeled data is added than when labeled data is added and after the peak the error decreases more slowly than when labeled data is added. The goal of this work is to describe and explain these effects. More specifically, we attempt to answer two questions:

1. What causes the performance in the semi-supervised setting to deteriorate faster than in the supervised case?
2. If we increase the amount of unlabeled data, will the performance of the semi-supervised learner converge to an error rate below the error rate of the supervised learner that does not take the additional unlabeled data into account?

To answer these questions, we first revisit the supervised peaking phenomenon and explain its causes in Sect. 2. In Sect. 3 we show how the results from Sect. 2 relate to the least squares classifier and how we specifically adapt this classifier to the semi-supervised setting. In Sects. 4 and 5 we attempt to answer our two questions in two ways: firstly by adapting the learning curve approximation of Raudys and Duin [12] and secondly through simulation studies. We end with an investigation of the semi-supervised peaking phenomenon on some benchmark datasets.

2 Supervised Peaking

Raudys and Duin [12] attempt to explain the peaking phenomenon in the supervised case by constructing an asymptotic approximation of the learning curve and decomposing this approximation into several terms that explain the effect of adding labeled data on the learning curve. The classifier they consider is the Fisher linear discriminant, whose normal to the decision boundary is defined as the direction that maximizes the between-class variance while minimizing the within-class variance:

$$\arg\max_{w} \frac{(w^\top m_1 - w^\top m_2)^2}{w^\top W w},$$ (1)

where m_c is the sample mean of class c and $W = \frac{1}{n} \sum_{c=1}^{2} \sum_{i=1}^{N_c} (x_{ci} - m_c)(x_{ci} - m_c)^\top$ is the sample within-class scatter matrix. The solution is given by

$$w = W^{-1}(m_1 - m_2).$$ (2)

The intercept (or threshold value) that we consider in actual classification is right in between the two class means: $-\frac{1}{2}(m_1 + m_2)^\top w$. The peaking phenomenon occurs when $n = 2N < p$, where N is the number of (labeled) objects per class and p is the dimensionality of the data. In this case, a pseudo-inverse needs to be applied instead of the regular inverse of W. This is equivalent to removing directions with an eigenvalue of 0 and training the classifier in a lower dimensional subspace, a subspace whose dimensionality increases as more training data is added.

   The goal of the analysis in [12] is to construct an approximation of the learning curve, which decomposes the error into different parts. These parts relate the observed peaking behaviour to different individual effects of increasing the number of training objects. To do this they construct an asymptotic approximation where both the dimensionality and the number of objects grows to infinity. An important assumption in the derivation, and the setting we also consider in our analysis, is that the data are generated from two Gaussian distributions corresponding to two classes, with true variance $\mathbf{I}$ and a Euclidean distance between the true means μ_1 and μ_2 of δ. Lastly, objects are sampled in equal amounts from both classes.

The approximation of the learning curve is then given by[1]

$$e(N, p, \delta) = \Phi \left\{ -\frac{\delta}{2} T_r \sqrt{(1 + \gamma^2) T_\mu + \gamma^2 \frac{3\delta^2}{4p}}^{-1} \right\},$$

where Φ is the cumulative distribution function of a standard normal distribution and N is the number of objects per class. The main quantities introduced are T_μ, T_r, and γ and [12] notes that the approximation of the learning curve can be broken down to depend on exactly these three quantities all with their own specific interpretation:

$$T_\mu = 1 + \frac{1}{N} + \frac{2p^2}{\delta^2(2N - 2)N} + \frac{p^2}{\delta^2(2N - 2)N^2},$$

relates to how well we can estimate the means, $T_r = \sqrt{\frac{2N-2}{p}}$ relates to the reduction in features brought about by using the pseudo-inverse and γ is a term related to the estimation of the eigenvalues or W. The T_μ and T_r terms lead to a decrease in the error rate as N, the number of objects per class increases. This is caused by the improved estimates of the means and the increasing dimensionality. The γ term increases the generalization error as N increases, which is caused by the fact that the smallest eigenvalues are difficult to accurately estimate but can have a large effect on the computation of the pseudo-inverse.

When $n > p$ the pseudo-inverse is no longer necessary and other approximations of the learning curve can be applied. The comparison of these approximations in [16] shows that the approximation

$$e(N, p, \delta) = \Phi \left\{ -\frac{\delta}{2} \sqrt{T_\mu T_\Sigma}^{-1} \right\},$$

with $T_\mu = 1 + \frac{2p}{\delta^2 N}$ and $T_\Sigma = 1 + \frac{p}{2N-p}$ works reasonably well. The former term again relates to the estimation of the means while the latter term relates to the estimation of the within scatter matrix W. Figure 2a shows these approximations and the empirical learning curve on a simple dataset with 2 Gaussian classes, with a distance between the means of $\delta = 4.65$.

---

[1] While going through the derivation we found a different solution than the one reported in [12], which renders the last term in the formulation independent of N. This slightly changes the expressions in the explanation of the peaking behaviour.

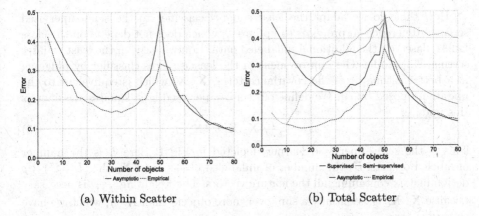

(a) Within Scatter   (b) Total Scatter

Fig. 2. Empirical learning curves and their asymptotic approximations for different classifiers. (a) Supervised learning curve corresponding to the formulation in Eq. (2). (b) Supervised and semi-supervised learning curves corresponding to the formulations in Eqs. (3), (4) and (5). Semi-supervised uses 5 labeled objects per class and the rest as unlabeled objects.

3   Semi-supervised Classifier

Unfortunately for our analysis, the classifier studied by Raudys and Duin does not correspond directly to the least squares classifier we wish to study, nor is it directly clear how their classifier can be extended to the semi-supervised setting. We therefore consider a slightly different version in which we follow [2,4]:

$$w = T^{-1}(m_1 - m_2). \tag{3}$$

This leads to the same classifier as Eq. (2) when $n > p$ [2]. Moreover, when the data are centered ($m = 0$) and the class priors are exactly equal it is equivalent to the solution of the least squares classifier, which minimizes the squared loss $(x_i^\top w - y_i)^2$ and whose solution is given by

$$w = (\mathbf{X}^\top \mathbf{X})^{-1} \mathbf{X}^\top y, \tag{4}$$

where y is a vector containing a numerical encoding of the labels and $\mathbf{X}$ is the $L \times p$ design matrix containing the L labeled feature vectors x_i.

While Eq. (3) is equivalent to Eq. (2) when $n > p$, this solution is not necessarily the same in the scenario where $n < p$ (compare the dashed black lines in Fig. 2a and b). This makes it impossible to apply the results from [12] directly to get a quantitatively good estimator for the learning curve. Moreover, their proof is not easily adapted to this new classifier. This is caused by dependencies that are introduced between the total scatter matrix T (which is proportional to $\mathbf{X}^\top \mathbf{X}$ in case $m = 0$) and the mean vectors m_c that complicate the derivation of the approximation. Their result does, however, offer a qualitative explanation of the peaking phenomenon in the semi-supervised setting—as we will see in Sect. 4.1.

How then do we adapt the least squares classifier to the semi-supervised setting? Reference [4] proposes to update T, which does not depend on the class labels, based on the additional unlabeled data. Equivalently, in the least squares setting, [13] studies the improvement in the least squares classifier by plugging in a better estimator of the covariance term, $\mathbf{X}^\top\mathbf{X}$, which is equivalent to the update proposed by [4]. We define our semi-supervised least squares classifier as this update:

$$w = (\tfrac{L}{L+U}\mathbf{X}_e^\top\mathbf{X}_e)^{-1}\mathbf{X}^\top y. \tag{5}$$

This is the semi-supervised learner depicted in Fig. 1. Here L is the number of labeled objects, U, the number of unlabeled objects and $\mathbf{X}_e$ the $(L+U) \times p$ design matrix containing all the feature vectors. The weighting $\frac{L}{L+U}$ is necessary because $\mathbf{X}_e^\top\mathbf{X}_e$ is essentially a sum over more objects than $\mathbf{X}^\top y$, which we have to correct for.

4 Why Peaking is More Extreme Under Semi-supervision

One apparent feature of the semi-supervised peaking phenomenon is that before the peak occurs, the learning curve rises more steeply when unlabeled data are added vs. when labeled data are.

4.1 Asymptotic Approximation

To explain this behaviour using the learning curve approximation, we hold the term that relates to the increased accuracy of the estimate of the means, T_μ, constant and consider the change in the approximation. As we noted before, the learning curve approximation is for a slightly different classifier, yet it might offer a qualitative insight as to the effect of only adding unlabeled data. Looking at the resulting curve in Fig. 2b, we indeed see that the semi-supervised approximation rises more quickly than the supervised approximation due to the lack of labeled data to improve the estimates of the mean. After the peak we see that the curve drops off less quickly for the same reason. The approximation, however, is not a very accurate reflection of the empirical learning curve.

4.2 Simulation of Contributions

Because the approximation used does not approximate the empirical learning curve very well, the question remains whether the lack of the updating of the means based on new data fully explains the increase in the semi-supervised learning curve over the supervised learning curve. To explore this, we decompose the change in the supervised learning curve into separate components by calculating the change in the error rate from adding data to improve respectively the estimator of the total covariance, T, the means or both at the same time. The result is shown in Fig. 3.

To do this we compare the difference in error of the semi-supervised classifier that has two additional unlabeled objects available to the supervised classifier

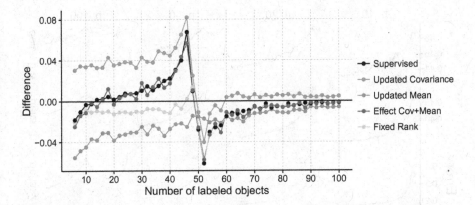

Fig. 3. Average gain in error rate by adding 2 additional objects to the training set, either in a supervised way, by adding labeled objects, or by only using them to improve the estimator of the total scatter (covariance).

that does not have these unlabeled data available. We see that adding these objects typically increases the error rate when $n < p$. We then compare the error of the supervised classifier to the one where we remove 2 labels and the classifier where we do not remove these labels. By negating this difference we get the value of having two additional labels. We see that for this dataset this effect always decreases the error. Adding up the effect of adding unlabeled objects to the effect of having additional labels, we find this approximates the total effect of adding labeled objects very well. It seems, therefore, that in the semi-supervised setting, by not having additional labels, the positive effect of these labels as shown Fig. 3 is removed, explaining the difference between the supervised and semi-supervised setting.

It is also clear from these results that peaking is caused by the estimation of the inverse of the covariance matrix, which leads to an increase in the error before $n > p$. To understand why this happens, consider the "Fixed rank" curve in Fig. 3. This curve shows the change in terms of the error rate when we add two labeled objects but leave the rank of the covariance matrix unchanged during the calculation of the inverse, merely considering the largest n eigenvectors of the newly obtained covariance matrix that was estimated using $n + 2$ objects. Since this tends to decrease the error rate, the error increase for the other curve may indeed stem from the actual growth of the rank. Especially when n is close to p, the eigenvalues of the dimensions that are added by increasing the rank become increasingly hard to estimate. This is similar to the γ term in the approximation, which captures the difficulty of estimating the eigenvalues for these directions.

5   Convergence to a Better Solution than the Base Learner?

The slow decline of the error rate after the peak in the learning curve begs the question whether the semi-supervised learner's error will ever drop below the

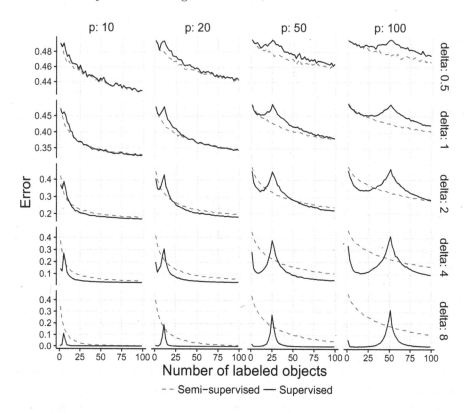

Fig. 4. Learning curves for the supervised learner and the semi-supervised learner with infinite amounts of unlabeled data for different dimensionalities, p, and distances between the means, δ.

error of the original supervised learner. If not, it would be worthwhile to refrain from using the semi-supervised learner in these settings. The approximation in Fig. 2b indicates that the learning curve will decline more slowly when $n \gtrsim p$ when unlabeled data are added. From this approximation, however, it is not clear if and under which circumstances the error of the semi-supervised classifier will improve over the base learner if larger amounts of unlabeled data become available.

To investigate this issue we consider, for the two-class Gaussian problem with different dimensionalities, p, and different distances between the means, δ, whether adding infinite unlabeled data improves over the supervised learner, for different amounts of limited labeled data. We can simulate this by setting the true means as $\boldsymbol{\mu}_1 = -\frac{\delta}{2\sqrt{p}}\mathbf{1}$ and $\boldsymbol{\mu}_2 = +\frac{\delta}{2\sqrt{p}}\mathbf{1}$. In this case, when the amount of unlabeled data increases, the total scatter matrix will converge to

$$T = I + \mathbf{1}\mathbf{1}^\top \frac{1}{4}\frac{\delta^2}{p}.$$

Using this we can calculate the semi-supervised classifier based on an infinite unlabeled sample and with a finite amount of labeled data. The results are shown in Fig. 4.

We observe that the dimensionality of the data does not have a large effect on whether the semi-supervised learner can outperform the supervised learner. It merely shifts the peak while qualitatively the differences between the supervised and semi-supervised curves remain the same. If we decrease the Bayes error by moving the means of the classes further apart, however, there are clear changes. For small distances between the means, the semi-supervised learner generally does increase performance for a larger range of sizes of the labeled set, while for larger distances this is no longer the case and the semi-supervised solution is typically worse than the supervised solution that does not take the unlabeled data into account.

6 Observations on Benchmark Datasets

The goal of this section is to observe the semi-supervised peaking phenomenon on several benchmark datasets (taken from [1,8]) and relate these observations to the results in the previous sections. We generate semi-supervised learning curves for eight benchmark datasets as follows. We select $L = \lceil p/2 \rceil$ where p is the dimensionality of the dataset after applying principal component analysis and retaining as many dimensions as required to retain 99 % of the variance.

We then randomly, with replacement, draw additional training samples, with a maximum of 100 for the smaller datasets and 1000 for the larger datasets. We also sample a separate set of 1000 objects with replacement to form the test set. The additional training samples are used as labeled examples by the supervised learner and as unlabeled samples for semi-supervised learning. We repeat this process 100 times and average the results. These averaged learning curves are shown in Fig. 5.

Both behaviours studied in the previous sections, the steeper ascent in the semi-supervised setting before the peak and the slower decline after the peak, are apparent on these example datasets. We also notice that for most of these datasets it seems unlikely that the semi-supervised learning will improve over the base classifier. This may suggest we are in a scenario similar to the large difference between the means in Fig. 4. The exception is the SPECT and SPECTF datasets, where the situation is more similar to the smaller δ. Notice that for all datasets it is still possible we are in a situation similar to $\delta = 2$ in Fig. 4: while adding unlabeled data does not help with the given amount of labeled examples, this effect might reverse if a few more labeled objects become available.

7 Discussion and Conclusion

In this work, we have studied the behaviour of the learning curve for one particular semi-supervised adaptation of the least squares classifier. This adaptation, based on the ideas from [4,13], was amenable to analysis. It is an open question

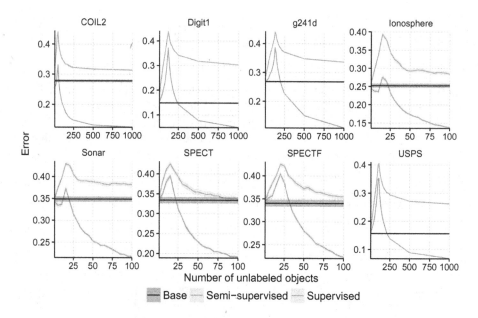

Fig. 5. Learning curves on benchmark datasets. The number of labeled objects is equal to $\lceil p/2 \rceil$. For the semi-supervised curve we add more unlabeled data, for the supervised curve more labeled data. "Base" shows performance of the supervised classifier using only the original $\lceil p/2 \rceil$ objects. For each dataset 100 curves where generated and averaged. Shaded area (small) indicates the standard error around the mean.

what the typical learning curve for other semi-supervised least squares adaptations looks like, such as self-learning or the constraint based approach in [7] where we first noticed this behaviour and which inspired us to look into this phenomenon. The lack of a closed form solution in these cases makes it more difficult to subject them to a similar analysis. Nevertheless, the current study does provide insight in the additional problems that small samples entail in the semi-supervised setting and largely explains the learning curve behaviour, at least for the specific semi-supervised learner considered.

Acknowledgements. This work was funded by project P23 of the Dutch public/private research network COMMIT.

References

1. Chapelle, O., Schölkopf, B., Zien, A.: Semi-Supervised Learning. MIT press, Cambridge (2006)
2. Duin, R.P.W.: Small sample size generalization. In: Proceedings of the Scandinavian Conference on Image Analysis, pp. 957–964 (1995)
3. Duin, R.P.: Classifiers in almost empty spaces. In: Proceedings of the 15th International Conference on Pattern Recognition, pp. 1–7 (2000)

4. Fan, B., Lei, Z., Li, S.Z.: Normalized LDA for semi-supervised learning. In: International Conference on Automatic Face & Gesture Recognition, pp. 1–6 (2008)
5. Hughes, G.F.: On the mean accuracy of statistical pattern recognizers. IEEE Trans. Inf. Theor. **14**(1), 55–63 (1968)
6. Jain, A.K., Chandrasekaran, B.: Dimensionality and sample size considerations in pattern recognition practice. In: Krishnaiah, P.R., Kanal, L. (eds.) Handbook of Statistics, vol. 2, pp. 835–855. North-Holland Publishing Company, Amsterdam (1982)
7. Krijthe, J.H., Loog, M.: Implicitly constrained semi-supervised least squares classification. In: Fromont, E., Bie, T., Leeuwen, M. (eds.) IDA 2015. LNCS, vol. 9385, pp. 158–169. Springer, Heidelberg (2015). doi:10.1007/978-3-319-24465-5_14
8. Lichman, M.: UCI Machine Learning Repository (2013). http://archive.ics.uci.edu/ml
9. Loog, M., Duin, R.P.W.: The dipping phenomenon. In: Gimefarb, G., Hancock, E., Imiya, A., Kuijper, A., Kudo, M., Omachi, S., Windeatt, T., Windeatt, T., Yamada, K. (eds.) SSPR & SPR 2012. LNCS, vol. 7626, pp. 310–317. Springer, Heidelberg (2012)
10. Opper, M.: Learning to generalize. In: Baltimore, D. (ed.) Frontiers of Life: Intelligent Systems, pp. 763–775. Academic Press, Cambridge (2001)
11. Opper, M., Kinzel, W.: Statistical mechanics of generalization. In: Domany, E., Hemmen, J.L., Schulten, K. (eds.) Physics of Neural Networks III, pp. 151–209. Springer, New York (1995)
12. Raudys, S., Duin, R.P.W.: Expected classification error of the Fisher linear classifier with pseudo-inverse covariance matrix. Pattern Recogn. Lett. **19**(5–6), 385–392 (1998)
13. Shaffer, J.P.: The Gauss-Markov theorem and random regressors. Am. Stat. **45**(4), 269–273 (1991)
14. Skurichina, M., Duin, R.P.W.: Stabilizing classifiers for very small samplesizes. In: Proceedings of the 13th International Conference on Pattern Recognition, pp. 891–896 (1996)
15. Skurichina, M., Duin, R.P.W.: Regularisation of linear classifiers by adding redundant features. Pattern Anal. Appl. **2**(1), 44–52 (1999)
16. Wyman, F.J., Young, D.M., Turner, D.W.: A comparison of asymptotic error rate expansions for the sample linear discriminant function. Pattern Recogn. **23**(7), 775–783 (1990)

Walker-Independent Features for Gait Recognition from Motion Capture Data

Michal Balazia$^{(\boxtimes)}$ and Petr Sojka

Faculty of Informatics, Masaryk University,
Botanická 68a, 60200 Brno, Czech Republic
xbalazia@mail.muni.cz, sojka@fi.muni.cz

Abstract. MoCap-based human identification, as a pattern recognition discipline, can be optimized using a machine learning approach. Yet in some applications such as video surveillance new identities can appear on the fly and labeled data for all encountered people may not always be available. This work introduces the concept of learning walker-independent gait features directly from raw joint coordinates by a modification of the Fisher's Linear Discriminant Analysis with Maximum Margin Criterion. Our new approach shows not only that these features can discriminate different people than who they are learned on, but also that the number of learning identities can be much smaller than the number of walkers encountered in the real operation.

1 Introduction

Recent rapid improvement in motion capture (MoCap) sensor accuracy brought affordable technology that can identify walking people. MoCap technology provides video clips of walking individuals containing structural motion data. The format keeps an overall structure of the human body and holds estimated 3D positions of major anatomical landmarks as the person moves. MoCap data can be collected online by a system of multiple cameras (Vicon) or a depth camera (Microsoft Kinect). To visualize motion capture data (see Fig. 1), a simplified stick figure representing the human skeleton (graph of joints connected by bones) can be recovered from body point spatial coordinates.

Recognizing a person by walk involves capturing and normalizing their walk sample, extracting gait features to compose a template, and finally querying a central database for a set of similar templates to report the most likely identity. This work focuses on extracting robust and discriminative gait features from raw MoCap data.

Many geometric gait features have been introduced over the past few years. They are typically combinations of static body parameters (bone lengths, person's height) [12] with dynamic gait features such as step length, walk speed, joint angles and inter-joint distances [1,4,12,15], along with various statistics (mean, standard deviation or maximum) of their signals [3]. Clearly, these features are schematic and human-interpretable, which is convenient for visualizations and for intuitive understanding, but unnecessary for automatic gait

© Springer International Publishing AG 2016
A. Robles-Kelly et al. (Eds.): S+SSPR 2016, LNCS 10029, pp. 310–321, 2016.
DOI: 10.1007/978-3-319-49055-7_28

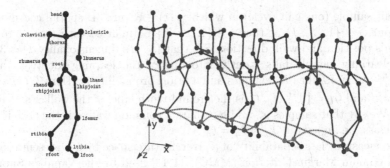

Fig. 1. Motion capture data. Skeleton is represented by a stick figure of 31 joints (only 17 are drawn here). Seven selected video frames of a walk sequence contain 3D coordinates of each joint in time. The red and blue lines track trajectories of hands and feet. (Color figure online)

recognition. Instead, this application prefers learning features that maximally separate the identity classes and are not limited by such dispensable factors.

Methods for 2D gait recognition extensively use machine learning models for extracting gait features, such as principal component analysis and multi-scale shape analysis [8], genetic algorithms and kernel principal component analysis [17], radial basis function neural networks [19], or convolutional neural networks [7]. All of those and many other models are reasonable to be utilized also in 3D gait recognition.

In the video surveillance environment data need to be acquired without walker's consent and new identities can appear on the fly. Here and also in other applications where labels for all encountered people may not always be available, we value features that have a high power in distinguishing all people and not exclusively who they were learned on. We call these walker-independent features. The main idea is to statistically learn what aspects of walk people generally differ in and extract those as gait features. The features are learned in a supervised manner, as described in the following section.

2   Learning Gait Features

In statistical pattern recognition, reducing space dimensionality is a common technique to overcome class estimation problems. Classes are discriminated by projecting high-dimensional input data onto low-dimensional sub-spaces by linear transformations with the goal of maximizing the class separability. We are interested in finding an optimal feature space where a gait template is close to those of the same walker and far from those of different walkers.

Let the model of a human body have J joints and all samples be linearly normalized to their average length T. Labeled learning data in the measurement space $\mathcal{G}_L$ are in the form $\{(\mathbf{g}_n, \ell_n)\}_{n=1}^{N_L}$ where

$$\mathbf{g}_n = \left[\left[\gamma_1(1) \cdots \gamma_J(1) \right]^\mathsf{T} \cdots \left[\gamma_1(T) \cdots \gamma_J(T) \right]^\mathsf{T} \right]^\mathsf{T} \tag{1}$$

is a gait sample (one gait cycle) in which $\gamma_j(t) \in \mathbb{R}^3$ are 3D spatial coordinates of a joint $j \in \{1, \ldots, J\}$ at time $t \in \{1, \ldots, T\}$ normalized with respect to the person's position and walk direction. See that $\mathcal{G}_L$ has dimensionality $D = 3JT$. Each learning sample falls strictly into one of the learning identity classes $\{\mathcal{I}_c\}_{c=1}^{C_L}$. A class $\mathcal{I}_c$ has N_c samples and a priori probability p_c. Here $\mathcal{I}_c \cap \mathcal{I}_{c'} = \emptyset$ for $c \neq c'$ and $\mathcal{G}_L = \bigcup_{c=1}^{C_L} \mathcal{I}_c$. ℓ_n is the ground-truth label of the walker's identity class. We say that samples $\mathbf{g}_n$ and $\mathbf{g}_{n'}$ share a common walker if they are in the same class, i.e., $\mathbf{g}_n, \mathbf{g}_{n'} \in \mathcal{I}_c \Leftrightarrow \ell_n = \ell_{n'}$.

We measure class separability of a given feature space by a representation of the Maximum Margin Criterion (MMC) [11,13] used by the Vapnik's Support Vector Machines (SVM) [18]

$$\mathcal{J} = \frac{1}{2} \sum_{c,c'=1}^{C_L} p_c p_{c'} \left((\mu_c - \mu_{c'})^\top (\mu_c - \mu_{c'}) - \mathrm{tr}\,(\Sigma_c + \Sigma_{c'}) \right) \tag{2}$$

which is actually a summation of $\frac{1}{2} C_L(C_L - 1)$ between-class margins. The margin is defined as the Euclidean distance of class means minus both individual variances (traces of scatter matrices $\Sigma_c = \sum_{n=1}^{N_c} p_c \left(\mathbf{g}_n^{(c)} - \mu_c \right) \left(\mathbf{g}_n^{(c)} - \mu_c \right)^\top$ and similarly for $\Sigma_{c'}$). For the whole labeled data, we denote the between- and within-class and total scatter matrices

$$\Sigma_{\mathrm{B}} = \sum_{c=1}^{C_L} p_c (\mu_c - \mu)(\mu_c - \mu)^\top$$

$$\Sigma_{\mathrm{W}} = \sum_{c=1}^{C_L} p_c \sum_{n=1}^{N_c} \left(\mathbf{g}_n^{(c)} - \mu_c \right) \left(\mathbf{g}_n^{(c)} - \mu_c \right)^\top \tag{3}$$

$$\Sigma_{\mathrm{T}} = \sum_{c=1}^{C_L} p_c \sum_{n=1}^{N_c} \left(\mathbf{g}_n^{(c)} - \mu \right) \left(\mathbf{g}_n^{(c)} - \mu \right)^\top = \Sigma_{\mathrm{B}} + \Sigma_{\mathrm{W}}$$

where $\mathbf{g}_n^{(c)}$ denotes the n-th sample in class $\mathcal{I}_c$ and μ_c and μ are sample means for class $\mathcal{I}_c$ and the whole data set, respectively, that is, $\mu_c = \frac{1}{N_c} \sum_{n=1}^{N_c} \mathbf{g}_n^{(c)}$ and $\mu = \frac{1}{N_L} \sum_{n=1}^{N_L} \mathbf{g}_n$. Now we obtain

$$\mathcal{J} = \frac{1}{2} \sum_{c,c'=1}^{C_L} p_c p_{c'} (\mu_c - \mu_{c'})^\top (\mu_c - \mu_{c'}) - \frac{1}{2} \sum_{c,c'=1}^{C_L} p_c p_{c'}\, \mathrm{tr}\,(\Sigma_c + \Sigma_{c'})$$

$$= \frac{1}{2} \sum_{c,c'=1}^{C_L} p_c p_{c'} (\mu_c - \mu + \mu - \mu_{c'})^\top (\mu_c - \mu + \mu - \mu_{c'}) - \sum_{c=1}^{C_L} p_c\, \mathrm{tr}\,(\Sigma_c) \tag{4}$$

$$= \mathrm{tr}\left(\sum_{c=1}^{C_L} p_c (\mu_c - \mu)(\mu_c - \mu)^\top \right) - \mathrm{tr}\left(\sum_{c=1}^{C_L} p_c \Sigma_c \right) = \mathrm{tr}\,(\Sigma_{\mathrm{B}}) - \mathrm{tr}\,(\Sigma_{\mathrm{W}}) = \mathrm{tr}\,(\Sigma_{\mathrm{B}} - \Sigma_{\mathrm{W}}).$$

Since $\mathrm{tr}\,(\Sigma_{\mathrm{B}})$ measures the overall variance of the class mean vectors, a large one implies that the class mean vectors scatter in a large space. On the other hand, a small $\mathrm{tr}(\Sigma_\mathrm{W})$ implies that classes have a small spread. Thus, a large $\mathcal{J}$ indicates that samples are close to each other if they share a common walker but are far from each other if they are performed by different walkers. Extracting features, that is, transforming the input data in the measurement space into a feature space of higher $\mathcal{J}$, can be used to link new observations of walkers more successfully.

Feature extraction is given by a linear transformation (feature) matrix $\Phi \in \mathbb{R}^{D \times \widehat{D}}$ from a D-dimensional measurement space $\mathcal{G} = \{\mathbf{g}_n\}_{n=1}^N$ of not necessarily labeled gait samples to a $\widehat{D}$-dimensional feature space (manifold) $\widehat{\mathcal{G}} = \{\widehat{\mathbf{g}}_n\}_{n=1}^N$ of gait templates where $\widehat{D} < D$ and each gait sample $\mathbf{g}_n$ is transformed into a gait template $\widehat{\mathbf{g}}_n = \Phi^\top \mathbf{g}_n$. The objective is to learn a transform Φ that maximizes MMC in the feature space

$$\mathcal{J}(\Phi) = \mathrm{tr}\left(\Phi^\top (\Sigma_\mathrm{B} - \Sigma_\mathrm{W})\,\Phi\right). \tag{5}$$

Once the transformation is found, all measured samples are transformed to the feature space (to become gait templates) along with the class means and covariances. Assigning a new sample to a walker can be done using the same transformation and comparing against the gallery templates with a distance measure such as the Mahalanobis distance

$$\widehat{\delta}(\widehat{\mathbf{g}}_n, \widehat{\mathbf{g}}_{n'}) = \sqrt{(\widehat{\mathbf{g}}_n - \widehat{\mathbf{g}}_{n'})^\top \widehat{\Sigma}_\mathrm{T}^{-1} (\widehat{\mathbf{g}}_n - \widehat{\mathbf{g}}_{n'})}. \tag{6}$$

We show that solution to the optimization problem in Eq. (5) can be obtained by eigendecomposition of the matrix $\Sigma_\mathrm{B} - \Sigma_\mathrm{W}$. An important property to notice about the objective $\mathcal{J}(\Phi)$ is that it is invariant w.r.t. rescalings $\Phi \to \alpha\Phi$. Hence, we can always choose $\Phi = \mathbf{f}_1 \| \cdots \| \mathbf{f}_{\widehat{D}}$ such that $\mathbf{f}_{\widehat{d}}^\top \mathbf{f}_{\widehat{d}} = 1$, since it is a scalar itself. For this reason we can reduce the problem of maximizing $\mathcal{J}(\Phi)$ into the constrained optimization problem

$$\max \quad \sum_{\widehat{d}=1}^{\widehat{D}} \mathbf{f}_{\widehat{d}}^\top (\Sigma_\mathrm{B} - \Sigma_\mathrm{W}) \mathbf{f}_{\widehat{d}} \tag{7}$$

$$\text{subject to} \quad \mathbf{f}_{\widehat{d}}^\top \mathbf{f}_{\widehat{d}} - 1 = 0 \qquad \forall \widehat{d} = 1, \ldots, \widehat{D}.$$

To solve the above optimization problem, let us consider the Lagrangian

$$\mathcal{L}\left(\mathbf{f}_{\widehat{d}}, \lambda_{\widehat{d}}\right) = \sum_{\widehat{d}=1}^{\widehat{D}} \mathbf{f}_{\widehat{d}}^\top (\Sigma_\mathrm{B} - \Sigma_\mathrm{W}) \mathbf{f}_{\widehat{d}} - \lambda_{\widehat{d}} \left(\mathbf{f}_{\widehat{d}}^\top \mathbf{f}_{\widehat{d}} - 1\right) \tag{8}$$

with multipliers $\lambda_{\widehat{d}}$. To find the maximum, we derive it with respect to $\mathbf{f}_{\widehat{d}}$ and equate to zero

$$\frac{\partial \mathcal{L}\left(\mathbf{f}_{\widehat{d}}, \lambda_{\widehat{d}}\right)}{\partial \mathbf{f}_{\widehat{d}}} = \left((\Sigma_\mathrm{B} - \Sigma_\mathrm{W}) - \lambda_{\widehat{d}}\mathbf{I}\right) \mathbf{f}_{\widehat{d}} = 0 \tag{9}$$

which leads to

$$(\Sigma_\mathrm{B} - \Sigma_\mathrm{W}) \mathbf{f}_{\widehat{d}} = \lambda_{\widehat{d}} \mathbf{f}_{\widehat{d}} \tag{10}$$

and, putting it all together, we get the Generalized Eigenvalue Problem (GEP)

$$(\Sigma_B - \Sigma_W) \, \Phi = \Lambda \Phi \tag{11}$$

where $\Lambda = \text{diag}\left(\lambda_1, \ldots, \lambda_{\widehat{D}}\right)$ is the generalized eigenvalue matrix. Therefore,

$$\mathcal{J}(\Phi) = \text{tr}\left(\Phi^\top (\Sigma_B - \Sigma_W) \, \Phi\right) = \text{tr}\left(\Phi^\top \Lambda \Phi\right) = \text{tr}(\Lambda) \tag{12}$$

is maximized when Λ has $\widehat{D}$ largest eigenvalues and Φ contains the corresponding leading eigenvectors of $\Sigma_B - \Sigma_W$.

In the following we discuss how to calculate the eigenvectors of $\Sigma_B - \Sigma_W$ and to determine the optimal dimensionality $\widehat{D}$ of the feature space. First, we rewrite $\Sigma_B - \Sigma_W = 2\Sigma_B - \Sigma_T$. Note that the null space of Σ_T is a subspace of that of Σ_B since the null space of Σ_T is the common null space of Σ_B and Σ_W. Thus, we can simultaneously diagonalize Σ_B and Σ_T to some Δ and I

$$\begin{aligned} \Psi^\top \Sigma_B \Psi &= \Delta \\ \Psi^\top \Sigma_T \Psi &= I \end{aligned} \tag{13}$$

with the $D \times \text{rank}(\Sigma_T)$ eigenvector matrix

$$\Psi = \Omega \Theta^{-\frac{1}{2}} \Xi \tag{14}$$

where Ω and Θ are the eigenvector and eigenvalue matrices of Σ_T, respectively, and Ξ is the eigenvector matrix of $\Theta^{-1/2}\Omega^\top \Sigma_B \Omega \Theta^{-1/2}$. To calculate Ψ, we use a fast two-step algorithm in virtue of Singular Value Decomposition (SVD). SVD expresses a real $r \times s$ matrix A as a product $A = UDV^\top$ where D is a diagonal matrix with decreasing non-negative entries, and U and V are $r \times \min\{r,s\}$ and $s \times \min\{r,s\}$ eigenvector matrices of $AA^\top$ and $A^\top A$, respectively, and the non-vanishing entries of D are square roots of the non-zero corresponding eigenvalues of both $AA^\top$ and $A^\top A$. See that Σ_T and Σ_B can be expressed in the forms

$$\begin{aligned} \Sigma_T &= XX^\top \text{ where } X = \frac{1}{\sqrt{N_L}}\left[(g_1 - \mu) \cdots (g_{N_L} - \mu)\right] \text{ and} \\ \Sigma_T &= \Upsilon\Upsilon^\top \text{ where } \Upsilon = \left[\sqrt{p_1}(\mu_1 - \mu) \cdots \sqrt{p_{C_L}}(\mu_{C_L} - \mu)\right], \end{aligned} \tag{15}$$

respectively. Hence, we can obtain the eigenvectors Ω and the corresponding eigenvalues Θ of Σ_T through the SVD of X and analogically Ξ of $\Theta^{-1/2}\Omega^\top \Sigma_B \Omega \Theta^{-1/2}$ through the SVD of $\Theta^{-1/2}\Omega^\top \Upsilon$. The columns of Ψ are clearly the eigenvectors of $2\Sigma_B - \Sigma_T$ with the corresponding eigenvalues $2\Delta - I$. Therefore, to constitute the transform Φ by maximizing the MMC, we should choose the eigenvectors in Ψ that correspond to the eigenvalues of at least $\frac{1}{2}$ in Δ. Note that Δ contains at most $\text{rank}(\Sigma_B) = C_L - 1$ positive eigenvalues, which gives an upper bound on the feature space dimensionality $\widehat{D}$. Algorithm 1 [5] provided below is an efficient way of learning the transform Φ for MMC on given labeled learning data $\mathcal{G}_L$.

Algorithm 1. LearnTransformationMatrixMMC($\mathcal{G}_L$)

1: split $\mathcal{G}_L = \{(\mathbf{g}_n, \ell_n)\}_{n=1}^{N_L}$ into identity classes $\{\mathcal{I}_c\}_{c=1}^{C_L}$
   such that $\mathbf{g}_n, \mathbf{g}_{n'} \in \mathcal{I}_c \Leftrightarrow \ell_n = \ell_{n'}$ and set p_c (we set $p_c = N_c/N_L$)
2: compute overall mean $\mu = \frac{1}{N_L} \sum_{n=1}^{N_L} \mathbf{g}_n$
   and individual class means $\mu_c = \frac{1}{N_c} \sum_{n=1}^{N_c} \mathbf{g}_n^{(c)}$
3: compute $\boldsymbol{\Sigma}_{\mathrm{B}} = \sum_{c=1}^{C_L} p_c \left(\mu_c - \mu\right)\left(\mu_c - \mu\right)^{\top}$
4: compute $\mathbf{X} = \frac{1}{\sqrt{N_L}} [(\mathbf{g}_1 - \mu) \cdots (\mathbf{g}_{N_L} - \mu)]$
5: compute $\boldsymbol{\Upsilon} = \left[\sqrt{p_1}\left(\mu_1 - \mu\right) \cdots \sqrt{p_{C_L}}\left(\mu_{C_L} - \mu\right)\right]$
6: compute eigenvectors $\boldsymbol{\Omega}$ and corresponding eigenvalues $\boldsymbol{\Theta}$ of $\boldsymbol{\Sigma}_{\mathrm{T}} = \mathbf{X}\mathbf{X}^{\top}$ through SVD of $\mathbf{X}$
7: compute eigenvectors $\boldsymbol{\Xi}$ of $\boldsymbol{\Theta}^{-1/2}\boldsymbol{\Omega}^{\top}\boldsymbol{\Sigma}_{\mathrm{B}}\boldsymbol{\Omega}\boldsymbol{\Theta}^{-1/2}$ through SVD of $\boldsymbol{\Theta}^{-1/2}\boldsymbol{\Omega}^{\top}\boldsymbol{\Upsilon}$
8: compute eigenvectors $\boldsymbol{\Psi} = \boldsymbol{\Omega}\boldsymbol{\Theta}^{-1/2}\boldsymbol{\Xi}$
9: compute eigenvalues $\boldsymbol{\Delta} = \boldsymbol{\Psi}^{\top}\boldsymbol{\Sigma}_{\mathrm{B}}\boldsymbol{\Psi}$
10: return transform $\boldsymbol{\Phi}$ as eigenvectors in $\boldsymbol{\Psi}$ that correspond to the eigenvalues of at least $\frac{1}{2}$ in $\boldsymbol{\Delta}$

---

3 Experiments and Results

3.1 Database

For the evaluation purposes we have extracted a large number of samples from the general MoCap database from CMU [9] as a well-known and recognized database of structural human motion data. It contains numerous motion sequences, including a considerable number of gait sequences. Motions are recorded with an optical marker-based Vicon system. People wear a black jumpsuit and have 41 markers taped on. The tracking space of $30\,\mathrm{m}^2$, surrounded by 12 cameras of sampling rate of $120\,\mathrm{Hz}$ in the height from 2 to $4\,\mathrm{m}$ above ground, creates a video surveillance environment. Motion videos are triangulated to get highly accurate 3D data in the form of relative body point coordinates (with respect to the root joint) in each video frame and stored in the standard ASF/AMC data format. Each registered participant is assigned with their respective skeleton described in an ASF file. Motions in the AMC files store bone rotational data, which is interpreted as instructions about how the associated skeleton deforms over time.

These MoCap data, however, contain skeleton parameters pre-calibrated by the CMU staff. Skeletons are unique for each walker and even a trivial skeleton check could result in 100 % recognition. In order to use the collected data in a fairly manner, a prototypical skeleton is constructed and used to represent bodies of all subjects, shrouding the unique skeleton parameters of individual walkers. Assuming that all walking subjects are physically identical disables the skeleton check as a potentially unfair classifier. Moreover, this is a skeleton-robust solution as all bone rotational data are linked with a fixed skeleton. To obtain realistic parameters, it is calculated as the mean of all skeletons in the provided ASF files.

We calculate 3D joint coordinates of joints using bone rotational data and the prototypical skeleton. One cannot directly use raw values of joint coordinates, as they refer to absolute positions in the tracking space, and not all potential methods are invariant to person's position or walk direction. To ensure such invariance, the center of the coordinate system is moved to the position of root joint $\gamma_{\text{root}}(t) = [0, 0, 0]^\top$ for each time t and axes are adjusted to the walker's perspective: the X axis is from right (negative) to left (positive), the Y axis is from down (negative) to up (positive), and the Z axis is from back (negative) to front (positive). In the AMC file structure notation it means setting the root translation and rotation to zero (**root 0 0 0 0 0 0**) in all frames of all motion sequences.

Since the general motion database contains all motion types, we extracted a number of sub-motions that represent gait cycles. First, an exemplary gait cycle was identified, and clean gait cycles were then filtered out using the DTW distance over bone rotations. The similarity threshold was set high enough so that even the least similar sub-motion still semantically represents a gait cycle. Finally, subjects that contributed with less than 10 samples were excluded. The final database has 54 walking subjects that performed 3,843 samples in total, which makes an average of about 71 samples per subject.

3.2   Evaluation Set-Ups and Metrics

Learning data $\mathcal{G}_L = \{(\mathbf{g}_n, \ell_n)\}_{n=1}^{N_L}$ of C_L identities and evaluation data $\mathcal{G}_E = \{(\mathbf{g}_n, \ell_n)\}_{n=1}^{N_E}$ of C_E identity classes have to be disjunct at all times. Evaluation is performed exclusively on the evaluation part, taking no observations of the learning part into account. In the following we introduce two set-ups of data separation: homogeneous and heterogeneous. The homogeneous set-up learns the transformation matrix on $1/3$ samples of C_L identities and is evaluated on templates derived from other $2/3$ samples of the same $C_E = C_L$ identities. The heterogeneous set-up learns the transform on all samples in C_L identities and is evaluated on all templates derived from other C_E identities. For better clarification we refer to Fig. 2. Note that unlike in homogeneous set-up, in heterogeneous set-up there is no walker identity ever used for both learning and evaluation at the same time.

Homogeneous set-up is parametrized by a single number $C_L = C_E$ of learning-and-evaluation identity classes, whereas the heterogeneous set-up has the form (C_L, C_E) specifying how many learning and how many evaluation identity classes are randomly selected from the database. Evaluation of each set-up is repeated 3 times, selecting new random C_L and C_E identity classes each time and reporting the average result. Please note that in the heterogeneous set-up the learning identities are disjunct from the evaluation identities, that is, there is no single identity used for both learning and evaluation.

Correct Classification Rate (CCR) is a standard qualitative measure; however, if a method has a low CCR, we cannot directly say if the system is failing because of bad features or a bad classifier. Providing an evaluation in terms of class separability of the feature space gives an estimate on the recognition potential of the extracted features and do not reflect eventual combination with

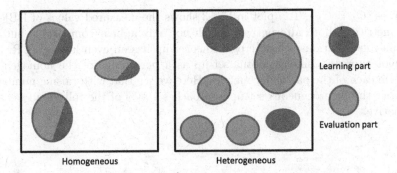

Fig. 2. Abstraction of data separation for homogeneous set-up of $C_L = C_E = 3$ learning-and-evaluation classes (left) and for heterogeneous set-up of $C_L = 2$ learning classes and $C_E = 4$ evaluation classes (right). Black square represents a database and ellipses are identity classes. (Color figure online)

an unsuitable classifier. Quality of features extraction algorithms is reflected in the Davies-Bouldin Index (DBI)

$$\text{DBI} = \frac{1}{C_E} \sum_{c=1}^{C_E} \max_{1 \leq c' \leq C_E, c' \neq c} \frac{\sigma_c + \sigma_{c'}}{\widehat{\delta}(\widehat{\mu}_c, \widehat{\mu}_{c'})} \tag{16}$$

where $\sigma_c = \frac{1}{N_c} \sum_{n=1}^{N_c} \widehat{\delta}(\widehat{\mathbf{g}}_n, \widehat{\mu}_c)$ is the average distance of all templates in identity class $\mathcal{I}_c$ to its centroid, and similarly for $\sigma_{c'}$. Templates of low intra-class distances and of high inter-class distances have a low DBI. DBI is measured on the full evaluation part, whereas CCR is estimated with 10-fold cross-validation taking one dis-labeled fold as a testing set and other nine as gallery. Test templates are classified by the winner-takes-all strategy, in which a test template $\widehat{\mathbf{g}}^{\text{test}}$ gets assigned with the label $\ell_{\arg\min_i \widehat{\delta}(\widehat{\mathbf{g}}^{\text{test}}, \widehat{\mathbf{g}}_i^{\text{gallery}})}$ of the gallery's closest identity class.

Based on Sect. 3.1, our database has 54 identity classes in total. We performed the series of experiments **A, B, C, D** below. The experiments **A** and **B** are to compare the homogeneous and heterogeneous set-up, whereas **C** and **D** examine how performance of the system in the heterogeneous set-up improves with increasing number of learning identities. The results are illustrated in Fig. 3 and in Fig. 4 in the next section.

A homogeneous set-up with $C_L = C_E \in \{2, \ldots, 27\}$;
B heterogeneous set-up with $C_L = C_E \in \{2, \ldots, 27\}$;
C heterogeneous set-up with $C_L \in \{2, \ldots, 27\}$ and $C_E = 27$;
D heterogeneous set-up with $C_L \in \{2, \ldots, 52\}$ and $C_E = 54 - C_L$.

3.3 Results

Experiments **A** and **B** compare homogeneous and heterogeneous set-ups by measuring the drop in performance on an identical number of learning and evaluation identities $(C_L = C_E)$. Top plot in Fig. 3 shows the measured values of DBI and CCR metrics in both alternatives, which not only appear comparable but also in some configurations the heterogeneous set-up has an even higher CCR. Bottom plot expresses heterogeneous set-up as a percentage of the homogeneous set-up in each of the particular metrics. Here we see that with raising number of identities the heterogeneous set-up approaches 100 % of the fully homogeneous alternative.

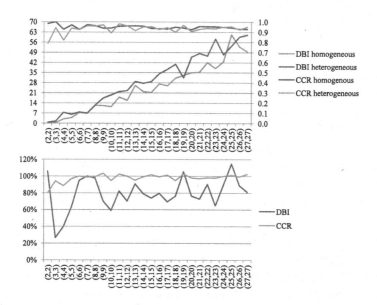

Fig. 3. DBI (left vertical axis) and CCR (right vertical axis) for experiments **A** of homogeneous set-up and **B** of heterogeneous set-up (top) with (C_L, C_E) configurations (horizontal axes) and their percentages (bottom).

Experiments **C** and **D** investigate on the impact of the number of learning identities in the heterogeneous set-up. Observing from the Fig. 4, the performance grows quickly on the first configurations with very few learning identities, which we can interpret as an analogy to the Pareto (80–20) principle. Specifically, the results of experiment **C** say that 8 learning identities achieve almost the same performance (66.78 DBI and 0.902 CCR) to as if learned on 27 identities (68.32 DBI and 0.947 CCR). The outcome of experiment **D** indicates a similar growth of performance and we see that yet 14 identities can be enough to learn the transformation matrix to distinguish 40 completely different people (0.904 CCR).

The proposed method and seven other state-of-the-art methods [2,4,6,10,12, 14,16] have been subjected to extensive simulations on homogeneous set-up in our recent research paper [5]. A variety of class-separability coefficients and classification metrics allows insights from different statistical perspectives. Results

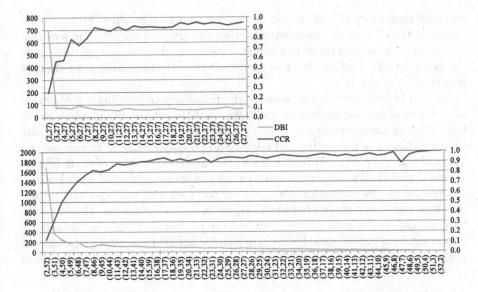

Fig. 4. DBI (left vertical axes) and CCR (right vertical axes) for experiments **C** (top) and **D** (bottom) on heterogeneous set-up with (C_L, C_E) configurations (horizontal axes).

indicate that the proposed method is a leading concept for rank-based classifier systems: lowest Davies-Bouldin Index, highest Dunn Index, highest (and exclusively positive) Silhouette Coefficient, second highest Fisher's Discriminant Ratio and, combined with rank-based classifier, the best Cumulative Match Characteristic, False Accept Rate and False Reject Rate trade-off, Receiver Operating Characteristic (ROC) and recall-precision trade-off scores along with Correct Classification Rate, Equal Error Rate, Area Under ROC Curve and Mean Average Precision. We interpret the high scores as a sign of robustness. Apart from performance merits, the MMC method is also efficient: low-dimensional templates ($\widehat{D} \leq C_L - 1 = C_E - 1 = 53$) and Mahalanobis distance ensure fast distance computations and thus contribute to high scalability.

4   Conclusions

Despite many advanced optimization techniques used in statistical pattern recognition, a common practice of state-of-the-art MoCap-based human identification is still to design geometric gait features by hand. As the first contribution of this paper, the proposed method does not involve any ad-hoc features; on the contrary, they are computed from a much larger space beyond the limits of human interpretability. The features are learned directly from raw joint coordinates by a modification of the Fisher's LDA with MMC so that the identities are maximally separated. We believe that MMC is a suitable criterion for optimizing gait features; however, our future work will continue with research on further potential optimality criterions and machine learning approaches. Furthermore, we are in the process of developing an evaluation framework with implementation details and source codes of all related methods, data extraction drive from the general CMU MoCap database and the evaluation mechanism to support reproducible research.

Second contribution lies in showing the possibility of building a representation on a problem and using it on another (related) problem. Simulations on the CMU MoCap database show that our approach is able to build robust feature spaces without pre-registering and labeling all potential walkers. In fact, we can take different people (experiments **A** and **B**) and just a fraction of them (experiments **C** and **D**). We have observed that with an increasing volume of identities the heterogeneous evaluation set-up is on par with the homogeneous set-up, that is, it does not matter what identities we learn the features on. One does not have to rely on the availability of all walkers for learning. This is particularly important for a system to aid video surveillance applications where encountered walkers never supply labeled data. Multiple occurrences of individual walkers can now be linked together even without knowing their actual identities.

Acknowledgments. Authors thank to the anonymous reviewers for their detailed commentary and suggestions. Data used in this project was created with funding from NSF EIA-0196217 and was obtained from http://mocap.cs.cmu.edu. Our extracted database is available at https://gait.fi.muni.cz/ to support results reproducibility.

References

1. Ahmed, F., Paul, P.P., Gavrilova, M.L.: DTW-based kernel and rank-level fusion for 3D gait recognition using Kinect. Vis. Comput. **31**(6), 915–924 (2015). http://dx.doi.org/10.1007/s00371-015-1092-0
2. Ahmed, M., Al-Jawad, N., Sabir, A.: Gait recognition based on Kinect sensor. Proc. SPIE **9139**, 91390B–91390B-10 (2014). http://dx.doi.org/10.1117/12.2052588
3. Ali, S., Wu, Z., Li, X., Saeed, N., Wang, D., Zhou, M.: Applying geometric function on sensors 3D gait data for human identification. In: Gavrilova, M.L., Kenneth Tan, C.J., Iglesias, A., Shinya, M., Galvez, A., Sourin, A. (eds.) Transactions on Computational Science XXVI. LNCS, vol. 9550, pp. 125–141. Springer, Heidelberg (2016). http://dx.doi.org/10.1007/978-3-662-49247-5_8
4. Andersson, V., Araujo, R.: Person identification using anthropometric and gait data from Kinect sensor (2015). http://www.aaai.org/ocs/index.php/AAAI/AAAI15/paper/view/9680
5. Balazia, M., Sojka, P.: Learning robust features for gait recognition by Maximum Margin Criterion. In: Proceedings of 23rd International Conference on Pattern Recognition, ICPR 2016, p. 6, December 2016. (preprint: arXiv:1609.04392)
6. Ball, A., Rye, D., Ramos, F., Velonaki, M.: Unsupervised clustering of people from 'skeleton' data. In: Proceedings of the Seventh Annual ACM/IEEE International Conference on Human-Robot Interaction, HRI 2012, pp. 225–226. ACM, New York (2012). http://doi.acm.org/10.1145/2157689.2157767
7. Castro, F.M., Marín-Jiménez, M.J., Guil, N., de la Blanca, N.P.: Automatic learning of gait signatures for people identification (2016). CoRR abs/1603.01006. http://arxiv.org/abs/1603.01006

8. Choudhury, S.D., Tjahjadi, T.: Robust view-invariant multi-scale gait recognition. Pattern Recogn. **48**(3), 798–811 (2015). http://www.sciencedirect.com/science/article/pii/S0031320314003835
9. CMU Graphics Lab: Carnegie-Mellon Motion Capture (MoCap) Database (2003). http://mocap.cs.cmu.edu
10. Dikovski, B., Madjarov, G., Gjorgjevikj, D.: Evaluation of different feature sets for gait recognition using skeletal data from Kinect. In: 37th International Convention on Information and Communication Technology, Electronics and Microelectronics, pp. 1304–1308, May 2014. http://ieeexplore.ieee.org/stamp/stamp.jsp?arnumber=6859769
11. Kocsor, A., Kovács, K., Szepesvári, C.: Margin maximizing discriminant analysis. In: Boulicaut, J.-F., Esposito, F., Giannotti, F., Pedreschi, D. (eds.) ECML 2004. LNCS, vol. 3201, pp. 227–238. Springer, Heidelberg (2004). http://dx.doi.org/10.1007/978-3-540-30115-8_23
12. Kwolek, B., Krzeszowski, T., Michalczuk, A., Josinski, H.: 3D gait recognition using spatio-temporal motion descriptors. In: Nguyen, N.T., Attachoo, B., Trawiński, B., Somboonviwat, K. (eds.) ACIIDS 2014. LNCS, vol. 8398, pp. 595–604. Springer, Heidelberg (2014). http://dx.doi.org/10.1007/978-3-319-05458-2_61
13. Li, H., Jiang, T., Zhang, K.: Efficient and robust feature extraction by Maximum Margin Criterion. IEEE Trans. Neural Netw. **17**(1), 157–165 (2006). http://dx.doi.org/10.1109/TNN.2005.860852
14. Preis, J., Kessel, M., Werner, M., Linnhoff-Popien, C.: Gait recognition with Kinect. In: 1st International Workshop on Kinect in Pervasive Computing, New Castle, UK, pp. 1–4, 18–22 June 2012. https://www.researchgate.net/publication/239862819_Gait_Recognition_with_Kinect
15. Reddy, V.R., Chakravarty, K., Aniruddha, S.: Person identification in natural static postures using Kinect. In: Agapito, L., Bronstein, M.M., Rother, C. (eds.) ECCV 2014 Workshops. LNCS, vol. 8926, pp. 93–808. Springer, Heidelberg (2015). http://dx.doi.org/10.1007/978-3-319-16181-5_60
16. Sinha, A., Chakravarty, K., Bhowmick, B.: Person identification using skeleton information from Kinect. In: Proceedings of the Sixth International Conference on Advances in CHI, ACHI 2013, pp. 101–108 (2013). https://www.thinkmind.org/index.php?view=article&articleid=achi_2013_4_50_20187
17. Tafazzoli, F., Bebis, G., Louis, S.J., Hussain, M.: Genetic feature selection for gait recognition. J. Electron. Imaging **24**(1), 013036 (2015). http://dx.doi.org/10.1117/1.JEI.24.1.013036
18. Vapnik, V.N.: The Nature of Statistical Learning Theory. Springer, New York (1995)
19. Zeng, W., Wang, C.: View-invariant gait recognition via deterministic learning. In: International Joint Conference on Neural Networks (IJCNN), pp. 3465–3472, July 2014. http://dx.doi.org/10.1109/IJCNN.2014.6889507

On Security and Sparsity of Linear Classifiers for Adversarial Settings

Ambra Demontis[(⊠)], Paolo Russu, Battista Biggio, Giorgio Fumera, and Fabio Roli

Department of Electrical and Electronic Engineering,
University of Cagliari, Piazza d'Armi, 09123 Cagliari, Italy
{ambra.demontis,paolo.russu,battista.biggio,fumera,roli}@diee.unica.it

Abstract. Machine-learning techniques are widely used in security-related applications, like spam and malware detection. However, in such settings, they have been shown to be vulnerable to *adversarial* attacks, including the deliberate manipulation of data at test time to evade detection. In this work, we focus on the vulnerability of linear classifiers to evasion attacks. This can be considered a relevant problem, as linear classifiers have been increasingly used in embedded systems and mobile devices for their low processing time and memory requirements. We exploit recent findings in robust optimization to investigate the link between regularization and security of linear classifiers, depending on the type of attack. We also analyze the relationship between the sparsity of feature weights, which is desirable for reducing processing cost, and the security of linear classifiers. We further propose a novel octagonal regularizer that allows us to achieve a proper trade-off between them. Finally, we empirically show how this regularizer can improve classifier security and sparsity in real-world application examples including spam and malware detection.

1 Introduction

Machine-learning techniques are becoming an essential tool in several application fields such as marketing, economy and medicine. They are increasingly being used also in security-related applications, like spam and malware detection, despite their vulnerability to *adversarial* attacks, *i.e.*, the *deliberate* manipulation of training or test data, to subvert their operation; *e.g.*, spam emails can be manipulated (at test time) to evade a trained anti-spam classifier [1–12].

In this work, we focus on the security of linear classifiers. These classifiers are particularly suited to mobile and embedded systems, as the latter usually demand for strict constraints on storage, processing time and power consumption. Nonetheless, linear classifiers are also a preferred choice as they provide easier-to-interpret decisions (with respect to nonlinear classification methods). For instance, the widely-used SpamAssassin anti-spam filter exploits a linear classifier [5,7][1]. Work in the adversarial machine learning literature has already

---

[1] See also http://spamassassin.apache.org.

A. Robles-Kelly et al. (Eds.): S+SSPR 2016, LNCS 10029, pp. 322–332, 2016.
DOI: 10.1007/978-3-319-49055-7_29

investigated the security of linear classifiers to evasion attacks [4,7], suggesting the use of more evenly-distributed feature weights as a mean to improve their security. Such a solution is however based on heuristic criteria, and a clear understanding of the conditions under which it can be effective, or even optimal, is still lacking. Moreover, in mobile and embedded systems, *sparse* weights are more desirable than evenly-distributed ones, in terms of processing time, memory requirements, and interpretability of decisions.

In this work, we shed some light on the security of linear classifiers, leveraging recent findings from [13–15] that highlight the relationship between classifier regularization and robust optimization problems in which the input data is potentially corrupted by noise (see Sect. 2). This is particularly relevant in adversarial settings as the aforementioned ones, since evasion attacks can be essentially considered a form of noise affecting the non-manipulated, initial data (*e.g.*, malicious code before obfuscation). Connecting the work in [13–15] to adversarial machine learning aims to help understanding what the optimal regularizer is against different kinds of adversarial *noise* (attacks). We analyze the relationship between the sparsity of the weights of a linear classifier and its security in Sect. 3, where we also propose an octagonal-norm regularizer to better tune the trade-off arising between sparsity and security. In Sect. 4, we empirically evaluate our results on a handwritten digit recognition task, and on real-world application examples including spam filtering and detection of malicious software (malware) in PDF files. We conclude by discussing the main contributions and findings of our work in Sect. 5, while also sketching some promising research directions.

2 Background

In this section, we summarize the attacker model previously proposed in [8–11], and the link between regularization and robustness discussed in [13–15].

2.1 Attacker's Model

To rigorously analyze possible attacks against machine learning and devise principled countermeasures, a formal model of the attacker has been proposed in [6,8–11], based on the definition of her goal (*e.g.*, evading detection at test time), knowledge of the classifier, and capability of manipulating the input data.

Attacker's Goal. Among the possible goals, here we focus on evasion attacks, where the goal is to modify a single malicious sample (*e.g.*, a spam email) to have it misclassified as legitimate (with the largest confidence) by the classifier [8].

Attacker's Knowledge. The attacker can have different levels of knowledge of the targeted classifier; she may have *limited* or *perfect* knowledge about the training data, the feature set, and the classification algorithm [8,9]. In this work, we focus on *perfect-knowledge* (worst-case) attacks.

Attacker's Capability. In evasion attacks, the attacker is only able to modify malicious instances. Modifying an instance usually has some cost. Moreover, arbitrary modifications to evade the classifier may be ineffective, if the resulting instance loses its malicious nature (*e.g.*, excessive obfuscation of a spam email could make it unreadable for humans). This can be formalized by an application-dependent constraint. As discussed in [16], two kinds of constraints have been mostly used when modeling real-world adversarial settings, leading one to define *sparse* (ℓ_1) and *dense* (ℓ_2) attacks. The ℓ_1-**norm** yields typically a sparse attack, as it represents the case when the cost depends on the number of modified features. For instance, when instances correspond to text (*e.g.*, the email's body) and each feature represents the occurrences of a given term in the text, the attacker usually aims to change as few words as possible. The ℓ_2-**norm** yields a dense attack, as it represents the case when the cost of modifying features is proportional to the distance between the original and modified sample in Euclidean space. For example, if instances are images, the attacker may prefer making small changes to many or even all pixels, rather than significantly modifying only few of them. This amounts to (slightly) blurring the image, instead of obtaining a salt-and-pepper noise effect (as the one produced by sparse attacks) [16].

Attack Strategy. It consists of the procedure for modifying samples, according to the attacker's goal, knowledge and capability, formalized as an optimization problem. Let us denote the legitimate and malicious class labels respectively with -1 and $+1$, and assume that the classifier's decision function is $f(\mathbf{x}) = \text{sign}\,(g(\mathbf{x}))$, where $g(\mathbf{x}) = \boldsymbol{w}^\top \boldsymbol{x} + b \in \mathbb{R}$ is a linear discriminant function with feature weights $\boldsymbol{w} \in \mathbb{R}^d$ and bias $b \in \mathbb{R}$, and $\mathbf{x}$ is the representation of an instance in a d-dimensional feature space. Given a malicious sample $\boldsymbol{x}_0$, the goal is to find the sample $\boldsymbol{x}^*$ that minimizes the classifier's discriminant function $g(\cdot)$ (*i.e.*, that is classified as legitimate with the highest possible confidence) subject to the constraint that $\boldsymbol{x}^*$ lies within a distance $d_{\max}$ from $\boldsymbol{x}_0$:

$$\boldsymbol{x}^* = \arg\min_{\boldsymbol{x}} g(\boldsymbol{x}) \tag{1}$$

$$\text{s.t.}\ \ d(\boldsymbol{x}, \boldsymbol{x}_0) \leq d_{\max}, \tag{2}$$

where the distance measure $d(\cdot, \cdot)$ is defined in terms of the cost of data manipulation (*e.g.*, the number of modified words in each spam) [1,2,8,9,12]. For sparse and dense attacks, $d(\cdot, \cdot)$ corresponds respectively to the ℓ_1 and ℓ_2 distance.

2.2 Robustness and Regularization

The goal of this section is to clarify the connection between regularization and input data uncertainty, leveraging on the recent findings in [13–15]. In particular, Xu *et al.* [13] have considered the following *robust* optimization problem:

$$\min_{\boldsymbol{w},b}\ \max_{\boldsymbol{u}_1,...,\boldsymbol{u}_m \in \mathcal{U}}\ \sum_{i=1}^{m} \left(1 - y_i(\boldsymbol{w}^\top(\boldsymbol{x}_i - \boldsymbol{u}_i) + b)\right)_+, \tag{3}$$

where $(z)_+$ is equal to $z \in \mathbb{R}$ if $z > 0$ and 0 otherwise, $\boldsymbol{u}_1, ..., \boldsymbol{u}_m \in \mathcal{U}$ define a set of bounded perturbations of the training data $\{\boldsymbol{x}_i, y_i\}_{i=1}^{m} \in \mathbb{R}^m \times \{-1, +1\}^m$, and the so-called *uncertainty set* $\mathcal{U}$ is defined as $\mathcal{U} \triangleq \{(u_1, \ldots, u_m) | \sum_{i=1}^{m} \|u_i\|^* \leq c\}$, being $\| \cdot \|^*$ the dual norm of $\| \cdot \|$. Typical examples of uncertainty sets according to the above definition include ℓ_1 and ℓ_2 balls [13,14].

Problem (3) basically corresponds to minimizing the hinge loss for a two-class classification problem under worst-case, bounded perturbations of the training samples x_i, *i.e.*, a typical setting in robust optimization [13–15]. Under some mild assumptions easily verified in practice (including non-separability of the training data), the authors have shown that the above problem is equivalent to the following non-robust, regularized optimization problem (*cf.* Theorem 3 in [13]):

$$\min_{w,b} \ c\|w\| + \sum_{i=1}^{m} \left(1 - y_i(w^\top x_i + b)\right)_+ . \tag{4}$$

This means that, if the ℓ_2 norm is chosen as the dual norm characterizing the uncertainty set $\mathcal{U}$, then w is regularized with the ℓ_2 norm, and the above problem is equivalent to a standard Support Vector Machine (SVM) [17]. If input data uncertainty is modeled with the ℓ_1 norm, instead, the optimal regularizer would be the ℓ_∞ regularizer, and vice-versa.[2] This notion is clarified in Fig. 1, where we consider different norms to model input data uncertainty against the corresponding SVMs; *i.e.*, the standard SVM [17], the Infinity-norm SVM [18] and the 1-norm SVM [19] against ℓ_2, ℓ_1 and ℓ_∞-norm uncertainty models, respectively.

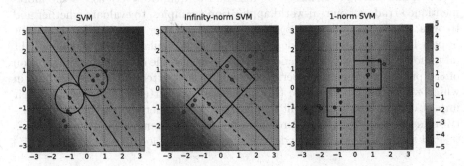

Fig. 1. Discriminant function $g(x)$ for SVM, Infinity-norm SVM, and 1-norm SVM (in colors). The decision boundary $(g(x) = 0)$ and margins $(g(x) = \pm 1)$ are respectively shown with black solid and dashed lines. Uncertainty sets are drawn over the support vectors to show how they determine the orientation of the decision boundary. (Color figure online)

---

[2] Note that the ℓ_1 norm is the dual norm of the ℓ_∞ norm, and vice-versa, while the ℓ_2 norm is the dual norm of itself.

3   Security and Sparsity

We discuss here the main contributions of this work. The result discussed in the previous section, similar to that reported independently in [15], helps understanding the security properties of linear classifiers in adversarial settings, in terms of the relationship between security and sparsity. In fact, what discussed in the previous section does not only confirm the intuition in [4,7], *i.e.*, that more uniform feature weighting schemes should improve classifier security by enforcing the attacker to manipulate more feature values to evade detection. The result in [13–15] also clarifies the meaning of *uniformity* of the feature weights $\boldsymbol{w}$. If one considers an ℓ_1 (sparse) attacker, facing a higher cost when modifying more features, it turns out that the optimal regularizer is given by the ℓ_∞ norm of $\boldsymbol{w}$, which tends to yield more uniform weights. In particular, the solution provided by ℓ_∞ regularization (in the presence of a strongly-regularized classifier) tends to yield weights which, in absolute value, are all equal to a (small) maximum value. This also implies that ℓ_∞ regularization does not provide a *sparse* solution.

For this reason we propose a novel *octagonal* (8gon) regularizer,[3] given as a linear (convex) combination of ℓ_1 and ℓ_∞ regularization:

$$\|\boldsymbol{w}\|_{\text{8gon}} = (1 - \rho)\|\boldsymbol{w}\|_1 + \rho\|\boldsymbol{w}\|_\infty \tag{5}$$

where $\rho \in (0, 1)$ can be increased to trade sparsity for security.

Our work does not only aim to clarify the relationships among regularization, sparsity, and *adversarial* noise. We also aim to quantitatively assess the aforementioned trade-off on real-world application examples, to evaluate whether and to what the extent the choice of a proper regularizer may have a significant impact in practice. Thus, besides proposing a new regularizer and shedding light on uniform feature weighting and classifier security, the other main contribution of the present work is the experimental analysis reported in the next section, in which we consider both dense (ℓ_2) and sparse (ℓ_1) attacks, and evaluate their impact on SVMs using different regularizers. We further analyze the weight distribution of each classifier to provide a better understanding on how sparsity is related to classifier security under the considered evasion attacks.

4   Experimental Analysis

We first consider dense and sparse attacks in the context of handwritten digit recognition, to visually demonstrate their blurring and salt-and-pepper effect on images. We then consider two real-world application examples including spam and PDF malware detection, investigating the behavior of different regularization terms against (*sparse*) evasion attacks.

---

[3] Note that octagonal regularization has been previously proposed also in [20]. However, differently from our work, the authors have used a pairwise version of the infinity norm, for the purpose of selecting (correlated) groups of features.

Handwritten Digit Classification. To visually show how evasion attacks work, we perform sparse and dense attacks on the MNIST digit data [21]. Each image is represented by a vector of 784 features, corresponding to its gray-level pixel values. As in [8], we simulate an adversarial classification problem where the digits 8 and 9 correspond to the legitimate and malicious class, respectively.

Spam Filtering. This is a well-known application subject to adversarial attacks. Most spam filters include an automatic text classifier that analyzes the email's body text. In the simplest case Boolean features are used, each representing the presence or absence of a given term. For our experiments we use the TREC 2007 spam track data, consisting of about 25000 legitimate and 50000 spam emails [22]. We extract a dictionary of terms (features) from the first 5000 emails (in chronological order) using the same parsing mechanism of SpamAssassin, and then select the 200 most discriminant features according to the information gain criterion [23]. We simulate a well-known (*sparse*) evasion attack in which the attacker aims to modify only few terms. Adding or removing a term amounts to switching the value of the corresponding Boolean feature [3,4,8,9,12].

PDF Malware Detection. Another application that is often targeted by attackers is the detection of malware in PDF files. A PDF file can host different kinds of contents, like Flash and JavaScript. Such third-party applications can be exploited by attacker to execute arbitrary operations. We use a data set made up of about 5500 legitimate and 6000 malicious PDF files. We represent every file using the 114 features that are described in [24]. They consist of the number of occurrences of a predefined set of keywords, where every keyword represents an action performed by one of the objects that are contained into the PDF file (*e.g.*, opening another document that is stored inside the file). An attacker cannot trivially remove keywords from a PDF file without corrupting its functionality. Conversely, she can easily add new keywords by inserting new object's operations. For this reason, we simulate this attack by only considering feature increments (decrementing a feature value is not allowed). Accordingly, the most convenient strategy to mislead a malware detector (classifier) is thus to insert as many occurrences of a *given* keyword as possible, which is a sparse attack.[4]

We consider different versions of the SVM classifier obtained by combining the hinge loss with the different regularizers shown in Fig. 2.

2-Norm SVM (SVM). This is the standard SVM learning algorithm [17]. It finds w and b by solving the following quadratic programming problem:

$$\min_{w,b} \frac{1}{2}\|w\|_2^2 + C\sum_{i=1}^{m}(1 - y_i g(x_i))_+ , \tag{6}$$

where $g(x) = w^\top x + b$ denotes the SVM's linear discriminant function. Note that ℓ_2 regularization does not induce sparsity on w.

---

[4] Despite no upper bound on the number of injected keywords may be set, we set the maximum value for each keyword to the corresponding one observed during training.

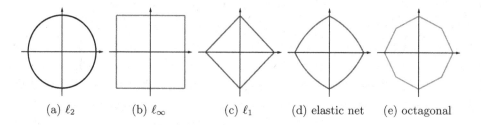

(a) ℓ_2 (b) ℓ_∞ (c) ℓ_1 (d) elastic net   (e) octagonal

Fig. 2. Unit balls for different norms.

Infinity-Norm SVM (∞-norm). In this case, the ℓ_∞ regularizer bounds the weights' maximum absolute value as $\|\boldsymbol{w}\|_\infty = \max_{j=1,\ldots,d} |w_j|$ [20]:

$$\min_{\boldsymbol{w},b} \|\boldsymbol{w}\|_\infty + C \sum_{i=1}^{m} (1 - y_i g(\boldsymbol{x}_i))_+ \ . \tag{7}$$

As the standard SVM, this classifier is not sparse; however, the above learning problem can be solved using a simple linear programming approach.

1-Norm SVM (1-norm). Its learning algorithm is defined as [19]:

$$\min_{\boldsymbol{w},b} \|\boldsymbol{w}\|_1 + C \sum_{i=1}^{m} (1 - y_i g(\boldsymbol{x}_i))_+ \ . \tag{8}$$

The ℓ_1 regularizer induces sparsity, while retaining convexity and linearity.

Elastic-net SVM (el-net). We use here the elastic-net regularizer [25], combined with the hinge loss to obtain an SVM formulation with tunable sparsity:

$$\min_{\boldsymbol{w},b} (1 - \lambda)\|\boldsymbol{w}\|_1 + \frac{\lambda}{2}\|\boldsymbol{w}\|_2^2 + C \sum_{i=1}^{m} (1 - y_i g(\boldsymbol{x}_i))_+ \ . \tag{9}$$

The level of sparsity can be tuned through the trade-off parameter $\lambda \in (0, 1)$.

Octagonal-Norm SVM (8gon). This novel SVM is based on our octagonal-norm regularizer, combined with the hinge loss:

$$\min_{\boldsymbol{w},b} (1 - \rho)\|\boldsymbol{w}\|_1 + \rho\|\boldsymbol{w}\|_\infty + C \sum_{i=1}^{m} (1 - y_i g(\boldsymbol{x}_i))_+ \ . \tag{10}$$

The above optimization problem is linear, and can be solved using state-of-the-art solvers. The sparsity of $\boldsymbol{w}$ can be increased by decreasing the trade-off parameter $\rho \in (0, 1)$, at the expense of classifier security.

Sparsity and Security Measures. We evaluate the degree of sparsity S of a given linear classifier as the fraction of its weights that are equal to zero:

$$S = \frac{1}{d}|\{w_j | w_j = 0, j = 1, \ldots, d\}| \ , \tag{11}$$

being $|\cdot|$ the cardinality of the set of null weights.

To evaluate security of linear classifiers, we define a measure E of *weight evenness*, similarly to [4,7], based on the ratio of the ℓ_1 and ℓ_∞ norm:

$$E = \frac{\|\boldsymbol{w}\|_1}{\mathsf{d}\|\boldsymbol{w}\|_\infty} \ , \tag{12}$$

where dividing by the number of features d ensures that $E \in \left[\frac{1}{\mathsf{d}}, 1\right]$, with higher values denoting more evenly-distributed feature weights. In particular, if only a weight is not zero, then $E = \frac{1}{\mathsf{d}}$; conversely, when all weights are equal to the maximum (in absolute value), $E = 1$.

Experimental Setup. We randomly select 500 legitimate and 500 malicious samples from each dataset, and equally subdivide them to create a training and a test set. We optimize the regularization parameter C of each SVM (along with λ and ρ for the Elastic-net and Octagonal SVMs, respectively) through 5-fold cross-validation, maximizing the following objective on the training data:

$$\mathrm{AUC} + \alpha E + \beta S \tag{13}$$

where AUC is the area under the ROC curve, and α and β are parameters defining the trade-off between security and sparsity. We set $\alpha = \beta = 0.1$ for the PDF and digit data, and $\alpha = 0.2$ and $\beta = 0.1$ for the spam data, to promote more secure solutions in the latter case. These parameters allow us to accept a marginal decrease in classifier security only if it corresponds to much sparser feature weights. After classifier training, we perform evasion attacks on all malicious test samples, and evaluate the corresponding performance as a function of the number of features modified by the attacker. We repeat this procedure five times, and report the average results on the original and modified test data.

Experimental Results. We consider first PDF malware and spam detection. In these applications, as mentioned before, only sparse evasion attacks make sense, as the attacker aims to minimize the number of modified features. In Fig. 3, we report the AUC at 10 % false positive rate for the considered classifiers, against an increasing number of words/keywords changed by the attacker. This experiment shows that the most secure classifier under sparse evasion attacks is

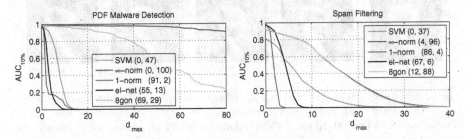

Fig. 3. Classifier performance under attack for PDF malware and spam data, measured in terms of $\mathrm{AUC}_{10\%}$ against an increasing number $d_{\max}$ of modified features. For each classifier, we also report (S, E) percentage values (Eqs. 11–12) in the legend.

the Infinity-norm SVM, since its performance degrades more gracefully under attack. This is an expected result, given that, in this case, infinity-norm regularization corresponds to the dual norm of the attacker's cost/distance function. Notably, the Octagonal SVM yields reasonable security levels while achieving much sparser solutions, as expected (*cf.* the sparsity values S in the legend of Fig. 3). This experiment really clarifies how much the choice of a proper regularizer can be crucial in real-world adversarial applications.

By looking at the values reported in Fig. 3, it may seem that the security measure E does not properly characterize classifier security under attack; *e.g.*, note how Octagonal SVM exhibits lower values of E despite being more secure than SVM on the PDF data. The underlying reason is that the attack implemented on the PDF data only considers feature increments, while E generically considers any kind of manipulation. Accordingly, one should define alternative security measures depending on specific kinds of data manipulation. However, the security measure E allows us to properly tune the trade-off between security and sparsity also in this case and, thus, this issue may be considered negligible.

Finally, to visually demonstrate the effect of sparse and dense evasion attacks, we report some results on the MNIST handwritten digits. In Fig. 4, we show the "9" digit image modified by the attacker to have it misclassified by the

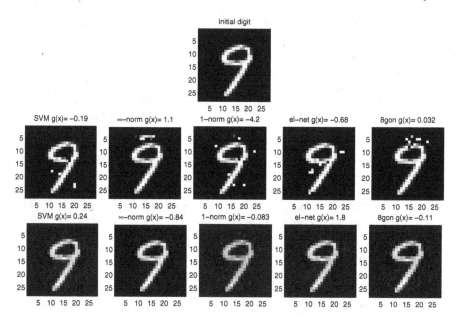

Fig. 4. Initial digit "9" (*first row*) and its versions modified to be misclassified as "8" (*second and third row*). Each column corresponds to a different classifier (from *left* to *right* in the second and third row): SVM, Infinity-norm SVM, 1-norm SVM, Elastic-net SVM, Octagonal SVM. *Second row*: sparse attacks (ℓ_1), with $d_{\max} = 2000$. *Third row*: dense attacks (ℓ_2), with $d_{\max} = 250$. Values of $g(\boldsymbol{x}) < 0$ denote a successful classifier evasion (*i.e.*, more vulnerable classifiers).

classifier as an "8". These modified digits are obtained by solving Problems (1)–(2) through a simple projected gradient-descent algorithm, as in [8][5]. Note how dense attacks only produce a slightly-blurred effect on the image, while sparse attacks create more evident visual artifacts. By comparing the values of $g(x)$ reported in Fig. 4, one may also note that this simple example confirms again that Infinity-norm and Octagonal SVM are more secure against sparse attacks, while SVM and Elastic-net SVM are more secure against dense attacks.

5 Conclusions and Future Work

In this work we have shed light on the theoretical and practical implications of sparsity and security in linear classifiers. We have shown on real-world adversarial applications that the choice of a proper regularizer is crucial. In fact, in the presence of sparse attacks, Infinity-norm SVMs can drastically outperform the security of standard SVMs. We believe that this is an important result, as (standard) SVMs are widely used in security tasks without taking the risk of adversarial attacks too much into consideration. Moreover, we propose a new octagonal regularizer that enables trading sparsity for a marginal loss of security under sparse evasion attacks. This is extremely useful in applications where sparsity and computational efficiency at test time are crucial. When dense attacks are instead deemed more likely, the standard SVM may be retained a good compromise. In that case, if sparsity is required, one may trade some level of security for sparsity using the Elastic-net SVM. Finally, we think that an interesting extension of our work may be to investigate the trade-off between sparsity and security also in the context of classifier poisoning (in which the attacker can contaminate the training data to mislead classifier learning) [9,26,27].

References

1. Dalvi, N., Domingos, P., Mausam, Sanghai, S., Verma, D.: Adversarial classification. In: 10th International Conference on Knowledge Discovery and Data Mining (KDD), Seattle, WA, USA, pp. 99–108. ACM (2004)
2. Lowd, D., Meek, C.: Adversarial learning. In: 11th International Conference on Knowledge Discovery in Data Mining (KDD), Chicago, IL, USA, pp. 641–647. ACM (2005)
3. Lowd, D., Meek, C.: Good word attacks on statistical spam filters. In: 2nd Conference on Email and Anti-Spam (CEAS), Mountain View, CA, USA (2005)
4. Kolcz, A., Teo, C.H.: Feature weighting for improved classifier robustness. In: 6th Conference on Email and Anti-Spam (CEAS), Mountain View, CA, USA (2009)
5. Nelson, B., Barreno, M., Chi, F.J., Joseph, A.D., Rubinstein, B.I.P., Saini, U., Sutton, C., Tygar, J.D., Xia, K.: Exploiting machine learning to subvert your spam filter. In: LEET 2008, Berkeley, CA, USA, pp. 1–9. USENIX Association (2008)

---

[5] Note also that, in the case of sparse attacks, Problems (1)–(2) amounts to minimizing a linear function subject to linear constraints. It can be thus formulated as a linear programming problem, and solved with state-of-the-art linear solvers.

6. Barreno, M., Nelson, B., Sears, R., Joseph, A.D., Tygar, J.D.: Can machine learning be secure? In: ASIACCS 2006, pp. 16–25. ACM, New York (2006)

7. Biggio, B., Fumera, G., Roli, F.: Multiple classifier systems for robust classifier design in adversarial environments. Int. J. Mach. Learn. Cybern. 1(1), 27–41 (2010)

8. Biggio, B., Corona, I., Maiorca, D., Nelson, B., Šrndić, N., Laskov, P., Giacinto, G., Roli, F.: Evasion attacks against machine learning at test time. In: Blockeel, H., Kersting, K., Nijssen, S., Železný, F. (eds.) ECML PKDD 2013. LNCS (LNAI), vol. 8190, pp. 387–402. Springer, Heidelberg (2013). doi:10.1007/978-3-642-40994-3_25

9. Biggio, B., Fumera, G., Roli, F.: Security evaluation of pattern classifiers under attack. IEEE Trans. Knowl. Data Eng. 26(4), 984–996 (2014)

10. Biggio, B., Fumera, G., Roli, F.: Pattern recognition systems under attack: design issues and research challenges. Int. J. Pattern Recognit. Artif. Intell. 28(7), 1460002 (2014)

11. Huang, L., Joseph, A.D., Nelson, B., Rubinstein, B., Tygar, J.D.: Adversarial machine learning. In: 4th ACM Workshop on Artificial Intelligence and Security (AISec), Chicago, IL, USA, pp. 43–57 (2011)

12. Zhang, F., Chan, P., Biggio, B., Yeung, D., Roli, F.: Adversarial feature selection against evasion attacks. IEEE Trans. Cybern. 46(3), 766–777 (2016)

13. Xu, H., Caramanis, C., Mannor, S.: Robustness and regularization of support vector machines. J. Mach. Learn. Res. 10, 1485–1510 (2009)

14. Sra, S., Nowozin, S., Wright, S.J.: Optimization for Machine Learning. MIT Press, Cambridge (2011)

15. Livni, R., Crammer, K., Globerson, A., Edmond, E.I., Safra, L.: A simple geometric interpretation of SVM using stochastic adversaries. In: AISTATS 2012. JMLR W&CP, vol. 22, pp. 722–730 (2012)

16. Wang, F., Liu, W., Chawla, S.: On sparse feature attacks in adversarial learning. In: IEEE International Conference on Data Mining (ICDM), pp. 1013–1018. IEEE (2014)

17. Cortes, C., Vapnik, V.: Support-vector networks. Mach. Learn. 20, 273–297 (1995)

18. Bennett, K.P., Bredensteiner, E.J.: Duality and geometry in SVM classifiers. In: 17th ICML, pp. 57–64. Morgan Kaufmann Publishers Inc. (2000)

19. Zhu, J., Rosset, S., Tibshirani, R., Hastie, T.J.: 1-norm support vector machines. In: Thrun, S., Saul, L., Schölkopf, B. (eds.) NIPS 16, pp. 49–56. MIT Press (2004)

20. Bondell, R.: Simultaneous regression shrinkage, variable selection, and supervised clustering of predictors with OSCAR. Biometrics 64, 115–123 (2007)

21. LeCun, Y., et al.: Comparison of learning algorithms for handwritten digit recognition. In: International Conference on Artificial Neural Networks, pp. 53–60 (1995)

22. Cormack, G.V.: TREC 2007 spam track overview. In: Voorhees, E.M., Buckland, L.P., (eds.) TREC. Volume Special Publication, pp. 500–274. NIST (2007)

23. Sebastiani, F.: Machine learning in automated text categorization. ACM Comput. Surv. 34, 1–47 (2002)

24. Maiorca, D., Giacinto, G., Corona, I.: A pattern recognition system for malicious PDF files detection. In: Perner, P. (ed.) MLDM 2012. LNCS (LNAI), vol. 7376, pp. 510–524. Springer, Heidelberg (2012). doi:10.1007/978-3-642-31537-4_40

25. Zou, H., Hastie, T.: Regularization and variable selection via the elastic net. J. R. Stat. Soc. Ser. B 67(2), 301–320 (2005)

26. Biggio, B., Nelson, B., Laskov, P.: Poisoning attacks against SVMs. In: Langford, J., Pineau, J., (eds.) 29th ICML, pp. 1807–1814. Omnipress (2012)

27. Xiao, H., Biggio, B., Brown, G., Fumera, G., Eckert, C., Roli, F.: Is feature selection secure against training data poisoning? In: Bach, F., Blei, D. (eds.) 32nd ICML. JMLR W&CP, vol. 37, pp. 1689–1698 (2015)

Semantic Segmentation via Multi-task, Multi-domain Learning

Damien Fourure[1], Rémi Emonet[1], Elisa Fromont[1(✉)], Damien Muselet[1],
Alain Trémeau[1], and Christian Wolf[2]

[1] Universite de Lyon, UJM, CNRS, Lab Hubert Curien UMR5516,
F-42000 Lyon, France
elisa.fromont@univ-st-etienne.fr

[2] Universite de Lyon, CNRS, INSA-Lyon, LIRIS, UMR5205, F-69621 Lyon, France

Abstract. We present an approach that leverages multiple datasets possibly annotated using different classes to improve the semantic segmentation accuracy on each individual dataset. We propose a new *selective loss* function that can be integrated into deep networks to exploit training data coming from multiple datasets with possibly different tasks (e.g., different label-sets). We show how the gradient-reversal approach for domain adaptation can be used in this setup. Thorough experiments on semantic segmentation applications show the relevance of our approach.

Keywords: Deep learning · Convolutional neural networks · Semantic segmentation · Domain adaptation · Multi-task learning

1 Introduction

Semantic scene parsing (a.k.a. semantic full scene labeling) from RGB images aims at segmenting an image into semantically meaningful regions, i.e. to provide a semantic class label for each pixel of an image—see Fig. 2 for examples of labels in an outdoor context. Semantic scene parsing is useful for a wide range of applications, for instance autonomous vehicles, automatic understanding and indexing of video databases, etc. Most semantic scene parsing methods use supervised machine learning algorithms and thus rely on densely labeled (manually annotated) training sets which are very tedious to obtain. Only a small amount of training data is currently available for this task, which makes this problem stand out from other problems in vision (as for instance object recognition and localization). This is a particularly stringent problem for the deep network models who are particularly needy in terms of training data.

Most datasets for scene parsing contain only several hundreds of images, some of them only several dozen [6,9,13–15,18,19,22,25]. Additionally, combining these datasets is a non-trivial task as target classes are often tailored to a custom application. One good example of such label-set diversity can be found within the KITTI Vision benchmark [4] which contains outdoor scene videos. Many research teams work on this dataset since its release in 2013, tackling computer vision tasks such as visual odometry, 3D object detection and 3D tracking

© Springer International Publishing AG 2016
A. Robles-Kelly et al. (Eds.): S+SSPR 2016, LNCS 10029, pp. 333–343, 2016.
DOI: 10.1007/978-3-319-49055-7_30

[9,13,14,18,19,22,25]. To tackle these tasks, several research teams have labeled parts of the original dataset, independently from the other teams and often for different goals (among the works listed above, semantic segmentation is the final goal only for [14,25]). In practice, the ground truth segmentation quality varies and both the granularity and the semantics of the labels differ. This inconsistency in the label-set is also true when using the Stanford Background [6] and SIFT Flow [15] datasets in combination with or in addition to KITTI.

The contributions of this paper are threefold: (I) we formalize a simple yet effective *selective loss* function that can be used in shared deep network to exploit training data coming from multiple datasets having different label-sets (different segmentation tasks). (II) we show that the gradient-reversal approach for domain adaptation [3] can be used but needs to be manipulated with care especially when datasets are unbalanced, (III) we run thorough scene segmentation experiments on 9 heterogeneously labeled datasets, underlining the impact of each part of the approach.

2   Related Works

In this section, we first discuss state-of-the-art methods dealing with multiple datasets, focusing in particular on feature and knowledge transfer methods in the context of deep networks. We then discuss semantic scene parsing with an emphasis on methods used for the KITTI benchmark.

Learning Across Datasets. Many recent papers [3,7,21,23,26] proposed methods to solve the problem of the transferability of deep features. Since CNNs require a lot of labeled data to provide good features, the usual methods exploit features learned on one big dataset and adapt them to other datasets and tasks [23]. In an extensive analysis about deep feature transfer, Yosinski et al. [23] show that it is better to initialize lower layers from features learned on a different (and maybe distant) task than using random weights. These transferred features improve generalization performance even after fine-tuning on a new task. Hinton et al. [7] propose another way to transfer (or distill) knowledge from one large network to a smaller one. The idea is for the small network to learn both the outputs of the large network as well as the correct labels of the data.

Considering the same task but datasets in multiple domains (different distributions of input data), the theory of domain adaptation tells us that the most similar the features are across domains [1,5,16], the better the adaptation will be. Ganin and Lempitsky [3] follow this principle by learning features that are invariant with respect to the shift between the source and target domains. In order to do that, they train a *domain classifier* and reverse the gradient during the backpropagation step in order to learn features that cannot help in discriminating the domains.

In this paper we show how domain adaptation techniques can be integrated in our proposed method to benefit from multiple heterogeneously labeled datasets without compromising the classification accuracy on each dataset.

Semantic Segmentation. Whereas the methods used for low level segmentation are diverse, high level semantic segmentation is dominated by machine learning. Learning methods range from random forests, to Support Vector Machines and deep learning who have been used in wide range of works [2,11,17]. Over years, segmentation algorithms have often been regularized through probabilistic graphical models like Conditional Random Fields (CRF). These methods have also been combined with deep networks [2,20]. For the 7 subsets of the KITTI dataset used in this paper [9,13,14,18,19,22,25], deep learning has never been used to tackle the semantic segmentation step. For example, [14] shows how to jointly classify pixels and predict their depth using a multi-class decision stumps-based boosted classifier. [25] uses a random forest classifier to classify segments of an image for different scales and sets of features. The aim of this paper is to show how to learn across datasets (with possibly different tasks) to improve the classification results. In particular, we do not optimize our method to produce the best accuracy for each of the used dataset. For example, while in KITTI many authors use rich features such as color, depth and temporal features, and complex post processing, we only use the RGB channels for our experiments and we do not take the label hierarchies into account. We also do not use any post-processing step.

3 Proposed Approach

Problem Statement. Given a set of images drawn from a set of K different datasets, pairs made of an input patch x_i^k and a target label y_i^k are grouped into sets $D_k = \{x_i^k, y_i^k\}$, where $k=1 \ldots K$ and i indexes patches. The label spaces are different over the datasets, therefore each y_i^k can take values in space $\mathcal{L}^k$.

Our goal is to learn a nonlinear mapping $y = \theta(x, \Theta)$ with parameters Θ which minimizes a empirical risk $\mathcal{R}[\Theta, D]$. The mapping θ is represented as a convolutional neural network, where each layer itself is a nonlinear mapping $f_l(\mathbf{W}_l \mathbf{h}_{l-1} + \mathbf{b}_l)$ where $\mathbf{h}_l$ is the lth hidden representation, $\mathbf{W}_l$ and $\mathbf{b}_l$ are the weights and bias of the lth layer and $f_l(.)$ is the activation function of the lth layer. We minimize the empirical risk, $\mathcal{R}[\Theta, D] = \frac{1}{N} \sum_{k=1}^{K} \sum_i J(x_i^k, y_i^k, \Theta)$, where N is the number of training samples and J is the loss function for each individual sample. We use the cross entropy loss $J(x_i^k, y_i^k, \Theta) = -\log \theta(x_i^k, \Theta)_{y_i^k}$, where $\theta(x_i^k, \Theta)_{y_i^k}$ is the network output for class y_i^k.

Limitations of Separate Training. Considering K different datasets, the classical baseline approach is to train K separate mappings (models) θ^k, each defined on its own label set $\mathcal{L}^k$. This baseline approach is illustrated in Fig. 1a. Unfortunately this basic approach presents the shortcoming that each mapping θ^k is trained on its own dataset D^k, which requires minimizing over separate sets of parameters Θ^k. In deep convolutional networks, the parameters Θ^k include all convolution's filters and the weights of all fully connected layers (overall, several millions of parameters). Learning such a large amount of parameters from limited (and generally small) amounts of training data is very challenging.

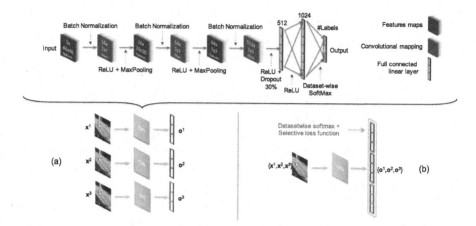

Fig. 1. Network used for our experiments with different learning strategies ((**a**) and (**b**)). (**a**) *No Fusion* is our baseline and consists in learning one network per dataset. (**b**) *Joint training* consists in learning a single network with our selective loss function.

Joint Feature Training with Selective Loss. We propose to tackle this shortcoming by exploiting the hierarchical nature of deep models. On most classical problems in computer vision, supervised training leads to a rising complexity and specificity of the features over the layers [24]. In our case, we propose to train a single deep network on the union of all individual datasets. This allows the network to decide at every layer which features should be generic and which ones should be task-specific. This joint training approach is illustrated in Fig. 1b. There is one output unit per label in the union of all label sets $\mathcal{L}^k$: the output layer thus produces predictions for all considered datasets.

In a traditional multi-class setting, the network output is computed using a soft-max function to produce a probability vector. However, with K different datasets, this is counter-productive: it maximizes the target class probability but minimizes the probability of all other classes, including the ones from different label sets. This minimization is problematic when there exists a correlation between labels across different datasets. For example, in the KITTI dataset (see Fig. 2 where all labels are reported) the class *Tree* of the dataset from He et al. [9] is correlated with the class *Vegetation* from the dataset labeled by Kundu et al. [13]. A plain softmax, optimizing the probability of the *Tree* class will implicitly penalize the probability of *Vegetation*, which is not a desired effect.

We thus define the *dataset-wise soft-max* (that produces a probabilities vector per dataset): for an input sample x_i^k, y_i^k from dataset k (denoted x, y for readability),

$$f(\theta(x, \Theta), y, k) = \frac{e^{\theta(x, \Theta)_y}}{\displaystyle\sum_{j' \in \mathcal{L}^k} e^{\theta(x, \Theta)_{j'}}} \tag{1}$$

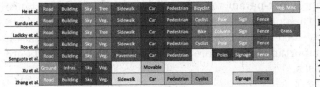

Data	Train	Val	Test	Total
He [9]	32	7	12	51
Ku [13]	28	7	15	50
La [14]	24	6	30	60
Ro [18]	80	20	46	146
Se [19]	36	9	25	70
Xu [22]	56	14	37	107
Zh [25]	112	28	112	252
Total	368	91	277	736

Fig. 2. The 68 labels (with the original colors) used by the different authors to annotate their subset of the KITTI benchmark as well as the number of images (and their train/validation/split decomposition, see details in Sect. 4) in each subset. (Color figure online)

During learning, the dataset-wise soft-max is combined with a *selective cross-entropy loss* function as follows:

$$J'(\theta(x,\Theta),y,k) = -\theta(x,\Theta)_y + \log(\sum_{j'\in\mathcal{L}^k} e^{\theta(x,\Theta)_{j'}}) \tag{2}$$

Gradients are null for parameters involving output units corresponding to labels from datasets l with $l \neq k$. This is equivalent to having a separate output layer for each dataset and intermediate layers that are shared across datasets.

Gradient Reversal for Domain Adaptation. So far, our method trains a single network on several datasets adapting for different label-sets, while ignoring potential shift in the input domain between datasets. This is not a problem in the case where the input data is sampled from a single distribution (e.g. for the different subsets of KITTI). In other cases, a non neglectable shift in input distribution does exist, for instance between the Stanford and SiftFlow data.

The theory of domain adaptation tells us that a better adaptation between source and target domains can be achieved when the feature distributions of both sets are closer to each other [1,5,16]. In the lines of Ganin and Lempitsky [3], this can be achieved using an additional classifier trained on the same features, which attempts to predict the domain of the input data. In the case of domain invariant features, this classifier will achieve high error. More precisely, our full mapping $y = \theta(x,\Theta)$ is conceptually split into two parts: the *feature extractor* $f = \theta_f(x,\Theta_f)$, which corresponds to the first convolutional layers and results in features f, and the *task classifier* $y = \theta_t(f,\Theta_t)$, which corresponds to the later fully connected layers including the selective loss function. The *domain classifier* is an additional mapping, which maps the features f to an estimated domain d, given as $d = \theta_d(f,\Theta_d)$ The goal here is to *minimize* the loss of the domain classifier θ_d over its parameters Θ_d in order to train a meaningful classifier, and to *maximize* the same loss over the features f, i.e. over the parameters Θ_f of the feature extractor θ_f, in order to create domain invariant features. In the lines of [3] this is achieved through a single backpropagation pass over the full network implementing θ_f, θ_t and θ_d, inversing the gradients between the domain classifier θ_d and the feature extractor θ_f. In practice, the gradients are not only inversed, they are multiplied with a hyper-parameter $-\lambda$, which controls the importance of the task classifier and the domain one, as in [3].

Our experiments described in Sect. 4 show, that this domain adaptation step is also useful and important in our more general setting where results are requested for different tasks.

4   Experimental Results

Training Details. For all experiments we used a network architecture illustrated at the top of Fig. 1. This architecture is a modified version of Farabet et al. [2]. The first two convolutional layers are composed by a bank of filters of size 7×7 followed by ReLU [12] units, 2×2 maximum pooling and batch normalization [10] units. The last convolutional layer is a filter bank followed by a ReLU unit, a batch normalization unit and dropout [8] with a drop factor of 30 %. The first fully connected linear layer is then followed by a ReLU unit and the last layer is followed by our dataset-wise softmax unit. To train the network, each RGB image is transformed into YUV space. A training input example is composed of a patch x_i of size 46×46 cropped from an image, the dataset k from which the image comes from, and y_i^k, the label of the center pixel of the patch x_i. Stochastic gradient descent with a mini-batch of size 128 was used to update the parameters. We used early stopping on a validation set in order to stop the training step before over-fitting. The only data augmentation strategy that we used is an horizontal flip of patches.

Datasets Details. The KITTI dataset [4] has been partially labeled by seven research groups [9,13,14,18,19,22,25] resulting in 736 labeled images (with almost no images in common) that are split into: a train set, a validation set and a test set. When the information was given by the author, we used the same train/test set as them, otherwise we randomly split them into approximately 70 % of data for the training and validation set and 30 % data for the test set, ensuring that any two images from the same video sequence end up in the same split. The labels used in the different subsets of the KITTI dataset are summarized in Fig. 2. We sample on average 390.000 patches in each video frame (depending on its size). This results into a dataset of about 280 million patches suitable to train a deep learning architecture. As mentioned in Sect. 2, the different labels provided by the different teams are not always consistent. As illustrated in Fig. 2, we can see that the granularity and the semantics of the labels may be very different from one labeling to another. For example, Ladicky et al. separate the *Trees* from the *Grass*. However, this might correspond to the *Vegetation* labels in the subset from Xu et al. but might also correspond (in the case of *Grass*) to the labels *Ground*. He et al. have not used the labels *Pole*, *Sign* or *Fence* used in most other labelings. These labels are likely to overlap with the label *Building* of He et al. but then, this *Building* class cannot be consistent anymore with the other labelings that contain the label *Building* in other subsets. Some groups have used the label *Bike* and some others have used the label

Cyclist. Those two labels are likely to overlap but in one case a team has focused on the entire entity "cyclist on a bike" whereas another has only focused on the bike device.

In addition to the KITTI dataset, we use two other scene labeling datasets: STANFORD BACKGROUND [6] and SIFT FLOW [15]. The Stanford Background dataset contains 715 images of outdoor scenes having 9 classes. Each image has a resolution of 320×240 pixels. We randomly split the images to keep 80 % of them for training and 20 % for testing. From these images, we extract a total of 40 millions patches. The SIFT Flow dataset contains 2688 manually labeled images of 256×256. The dataset has 2488 training and 200 test images containing 33 classes of objects. From this we extract 160 millions patches.

4.1 Segmentation Results with Different Training Strategies

Table 1 and Fig. 3 show the results obtained for all our training strategies. We report *global accuracy, average accuracy* and the *Intersection over Union* (IoU) measure. *Global* is the number of correctly classified pixels (*True Positive*) over the total number of pixels (also called *recall* or *pixel accuracy*), *Average* is the average of this recall per class (also called the *class accuracy*) and *IoU* is the ratio $TP/(TP + FP + FN)$ where TP is the number of *True Positive* and FP, FN are respectively the *False Positive* and *False Negative* averaged across all classes All results (global, average and IoU) are averaged over the considered datasets, taking into account their relative number of patches.

Table 1. Pixel (*Global*) and class (*Avg.*) accuracies and Intersection over Union (IoU) results for the combinations of each pair of datasets: 7 KITTI (7 heterogeneously labeled subsets of the KITTI dataset), Stanford and Siftflow with different training strategies: NF=No Fusion (see Fig. 1a); JT= Joint training (see Fig. 1b); DE=Dataset Equilibrium; GR=Gradient Reversal. Best results are highlighted in bold.

	7 KITTI			Stanford + SiftFlow			7 KITTI + Stanford			7 KITTI + SiftFlow		
Methods	Global	Avg	IoU	Global	Avg	IoU	Global	Avg	IoU	Global	Avg	IoU
No Fusion	77.35	54.51	41.99	72.66	33.24	25.84	77.09	55.64	43.10	76.72	45.18	34.73
JT	80.71	58.62	46.21	73.99	36.52	28.31	79.64	58.97	46.43	79.1	48.08	37.52
JT + GR	-	-	-	73.26	34.07	26.37	79.82	58.82	46.43	79.65	48.08	**37.76**
JT + DE	**80.84**	**58.89**	**46.32**	73.51	**37.08**	**28.71**	80.02	**59.23**	46.78	78.37	45.51	35.84
JT + GR + DE	-	-	-	**74.17**	35.89	27.60	**80.46**	59.02	**46.91**	**79.65**	**48.33**	37.62

Fig. 3. (a): ground truth image from [18]. **(b), (c):** our pixel classification results for the corresponding image using the JT+DE strategy (see Table 1) for the label-set outputs corresponding respectively to [18] and to [13].

No Fusion. The first learning strategy implemented consists in learning one network per dataset with the architecture described in Sect. 4 and illustrated in Fig. 1a. This is our baseline, and the results for this strategy are shown in the rows described as *No Fusion*. State of the art performances for the different KITTI sub datasets are (respectively for global and average accuracies): (92.77, 68.65) for He et al. [9]; (97.20, non reported) for Kundu et al. [13]; (82.4, 72.2) for Ladicky et. al. [14]; (51.2, 61.6) for Ros et al. [18]; (90.9, 92.10) for Sengupta et al. [19]; (non reported, 61.6) for Xu et al. [22]; and (89.3, 65.4) for Zhang et al. [25]. For Stanford with 8 classes (resp. SIFT Flow), [20] reports a global accuracy of 82.3 (resp. 80.9), a pixel accuracy of 79.1 (resp. 39.1) and an IoU of 64.5 (resp. 30.8). These isolated results are better (except for Ros et al. in Table 1) than the averages reported in our tables. This can be explained by the fact that: [13, 14, 18–20] only show results computed from a subset of their labels (e.g. the label *pedestrian* is ignored in [14, 18, 19]); the features used by all methods on KITTI are richer (e.g. depth and time); and the proposed methods always combine multiple classifiers tuned on one particular sub-dataset. To assess our contributions, we believe that *No Fusion* is the fairest baseline.

Joint Training (JT). The second alternative strategy consists in learning one single network (illustrated in Fig. 1b) with all the datasets using the selective loss function detailed in Sect. 3. We can see that this strategy gives better results than our baseline for all combination of datasets (for example, in Table 1, learning with all the subsets of KITTI gives, on average for all the 7 subsets, an improvement of +3.36 on the Global accuracy, of +4.11 on the Average accuracy and of +4.22 on the IoU). These results confirm that *Joint Training* allows us to increase the number of data used to train the network even if the datasets have different label-sets. For the sake of completeness and to evaluate the contribution of the selective loss over the mere augmentation of data, we trained our network with pairs of datasets where one dataset was used to initialize the weights of the convolutional part of the network and the other (usually the smaller one) was used to fine-tune the network. The results, not given here for lack of space, consistently show that this fine-tuning approach increases all the performance measures but at a lower extent than when using our method.

Gradient Reversal (GR) and Dataset Equilibrium (DE). As explained in Sect. 3, our selective loss does not full exploit datasets with different distributions. Thus we combine it with the gradient reversal techniques. Using gradient reversal for the KITTI dataset does not make sense since the sub-datasets all come from the same distribution. Table 1 shows that adding this gradient reversal does not always improve the performance (e.g., when learning with SIFT Flow and Stanford, the performance measures are worse than for *JT*). Starting from the intuition that unbalanced datasets can be an issue for *GR*, we experimented with weighting the contribution of each patch on the gradient computation depending on the size of the dataset this patch come from. The results show the importance of this *Dataset Equilibrium* step, even for the KITTI dataset alone. Most best results are obtained when the joint training approach is combined both with Dataset Equilibrium and with Gradient Reversal.

4.2 Detailed Analysis on Correlations Across Tasks

We computed a label correlation matrix for all sub-datasets of the KITTI dataset, shown in Fig. 4, by averaging the predictions made by the network for each target class label (from one of the 7 possible labelings). The (full) diagonal of the matrix gives the correlation rates between the expected target labels. In each line, the other non-zero values correspond to the labels correlated with the target label y_i^k. A diagonal is visible *inside* each block, including *off-diagonal blocks*, in particular for the first 5 labels of each block. This means that, as expected, these first 5 labels are all correlated across datasets. For instance, the label *Road* from He et al. is correlated with the label *Road* from Kundu et al. with

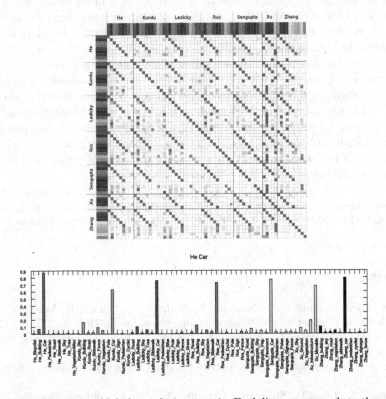

Fig. 4. Left: the empirical label correlation matrix. Each line corresponds to the average of the predictions of the network for a given target class (among the 68 labels given in Fig. 2). Darker cells indicate higher probabilities. Non-diagonal red cells correspond to labels highly correlated with the target (main diagonal) label. Below: detailed line of the correlation matrix corresponding to the label Car from the sub-dataset of He. (Color figure online)

the *Road* from Ladicky et al. etc. A second observation is that the correlation matrix is not symmetric. For example, the classes *Building, Poles, Signage* and *Fence* from Sengupta et al. have (as already discussed in Sect. 4) a high correlation with the class *Infrastructure* from Xu et al., meaning that these classes overlap. On the contrary, the class *Infrastructure* from Xu et al. has a very high correlation with the class *Building* from Sengupta et al. and a limited one with the classes *Poles, Signage* and *Fence*. This is due to the label distributions: the *Building* class from Sengupta et al. is more represented than the three other classes, so *Infrastructure* from Xu et al. is more correlated to *Building*.

5   Conclusion

In this paper, we considered the problem of multi-task multi-domain learning: we want to exploit multiple datasets that have related inputs (e.g. images) but that have been annotated for different tasks. This problem is important in two major situations: to fuse existing (small) datasets and to reuse existing dataset(s) for a related custom (new) task. We introduced a new *selective loss* function for deep networks that makes it possible to learn jointly across different tasks. We provide experimental results on semantic segmentation computer vision tasks with a total of 9 datasets. The results show that our approach allows to jointly learn from multiple datasets and to outperform per-task learning and classical fine-tuning based approaches. We also show that the domain adaptation methods (gradient reversal) can be applied for multi-task multi-domain learning but needs to be used with care and requires to balance the different datasets. An important perspective of this work is to design a strategy to practically take into account the correlations between labels highlighted by our method.

Acknowledgment. Authors acknowledge the support from the ANR project SoL-StiCe (ANR-13-BS02-0002-01). They also want to thank Nvidia for providing two Titan X GPU.

References

1. Ben-David, S., Blitzer, J., Crammer, K., Kulesza, A., Pereira, F., Vaughan, J.W.: A theory of learning from different domains. Mach. Learn. **79**(1–2), 151–175 (2010)
2. Farabet, C., Couprie, C., Najman, L., LeCun, Y.: Learning hierarchical features for scene labeling. IEEE TPAMI **35**(8), 1915–1929 (2013)
3. Ganin, Y., Lempitsky, V.: Unsupervised domain adaptation by backpropagation. In: ICML (2015)
4. Geiger, A., Lenz, P., Stiller, C., Urtasun, R.: Vision meets robotics: the KITTI dataset. Int. J. Robot. Res. **34**, 727–743 (2013)
5. Germain, P., Habrard, A., Laviolette, F., Morvant, E.: A new PAC-Bayesian perspective on domain adaptation. In: Proceedings of the 33rd International Conference on Machine Learning (ICML 2016), New York, NY, United States (2016). https://hal.archives-ouvertes.fr/hal-01307045

6. Gould, S., Fulton, R., Koller, D.: Decomposing a scene into geometric and seman-tically consistent regions. In: ICCV (2009)
7. Hinton, G., Vinyals, O., Dean, J.: Distilling the knowledge in a neural network. In: NIPS Deep Learning Workshop (2014)
8. Hinton, G., Srivastava, N., Krizhevsky, A., Sutskever, I., Salakhutdinov, R.: Improving neural networks by preventing co-adaptation of feature detectors arXiv:1207.0580 (2012)
9. Hu, H., Ben, U.: Nonparametric semantic segmentation for 3D street scenes. In: Intelligent Robots and Systems (IROS) (2013)
10. Ioffe, S., Szegedy, C.: Batch normalization: accelerating deep network training by reducing internal covariate shift. In: ICML (2015)
11. Kekec, T., Emonet, R., Fromont, E., Trémeau, A., Wolf, C.: Contextually con-strained deep networks for scene labeling. In: BMVC (2014)
12. Krizhevsky, A., Sutskever, I., Hinton, G.: Imagenet classification with deep convo-lutional neural networks. In: NIPS (2012)
13. Kundu, A., Li, Y., Dellaert, F., Li, F., Rehg, J.M.: Joint semantic segmentation and 3D reconstruction from monocular video. In: Fleet, D., Pajdla, T., Schiele, B., Tuytelaars, T. (eds.) ECCV 2014. LNCS, vol. 8694, pp. 703–718. Springer, Heidelberg (2014). doi:10.1007/978-3-319-10599-4_45
14. Ladicky, L., Shi, J., Pollefeys, M.: Pulling things out of perspective. In: CVPR (2014)
15. Liu, C., Yuen, J., Torralba, A.: Nonparametric scene parsing via label transfer. IEEE TPAMI 33(5), 978–994 (2011)
16. Mansour, Y., Mohri, M., Rostamizadeh, A.: Domain adaptation: learning bounds and algorithms. In: COLT 2009 - The 22nd Conference on Learning Theory, Mon-treal, Quebec, Canada, 18–21 June 2009
17. Pinheiro, P., Collobert, R.: Recurrent convolutional neural networks for scene label-ing. In: ICML (2014)
18. Ros, G., Ramos, S., Granados, M., Bakhtiary, A., Vazquez, D., Lopez, A.M.: Vision-based offline-online perception paradigm for autonomous driving. In: Winter Con-ference on Applications of Computer Vision (WACV) (2015)
19. Sengupta, S., Greveson, E., Shahrokni, A., Torr, P.H.: Urban 3D semantic mod-elling using stereo vision. In: ICRA (2013)
20. Sharma, A., Tuzel, O., Jacobs, D.W.: Deep hierarchical parsing for semantic seg-mentation. In: CVPR (2015)
21. Tzeng, E., Hoffman, J., Darrell, T., Saenko, K.: Simultaneous deep transfer across domains and tasks. In: ICCV (2015)
22. Xu, P., Davoine, F., Bordes, J.B., Zhao, H., Denoeux, T.: Information fusion on oversegmented images: an application for urban scene understanding. In: IAPR MVA (2013)
23. Yosinski, J., Clune, J., Bengio, Y., Lipson, H.: How transferable are features in deep neural networks? In: NIPS (2014)
24. Zeiler, M.D., Fergus, R.: Visualizing and understanding convolutional networks. In: Fleet, D., Pajdla, T., Schiele, B., Tuytelaars, T. (eds.) ECCV 2014. LNCS, vol. 8689, pp. 818–833. Springer, Heidelberg (2014). doi:10.1007/978-3-319-10590-1_53
25. Zhang, R., Candra, S.A., Vetter, K., Zakhor, A.: Sensor fusion for semantic seg-mentation of urban scenes. In: IEEE ICRA (2015)
26. Zhang, X., Yu, F.X., Chang, S., Wang, S.: Deep transfer network: unsupervised domain adaptation. arXiv (2015)

Sequential Labeling with Structural SVM Under an Average Precision Loss

Guopeng Zhang and Massimo Piccardi[✉]

Global Big Data Technologies Centre, University of Technology Sydney,
Ultimo, NSW, Australia
Guopeng.Zhang@student.uts.edu.au, Massimo.Piccardi@uts.edu.au

Abstract. The average precision (AP) is an important and widely-adopted performance measure for information retrieval and classification systems. However, owing to its relatively complex formulation, very few approaches have been proposed to learn a classifier by maximising its average precision over a given training set. Moreover, most of the existing work is restricted to i.i.d. data and does not extend to sequential data. For this reason, we herewith propose a structural SVM learning algorithm for sequential labeling that maximises an average precision measure. A further contribution of this paper is an algorithm that computes the average precision of a sequential classifier at test time, making it possible to assess sequential labeling under this measure. Experimental results over challenging datasets which depict human actions in kitchen scenarios (i.e., TUM Kitchen and CMU Multimodal Activity) show that the proposed approach leads to an average precision improvement of up to 4.2 and 5.7 % points against the runner-up, respectively.

Keywords: Sequential labeling · Structural SVM · Average precision · Loss-augmented inference

1 Introduction and Related Work

Choosing appropriate performance measures plays an important role in developing effective information retrieval and classification systems. Common figures include the false positive and false negative rates, the precision and recall, and the F-measure which can all assess the accuracy of a prediction by comparing the predicted labels with given ground-truth labels. However, in applications such as information retrieval, it is often important to assess not only the accuracy of the predicted labels, but also that of a complete ranking of the samples. In classification, too, it is often preferable to evaluate the prediction accuracy at various trade-offs of precision and recall, to ensure coverage of multiple operating points. For both these needs, the average precision (a discretised version of the area under the precision-recall curve) offers a very informative performance measure.

Amongst the various flavours of classification, sequential labeling, or tagging, refers to the classification of each of the measurements in a sequence. It is a very

© Springer International Publishing AG 2016
A. Robles-Kelly et al. (Eds.): S+SSPR 2016, LNCS 10029, pp. 344–354, 2016.
DOI: 10.1007/978-3-319-49055-7_31

important task in a variety of fields including video analysis, bioinformatics, financial time series and natural language processing [8]. Unlike the classification of independent samples, the typical sequential labeling algorithms such as Viterbi (including their n-best versions [7]) do not provide multiple predictions at varying trade-offs of precision and recall, and therefore the computation of their average precision is not trivial.

In the literature, a number of papers have addressed the average precision as a performance measure in the case of independent samples. For instance, [5] has studied the statistical behaviour of the average precision in the presence of relevance judgements. Yilmaz and Aslam in [15] have proposed an approximation of the average precision in retrieval systems with incomplete and imperfect judgements. Morgan et al. in [6] have proposed an algorithm for learning the weights of a search query with maximum average precision. Notably, Joachims et al. in [16] have proposed a learning algorithm that can efficiently train a support vector machine (SVM) under an average precision loss. However, all this work only considers independent and identically distributed (i.i.d.) samples, while very little work to date has addressed the average precision in sequential labeling and structured prediction. In [9], Rosenfeld et al. have proposed an algorithm for training structural SVM under the average precision loss. However, their algorithm assumes that the structured output variables can be ranked in a total order relationship which is generally restrictive.

For the above reasons, we propose a training algorithm that can train structural SVM for sequential labeling under an average precision loss. Our assumptions are very general and do not require ranking of the output space. The core component of our training algorithm is an inference procedure that returns sequential predictions at multiple levels of recall. The same inference procedure can also be used at test time, making it possible to evaluate the average precision of sequential labeling algorithms and to compare it with that of i.i.d. classifiers.

Experiments have been conducted over two challenging sequential datasets: the TUM Kitchen and the CMU-MMAC activity datasets [1,11]. The results, reported in terms of average precision, show that the proposed method remarkably outperforms other performing classifiers such as standard SVM and structural SVM trained with conventional losses.

2   Background

2.1   Average Precision

The average precision (AP) is a de-facto standard evaluation in the computer vision community since the popular PASCAL VOC challenges [2]. It is defined as the average of the precision at various levels of recall and is a discretised version of the area under the precision-recall curve (AUC). The AP is a very informative measure since it assesses the classification performance at different trade-offs of precision and recall, reflecting a variety of operating conditions. Its formal definition is:

$$AP = \frac{1}{R} \sum_r p_{@r} \tag{1}$$

where $p_{@r}$ is the precision at level of recall r, and R is the number of levels. The recall ranges between 0 and 1, typically in 0.1 steps. At its turn, the precision at a chosen value of recall, $p_{@r}$, is defined as:

$$p_{@r} = \frac{TP}{TP + FP} \quad s.t. \quad \frac{TP}{TP + FN} = r \tag{2}$$

where TP, FP and FN are the number of true positives, the number of false negatives and the number of false positives, respectively, computed from the classification contingency table of the predicted and ground-truth labels.

In general, the precision tends to decrease as r grows. However, it is not a monotonically non-increasing function of r. To ensure monotonicity of the summand, Everingham et al. in [3] modified the definition of the AP as:

$$AP = \frac{1}{R} \sum_r \max_{l=0...r} p_{@l} \tag{3}$$

This way of computing the average precision has become commonplace in the computer vision and machine learning communities and it is therefore adopted in our experiments. However, the algorithm we describe in Sect. 3 can be used interchangeably for either (1) or (3). Given that the AP is bounded between 0 and 1, a natural definition for an AP-based loss is $\Delta_{AP} = 1 - AP$.

2.2   Sequential Labeling

Sequential labeling predicts a sequence of class labels, $y = (y_1, \ldots, y_t, \ldots, y_T)$, from a given measurement sequence, $x = (x_1, \ldots, x_t, \ldots, x_T)$, where x_t is a feature vector at sequence position t and y_t is a corresponding discrete label, $y_t \in 1 \ldots M$. In many cases, it is not restrictive to assume that y_t is a binary label (1: positive class; 0: negative class), obtaining multi-class classification from a combination of binary classifiers. Therefore, in the following we focus on the binary case. The most widespread model for sequential labeling is the hidden Markov model (HMM) which is a probabilistic graphical model factorising the joint probability of the labels and the measurements. By restricting the model to the exponential family of distributions and expressing the probability in a logarithmic scale, the score of an HMM can be represented as a generalised linear model:

$$\ln p(x, y) \propto w^\top \phi(x, y) = w_{init}^\top f(y_1) +$$
$$+ \sum_{t=2}^{T} w_{tran}^\top f(y_{t-1}, y_t) + \sum_{t=1}^{T} w_{em}^\top f(x_t, y_t) \tag{4}$$

where w_{init} are the first-frame parameters, w_{tran} are the transition parameters, w_{em} are the emission parameters, and functions $f(y_1)$, $f(y_{t-1}, y_t)$ and $f(x_t, y_t)$ are arbitrary feature functions of their respective arguments. The inference problem for this model consists of determining the best class sequence for a given measurement sequence:

$$\bar{y} = \operatorname*{argmax}_{y} w^\top \phi(x, y) \tag{5}$$

This problem can be efficiently solved in $O(T)$ time by the well-known Viterbi algorithm operating in a logarithmic scale [8].

2.3   Structural SVM

SVM has been extended from independent (measurement, label) pairs to the prediction of structured labels, i.e. multiple labels that have mutual dependencies in the form of sequences, trees and graphs and that co-depend on multiple measurements [10,12]. Given a set of N training instances $\{x^i, y^i\}$, $i = 1, \ldots, N$, structural SVM finds the optimal model's parameter vector w by solving the following convex optimisation problem:

$$
\min_{w, \xi} \frac{1}{2} \|w\|^2 + C \sum_{i=1}^{N} \xi^i \quad s.t.
$$

$$
w^\top \phi(x^i, y^i) - w^\top \phi(x^i, y) \geq \Delta(y^i, y) - \xi^i,
$$

$$
i = 1 \ldots N, \quad \forall y \in \mathcal{Y}
$$

(6)

As usual, term $\sum_{i=1}^{N} \xi^i$ places an upper bound over the total training error, while term $\|w\|^2$ regularises the solution to encourage generalisation. Parameter C is an arbitrary, positive coefficient that balances these two terms. In the constraints, function $\phi(x, y)$ is a feature function that computes structured features from the pair $\{x, y\}$ such that $w^\top \phi(x, y)$ can assign a score to the pair. The constraint for labeling $y = y^i$ guarantees that $\xi^i \geq 0$, and $\Delta(y^i, y)$ is the chosen, arbitrary loss function.

The problem with Eq. (6) is that the size of the constraint set, $\mathcal{Y}$, is exponential in the number of of the output variables and it is therefore impossible to satisfy the full constraint set. However, [12] has shown that it is possible to find ϵ-correct solutions with a constraint subset of polynomial size, consisting of only the "most violated" constraint for each sample, i.e. the labeling with the highest sum of score and loss:

$$
\xi^i = \max_{y} (-w^\top \phi(x^i, y^i) + w^\top \phi(x^i, y) + \Delta(y^i, y))
$$

$$
\rightarrow \bar{y}^i = \arg\max_{y} (w^\top \phi(x^i, y) + \Delta(y^i, y))
$$

(7)

This problem is commonly referred to as "loss-augmented inference" due to its resemblance to the usual inference of Eq. (5) and is the main step of structural SVM.

3   Training and Testing Sequential Labeling with the AP Loss

The loss functions used for training structural SVM commonly include the 0–1 loss and the Hamming loss. Under these losses, the loss-augmented inference can still be computed by a conventional Viterbi algorithm with adjusted weights. Instead, training with the average precision cannot be approached in the same way since it requires predicting either a ranking or multiple labelings. For this reason, we propose a different formulation of the structural SVM primal problem:

$$\min_{w,\xi} \frac{1}{2}\|w\|^2 + C\sum_{i=1}^{N}\xi^i \quad s.t.$$

$$w^\top\phi(x^i,y^i) - \frac{1}{R}\sum_r w^\top\phi(x^i,y^{[r]}) \tag{8}$$

$$\geq \Delta_{AP}(y^i,y^{[0]},\dots y^{[1]}) - \xi^i, \xi^i \geq 0, \quad i = 1\dots N,$$

$$r = 0, 0.1,\dots 1, \quad \forall y^{[0]}\dots y^{[1]} \in \mathcal{Y}_0 \times \dots \times \mathcal{Y}_1$$

The constraints in Eq. (8) state that the score assigned to the ground-truth labeling, y^i, must be greater than or equal to the average score of any set of R labelings at the appropriate levels of recall by at least their average precision loss. In this way, we retain the structural SVM principle of imposing a margin between the ground truth and the prediction that is equal to the chosen loss, while we constrain all the predictions at the prescribed levels of recall. At the same time, we cannot ensure that the hinge loss ξ^i is an upper bound for $\Delta_{AP}(y^i,y^{[0]},\dots y^{[1]})$, and therefore the minimisation of the loss over the training set is only heuristic.

For Eq. (8), the loss-augmented inference becomes:

$$\bar{y}^{[0]}\dots\bar{y}^{[1]} = \operatorname*{argmax}_{y^{[0]}\dots y^{[1]}} \left(\frac{1}{R}\sum_r w^\top\phi(x^i,y^{[r]}) + \Delta_{AP}(y^i,y^{[0]},\dots y^{[1]})\right)$$

$$= \operatorname*{argmax}_{y^{[0]}\dots y^{[1]}} \left(\frac{1}{R}\sum_r w^\top\phi(x^i,y^{[r]}) + \frac{1}{R}\sum_r \Delta_{P@r}(y^i,y^{[r]})\right)$$

$$= \operatorname*{argmax}_{y^{[0]}}(w^\top\phi(x^i,y^{[0]}) + \Delta_{P@0}(y^i,y^{[0]})),\dots\operatorname*{argmax}_{y^{[1]}}(w^\top\phi(x^i,y^{[1]}) + \Delta_{P@1}(y^i,y^{[1]}))$$

$$\tag{9}$$

where we have made use of the definition of average precision from Eq. (1). Equation (9) shows an important property: that the R most violating labelings can be found independently of each other using the precision loss at the required level of recall. This is the key property for the algorithm we propose in the following sub-section.

3.1   Inference and Loss-Augmented Inference

Once the model is trained, testing it to report its AP requires, once again, the ability to produce a set of R predictions at the required levels of recall.

Therefore, the key problems for both training and testing can be summed up, respectively, as:

$$\operatorname*{argmax}_{y^{[r]}} (w^\top \phi(x^i, y^{[r]}) + \Delta_{p_{@r}}(y^i, y^{[r]})) \tag{10}$$

$$\operatorname*{argmax}_{y^{[r]}} w^\top \phi(x^i, y^{[r]}) \tag{11}$$

The algorithm we propose hereafter works interchangeably for both Eqs. (10) and (11), and also for the modified AP loss of Eq. (3). Given any ground-truth label sequence, y^i, the degrees of freedom of the precision loss are only the number of false positives, FP, and false negatives, FN. By making a prediction in left-to-right order along the sequence, the running values of FP and FN can only increment or remain unchanged. We can thus still approach the solution of Eq. (10) by dynamic programming, extending the state of a partial solution to include: (a) the ground-truth label of the current frame, y_t, as in conventional Viterbi; (b) the number of false positives, FP, in sub-sequence $y_{1:t}$; and (c) the number of false negatives, FN, in sub-sequence $y_{1:t}$. We use notation $\psi(FP, FN, y_t)$ to indicate the $y_{1:t}$ sub-sequence with the highest score for the given extended state, and $s(\psi)$ for its score. The generic induction step is as follows: at any time step, t, a partial solution is obtained by extending two of the partial solutions of time $t-1$ with the current prediction, y_t, and correspondingly incrementing either FP or FN if the prediction is incorrect, or neither if correct. After the final time step, T, Eq. (10) is computed over the stored sequences and the argmax returned. Algorithm 1 describes the solution formally.

4   Experiments

The proposed approach has been evaluated on two challenging datasets of human activities, TUM Kitchen and CMU Multimodal Activity (CMU-MMAC). Descriptions and results for these two datasets are reported in the following subsections. The compared algorithms include: (a) the proposed method based on the AP loss; (b) structural SVM using the common 0–1 loss and Hamming loss, and (c) a baseline offered by a standard SVM that classifies each frame separately. For SVM training, we have used constant $C = 0.1$ (based on a preliminary cross-validation), the RBF kernel (for non-linearity), and, for SSVM, convergence threshold $\epsilon = 0.01$ (default). For the AP loss, given the greater computational complexity of the loss-augmented inference (approximately quadratic for sequences with sparse positives), we decode each sequence in sub-sequences of 300 frames each. To develop the software, we have used the SVM^{struct} package and its MATLAB wrapper [4,13]. All experiments have been performed on a PC with an Intel i7 2.4GHz CPU with 8 GB RAM.

4.1   Results on the TUM Kitchen Dataset

The TUM Kitchen dataset is a collection of activity sequences recorded in a kitchen equipped with multiple sensors [11]. In the kitchen environment,

---

Algorithm 1. Algorithm for computing the loss-augmented inference of Eq. (10).

Input: w, $x = (x_1, \ldots, x_T)$, $y^g = (y_1^g, \ldots, y_T^g)$ (ground-truth labels), r
Output: $\bar{y}^{[r]}$
Initialize: $FP_{max} = FN_{max} = 0$

// FP, FN: running variables for the number of false positives and false negatives
// pos, neg: number of positives and negatives in y^g
// $\psi(invalidarg) = NULL, s(NULL) = -\infty$, [ ] = string concatenation operator

ψ = FindHighestScoringSequences(w, x, y^g);
$\bar{y}^{[r]}$ = FindMostViolatingLabeling(ψ, r);
return $\bar{y}^{[r]}$

---

function FindHighestScoringSequences(w, x, y^g)
// Finds all highest-scoring sequences for any combinations of FP and FN:

if $y_t^g = 0$
  $FP_{max} = FP_{max} + 1$
  for $FP = 0 : FP_{max}, FN = 0 : FN_{max}, t = 1 : T$
    $\psi(FP, FN, y_t = 0) =$
      $\text{argmax}(s([\psi(FP, FN, y_{t-1} = 0), 0]), s([\psi(FP, FN, y_{t-1} = 1), 0]))$
    $\psi(FP, FN, y_t = 1) =$
      $\text{argmax}(s([\psi(FP - 1, FN, y_{t-1} = 0), 1]), s([\psi(FP - 1, FN, y_{t-1} = 1), 1]))$
else
  $FN_{max} = FN_{max} + 1$
  for $FP = 0 : FP_{max}, FN = 0 : FN_{max}, t = 1 : T$
    $\psi(FP, FN, y_t = 0) =$
      $\text{argmax}(s([\psi(FP, FN - 1, y_{t-1} = 0), 0]), s([\psi(FP, FN - 1, y_{t-1} = 1), 0]))$
    $\psi(FP, FN, y_t = 1) =$
      $\text{argmax}(s([\psi(FP, FN, y_{t-1} = 0), 1]), s([\psi(FP, FN, y_{t-1} = 1), 1]))$
return ψ
end function

---

function FindMostViolatingLabeling(ψ, r)
// Finds the labeling maximising the sum of score and loss:

$FN^* = round(pos\,(1 - r))$ // sets the desired recall level

find $\text{argmax}_{\bar{y}^{[r]}}\, s(\bar{y}^{[r]})$ over $FP = 0 : neg, FN = FN^*$
  $\bar{y}^{[r]} = \text{argmax}_\psi$
    $[s(\psi(FP, FN^*, y_T = 0)) + \Delta_{p@r}(pos, FP, FN),$
    $s(\psi(FP, FN^*, y_T = 1)) + \Delta_{p@r}(pos, FP, FN)]$

// for Eq. (11), just remove $\Delta_{p@r}$
// for the modified AP loss of Eq. (3), set $FN = 0 : FN^*$

return $\bar{y}^{[r]}$
end function

---

various subjects were asked to set a table in different ways, performing 9 actions, *Reaching, TakingSomething, Carrying, LoweringAnObject, Releasing-Grasp, OpeningADoor, ClosingADoor, OpeningADrawer* and *ClosingADrawer*. For our experiments, we have chosen to use the motion capture data from the left and right hands. These data consist of 19 sequences for each hand, each ranging in length between 1,000 and 6,000 measurements. The first 6 sequences were used for training and the remaining for testing. Each measurement is a 45-D vector of 3D body joint locations. Figure 1a shows a scene from this dataset.

Table 1 reports the results for activity recognition from the left and right hand sequences. The table shows that the mean of the AP over the nine classes is the highest for the proposed technique, with an improvement of 4.2 % points

Table 1. Comparison of the average precision over the TUM Kitchen dataset. SVM: standard SVM baseline; 0–1 loss and Hamming loss: structural SVM with conventional loss functions; AP loss: proposed technique.

	Average precision (%)			
Left hand sequences	SVM	0–1 loss	Hamming loss	AP loss
Reaching	24.5	44.8	18.5	**50.1**
TakingSomething	31.1	79.7	20.0	**80.7**
LoweringAnObject	19.3	44.6	16.9	**49.9**
ReleasingGrasp	18.1	53.2	25.0	**54.4**
OpeningADoor	10.9	9.1	9.1	**15.5**
ClosingADoor	9.2	9.1	9.1	**11.5**
OpeningADrawer	10.5	14.8	11.8	**20.6**
ClosingADrawer	10.9	9.1	9.1	**15.5**
Carrying	62.3	75.6	51.9	**80.2**
Mean	21.9	37.8	19.0	**42.0**
Right hand sequences	SVM	0–1 loss	Hamming loss	AP loss
Reaching	18.0	65.5	18.3	**68.9**
TakingSomething	12.8	**91.6**	14.1	90.9
LoweringAnObject	13.7	43.1	15.1	**47.7**
ReleasingGrasp	17.9	40.8	18.8	**45.4**
OpeningADoor	29.1	68.5	16.3	**73.9**
ClosingADoor	13.2	36.4	15.6	**41.3**
OpeningADrawer	14.7	26.8	13.8	**30.2**
ClosingADrawer	12.3	30.7	13.0	**38.0**
Carrying	58.7	85.4	63.1	**89.9**
Mean	21.3	54.3	20.8	**58.5**

(a)   (b)

Fig. 1. Sample frames from (a) the TUM Kitchen dataset and (b) the CMU-MMAC dataset.

over the runner-up for both the left and right hand sequences. In addition, the proposed technique reports the highest average precision in all the classes with the left hand sequences, and in 8 classes out of 9 with the right hand sequences. In addition, the average precision of the proposed technique is about double that of the standard SVM baseline that does not leverage sequentiality.

4.2   Results on the CMU Multimodal Activity Dataset

The CMU Multimodal Activity (CMU-MMAC) dataset contains multimodal measurements of the activities of 55 subjects preparing 5 different recipes: "brownies", a salad, a pizza, a sandwich and scrambled eggs [1]. For our experiments, we have chosen to use the video clips of the 12 subjects preparing brownies from a dry mix box. The actions performed by the subjects are very realistic and are divided over 14 basic activities. The length of the 12 video clips ranges from 8,000 to 20,000 frames. For the experiments, we have used the first 8 videos for training and the remaining 4 for testing. For the feature vector of each frame, we have first extracted dense SIFT features at a 32-pixel step and used k-means with 32 clusters to generate a codebook. Then, the descriptors of each frame have been encoded into a 4,096-D VLAD vector [14]. Figure 1b displays a scene from this dataset, showing that the kitchen environment and camera view are significantly different from TUM's.

Table 2 reports the results for activity recognition over this dataset. The table shows that the mean of the AP is the highest for the proposed technique, with an improvement of 5.7 % points over the runner-up. In addition, the proposed technique reports the highest average precision for 12 classes out of 14, and more than doubles the SVM baseline.

Table 2. Comparison of the average precision over the CMU-MMAC dataset. SVM: standard SVM baseline; 0–1 loss and Hamming loss: structural SVM with conventional loss functions; AP loss: proposed technique.

	Average Precision (%)			
	SVM	0–1 loss	Hamming loss	AP loss
Close	9.9	**16.8**	9.0	16.2
Crack	11.4	23.1	8.3	**28.5**
None	30.9	46.6	29.8	**54.1**
Open	15.8	**33.3**	16.1	29.0
Pour	30.6	50.0	27.4	**61.4**
Put	13.1	27.8	11.4	**34.3**
Read	9.1	10.9	11.8	**16.5**
Spray	14.8	25.6	10.2	**28.4**
Stir	28.0	39.4	23.5	**45.5**
Switch-on	11.3	27.6	9.9	**32.9**
Take	22.5	47.1	19.8	**60.8**
Twist-off	10.1	25.5	7.9	**30.0**
Twist-on	9.8	19.0	8.1	**27.6**
Walk	10.4	23.2	9.6	**29.7**
Mean	16.3	29.7	14.5	**35.4**

5 Conclusion

The average precision has become a reference evaluation measure for its ability to assess performance at multiple operating points. However, the typical sequential labeling algorithms such as Viterbi do not allow the computation of the average precision. For this reason, in this paper, we have proposed an inference procedure that infers a set of predictions at multiple levels of recall and allows measuring the average precision of a sequential classifier. In addition, we have proposed a structural SVM training algorithm for sequential labeling that minimises an average precision loss. Experiments conducted over two challenging activity datasets - TUM Kitchen and CMU-MMAC - have shown that the proposed approach significantly outperforms all of the other compared techniques and more than doubles the performance of a baseline. Moreover, while we have only focused on sequential labeling in this paper, the proposed approach could readily be employed for more general structures such as trees and graphs.

References

1. De la Torre, F., Hodgins, J.K., Montano, J., Valcarcel, S.: Detailed human data acquisition of kitchen activities: the CMU-multimodal activity database (CMU-MMAC). In: CHI 2009 Workshops, pp. 1–5 (2009)

2. Everingham, M., Van Gool, L., Williams, C.K.I., Winn, J., Zisserman, A.: The PASCAL Visual Object Classes Challenge 2007 (VOC 2007) Results (2007). http://www.pascal-network.org/challenges/VOC/voc2007/workshop/index.html
3. Everingham, M., Van Gool, L., Williams, C.K.I., Winn, J., Zisserman, A.: The Pascal visual object classes (VOC) challenge. IJCV **88**(2), 303–338 (2010)
4. Joachims, T.: SVM^{struct}: Support vector machine for complex output 3.10 (2008). http://www.cs.cornell.edu/people/tj/svm_light/svm_struct.html
5. Kishida, K.: Property of Average Precision and Its Generalization: An Examination of Evaluation Indicator for Information Retrieval Experiments. NII Technical report, National Institute of Informatics (2005)
6. Morgan, W., Greiff, W., Henderson, J.: Direct maximization of average precision by hill-climbing, with a comparison to a maximum entropy approach. In: HLT-NAACL 2004, HLT-NAACL-Short 2004, pp. 93–96 (2004)
7. Nilsson, D., Goldberger, J.: Sequentially finding the N-best list in hidden Markov models. In: IJCAI 2001, pp. 1280–1285 (2001)
8. Rabiner, L.: A tutorial on hidden Markov models and selected applications in speech recognition. IEEE Proc. **77**, 257–286 (1989)
9. Rosenfeld, N., Meshi, O., Tarlow, D., Globerson, A.: Learning structured models with the AUC loss and its generalizations. In: AISTATS 2014, pp. 841–849 (2014)
10. Taskar, B., Guestrin, C., Koller, D.: Max-margin Markov networks. In: NIPS, pp. 25–32 (2003)
11. Tenorth, M., Bandouch, J., Beetz, M.: The TUM kitchen data set of everyday manipulation activities for motion tracking and action recognition. In: IEEE International Workshop on Tracking Humans for the Evaluation of their Motion in Image Sequences (THEMIS), in Conjunction with ICCV 2009, pp. 1089–1096 (2009)
12. Tsochantaridis, I., Joachims, T., Hofmann, T., Altun, Y.: Large margin methods for structured and interdependent output variables. JMLR **6**, 1453–1484 (2005)
13. Vedaldi, A.: A MATLAB wrapper of SVM^{struct} (2011). http://www.vlfeat.org/~vedaldi/code/svm-struct-matlab.html
14. Vedaldi, A., Fulkerson, B.: VLFeat: An open and portable library of computer vision algorithms (2008). http://www.vlfeat.org/index.html
15. Yilmaz, E., Aslam, J.A.: Estimating average precision with incomplete and imperfect judgments. In: ACM CIKM 2006, pp. 102–111 (2006)
16. Yue, Y., Finley, T., Radlinski, F., Joachims, T.: A support vector method for optimizing average precision. In: SIGIR, pp. 271–278 (2007)

Shape Analysis

Forensic Analysis of Manuscript Authorship: An Optimized Computational Approach Based on Texture Descriptors

Jean Felipe Felsky, Edson J.R. Justino[✉], and Jacques Facon

Pontifícia Universidade Católica do Paraná (PUCPR) Rua Imaculada Conceicao,
1155, Prado Velho, Curitiba, Parana, Brazil
jean.felsky@gmail.com, {justino,facon}@ppgia.pucpr.br

Abstract. This paper presents an optimized method for establishing the authorship of questioned handwritten documents, on the basis of a forensic analysis and a computational model using texture descriptors. The proposed method uses two classes of texture descriptors: model-based, using fractal geometry, and statistical, using GLCM (Gray-Level Co-occurrence Matrix) and Haralick's descriptors. The proposed method also uses an SVM (Support Vector Machine) as a classifier and generator of the writer-independent training. The results demonstrate the robustness of the writer-independent obtained from the features by using texture descriptors and robustness in the amount low of samples used as references for comparison and the number of feature used. The results appear promising, in the order of 97.7 %, and are consistent with those obtained in other studies that used the same database.

Keywords: Handwritten · Document · Classifier · Texture · Descriptors

1 Introduction

In forensic science, graphoscopy covers, among other aspects of writing, the finding of the authorship of handwritten documents, that is, the association of a questioned text to a specific writer. Further, its complexity is in the subjectivity used by the expert in the establishment of measures of similarities between writing patterns found in the reference models of a writer and the questioned document, allegedly belonging to the same writer.

From a computational point of view, establishing the authorship of handwritten documents is a major challenge, as the transposition of graphometric features to a computational model can present a high complexity.

In recent decades, computational solutions have been presented in the literature. These solutions can be classified into two types of basic segmentation approaches: contextual and non-contextual [2]. Contextual approaches generally use segmentation processes based on the text's content [3, 4, 7, 8]. The segmentation of the text into lines, words, letters, and letter segments is the approach more preferably used. The results generated by this segmentation are subsequently subjected to different feature extraction techniques, which may or may not be associated with the handwritten

A. Robles-Kelly et al. (Eds.): S+SSPR 2016, LNCS 10029, pp. 357–367, 2016.
DOI: 10.1007/978-3-319-49055-7_32

features [1, 2]. Approaches that use contextual models feature high complexity and may not meet all writing styles. This results from the high intrapersonal variability and graphical anomalies presented by some writers. For example, there are cases where the contents of adjacent lines overlap (model behavior – Fig. 1(a), their slant is high and they occur irregularly (basic behavior – Fig. 1(b), or the text is unreadable, preventing the identification of connectivity points between letters (moments) and the spacing between words (spacing – Fig. 1(c). Because of these factors, the computational solutions usually presented are based on manual or semi-automatic segmentation.

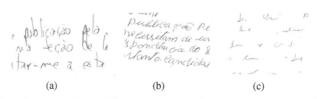

(a)   (b)   (c)

Fig. 1. (a) Text with related content; (b) Text with irregular alignment; (c) Illegible text.

Non-contextual approaches are based on the morphology of the extracted information and not on its meaning. Therefore, these approaches do not take into account the significance of the information obtained, such as letters and words. In some cases, they seek only stroke segments [1, 2]. The advantage of these approaches, as compared to those mentioned above, is the simplicity of the techniques of segmentation and features extraction, which enable the use of computational applications for forensic purposes.

In both contextual and non-contextual computational approaches, the analysis of the texture obtained from the handwritten document has proven to be a promising alternative. This is due in part to the minimization of the complexity of the features extraction process.

Non-contextual methods offer simplicity and efficiency in computer applications. They allow the use of manuscripts with different contents, in the set of documents to be analyzed, and the use of manuscripts in any language, even among the documents of the analysis set.

Based on the generation of textures, various classification and features extraction methods have been proposed in the literature; the most popular ones among them are Gabor's filters, Wavelets [9], GLCM (Gray-Level Co-occurrence Matrix) [2], LBP (Local Binary Patterns), and LPQ (Local Phase Quantization) [1].

This paper presents a non-contextual approach for determining the authorship of handwritten documents. The segmentation method used was proposed by Hanuziak [2], and the features were obtained from two distinct classes of texture descriptors, the statistical – GLCM [2] – and the model-based – based on fractals [10]. For classification, a protocol based on an SVM (Support Vector Machine) [1, 2, 7] classifiers was used, and measures of dissimilarities [1, 2, 10] were used as metrics. The results, around 97.8 % witch reduced number of feature and number of references samples per writer are promising.

2   Graphometric Expertise in Handwritten Documents

Given a set of reference manuscripts K (of known authorship) and the questioned manuscript Q (of unknown authorship), from the point of view of an expert, the result of expert analysis is based on the disclosure of a minimum set of graphometric features and the respective degrees of similarity found. The authorship of a questioned document is determined on the basis of the similarities observed during the review process of these features. This set consists of a large group of features such as attacks, pressure, and endings [2], as listed in Table 1.

Table 1. Genetic and generic attributes.

Generic features	Genetic features
Caliber	Pressure
Spacing	Progression
Model behavior	Attacks
Basic behavior	Endings
Angle values	Development
Curvilinear values	Axial slant
Moments	Graphic minimums
Proportionality	

Graphometric features are divided into two major classes, generic and genetic features of writing [1]. Generic features have elements related to the general shape of the text. Genetic features represent the genesis of the author's writing, presenting unconsciously produced aspects of the stroke. These are difficult for the writer to hide and for others to reproduce. In an expert examination, each feature can be classified into minimum, medium, or maximum convergence or divergence.

One aspect relevant to the expert analysis process is the set of documents used as $K_{1...n}$ models or references, n being the number of reference documents. The number of these documents should be sufficient to allow an accurate analysis of the similarities between the features of K_i models and the Q questioned document. The expert observes a set of graphometric features in the n reference samples and looks for them in the analysis of the questioned document. By the end of the analysis, he provides a similarity index between them (questioned and reference documents) and subsequently, makes a decision. The resulting expert report depends on the sum of the results obtained from individual comparisons of pairs of documents (reference/questioned).

2.1   Textures and Graphometric Features

One of the texture's most important properties is the preservation of the dynamic behavior of the writer's stroke. Table 2 shows that generic and genetic attributes can be observed closely by watching a texture image generated from a manuscript. Therefore, in this study, we seek to associate the observable properties of texture to graphometric

Table 2. Relationship between graphometric features and the observable behavior of compressed blocks of handwritten text in texture form.

Graphometric Feature	Handwriting	Texture	
Onrush and upshot			
Development or progression			
Slant			
Caliber			
Curvilinear and angular values			
Proportion			
Pressure			
Moments			

features. Thus, two different approaches were used to obtain texture descriptors, namely, model-base and statistical.

2.2   GLCM and Haralick's Descriptors

The descriptors proposed by Haralick [14] are statistical descriptors whose properties seek to model the representation of irregular texture patterns. The set consists of 13 descriptors. They seek to statistically determine the distribution properties and the relationship between the grayscale contained in a texture. The descriptors are calculated by using GLCM obtained from the image of the texture under analysis.

The works based on GLCM demonstrated that among the various descriptors of Haralick [14], entropy has stood out in the approaches where the goal is to find authorship. Entropy describes the randomness of the pixel distribution in the texture. In manuscripts, this randomness is closely related to the writer's habits during the writing process.

Entropy E can be defined by the following equation:

$$E = -\sum_{i=0}^{n} \sum_{j=0}^{m} p(i,j).\log(p(i,j)), \tag{1}$$

where $p(i, j)$ denotes the matrix generated by GLCM and $n\neg$ and m represent the maximum dimensions of the matrix.

2.3 Fractal Geometry

Fractals are elementary geometric shapes, whose pattern replicates indefinitely, creating complex objects that preserve, in each of their parts, features of the whole. Two of the most important properties of fractals are self-similarity and non-integer dimension [13]. Self-similarity is evident when a portion of an object can be seen as a replica of the whole, on a smaller scale. The non-integer dimension is equivalent to modeling phenomena only representable in geometric scales in the range between integer values.

The property that combines fractal geometry with graphometric features is the possibility of mapping the structural behavior of the writer's stroke, such as the slant The relationship between different dimensions S can be defined as follows:

$$S = L.D, \tag{2}$$

where L denotes the linear scale and D represents the fractal dimension.

The relationship between dimension D and linear scaling in a result of increasing size S can be generalized as follows:

$$D = \frac{log(S)}{log(L)} \tag{3}$$

Samarabandu [11] presents a fractal model based on mathematical morphology. Developed by Serra [12], this model uses a series of morphological transformations to analyze an image. The method proposed by Samarabandu was chosen because of its simple implementation and performance compared with other techniques proposed in the literature.

Consider a grayscale image X as a surface in three dimensions, whose height represents the scale of gray at each point. The surface area in different scales is estimated using a number of dilations of this surface by a structuring element Y, whose size determines the range ρ. Thus, one can define the surface area of a compact set X regarding a structural element Y, symmetrical to its origin, by using the following function:

$$S(X,Y) = lim_{\rho \to 0} \frac{V(\vartheta X \oplus \rho Y)}{2\rho} \tag{4}$$

where V denotes the surface, ϑX denotes the edge of set X and $\oplus$ represents the dilation of the edge of set X by the structural element Y, in the ρ scale.

3   Materials and Methods

The proposed method is divided into two main stages, namely, the training procedure and the expert examination procedure, as shown in Fig. 2. Each stage is divided into processes that can be repeated in both of them, such as compression of the manuscript image and extraction of texture descriptors, both detailed in the following sections.

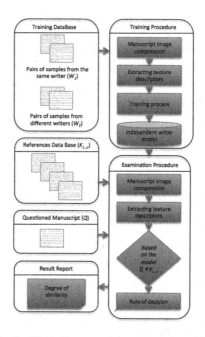

Fig. 2. Block diagram of the proposed method.

3.1   Acquisition and Preparation of Manuscript Database

The database chosen for this work was the IAM Handwriting Database [1, 3–5, 7, 8]. The IAM database was chosen because it is an international database widely used in works related to manuscripts.

The abovementioned database includes 1,539 documents from 657 different authors. One document sample from the database is shown in Fig. 3. The number of letters by author varies between 1 and 59 copies. For this work, the header and the footer were removed, as displayed in the rectangles in Fig. 3.

Fig. 3. Example of manuscript from the IAM database.

The documents were then submitted to the compression process presented by Hanusiak [2]. The resulting image was fragmented into rectangles of 128 × 128 pixels, as shown in Fig. 4.

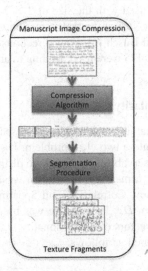

Fig. 4. Example of manuscript from the IAM database.

Each set of manuscripts by an author generated an average of 24 fragments measuring 128 × 128 pixels. These images were then separated into training and testing databases and used in the process of obtaining the texture descriptors.

For generation of the model, the writer-independent approach [1, 2] was used. This implied that the feature vectors generated from the combination of two samples of the same writer formed the sample set called "authorship" (W_1). Vectors generated from the combination of samples of different writers formed the set referred to as "non-authorship" (W_2). The writers who participated in training the model were not used in the test. This model has the advantage of not requiring new training when a new writer is inserted into the writer database. Sets of 50, 100, 200, and 410 authors were tested in the training.

Samples of the test base, consisting of 240 writers, were divided into two sets. The first was allocated to the reference set $K_{1...n}$, with n ranging from 3 to 9 fragments. The remaining author samples were allocated to the test, appearing as questioned document Q.

3.2   Extraction of Texture Descriptors

The features used are based on fractal geometry, using fractal dimension and GLCM through entropy.

In this stage, the calculation of the fractal dimension was applied on the fragments of compressed manuscript images. The calculation was repeated on the images in different gray levels to thereby form the feature vector. Levels 2, 4, 8, 16, 32, 64, 128, and 256 were used, leading to an array of eight components.

The Samarabandu [11] algorithm was used with $\rho = 6$ and structuring element Y crosswise with initial size 1. These values were chosen in the preliminary tests.

For entropy, GLCM was used with the size of 2×2, applied to a fragment of the image in black and white. The size of the matrix was chosen on the basis of the results obtained by Hanuziak [2] and Busch [6]. Four directions with $q = 0°, 45°, 90°, 135°$, and five distances $d = 1, 2, 3, 4, 5$ were used, leading to an m vector of 20 components.

3.3   Calculation of Dissimilarity

The dissimilarity between feature vectors was proposed by Cha [9] to transform a problem of multiple classes into a two-class problem. Dissimilarity was obtained from the difference between the observable features in two distinct fragments of text. For the calculation of this difference, a measure of distance was used, such as the Euclidean distance, Cityblock, Chebychev, Correlation, and Spearman [1, 2]. For this work, the Euclidean distance was chosen $D_{x,y}$. The choice was based on preliminary tests.

X and Y are two feature vectors of dimensionality m. The dissimilarity vector $D_{x,y}$ between X and Y can be defined as follows:

$$D_{x,y} = \bigcup_{i=1}^{m} \left(\sqrt{(x_i - y_i)^2} \right) \tag{5}$$

3.4   Rule of Decision

The concept of similarity establishes that the distance between the measurements of the features of the same writer tends to zero (similarity), while the distance between the measurements of the features of different writers tends to a value far from zero (dissimilarity). Consequently, the feature vector generated by these distances can be used by the SVM for classification purposes. Therefore, on the basis of the training model, the SVM is then able to classify it into authorship and non-authorship.

Each result of the comparison of the questioned manuscript Q with one of the references K is then combined in a process of merging the results. With the number of references greater than 1 ($K \geq 1$), it is possible to define a decision criterion that takes into account the results from the individual comparison of each reference to the questioned sample. The concept presented here is the same used by an expert. There are several decision rules proposed in the literature, such as majority vote, the sum of the results, and the average of the results [1, 2]. In this work, a combination of the linear kernel in the SVM and the sum of the results as the decision rule were used, as they presented the best results in the preliminary tests.

4 Results

The following are the results of the tests performed on the IAM database.

The first test (Table 3) aimed to observe the behavior of the variation of the training set, given that this is a writer-independent model. Under these conditions, the result was expected to show significant improvement with an increase in the number of training samples in both classes, W_1 and W_2. However, the results showed that even when a significantly small number of writers were used in the training, the results ranged around 1.5 %. This was attributed in part to the dimensionality of the feature vector used, which provided stable performance even in small sets of training samples.

Table 3. Results obtained with the proposed method.

Number of references K	Performance (%) writers for training			
	50	100	200	410
3	84.8	94.6	93.8	94.0
5	96.2	96.8	96.3	96.7
7	96.4	96.8	97.1	**97.7**
9	96.3	97.4	97.1	97.2

The factor that most influenced the results was the number of samples used as reference for each writer K. The use of $K = 3$ presented the least significant result. The values of $K = 7$ and 9 had the greatest stability of the tested set. This behavior was also observed in the analysis conducted by the expert. That is, the more reference samples he evaluated, the higher was his capacity to distinguish writer variability and, therefore, the lower was the likelihood of mistakes.

The comparison of the proposed method with others presented in the literature is always an approximate approach, even when using the same database. Various aspects can interfere with this or that result, such as the classifier used, the training and testing protocol, the texture fragment size, the number of references, the feature vector dimensionality, and the text compression process. However, it is possible to establish behavior trends showing how promising the proposed method can be. Table 4 presents a comparative view of some authorship analysis methods.

Table 4. Comparison of results obtained with the methods found in the literature and the proposed method.

Method	Feature	Number of writers	Fragment size/feature	Classifier	Performance (%)
Schlapbach and Bunke [5]	Geometric	120	-/9	HMM	97.5
Bensefia et al. (2005) [7]	Graphemes	150	-/50	VSM	96.0
Schlapbach and Bunke [8]	Geometric	100	-/15	HMM	97.5
Siddiqi and Vincent [3]	Global and Local	650	-/586	Dist.X2	97.7
Bertolini et al. [1]	Texture (LPQ)	240	$256 \times 128/59$	SVM	99.6
Proposed Method	Entropy and Fractal	240	$\mathbf{128 \times 128/28}$	SVM	97.7

5 Conclusion and Future Studies

In this paper, we presented an optimized method for establishing the authorship of questioned handwritten documents by using texture descriptors. Two approaches to obtain the descriptors were used: fractal geometry, using fractal dimension; and entropy, proposed by Haralick, using GLCM. The results show the stability of the writer-independent model with respect to the number of writers used for training, the reduced number of features and the low number of samples, compared to other works.

Two other important aspects are as follows: (1) the texture-based method uses descriptors that provide a direct relationship with graphometric features and (2) the use of a low-dimensional feature vector allows training with a small number of samples to not significantly affect the performance of the method.

The results of 97.7 % obtained by using the proposed method were promising and consistent with those obtained by other studies.

In future studies, the intention is to incorporate other texture descriptors that can describe the curvilinear behavior of strokes, such as curvelets.

References

1. Bertolini, D., Oliveira, L.S., Justino, E., Sabourin, R.: Texture-based descriptors for writer identification and verification. Expert Syst. Appl. **40**, 2069–2080 (2013)
2. Hanusiak, R.K., Oliveira, L.S., Justino, E., Sabourin, R.: Writer verification using texture-based features. Int. J. Doc. Anal. Recogn. **15**, 213–226 (2012)

3. Siddiqi, I., Vincent, N.: Text independent writer recognition using redundant writing patterns with contour-based orientation and curvature features. Pattern Recogn. **43**, 3853–3865 (2010)
4. Shomaker, L.: Advances in writer identification and verification. In: Proceedings of the International Conference on Document Analysis and Recognition (ICDAR), Curitiba, Brazil (2007)
5. Schlapbach, A., Bunke, H.: A writer identification and verification system using HMM based recognizers. Pattern Anal. Appl. **10**, 33–43 (2007)
6. Bush, A., Boles, W., Sridharan, S.: Texture for script identification. IEEE Trans. Pattern Anal. Mach. Intell. **27**(11), 1721–1732 (2005)
7. Bensefia, A., Paquet, T., Heutte, L.: A writer identification and verification system. Pattern Recogn. Lett. **26**(13), 2080–2092 (2005)
8. Schlapbach, A., Bunke, H.: Using HMM based recognizers for writer identification and verification. In: Ninth International Workshop on Frontiers in Handwriting Recognition, pp. 67–172 (2004)
9. Bulacu, M., Shomaker, L.: Writer identification using edge-based directional features. In: Proceedings of 7th International Conference on Document Analysis and Recognition (ICDAR). IEEE Computer Society (2003)
10. Cha, S.H.: Use of the distance measures in handwriting analysis. Doctor Theses. State University of New York at Buffalo, EUA, 208 (2001)
11. Samarabandu, J., et al.: Analysis of bone X-rays using morphological fractals. IEEE Trans. Med. Imaging **12**(3), 466–470 (1993)
12. Serra, J.: Image Analysis and Mathematical Morphology, vol. 1. Academic Press, Cambridge (1982)
13. Mandelbrot, B.B.: The Fractal Geometry of Nature. W.H. Freeman, London (1977)
14. Haralick, R.M., Shanmugam, K., Dinstein, I.: Textural features for image classification. IEEE Trans. Syst. Cyber. **smc-3**(6), 610–621 (1973)

Information Theoretic Rotationwise Robust Binary Descriptor Learning

Youssef El Rhabi[2]([⊠]), Loic Simon[1], Luc Brun[1], Josep Llados Canet[3], and Felipe Lumbreras[3]

[1] Groupe de Recherche en Informatique, Image,
Automatique et Instrumentation de Caen Normandie Univ, UNICAEN,
ENSICAEN, CNRS, GREYC, 14000 Caen, France
{loic.simon,luc.brun}@ensicaen.fr
[2] 44screens, Paris, France
yer@44screens.com
[3] Computer Vision Center Dep. Informàtica, Universitat Autònoma de Barcelona,
08193 Bellaterra (Barcelona), Spain
{josep,felipe}@cvc.uab.es

Abstract. In this paper, we propose a new data-driven approach for binary descriptor selection. In order to draw a clear analysis of common designs, we present a general information-theoretic selection paradigm. It encompasses several standard binary descriptor construction schemes, including a recent state-of-the-art one named BOLD. We pursue the same endeavor to increase the stability of the produced descriptors with respect to rotations. To achieve this goal, we have designed a novel offline selection criterion which is better adapted to the online matching procedure. The effectiveness of our approach is demonstrated on two standard datasets, where our descriptor is compared to BOLD and to several classical descriptors. In particular, it emerges that our approach can reproduce equivalent if not better performance as BOLD while relying on twice shorter descriptors. Such an improvement can be influential for real-time applications.

1 Introduction

Since the advent of SIFT [12], extracting local descriptors has become a common practice in order to assess the similarity of image regions. Applications of local descriptors have been considerable, such as image stitching to build panoramas [5], context-based image retrieval, visual odometry or multi-view 3D reconstruction [15]. As a result of its success, this line of research has greatly impacted our everyday behaviour, be it by our use of efficient exemplar based image search engine, or the pervasive introduction of computer vision in mobile devices. Due to this important economical and societal repercussions, the design of ever improving descriptors has drawn a strong interest [4,14]. One of the main enhancements relates to data-driven construction schemes, where a typical database of image correspondences is leveraged to learn an efficient descriptor [8,21].

© Springer International Publishing AG 2016
A. Robles-Kelly et al. (Eds.): S+SSPR 2016, LNCS 10029, pp. 368–378, 2016.
DOI: 10.1007/978-3-319-49055-7_33

In particular, recent approaches based on deep learning techniques [25] have shown a strong improvement on the state of the art.

However, some kind of "no free lunch" principle applies in that quest. Depending on the targeted application, the desired properties of the descriptor may differ significantly, leading to several trade-offs and design principles. Among others, the following questions are recurrent. Is the computational complexity of paramount importance? Does accuracy matter more than completeness? What class of invariance is required? For instance, in context-based image retrieval, a query image is proposed to a system that should propose several images similar to the query. But often semantic similarity is more crucial than purely visual resemblance. On a different note, perspective or affine invariance is a desirable asset in a multi-view reconstruction system but not in a tracking scenario. These central questions become further more complicated in practice, since descriptors are no more than a brick in a complex pipeline. Therefore, some properties of the descriptors can be destroyed or corrected by other parts of the systems. For instance, invariance can be embedded in the design of a descriptor or provided by detecting an orientation and scale before computing a non invariant descriptor. A more sophisticated case is exposed in [7], where the authors acknowledge the benefit of binary descriptors for real-time applications but claim that in a 3D reconstruction system, typical descriptors like SIFT provide a better compromise between accuracy and run time, in particular when matching is accelerated thanks to adapted data structures [16].

In this article, we intend to improve the state-of-the-art of descriptors with real-time applications in mind (e.g. SLAM). We therefore focus on low-complexity binary descriptors based on image intensity comparisons [1,6]. This active line of research lies at the crossroad of several intertwined areas such as feature selection [18,20] and hashing [9]. Our contributions include a clear exposition of a generic framework encompassing the typical state-of-the-art descriptor pipelines, as well as the design of an elegant information theoretic criterion used in the feature selection process. It yields a consensus between the discriminative power of the descriptor and its resilience to rotations. This contribution is evaluated on classical benchmarks and decreases the time and space complexity by a factor 2 compared to a recent state-of-the-art technique [3].

2 State of the Art

Binary descriptors have been in the spotlight during the past decade. Indeed these descriptors come with two central properties: low memory footprint and efficient computation. As a rule of thumb, binary descriptors require up to 512 bit storage, while full-spectrum descriptors typically involve 512 floating point values (32 times larger memory). To reduce the memory requirement one may apply dimensionality reduction and hashing techniques to a full-spectrum descriptor [10,21]. On the contrary, binary descriptors skip the full-spectrum descriptor and produce directly a reduced number of simple binary features (a.k.a tests). As a result, they are not only cheap to store, but are also faster to compute.

The two key strengths of binary descriptors come at a price, namely a lower distinctness. Therefore the main line of research in this area aims at increasing the expressive power of binary descriptors while maintaining a good trade-off in terms of memory and complexity. Attempts in that direction are numerous, and we will hereby extract a few representative ones. Two early instances of binary descriptors are CENSUS and LBP [17,24]. They are based on the systematic comparison of the pixels in a neighborhood to the central pixel. The two methods differ by the shape of the neighborhood: a full-square for CENSUS and a ring for LBP. Such procedures produce a binary string whose length depends directly on the size of the neighborhood. Therefore, in order to remain short and fast, the descriptor must be computed on a small neighborhood, which in turns restricts its distinctness.

BRIEF [6] is a recent approach to tackle the trade-off between locality and efficiency. It is built upon ideas of Locally Sensitive Hashing [9]. It relies on a random pre-selection of pairs of pixels within a large neighborhood. Afterward, the descriptor of a patch is computed by aggregating the results of the binary comparison applied on each pixel pair. In that way, the size of the neighborhood and the length of the descriptor can be chosen independently (e.g. 32×32 and 512). The authors of ORB [19] argue that the selection mechanism should account for the typical data distribution. They propose a principled scheme to select good binary features. Their approach operates in a greedy manner allowing to select uncorrelated features of maximum variance. In Sect. 3, we will give an interpretation of this procedure as a maximization of the overall information quantity. What matters most is that the variance and correlation are estimated on a representative database. In that way, the trade-off between the descriptor complexity and its expressive power is sought according to the data distribution. In addition, some authors guide the feature selection thanks to other principles. For instance, in BRISK [11], a set of concentric sampling points are selected and pairs are created from any two points that are not too far apart. Similarly, FREAK [1] designs a concentric pattern that mimics the retinal layout by having higher sampling density near the center. Then the greedy selection from ORB is used to select good pairs.

Making binary descriptors invariant to natural transformations represents also an important task. By construction, descriptors based on local differences are invariant to contrast changes. On the contrary, noise is by default badly tackled. This can be compensated by pre-smoothing the image patch before computing the descriptor. More sophisticated binary tests were also designed such as in LDB [23] where pixel data are aggregated on cell grids. As for geometric transformations, their impact can be efficiently neutralized if the main orientation (and scale) of the feature is estimated. This is the case for instance in ORB, or in AKAZE [2]. More recently, the authors of BOLD [3] have proposed an alternative where the robustness is introduced by using an online feature selection. The results are compelling and motivated our own work.

In Sect. 3, we establish the general framework underlying our approach and present it in details. Then, we provide an in-depth analysis of the observations that have led to our formulation. In Sect. 4, we demonstrate the benefits of our contributions on several standard benchmarks. We first compare our descriptor with BOLD, its most direct challenger. In a nutshell, our contributions allows us to achieve similar performance while using half as many features. In addition, we provide also a comparison with a larger collection of classical descriptors such as SIFT [12], SURF [4], LIOP [22] and BRISK [11].

3   Binary Descriptors Construction Scheme

In this section we describe a generic information theoretic framework that encompasses most binary descriptors based on feature selection. In particular, we illustrate how BOLD descriptor can be recovered as a special case, and we analyse its limitations. Based on this analysis, we lay out a possible extension. In what follows we denote patches by the letter $w \in \mathbb{R}^{s \times s}$. A test is a function $t : \mathbb{R}^{s \times s} \to \{0, 1\}$ that maps any patch to a binary value $t(w)$. A database is a collection of patches $\mathcal{D} = \{w_1, \cdots, w_d\}$ drawn from a common (unknown) distribution. Given N tests $t_1, \cdots, t_N$, we denote $x_{k,i} = t_i(w_k)$ the collection of binary samples obtained by applying each individual test to all the patches. For convenience, we denote generically by an upper case W a random variable following the underlying patch distribution, and $X_t = t(W)$ the Bernoulli random variable induced by the test t.

3.1   Global Framework

Our main purpose is, given two patches w_1, w_2, to decide if they are similar up to some allowed transformations. For that we compute a distance d between w_1 and w_2 and match them based on a hard threshold. In practice, binary descriptors x_1 and x_2 are computed for both patches, and the distance is computed between these descriptors. Learning a good metric boils down to selecting good features which is typically done in an offline procedure. In this phase the main goal is to choose a fixed number N of tests which bring as much information as possible. Ideally those tests will be chosen by optimally learning them on a representative database. For that purpose a greedy approach is convenient, where tests are selected iteratively by maximizing a measure tied to the information quantity of the new test given the previously selected ones. This procedure is detailed in Algorithm 1. Overall, the metric construction scheme relies on an offline and an online procedure. Each of them is driven by a criterion: $J_{\mathcal{D}}(t|\mathcal{S})$ for the offline procedure and $d(x_1, x_2)$ for the online one. In Sect. 3.2 we present some common ways to evaluate those two criteria.

3.2   Usual Selection Mechanisms

In order to proceed to the offline test selection we need to set a criterion $J_{\mathcal{D}}(t|\mathcal{S})$. $J_{\mathcal{D}}(t|\mathcal{S})$ can be an estimate of the conditional entropy $H(X_t|X_{\mathcal{S}})$ and $X_{\mathcal{S}}$ is the collection of random variables induced by the current set of selected tests. However, computing the conditional entropy requires estimates of joint probabilities which can be unreliable. Some conditional independence assumptions (*e.g.*

Algorithm 1. Offline algorithm

---

    input   : Image patches dataset $\mathcal{D}$
    output : $\mathcal{S}$ selected tests
1 generate a pool $\mathcal{P} = \{t_1, \cdots, t_M\}$ of M random tests   // $M \gg N$
2 $\mathcal{S} = \varnothing$
3 **for** $i = 1..N$ **do**
4 | $t_i^* = \arg\max_{t \in \mathcal{P}}(J_{\mathcal{D}}(t|\mathcal{S}))$
5 | $\mathcal{S} = \mathcal{S} \cup \{t_i^*\}$

---

pairwise dependence) can make this task more scalable but it remains computationally intensive. Practitioners [19] often prefer to fall back to related criteria of the form: $J_{\mathcal{D}}(t|\mathcal{S}) = J_{\mathcal{D}}(t) - \infty \mathbb{1}_{|max_{s \in \mathcal{S}}(\text{corr}(X_t, X_s))| > \tau}$. Such a criterion lends itself to an efficient implementation. One may, for example, maximize the first part $J_{\mathcal{D}}(t)$ among the tests that comply with the hard correlation thresholding constraints. $J_{\mathcal{D}}(t)$ can be the entropy, but other measures exist. For instance $J_{\mathcal{D}}(t) = \text{var}(X_t)$ or $J_{\mathcal{D}}(t) = |\mathbb{E}(X_t) - 0.5|$ are preferred. Besides, with Bernoulli variables all those measures are equivalent in terms of maximisation.

In order to compute the online distance between patches w_1 and w_2, the typical matching procedure starts by computing the Hamming distance, $d_{ham}(x_1, x_2)$, between the test results $x_1, x_2 \in \{0,1\}^N$:

$$d_{ham}(x_1, x_2) = \sum_{i=1}^{N} x_{1,i} \oplus x_{2,i} \tag{1}$$

where $\oplus$ is the *XOR* operator and the sum can be efficiently computed thanks to the *popcount* routine.

BOLD improves the robustness to natural transformations by using an online-selection strategy leading to the derivation of a masked Hamming distance. This is done by computing p transformed versions $w_k^1, \cdots, w_k^p$ of each patch w_k ($k \in \{1,2\}$). Then a mask $y_k \in \{0,1\}^N$ allowing to filter out non robust test bits is built as follows[1]:

$$y_{k,i} = \bigoplus_{j=1}^{p} x_{k,i}^j \text{ with } x_{k,i}^j = t_i(w_k^j) \tag{2}$$

Based on this criterion a masked Hamming distance is constructed taking into account only those tests that are robust to the chosen deformations:

$$d_{masked}(x_1, x_2; y_1, y_2) = \sum_{i=1}^{N} \lambda_1 y_{1,i} \wedge (x_{1,i} \oplus x_{2,i}) + \lambda_2 y_{2,i} \wedge (x_{1,i} \oplus x_{2,i}) \tag{3}$$

where the weights[2] are given by $\lambda_k = \frac{|y_k|}{|y_1| + |y_2|}$ with $|y_k|$ the number of 1's in y_k.

---

[1] In the formula, $\bigoplus$ denotes the n-ary *XOR* (true when all its arguments are equal).
[2] The formula corresponds to the implementation provided by the authors of BOLD and the weights are different from those exposed in their article.

3.3   Analysis of Information Distribution

Data-driven selection methods implicitly rely on the fact that certain distribution estimates generalize well across databases. In particular, in order to preserve tests carrying much information, it is important that the probability estimates p_{x_t} of success of test t should remain consistent independently of the dataset. In Table 1-(a), we present the linear regression between $p_{x_t}^{\mathcal{D}}$ and $p_{x_t}^{\mathcal{D}'}$ for several pairs of databases $(\mathcal{D}, \mathcal{D}')$ from [8]. It shows with high level of confidence that their relationship is well approximated by the identity function.

Setting the number N of selected tests leads to a trade-off between performance and computation time. Besides after a critical number of tests, a saturation point is typically observed, where performance stalls and eventually worsen. Such a phenomenon is shared by data-driven approaches beyond binary descriptors. As an example, in the dimensionality reduced GLOH descriptor [14], the results are worse for a 272 dimension descriptor than for the 128 alternative. This saturation can be observed in Fig. 1-(a) for an offline selection maximizing the test variance. At the saturation, BOLD gets better performance by ignoring bits that are not resilient to some natural transformations. This feature selection is done entirely online because the resilience of a test depends on the chosen patch. A saturation phenomenon can still be observed, as shown in Fig. 1-(b). In this figure, once $N = 512$ tests are selected then no gain in performance is noticeable.

Table 1. Linear regressions (with 95 % confidence intervals) in the form $p^{\mathcal{D}'} = ap^{\mathcal{D}} + b$ with Yosemite as $\mathcal{D}$ and two alternative datasets $\mathcal{D}'$.

$\mathcal{D}'$		estimate	l.b. 95%	u.b. 95%	$\mathcal{D}'$		estimate	l.b. 95%	u.b. 95%
ND	a	1.083	1.078	1.088	**ND**	a	1.007	1.001	1.014
	b	-0.043	-0.046	-0.040		b	0.016	0.012	0.020
Liberty	a	1.004	0.997	1.011	**Liberty**	a	0.938	0.930	0.946
	b	-0.0002	-0.004	0.004		b	0.036	0.031	0.041

$$(a)\ p_{x_t}^{\mathcal{D}}\ \text{vs}\ p_{x_t}^{\mathcal{D}'} \qquad\qquad (b)\ p_{y_t}^{\mathcal{D}}\ \text{vs}\ p_{y_t}^{\mathcal{D}'}$$

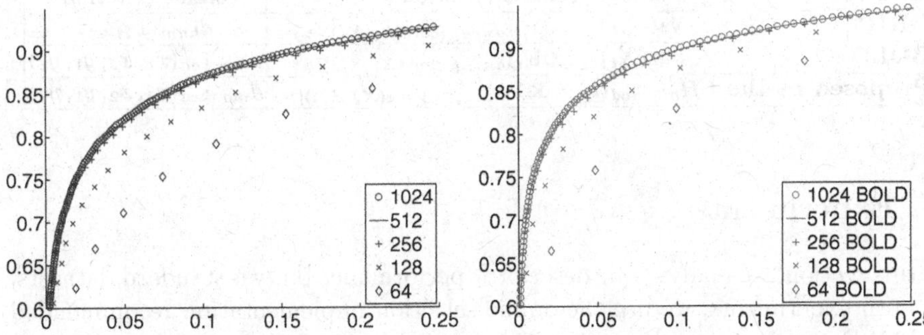

(a) Offline selection   (b) BOLD online selection

Fig. 1. Saturation on the ROC curves for the offline selection and BOLD.

3.4 Proposed Approach

Even though the online selection performs on a per patch basis, it is interesting to note that some tests are statistically more robust to geometric transforms than others. This fact becomes manifest when observing the generalization of the probabilities p_{y_t} that a test t is kept by the online selection (see Table 1-(b) for linear regressions across datasets). Since p_{y_t} generalizes well, tests that are robust to geometric transforms can also be learnt offline. We want to take this online filtering into account in the offline selection. Otherwise, information quantity as estimated offline misrepresents the actual information kept during the online phase. We propose a modified information quantity measure $H_{masked}(t)$. It serves as another way to define $J_\mathcal{D}(t)$. Therefore $H_{masked}(t)$ is an offline criterion but it is designed to take into account the online selection proposed by BOLD:

$$H_{masked}(t) = -[p_{x_t} \log(p_{x_t}) + (1 - p_{x_t}) \log(1 - p_{x_t})] \times p_{y_t} \qquad (4)$$

where $p_{x_t} = p(X_t = 1)$ is the estimated probability that test t is successful. The interpretation of Eq. 4 is straightforward. On the one hand, $-(p_{x_t} \log(p_{x_t}) + (1 - p_{x_t}) \log(1 - p_{x_t}))$ represents the expected information quantity for bit t irrespectively of the online selection. On the other hand, it is multiplied by p_{y_t} since information will be thrown away with probability $1 - p_{y_t}$. A similar definition of a conditional information quantity is possible. Nonetheless, for computational purpose, we choose to rely on a hard decorrelation scheme, and use only the marginal definition which is set as a substitute for $J_\mathcal{D}(t)$. This measure is intended to strengthen the overall information flow after the online selection. This asset shall be confirmed by experiments hereafter.

In this section we have set a generic framework that encompasses most of the state of the art test selection based descriptors. In Table 2 we show how online and offline criteria can be combined so as to retrieve some state of the art detectors as well as ours.

Table 2. Description of several methods that derivates from the generic framework

method	offline criterion	online criterion
ORB	$var(X_t) - \infty \mathbb{1}_{|max_{s \in S}(corr(X_t, X_s))| > \tau}$	$d_{ham}(x_1, x_2)$
BOLD	$var(X_t) - \infty \mathbb{1}_{|max_{s \in S}(corr(X_t, X_s))| > \tau}$	$d_{masked}(x_1, x_2; y_1, y_2)$
Proposed method	$H_{masked}(t) - \infty \mathbb{1}_{|max_{s \in S}(corr(X_t, X_s))| > \tau}$	$d_{masked}(x_1, x_2; y_1, y_2)$

4 Experiments

In this section, we analyse our descriptor performance on two standard datasets. In both experiments, we use the online selection implementation recommended in BOLD: we draw $p = 3$ rotations (up to $20°$) in Eq. 2.

Photo Tourism Dataset: First, we present the evaluation results on the dataset proposed in [8] with the evaluation protocol of [21]. This protocol uses 3

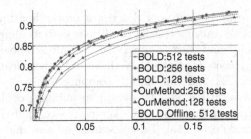

Fig. 2. ROC curves for our descriptor and BOLD under different regimes. (Color figure online)

datasets (Liberty, Notre Dame, Yosemite). The groundtruth on these datasets is encoded through correspondences between pairs of patches. Half of those correspondences are correct matches while the other half correspond to non matches. Interest points are detected with usual detectors (such as differences of Gaussians) and matches are found using a multi-view stereo algorithm as detailed in [21]. In this evaluation We compare our method specifically to the BOLD descriptor (based on their original implementation). In Fig. 2 descriptors were trained on the Yosemite dataset and were tested on 130 k patches of size 32×32 from Notre Dame. This figure highlights that our method with 256 tests, yields as good results as BOLD with 512 tests. On the contrary, BOLD with 256 tests yields lower results. In addition, the first tests selected by our approach are more informative than the ones produced by BOLD. Indeed, we can observe a substantial gap between both approaches when 128 tests are aggregated. Our descriptor performs close to the online saturation level, while BOLD is closer to the offline selection regime (in green).

We tested our approach and BOLD under different configurations (training/testing combinations). We reported area under ROC curves in Tables 3a, b and c. In all configurations, we obtain as good if not better results than BOLD. In particular, with short descriptors our approach is significantly superior to BOLD. Also BOLD reaches the saturation at a slower rate. In a nutshell, saturation occurs around 256 tests for our descriptor against 512 for BOLD.

Table 3. Area under PR curves (values are rounded at 3 decimals).

bits	Notre Dame Bold	Notre Dame Us	Liberty Bold	Liberty Us	Yosemite Bold	Yosemite Us	Liberty Bold	Liberty Us	Notre Dame Bold	Notre Dame Us	Yosemite Bold	Yosemite Us
1024	0.959	0.959	0.941	0.941	0.951	0.952	0.939	0.940	0.957	0.959	0.952	0.954
512	0.958	0.959	0.941	0.942	0.952	0.954	0.939	0.943	0.956	0.960	0.953	0.955
256	0.957	0.959	0.941	0.942	0.951	0.955	0.939	0.943	0.955	0.958	0.952	0.954
128	0.950	0.954	0.934	0.939	0.946	0.951	0.933	0.940	0.947	0.953	0.944	0.949
64	0.932	0.943	0.916	0.929	0.930	0.941	0.916	0.931	0.933	0.940	0.931	0.938

(a) Training on Yosemite   (b) Training on Notre Dame   (c) Training on Liberty

Vgg Dataset: Here we evaluate our descriptor on the benchmark proposed in [14]. This dataset offers varying testing conditions such as illumination changes, rotations, zooms, blur and jpeg compression. We also compare our descriptor with standard and recent descriptors. We have used the vlfeat implementation of SIFT and LIOP [22], and the Matlab computer vision toolbox implementation of BRISK and SURF [4]. To obtain meaningful comparisons, all the descriptors are extracted from the same key-points computed with the multiscale Harris-Laplace detector [13]. Table 4 shows the area under ROC curve for several descriptors and image pairs in a nearest neighbor matching scenario. Since our contributions relate to binary feature selection and additional robustness with respect to rotations, we have organised the table as follows. Vertically, columns are ordered according to the level of orientation change in the image pair[3]. We have extracted 3 characteristic rotation regimes. Then horizontally, the binary descriptors are separated from full-spectrum ones. Among the binary descriptors, we consider a recent handcrafted method (BRISK), as well as a few data-driven variants lying within our framework. The chosen variants implement several mechanisms to handle rotations. The first two variations correspond to our descriptor (with 512 or 256 tests) and to BOLD. Then the two remaining variants rest on the offline selection only. The first one, compensates for the orientation of the patch while the other is applied directly on the patch. All full-spectrum descriptors rely by design on explicit rotation compensation. We have highlighted in bold the best results among binary and full-spectrum descriptors.

In the first regime (very small orientation changes), the online selection (BOLD, Us512, Us256) decreases but slightly the performance as compared to the offline-only selection. Explicit compensation of the rotation exacerbates the result deterioration. This observation echoes the fact that enforcing unneeded invariance can be harmful. The second regime (medium angles roughly below

Table 4. Area under ROC curves with a nearest neighbour matching scenario.

pair angle	ubc 1:5 0°	bikes 1:5 6°	leuven 1:5 7°	boat 1:5 8°	boat 1:2 14°	bark 1:2 31°	boat 1:4 79°	bark 1:4 120°
Us512	0.880	0.828	0.820	**0.619**	**0.720**	0.179	0	0
Us256	0.877	0.832	0.827	0.597	0.703	0.194	0	0
Bold	0.878	0.827	**0.840**	0.605	0.719	0.157	0	0
offline oriented	0.859	0.808	0.796	0.592	0.676	**0.705**	**0.592**	**0.677**
offline	**0.883**	**0.839**	0.830	0.479	0.359	0.012	0	0
BRISK	0.779	0.648	0.647	0.406	0.594	0.668	0.382	0.574
SIFT	**0.865**	**0.859**	**0.880**	**0.749**	0.698	0.801	**0.707**	**0.807**
SURF	0.711	0.604	0.553	0.404	0.516	0.582	0.381	0.418
LIOP	0.815	0.792	0.754	0.588	0.662	**0.804**	0.539	0.695

---

[3] Apart from JPEG experiments (UBC pair), all the image pairs correspond to two independent camera shots and present varying degrees of geometric changes.

20°) is the one for which the online selection was designed. As a matter of fact, this is the mode where BOLD and our descriptors excel among binary ones. They even compete favorably with full-spectrum descriptors. Looking more closely, the orientation compensation is less efficient than the online selection. Besides, our method in this regime performs better than BOLD thanks to the modified entropy proposed in Eq. 4. In the third regime, the online selection cannot tackle the intensity of the underlying rotations. Here only explicit rotation compensation is fruitful, with a large advantage for SIFT. Apart from a single exception, the comparison with BOLD is uniformly at our advantage. This was confirmed by synthetic experiments we carried out on pure rotations, especially for ones between 15° and 30°.

5   Conclusion

In this article we have developed a novel binary image descriptor that finds its roots in the context of real time applications. To construct this descriptor, we have laid a common foundation based on feature selection. This framework covers most of the recent data-driven binary descriptors including a recent one called BOLD. We have also complemented the online selection mechanism proposed in BOLD by an adapted offline criterion applied beforehand. This new mechanism presents an elegant information theoretic interpretation and above all a perceptible practical influence. The immediate comparison to BOLD conveys that in most cases, our descriptor carries as much useful information while being twice more compact. Such an asset is an important benefit in the considered applications. Comparisons to a few other classical descriptor show that our approach obtains favorable results under mild geometric transforms. This situation arises easily in applications on mobile devices where guessing a rough estimate is often possible thanks to additional sensors. Our descriptor is therefore a perfect fit for real-time applications.

References

1. Alahi, A., Ortiz, R., Vandergheynst, P.: Freak: fast retina keypoint. In: CVPR, pp. 510–517. IEEE (2012)
2. Alcantarilla, P.F., Nuevo, J., Bartoli, A.: Fast explicit diffusion for accelerated features in nonlinear scale spaces. In: BMVC (2013)
3. Balntas, V., Tang, L., Mikolajczyk, K.: Bold-binary online learned descriptor for efficient image matching. In: CVPR, pp. 2367–2375 (2015)
4. Bay, H., Tuytelaars, T., Gool, L.: SURF: speeded up robust features. In: Leonardis, A., Bischof, H., Pinz, A. (eds.) ECCV 2006. LNCS, vol. 3951, pp. 404–417. Springer, Heidelberg (2006). doi:10.1007/11744023_32
5. Brown, M., Lowe, D.G.: Automatic panoramic image stitching using invariant features. IJCV 74(1), 59–73 (2007)
6. Calonder, M., Lepetit, V., Strecha, C., Fua, P.: BRIEF: binary robust independent elementary features. In: Daniilidis, K., Maragos, P., Paragios, N. (eds.) ECCV 2010. LNCS, vol. 6314, pp. 778–792. Springer, Heidelberg (2010). doi:10.1007/978-3-642-15561-1_56

7. Fan, B., Kong, Q., Sui, W., Wang, Z., Wang, X., Xiang, S., Pan, C., Fua, P.: Do we need binary features for 3D reconstruction? arXiv:1602.04502 (2016)
8. Hua, G., Brown, M., Winder, S.: Discriminant embedding for local image descriptors. In: ICCV, pp. 1–8. IEEE (2007)
9. Indyk, P., Motwani, R.: Approximate nearest neighbors: towards removing the curse of dimensionality. In: Proceedings of 30th Annual ACM Symposium on Theory of Computing, pp. 604–613. ACM (1998)
10. Ke, Y., Sukthankar, R.: PCA-SIFT: a more distinctive representation for local image descriptors. In: CVPR, vol. 2, pp. II-506. IEEE (2004)
11. Leutenegger, S., Chli, M., Siegwart, R.Y.: Brisk: binary robust invariant scalable keypoints. In: ICCV, pp. 2548–2555. IEEE (2011)
12. Lowe, D.G.: Object recognition from local scale-invariant features. In: ICCV, vol. 2, pp. 1150–1157. IEEE (1999)
13. Mikolajczyk, K., Schmid, C.: Scale & affine invariant interest point detectors. IJCV 60(1), 63–86 (2004)
14. Mikolajczyk, K., Schmid, C.: A performance evaluation of local descriptors. PAMI 27(10), 1615–1630 (2005)
15. Moulon, P., Monasse, P., Marlet, R.: Global fusion of relative motions for robust, accurate and scalable structure from motion. In: ICCV, pp. 3248–3255 (2013)
16. Muja, M., Lowe, D.G.: Fast matching of binary features. In: CRV. IEEE (2012)
17. Ojala, T., Pietikäinen, M., Harwood, D.: A comparative study of texture measures with classification based on featured distributions. Pattern Recogn. 29(1), 51–59 (1996)
18. Peng, H., Long, F., Ding, C.: Feature selection based on mutual information criteria of max-dependency, max-relevance, and min-redundancy. PAMI 27(8), 1226–1238 (2005)
19. Rublee, E., Rabaud, V., Konolige, K., Bradski, G.: ORB: an efficient alternative to SIFT or SURF. In: ICCV, pp. 2564–2571. IEEE (2011)
20. Sechidis, K., Nikolaou, N., Brown, G.: Information theoretic feature selection in multi-label data through composite likelihood. In: S+SSPR, pp. 143–152 (2014)
21. Trzcinski, T., Christoudias, M., Fua, P., Lepetit, V.: Boosting binary keypoint descriptors. In: CVPR, pp. 2874–2881 (2013)
22. Wang, Z., Fan, B., Wu, F.: Local intensity order pattern for feature description. In: ICCV, pp. 603–610. IEEE (2011)
23. Yang, X., Cheng, K.T.: Ldb: An ultra-fast feature for scalable augmented reality on mobile devices. In: ISMAR, pp. 49–57. IEEE (2012)
24. Zabih, R., Woodfill, J.: Non-parametric local transforms for computing visual correspondence. In: Eklundh, J.-O. (ed.) ECCV 1994. LNCS, vol. 801, pp. 151–158. Springer, Heidelberg (1994). doi:10.1007/BFb0028345
25. Zagoruyko, S., Komodakis, N.: Learning to compare image patches via convolutional neural networks. In: CVPR, pp. 4353–4361 (2015)

Markov–Gibbs Texture Modelling with Learnt Freeform Filters

Ralph Versteegen[⊠], Georgy Gimel'farb, and Patricia Riddle

Department of Computer Science, The University of Auckland,
Auckland, New Zealand
rver017@aucklanduni.ac.nz, {g.gimelfarb,p.riddle}@auckland.ac.nz

Abstract. Energy-based Markov–Gibbs random field (MGRF) image models describe images by statistics of localised features; hence selecting statistics is crucial. This paper presently a procedure for searching much broader than typical families of linear-filter-based statistics, by alternately optimising in continuous parameter space and discrete graphical structure space. This unifies and extends the divergent models deriving from the well-known Fields of Experts (FoE), which learn parametrised features built on small linear filters, and the constrasting FRAME (exponential family) approach which iteratively selects large filters from a fixed set. While FoE is limited by computational cost to small filters, we use large sparse (non-contiguous) filters with arbitrary shapes which can capture long-range interactions directly. A filter pre-training step also improves speed and results. Synthesis of a variety of textures shows promising abilities of the proposed models to capture both fine details and larger-scale structure with a low number of small and efficient filters.

1 Introduction

An increasing variety of energy-based models have been proposed for image and texture modelling, motivated by their applicability to both generation (e.g. image synthesis and inpainting) and inference (e.g. classification and segmentation), and as building blocks of higher-level computer vision systems. These models estimate the density function of the data and depend on the selection of features whose statistics identify relevant data attributes. In particular, traditional statistical texture models are maximum-entropy Markov–Gibbs random fields (MGRFs), i.e. MRFs with Gibbs probability distributions, with parameters adjusting the strength of the Gibbs factors/potentials which are learnt by maximum likelihood estimation (MLE). Recently a number of non-max-entropy models have been proposed with much more complex parameterised potentials (usually interpretable as compositions of linear filters) and which often also incorporate implicit or explicit latent variables. Fields-of-Experts (FoE) [17] and restricted Boltzmann machines (RBMs) [9] are influential models of this class. Although not the max-entropy solutions, the MLEs are computed identically.

Learning of MGRFs by "model nesting" [2,21,23], also known as the minimax entropy principle [22,23], is the application of the max-entropy principle to iterative model selection by repeatedly adding features/potentials to a base model.

© Springer International Publishing AG 2016
A. Robles-Kelly et al. (Eds.): S+SSPR 2016, LNCS 10029, pp. 379–389, 2016.
DOI: 10.1007/978-3-319-49055-7_34

Each iteration selects features estimated to provide the most additional information (about disagreements between the model and data), then learns parameters that encode that information. We generalise this procedure, moving beyond traditional MGRF models, which are parametrised only with "natural parameters" which select the model with maximum entropy among those with given sufficient statistics, to consider those with potentials additionally parametrised with "feature parameters", such as filter coefficients. Both are optimised simultaneously with MLE, as in FoE. This unifies FoE (which pre-specified the graphical structure, i.e. the shapes of the filters) with the well-known FRAME [23] texture models (which used pre-specified filters) and other MGRFs using iterative selection of arbitrary features, e.g. local binary patterns [21].

Below, we apply nesting to learn MGRF texture models which capture the marginal distributions of learnt filters. Taking advantage of feature learning, we allow non-contiguous (sparse) filters, proposed to efficiently capture large-scale texture-specific visual features in a much simpler way than a multi-scale or latent-variable model. We follow FRAME and other earlier MGRFs by describing filter outputs non-parametrically with histograms (nearly entirely abandoned since) because the marginals become increasingly complex and multimodal as one moves even a short way away from the simplest near-regular textures.

This paper extends our previous work [20] which introduced model nesting with square or non-contiguous filters selected by a pre-training procedure (see Sect. 3.3) rather than MLE filter optimisation (as in FoE and RBMs), but only picked filter shapes heuristically. We present below a better method for learning non-contiguous filters, even adjusting their shapes during gradient descent, by using regularisation. We retain the cheap pre-training step, finding that it improves results compared to initialising to noise. Texture synthesis experiments with varied and difficult types of textures qualitatively compare the efficacy of different modelling frameworks and show that our models can perform at least as well as others on many textures, while only using sparse filters, minimising the cost of sampling, and without latent variables.

2   Related Work

Model nesting was described independently in [2,22] and applied to text and texture modelling respectively. Della Pietra et al. [2] used features composed from previously selected ones, testing for the occurrence of growing patterns of characters. The FRAME models [22,23] use histograms of responses of filters of varying sizes, selected by model nesting from a fixed hand-selected bank. Nesting can learn heterogeneous models; as shown in [21], texture models containing both long-range grey-level differences (GLDs) $f_{GLD}(x_1, x_2) := x_2 - x_1$ (which cheaply capture second-order interactions) and higher-order local binary patterns are much more capable than either alone. Feature selection for max-entropy models is also commonly performed by adding sparsity inducing regularisation [12], which is a variant of nesting that can also remove unnecessary features.

The FoE model [17] is a spatially homogeneous (convolutional) MGRF composed of nonlinear 'expert' potentials, fed linear filter responses as input. The responses of linear filters (even random ones) averaged over natural images are usually heavy-tailed unimodal distributions, so unimodal experts were used in [17]. However as Heess et al. [8] argued, for specific texture classes the marginal distributions are more complicated, and hence introduced BiFoE models with three-parameter uni- or bi-modal expert functions. These were much more capable than FoE of modelling textures. However, as pointed out by Schmidt et al. [18] (and further by [3]), the graph of the ideal expert function need not correspond to the shape of the filter marginal; this misidentification meaning that "the original FoE model does not capture the filter statistics".

Many works on image modelling with learnt filters have found that many of the filters are zero nearly everywhere, e.g. [3,17,18]. This suggests that fixing filter sizes and shapes is inefficient. (As an exception, for regular textures with very short tessellation distance periodic filters are likely to be learnt, e.g [4].) Many researchers have also encountered difficulties in learning FoE filters and other parameters using gradient descent e.g. [3,8,11,17]. This is one reason why these models have been limited to only small filters, from 7×7 in [4,8] up to 11×11 in [10]. Compare this to the fixed filters used in FRAME of up to 31×31 which were necessary to capture long range interactions, at a very high computational cost.

Many recent MGRF image models are hierarchical, generalising the FoE by adding latent variables, e.g. [16,18], including several works on texture modelling [4,7,10,13]. With the right design, these can be marginalised out easily or used for efficient block-Gibbs sampling; RBMs are very popular for the latter scheme. MGRFs with sophisticated higher-order structures including variables which locally modulate interactions, such as to represent edge discontinuities [16], pool features, or combine multiple texture models [10], are able to model complex textures. Building on [16], Luo et al. [13] stacked layers of latent variables to build convolutional deep belief networks (DBNs) for texture modelling, producing a quickly mixing sampler with leading synthesis results.

Portilla and Simoncelli [15] introduced a powerful texture synthesis algorithm which collects covariances of wavelet responses and uses iterated projections to produce images that match these statistics. However, these are not density models so are not as broadly applicable to different tasks. More recently Gatys et al. [5] introduced a similar algorithm using nearly a million covariances of features from a fixed 21 layer deep convolutional neural network, producing very high quality synthesis results. However these hugely complex summary statistics result in overfitting and computationally expensive texture synthesis.

3   Learning Markov–Gibbs Random Fields

Formulating as a general-form exponential family distribution, a generic MGRF (without explicit hidden variables) can be defined as

$$p(\mathbf{g}|\boldsymbol{\theta}, \boldsymbol{\lambda}) = \frac{1}{Z(\boldsymbol{\theta}, \boldsymbol{\lambda})} q(\mathbf{g}) \exp(-\boldsymbol{\theta} \cdot \boldsymbol{S}(\mathbf{g}|\boldsymbol{\lambda})) \qquad (1)$$

where $q(\mathbf{g})$ is the base distribution, $\mathbf{g}$ is an image, parameters are $\mathbf{\Lambda} := (\boldsymbol{\theta}, \boldsymbol{\lambda})$ where $\boldsymbol{\theta}$ are the *natural parameters*, and the *feature parameters* $\boldsymbol{\lambda}$ parametrise the *sufficient statistics* $\mathbf{S}$. $E(\mathbf{g}|\boldsymbol{\theta}, \boldsymbol{\lambda}) := -\boldsymbol{\theta} \cdot \mathbf{S}(\mathbf{g}|\boldsymbol{\lambda})$ is called the *energy function* of the distribution and Z is the normalising constant. $\mathbf{S}$ is a vector of sums of *feature functions* f_i over subsets of the image called cliques: $\mathbf{S}_i(\mathbf{g}|\boldsymbol{\lambda}) = \sum_c f_i(\mathbf{g}_c|\boldsymbol{\lambda}_i)$ where $\mathbf{g}_c$ denotes the values of the pixels of $\mathbf{g}$ in the clique c of f_i.

The gradient of the log-likelihood $l(\mathbf{\Lambda}|\mathbf{g}_{\mathrm{obs}}) = \log p(\mathbf{g}_{\mathrm{obs}}|\mathbf{\Lambda})$ for a training image $\mathbf{g}_{\mathrm{obs}}$ is given by [1]

$$\frac{\partial}{\partial \mathbf{\Lambda}_i} l(\mathbf{\Lambda}|\mathbf{g}_{\mathrm{obs}}) = \mathbb{E}_{p(\cdot|\mathbf{\Lambda})} \left[\frac{\partial E(\mathbf{g}|\mathbf{\Lambda})}{\partial \mathbf{\Lambda}_i} \right] - \frac{\partial E(\mathbf{g}_{\mathrm{obs}}|\mathbf{\Lambda})}{\partial \mathbf{\Lambda}_i} \tag{2}$$

The expectation is intractable so must be approximated, e.g. by MCMC.

If $\boldsymbol{\lambda}$ is empty or fixed, then the distribution is an exponential family distribution, and l is unimodal in $\mathbf{\Lambda}$ with gradient $\frac{\partial}{\partial \boldsymbol{\theta}_i} l(\boldsymbol{\theta}|\mathbf{g}_{\mathrm{obs}}) = \mathbb{E}_{p_i}[\mathbf{S}_i(\mathbf{g})] - \mathbf{S}_i(\mathbf{g}_{\mathrm{obs}})$. This shows that the MLE distribution p^* satisfies the constraint $\mathbb{E}_{p^*}[\mathbf{S}(\mathbf{g})] = \mathbf{S}(\mathbf{g}_{\mathrm{obs}})$. Further, the MLE solution is the maximum entropy (ME) distribution, i.e. it is the distribution meeting this constraint which deviates least from $q(\mathbf{g})$.

3.1   Model Nesting and Sampling

For completeness, model nesting is outlined briefly here (see [21] for details). Nesting iteratively builds a model by greedily adding potentials/features f (and corresponding constraints $\mathbb{E}_{p^*}[\mathbf{S}_f(\mathbf{g})] = \mathbf{S}_f(\mathbf{g}_{\mathrm{obs}})$) which will most rapidly move the model distribution closer to the training data, and then learning approximate MLE parameters to meet those constraints (i.e. repairing the difference in statistics). An arbitrary base distribution q can be used, as it does not even appear in the gradient (Eq. (2)). However, we do not need these constraints to be satisfied completely (by finding the MLE parameters), but only wish to improve the model by correcting some of the statistical difference on each iteration. Thus we can make the approximation of not drawing true samples from the model, but only finding images with which to approximate $\mathbb{E}_{p^*}[\mathbf{S}_f(\mathbf{g})]$.

Each iteration a set or space of candidate features is searched for one or more which might most increase the likelihood $p(\mathbf{g}_{\mathrm{obs}})$. This is normally estimated with a norm of the gradient (2) w.r.t. to the parameters $\boldsymbol{\theta}_f$ of the new potential f, i.e. $||\mathbb{E}_{p_i}[\mathbf{S}_f(\mathbf{g})] - \mathbf{S}_f(\mathbf{g}_{\mathrm{obs}})||_1$. This can be approximated using samples obtained from the model during the previous parameter learning step. Running time is quadratic in the number of nesting iterations.

Sampling. To rapidly learn parameters and obtain approximate 'samples' from the model we use a slight variant of persistent contrastive divergence (PCD) [19], earlier called CSA [6], starting from a small 'seed' image of random noise and interleaves Gibbs sampling steps with parameter gradient descent steps with a decaying learning rate, while slowly enlarging the seed. When feature parameters are fixed, the resulting image comes close to matching the desired statistics, which means this approximates Gibbs sampling from the model with optimal natural parameters [21]. This is sufficient to estimate the required expectations, even if still far from the MLE parameters. However a longer learning process with smaller steps is needed to get close to the MLE, or when the model has poorly initialised feature parameters in which case there are no real desired statistics to begin with. We used 200 PCD/Gibbs sampler steps for each nesting iteration, to produce a 100×100 image (after trimming boundaries), and 300 steps to synthesise the final images shown in the figures. Feature parameters were kept fixed while performing final synthesis, because there the intention is to match statistics, not tune the model. These hyperparameters were set to produce reasonable results quickly, rather than approach asymptotic performance.

3.2 Filter Learning

For filter w the marginal histogram is $S_w(g) := [\sum_c \text{bin}_i(w \cdot g_c) : i \in \{1, \dots, k\}]$, where $\text{bin}_i : \mathbb{R} \to [0, 1]$ is the non-negative band-pass function for the ith bin, with $\sum_i \text{bin}_i(x) = 1$. In experiments we used 32 bins stretched over the full possible range (usually less than half of the bins were nonempty). While performing Gibbs sampling we use binary-valued bins for speed but in order to have meaningful gradients or subgradients some smoothing is required. Hence for the purpose of computing the energy gradient we use a triangular bin function which linearly interpolates between the centres of each two adjacent bins.

By considering a sparse non-contiguous filter as a large square filter with most coefficients zero, filter shapes can be learnt by using sparsity-inducing regularisation. Calculating $S_w(g_{\text{samp}})$ or $\frac{\partial}{\partial \lambda_w} S_w(g_{\text{samp}})$, or performing one step of Gibbs sampling (with caching of filter responses) are all linear in the filter size, hence starting from a large square filter and waiting for it to become sparse would be slow. Instead, we start with a sparse filter (see below), and on every MLE gradient ascent iteration consider a subset of a larger set of available coefficients (we used an area of 25×25), looping through all possibilities every 10 iterations in a strided fashion. The majority of time is spent in Gibbs sampling, so considering extra coefficients only in the gradient calculation step has an insigificant cost. We apply l_1 regularisation to filter coefficients and limit the number of coefficients which are nonzero - when this constraint is exceeded, the regularisation amount is temporarily increased to force sufficient coefficients to zero. Once zero, a coefficient is removed from the filter until it randomly re-added to the active set. This directly imposed limit requires less finetuning than the indirect limit imposed by the l_1 penalty. Filters were constrained by projection to have zero mean and bounded coefficients.

3.3 Filter Pre-training

Optimising natural parameters and filters simultaneously according to (2) creates a non-convex objective. For example, since $\frac{\partial}{\partial \lambda_i} l(\Lambda | g_{\text{obs}}) \propto \theta_i$, if a potential has no influence then its filter will not be changed and it remains useless, the vanishing gradient problem. One way to simplify the learning procedure is to learn feature parameters and natural parameters separately, doing so in the feature selection step of nesting rather than the MLE step. That is, we attempt to find the coefficients of an ideal filter to add to the current model by maximising the expected gain in information by doing so. This objective function of for a new filter $\boldsymbol{w}$ is the error

$$e(\boldsymbol{w}) := ||\boldsymbol{S_w}(\mathbf{g}_{\mathrm{obs}}) - \boldsymbol{S_w}(\mathbf{g}_{\mathrm{samp}})||_1 \approx ||\frac{\partial}{\partial \boldsymbol{\theta_w}}\ell(\boldsymbol{\Lambda}|\mathbf{g}_{\mathrm{obs}})||_1 \qquad (3)$$

(presented for the simple case where we have a single sample $\mathbf{g}_{\mathrm{samp}}$ synthesised from the current model, from the previous MLE step). The new filters are indirectly pushed away from existing filters, which would provide little information. Figure 1 compares filters found with and without pre-training and MLE.

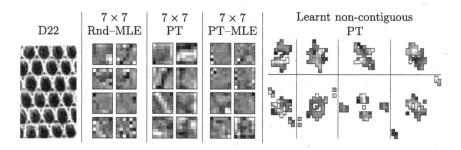

Fig. 1. Examples of learnt filters (GLD filters excluded) for texture D22. Rnd-MLE: starting from random noise then MLE-optimised. PT: pre-learnt filters kept fixed (only natural parameters adjusted during MLE). PT-MLE: pre-learnt then MLE-optimised.

We initialise filters to random noise, and follow the gradient while periodically doubling the number of empirical marginal bins, starting from 4. This allows the coefficients to be tuned to shift the mass of the empirical marginal around and gradually refine it, while overcoming the localised gradient associated with individual empirical marginals. It can be seen as a discrete version of a similar empirical marginal-smoothing technique used in [23].

4   Experimental Results

Models composed solely of GLD filters at appropriate offsets capture a majority of second order interactions and are relatively cheap to sample from. This frees other potentials to focus on more complex structure. The addition of GLD resulted in vastly improved models compared to those with only pairwise or only higher-order potentials. Hence in the majority of experiments we start with a 1st order potential to describe the grey-level histogram and then select 3 GLDs per nesting iteration, of offset up to 40 pixels, before adding filter potentials one at a time. For comparability with [21] and to avoid the randomness of a stopping rule we used the same fixed number of nesting iterations: 8 of GLDs and 8 of filters. Typical learning time for a nested model was 8–11 min (<1 min for nesting with only GLD potentials), mainly spent in the single-threaded Gibbs sampler.

Unfortunately quantitative evaluation of texture synthesis is very difficult, and most texture similarity measures implicitly make a choice of relevant statistics; We agree with Luo et al. [13] that the quantitative measures used in [8] are flawed, and since they are only applicable to highly regular textures as used in [8] and later, visual inspection of results on more challenging textures is necessary. Figure 2 provides comparison to all recent works which gave results for a set of 8 popular regular textures, following the same procedure of downscaling the original texture and using the top half for training; we quantised images to 16 grey levels so that Gibbs sampling could be used. For all other synthesis experiments we used 8 grey levels. Synthesis results comparing different methods of filter learning are shown in Fig. 3. We include a comparison (third column) to FoE-style unnested models by starting with 8 random noise

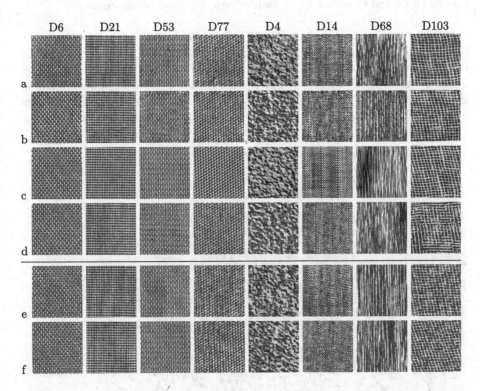

Fig. 2. Comparison of synthesis results against previously published works (images scaled, grey levels reversed and individually renormalised to allow comparison): (a) the eight original 98 × 98 Brodatz textures; results of (b) Multi-Tm [10] (a single model for all 8 textures); (c) a 2-layer TssDBN [13]; (d) nesting with jagstar-BP[13] (local binary pattern features) [21]; (e) our proposed nested models with 7×7 filters; (f) our proposed nested models with non-contiguous filters.

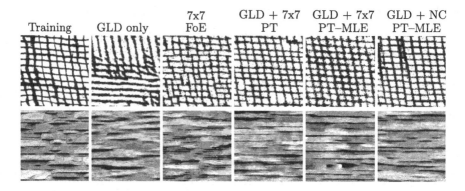

Fig. 3. Synthesis results comparing filter learning approaches. Columns are: *First*: Brodatz D103 and Simoncelli's stone-wall4 (http://www.cns.nyu.edu/~lcv/texture/). *Second*: nested models with 24 GLD potentials. *Third*: FoE-style model with eight 7×7 filter histogram potentials and no pre-training or nesting. *Following*: nested models with 24 GLDs and filters; see Fig. 1 for headings. NC: Non-contiguous filters.

7×7 filters and performing 800 PCD/Gibbs steps (restarting every 200 steps), so that the total computation was about the same as the others. These models without long-range GLD or filter potentials suffer badly when modelling any image detail or structure more than 7 pixels across, producing jumbled images. If the image is regular, adding GLD potentials works very well, but this fix does not work for irregular textures. Figure 4 shows further examples on a broader

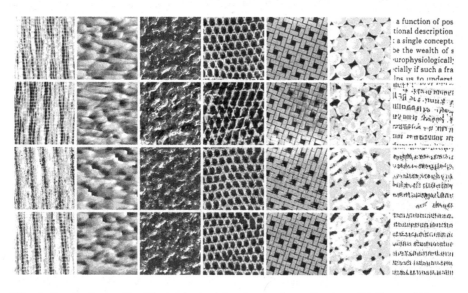

Fig. 4. Synthesis results. *First row*: training images from Brodatz, VisTex and Simoncelli. *Second*: Results by [15] as a baseline (quantised after synthesis). *Third row*: our synthesis results with nested MGRFs, 24 GLDs and square 7×7 filters. *Fourth row*: our results with 24 GLDs and non-contiguous filters.

range of challenging textures. Results for additional textures and models, and source code for the experiments are available on the accompanying website at http://www.ivs.auckland.ac.nz/texture_modelling/sparsefilt/.

5   Conclusions

Our synthesis results, using models with only at most 8 small filters plus simple GLD features and no latent variables, are comparable with those of other models (Fig. 2)—the best performing of which have up to hundreds of filters (in [13] and the multi-texture models of [10]) or up to 4096 parameters per potential [21]—and show ability on irregular textures which are seemingly too difficult for many previous works to attempt. The combination of learning the structure of a MGRF model (as in FRAME) and optimising the feature parameters (as in FoE) appears to allow complex textures to be reproduced more easily by considering a more general set of possible models, although suffering potentially longer learning times due to the iterated nesting procedure.

Filter pre-training instead of starting from random filters was essential for our potentials (results were poor without it), because the gradient can not push mass between histogram bins that are not adjacent, a major disadvantage of using histograms. This is likely a reason that learning sometimes failed to find sensible filters. An alternative parametrisation of the histogram with overlapping coarse and fine bins may provide a more navigable optimisation landscape. Performing MLE learning of filter coefficents gave mixed results, sometimes leading to worse or better results than using purely pre-trained filters, possibly due to the non-convexity of the optimisation problem and the use of histograms. Non-contiguous and contiguous filters also showed tradeoffs; in experiments we found that the noncontiguous ones are more able to handle larger-scale textures, and are more robust because models using square filters will produce bad results if the filters are too small. On the other hand for many textures with small localised details it appeared to be better to use traditional square filters instead. Ideally, such tradeoffs would be made by the nesting algorithm itself, by providing the right stopping rule and regularisation/prior. Unfortunately, this further increases the number of hyperparameters to be tuned, and there are already a relatively large number; there is a tradeoff between selecting model aspects manually (e.g. graphical structure), which may be easier but less robust, and having to tune hyperparameters to select them automatically.

More sophisticated image statistics in [5, 14, 15], notably filter response inter-dependencies/co-occurrences, have proven to be powerful texture descriptors when used for synthesis of complex and inhomogeneous textures. However, energy-based texture models with such sophisticated statistics are surprisingly yet to be investigated. This paper has focused on learning of linear filters as a case study, but future work should clearly attempt to bridge the gap between these fields by incorporating such kinds of powerful co-occurrence statistics.

References

1. Barndorff-Nielsen, O.: Information and Exponential Families in Statistical Theory. Wiley, Hoboken (1978)
2. Della Pietra, S.; Della Pietra, V., Lafferty, J.: Inducing features of random fields. IEEE Trans. Pattern Anal. Mach. Intell. **19**(4), 380–393 (1997)
3. Gao, Q., Roth, S.: How well do filter-based MRFs model natural images? In: Pinz, A., Pock, T., Bischof, H., Leberl, F. (eds.) DAGM/OAGM 2012. LNCS, vol. 7476, pp. 62–72. Springer, Heidelberg (2012). doi:10.1007/978-3-642-32717-9_7
4. Gao, Q., Roth, S.: Texture synthesis: from convolutional RBMs to efficient deterministic algorithms. In: Fränti, P., Brown, G., Loog, M., Escolano, F., Pelillo, M. (eds.) S+SSPR 2014. LNCS, vol. 8621, pp. 434–443. Springer, Heidelberg (2014). doi:10.1007/978-3-662-44415-3_44
5. Gatys, L.A., Ecker, A.S., Bethge, M.: Texture synthesis using convolutional neural networks. In: Advances in Neural Information Processing Systems, pp. 262–270 (2015)
6. Gimel'farb, G.: Image Textures and Gibbs Random Fields. Kluwer Academic Publishers, Dordrecht (1999)
7. Hao, T., Raiko, T., Ilin, A., Karhunen, J.: Gated Boltzmann machine in texture modeling. In: Villa, A.E.P., Duch, W., Érdi, P., Masulli, F., Palm, G. (eds.) ICANN 2012. LNCS, vol. 7553, pp. 124–131. Springer, Heidelberg (2012). doi:10.1007/978-3-642-33266-1_16
8. Heess, N., Williams, C.K.I., Hinton, G.E.: Learning generative texture models with extended fields-of-experts. In: Proceedings of British Machine Vision Conference (BMVC 2009), pp. 1–11 (2009)
9. Hinton, G.E.: Training products of experts by minimizing contrastive divergence. Neural Comput. **14**(8), 1771–1800 (2002)
10. Kivinen, J.J., Williams, C.: Multiple texture Boltzmann machines. In: Proceedings of 15th International Conference on Artificial Intelligence and Statistics, vol. 2, pp. 638–646 (2012)
11. Köster, U., Lindgren, J.T., Hyvärinen, A.: Estimating Markov random field potentials for natural images. In: Adali, T., Jutten, C., Romano, J.M.T., Barros, A.K. (eds.) ICA 2009. LNCS, vol. 5441, pp. 515–522. Springer, Heidelberg (2009). doi:10.1007/978-3-642-00599-2_65
12. Lee, S., Ganapathi, V., Koller, D.: Efficient structure learning of Markov networks using L1-regularization. In: Advances in Neural Information Processing Systems 19 (NIPS 2006), pp. 817–824 (2007)
13. Luo, H., Carrier, P.L., Courville, A., Bengio, Y.: Texture modeling with convolutional spike-and-slab RBMs and deep extensions. J. Mach. Learn. Res. **31**, 415–423 (2013)
14. Peyré, G.: Sparse modeling of textures. J. Math. Imaging Vis. **34**(1), 17–31 (2009)
15. Portilla, J., Simoncelli, E.P.: A parametric texture model based on joint statistics of complex wavelet coefficients. Int. J. Comput. Vis. **40**(1), 49–70 (2000)
16. Ranzato, M., Mnih, V., Hinton, G.E.: Generating more realistic images using gated MRFs. In: Advances in Neural Information Processing Systems 23 (NIPS 2010), pp. 2002–2010. Curran Associates (2010)
17. Roth, S., Black, M.J.: Fields of experts. Int. J. Comput. Vis. **82**(2), 205–229 (2009)
18. Schmidt, U., Gao, Q., Roth, S.: A generative perspective on MRFs in low-level vision. In: IEEE Conference on Computer Vision and Pattern Recognition (CVPR 2010), pp. 1751–1758 (2010)

19. Tieleman, T.: Training restricted Boltzmann machines using approximations to the likelihood gradient. In: Proceedings of 25th International Conference on Machine Learning (ICML 2008), pp. 1064–1071. ACM (2008)
20. Versteegen, R., Gimel'farb, G., Riddle, P.: Texture modelling with non-contiguous filters. In: Proceedings of 30th International Conference on Image and Vision Computing New Zealand (IVCNZ 2015) (2015)
21. Versteegen, R., Gimel'farb, G., Riddle, P.: Texture modelling with nested high-order Markov-Gibbs random fields. Comput. Vis. Image Underst. **143**, 120–134 (2016)
22. Zhu, S.C., Wu, Y., Mumford, D.: Minimax entropy principle and its application to texture modeling. Neural Comput. **9**(8), 1627–1660 (1997)
23. Zhu, S.C., Wu, Y., Mumford, D.: Filters, random fields and maximum entropy (FRAME): towards a unified theory for texture modeling. Int. J. of Comput. Vis. **27**(2), 107–126 (1998)

Shape Normalizing and Tracking Dancing Worms

Carmine Sansone[1,2], Daniel Pucher[1(✉)], Nicole M. Artner[1(✉)],
Walter G. Kropatsch[1], Alessia Saggese[2], and Mario Vento[2]

[1] Pattern Recognition and Image Processing Group, TU Wien, Vienna, Austria
{daniel.pucher,artner,krw}@prip.tuwien.ac.at
[2] Department of Information Engineering, Electrical Engineering and Applied
Mathematics (DIEM), Faculty of Engineering, University of Salerno, Fisciano, Italy

Abstract. During spawning, the marine worms Platynereis dumerilii exhibit certain swimming behaviors, which are described as *nuptial dance*. To address the hypothesis that characteristic male and female spawning behaviors are required for successful spawning and fertilization, we propose a 2D tracking approach enabling the extraction of spatio-temporal data to quantify gender-specific behaviors. One of the main issues is the complex interaction between the worms leading to collisions, occlusions, and interruptions of their continuous trajectories. To maintain the individual identities under these challenging interactions a combined tracking and re-identification approach is proposed. The re-identification is based on a set of features, which take into account position, shape and appearance of the worms. These features include the *normalized shape* of a worm, which is computed using a novel approach based on its distance transform and skeleton.

Keywords: Object tracking · Appearance models · Shape normalization · Shape analysis

1 Introduction

Platynereis dumerilii [19] are marine worms, who reproduce through external fertilization when sexually mature. During spawning they exhibit particular swimming behaviors, which differ based on their gender and spawning phase [6]. The aim of the biologists is to verify that characteristic male and female spawning behaviors are required for successful spawning and fertilization. Therefore, they want to analyze the spawning behaviors in a quantitative manner and characterize and compare male- and female-specific behaviors. To achieve this, the aim is to develop methods that enable the 2D tracking of spawning worms and extract features describing their appearance (skeleton, curvature, normalized shape, etc.) and motion (trajectories of head and tail). These features are described and discussed in more detail in [13].

© Springer International Publishing AG 2016
A. Robles-Kelly et al. (Eds.): S+SSPR 2016, LNCS 10029, pp. 390–400, 2016.
DOI: 10.1007/978-3-319-49055-7_35

For the proposed spatio-temporal analysis, the nuptial dance [3] of the worms is recorded with the help of a special setup. In nature, spawning happens in the sea at night around new moon. Therefore, it is necessary to record the videos inside a light-tight box with an infrared camera. During the recordings, the worms are placed into a shallow, cylindrical bowl filled with sea water referred to as *arena*. On the water surface, the arena has a diameter of 7 cm. The average length of a worm is between 1 and 2 cm.

When the worms are not close to each other, analysis and tracking is straightforward. As the camera is fixed, the worms can be reliably segmented by subtracting a background image (showing the empty arena) and removing noise (morphological filters). The resulting binary regions of both worms are represented by their center line (extracted based on skeleton) limited by head and tail position. Tracking of the worms is done by employing a Kalman filter, which finds the correspondences of head and tail for both worms in each frame.

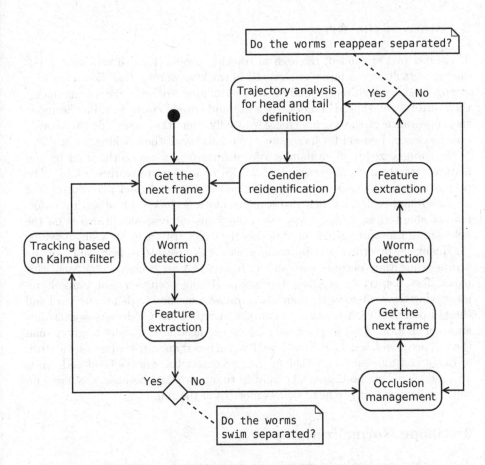

Fig. 1. Flow chart of the proposed method

The contribution and focus of this paper lies in the active phase of the spawning process, where the worms interact. During this active phase the worms collide and occlude each other in the 2D image, thus leading to ambiguities in segmentation and tracking. As the worms frequently change their appearance and motion, it is a challenging task to correctly identify them throughout the whole video sequence. The main aim of this paper is to re-identify the worms after their frequent interactions. This ensures that the collected spatio-temporal data is associated with the correct label (male or female) and is a reliable source of information for the following analysis of the biologists. Figure 1 gives an overview of the overall tracking pipeline.

The remaining part of this paper is organized as follows: Section 2 discusses existing tracking methods, Sect. 3 explains shape normalization in detail, Sect. 4 presents the tracking and re-identification method. Experimental results are shown in Sect. 5 and Sect. 6 concludes the paper.

2   State of the Art

There is a vast amount of research in tracking people [11] and vehicles [5], but only modest attention has been devoted to tracking worms. Traditional tracking methods cannot be directly used to track swimming worms as the worms change their direction and speed of motion in a fast and unpredictable way. Furthermore, the appearance of the worms changes rapidly, thus the history of the moving objects cannot be used for increasing the reliability of the tracking, as in [11].

As summarized in [9], a number of worm trackers focus on the tracking and feature extraction of one worm [2,15,17,18] or a group of worms [14,16]. The proposed methodologies use background subtraction and track the centroids of all foreground regions. When two or more blobs occlude each other, the information about these blobs is lost since the focus of these algorithms is on the behavior of the worm groups and not on the exact trajectory of a single worm.

Huang [8] describes an approach, where it is possible to keep track of the worms while they occlude each other. In contrast to Huang's approach, this paper does not try to segment the worms during occlusion, but aims at re-identifying them afterwards. They also propose a method to define the head and the tail points of each worm based on the fact that the C. elegans worms have an accumulation of fat in the head making the head area usually brighter than the tail area, which does not hold for Platynereis dumerilii worms. In the work of Hoshi and Shingai [7] another method to detect the worm's head and tail is proposed. This method cannot be applied to our problem because it is based on the assumption that the head swings more than the tail.

3   Shape Normalization

As the worms are highly deformable, it is necessary to come up with a description of their shape, which is independent of deformations. For this, we follow a recent strategy, which is known as co-registration, where shapes are first straightened or

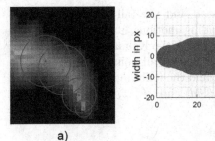
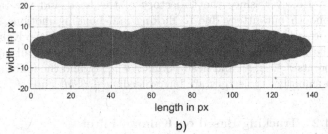

a)   b)

Fig. 2. (a) Four circles (red) representing the shape at four selected positions in the skeleton (red points). Yellow pixels represent the skeleton. Gray values visualize the distance to the boundary. (b) Normalized shape representation of a male worm. (Color figure online)

flattened to then register different views/deformations of the same normalized shape [1]. The shape normalization is based on the distance transform of the binary region and the skeleton of the shape. The skeleton is restricted to a line delimited by two end-points without any branches. For every point of the skeleton, the corresponding value of the distance transform holds the Euclidean distance to the nearest boundary pixel of the binary region. These distances hold information on the thickness of the shape and also serve as radii of circles used to draw the normalized shape representation. Figure 2a shows the circles of four points on the skeleton of a worm.

For every pixel of the skeleton a corresponding circle can be drawn. The union of these circles covers the original worm shape. By arranging the skeleton points in a straight line, the representation becomes independent of the deformations of the worm and gives the normalized shape (See Fig. 2b). To maintain the correspondence between deformed and normalized shape, the geodesic distance along the deformed skeleton is mapped to the Euclidean distance in the normalized shape. This one-to-one mapping also allows to return to the image space. In this paper, the normalized shape is used to compare worm shapes before and after an occlusion occurs.

4   Worm Tracking

4.1   Representation of Worms

Let $\omega(\tau, z, id)$ be a function that defines the position of a point along the worm's skeleton, where $\tau \in [0, 1]$ is the percentage of the worm length that defines the normalized distance along the skeleton between the head point $\omega(1, z, id)$ and the tail point $\omega(0, z, id)$. z represents the frame number of the video and id is the worm's identifier. In case of two worms $id \in [male, female]$. For a fixed frame $\hat{z}$ of the video and for a fixed worm $\hat{id}$, the position of the tail point is $\omega(0, \hat{z}, \hat{id})$ and the position of the head point is $\omega(1, \hat{z}, \hat{id})$.

However, since the trajectory of the head and the tail of a worm is not smooth but jittery due to flicking and bending motions, it is not possible to correctly predict the position of these points using the Kalman filter [10]. Thus, the skeleton is partitioned into three parts, namely tail points $\tau \in [0, \frac{1}{3}]$, body points $\tau \in [\frac{1}{3}, \frac{2}{3}]$, and head points $\tau \in [\frac{2}{3}, 1]$, and the following two points having a smoother trajectory are introduced: $\omega(\frac{1}{3}, z, id)$ and $\omega(\frac{2}{3}, z, id)$.

4.2   Tracking Based on Kalman Filter

During a preliminary detection step based on a traditional background subtraction algorithm, the connected regions associated to the worms are detected and their skeletons are extracted [13]. Each skeleton is then represented by the two points p_1 and p_2, having a geodesic distance of $\frac{1}{3}$ and $\frac{2}{3}$ from one of ending points of the skeleton, respectively. At the generic frame $\hat{z}+1$, the positions of the worms $w(\frac{1}{3}, \hat{z}, male)$, $w(\frac{2}{3}, \hat{z}, male)$, $w(\frac{1}{3}, \hat{z}, female)$ and $w(\frac{2}{3}, \hat{z}, female)$ are known from the previous frame $\hat{z}$. In the first frame of a video, the positions are defined manually. The tracking algorithms has to find the correct association among the above positions identified at the frame $\hat{z}$ (the output of the tracking at the previous frame) and the points p_1 and p_2 defined in frame $\hat{z}+1$ (the output of the detection).

The Kalman filter provides predictions for the positions of the points of frame $\hat{z}$ based on the trajectory of these points [10]. These predictions are compared with the points that describe the worms in frame $\hat{z}+1$. In the case of two worms, there are just eight possible associations. For every prediction, the corresponding error is calculated as the Euclidean distance between the position of the prediction and the position of the predicted point in frame $\hat{z}+1$. For every hypothesis four errors are taken in account: the errors of the head and tail predictions of the male worm and the errors of the head and tail predictions of the female worm. The hypothesis with the lowest root mean square error is chosen because the prediction errors follow a norm distribution [4]. Furthermore the evaluation showed that using the root mean square error brings better results.

4.3   Handling Occlusions by Re-Identification

An occlusion is defined as a set of frames in which the worms overlap each other and appear as a single connected region in the binary image. In the frames before an occlusion, a set of features is extracted to describe each worm, while during an occlusion it is not possible to reliably extract and associate these features. After an occlusion, there are again two connected regions: $region_1$ and $region_2$, each describing a worm, but their identities (male or female) are unknown. Furthermore, it is unknown which of the two end-points of each skeleton is the head and which is the tail. When the worms occlude each other, they often change the direction of their movement in an unpredictable way. Information about the movement before the occlusion can't be robustly used to predict the position of the worms after the occlusion.

As motion (speed and direction) is not a reliable indicator for the identity of the worms, this paper proposes an approach based on comparing the appearance immediately before and after the occlusion. For this comparison, the following features are taken into account:

- Normalized Shape, $f^s \in \mathbf{R}^N$
- Area in pixels, $f^a \in \mathbf{N}$
- Mean gray-scale value of skeleton, $f^m \in [0, 255]$
- Length of skeleton in pixels, $f^l \in \mathbf{R}$

A sliding window is used to analyze the worms over several frames to increase the reliability in the classification process. The features collected before the occlusion are used to build two models: $model_1$ for the male worm and $model_2$ for the female worm. To compare the worm models with the features extracted from the unidentified regions after the occlusion, a measure of similarity (see Eq. 1) and a method of comparison are necessary.

After the occlusion, a majority voting approach is used to combine the decisions taken in every frame of the temporal window to find the identity of the worms (male or female). The identity that receives the most votes is the final identity of each worm.

The size of the sliding window has to be chosen according to the rate in which the chosen features change over time. For Platynereis dumerilii worms, some of the features depend on how and where the worms are moving. When they swim in a straight line, the body area is bigger than when they are moving in circles. The luminosity also changes depending on where they swim. It is higher in the middle of the arena than at the border. Therefore a *temporal window* of five frames was chosen empirically in the current tracking setup.

To build the male and female model, the average value of the features in the last five frames before an occlusion is calculated. The models define how the worms should appear after the occlusion. For each of the five frames after the occlusion, the features of the two unidentified regions are compared with the male and female models. For the comparison, the following measures of similarity are computed: s_{ij}^s (normalized shape), s_{ij}^a (area), s_{ij}^m (mean gray scale value) and s_{ij}^l (length), where $i \in [1,2]$ is an index referring to the two unidentified regions and $j \in [1,2]$ is an index referring to the models. All these measures of similarity are normalized to one and combined according to Eq. 1.

$$s_{ij} = \sqrt{(s_{ij}^s)^2 + (s_{ij}^a)^2 + (s_{ij}^m)^2 + (s_{ij}^l)^2} \qquad (1)$$

Let x $\in$ {a, m, l} be a variable that represents the value of one of the following features: area, mean gray scale value and length. d_{ij}^x is the difference between the value of the feature x of region i and the value of the same feature in model j (see Eq. 2). s_{ij}^x is computed according to Eq. 3 where the features are normalized with the maximum feature value d_{max}^x which is defined empirically.

$$d_{ij}^x = f_i^x - f_j^x \qquad (2)$$

$$s_{ij}^x = 1 - \frac{|d_{ij}^x|}{d_{max}^x} \tag{3}$$

The normalized shape similarity s_{ij}^s is computed according to the Pearson correlation [12] (see Eq. 4), where N is an integer value computed as the minimum value between the geodesic length of the skeleton from $region_i$ and the geodesic length of the skeleton of $worm_j$. $r_i(n)$ and $r_j(n)$ are two functions that define the radius of the normalized shape for the $region_i$ and for the $worm_j$ of a fixed point n on the skeleton. Comparing normalized shapes 3D movements are managed.

$$s_{ij}^s = \frac{N \cdot \sum_{n=1}^N r_i(n) \cdot r_j(n) - \sum_{n=1}^N r_i(n) \cdot \sum_{n=1}^N r_j(n)}{\sqrt{N \cdot \sum_{n=1}^N r_i^2(n) - \left(\sum_{n=1}^N r_i(n)\right)^2} \cdot \sqrt{N \cdot \sum_{n=1}^N r_j^2(n) - \left(\sum_{n=1}^N r_j(n)\right)^2}}$$
$$\tag{4}$$

There are two possible associations: $region_1$ is the male worm and $region_2$ is the female worm (hypothesis Hp_1) or vice versa (hypothesis Hp_2). A measure of confidence is assigned to both hypotheses: $s_{11} + s_{22}$ to Hp_1 and $s_{12} + s_{21}$ to Hp_2. The association with the greater confidence value is chosen as result of the re-identification process.

From Eq. 1 it can be observed that all the features bring an equal contribution to the final decision. From the evaluation it has been observed that there is no need to weight the features.

4.4   Trajectory Analysis for Head and Tail Definition

After the re-identification, the regions are labeled as male and female, but the head and tail of each worm is not identified yet. To solve this open issue, a trajectory analysis is done. The basic idea is to analyze the direction of the motion of the worms, as they in general swim in a forward manner (tail point on skeleton is pulled from head point).

A new predefined temporal window after the occlusion is considered for the trajectory analysis. In each frame, the two end-points of the skeletons of both worms are tracked as shown in Fig. 3. In the synthetic example in Fig. 3, the Euclidean distance between the point $w(\frac{1}{3}, \hat{z} + 1, \hat{id})$ and the point $w(\frac{1}{2}, \hat{z}, \hat{id})$ is smaller than the Euclidean distance between the point $w(\frac{2}{3}, \hat{z} + 1, \hat{id})$ and the point $w(\frac{1}{2}, \hat{z}, \hat{id})$. Therefore, the point $w(0, \hat{z} + 1, \hat{id})$ votes for tail in this frame pair. This voting is repeated for every consecutive frame pair in the temporal window. Finally, the majority of votes decides on head and tail of each worm.

For the trajectory analysis it is assumed that the worms do not move backwards for more than half of the frames in the temporal window. The temporal window for the trajectory analysis has to be chosen considering the trade off between the probability of making the correct decision and the elaboration time. Therefore, a window of 30 frames was chosen empirically related to speed and longest backward movement of Platynereis dumerilii worms in the current tracking setup. Figure 4 shows an example of occlusion management.

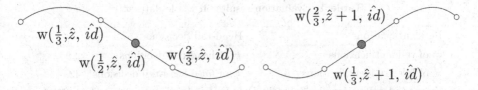

Fig. 3. Information taken in account to identify the head and the tail points in a frame pair (left and right).

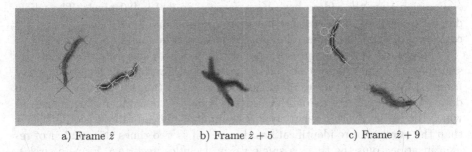

a) Frame $\hat{z}$ b) Frame $\hat{z}+5$ c) Frame $\hat{z}+9$

Fig. 4. Three frames taken out of a sequence of ten frames. (a) Male (blue) and female (pink) worm before occlusion. Small crosses identify the points $w(0,\hat{z},id)$, $w(\frac{1}{3},\hat{z},id)$ and $w(\frac{2}{3},\hat{z},id)$. Circles identify the predictions of the points $w(\frac{1}{3},\hat{z},id)$ and $w(\frac{2}{3},\hat{z},id)$. The big cross on both worms identifies the head point $w(1,\hat{z},id)$. (b) interactions during occlusion. (c) Re-identification after occlusion. (Color figure online)

5   Evaluation

5.1   Normalized Shape

The normalized shape representation has been evaluated in [13] on a dataset containing 100 images selected from 7 different video sequences. This section gives an overview of the results. For the evaluation, the original binary worm region is compared with its normalized shape, which is projected into the image space to match the original worm region. Errors occur when pixels are missed from the original shape. Non-original pixels are not added when projecting the normalized shape, since the circles are always inside the original shape. Comparing the projected normalized shape to the original worm region, the minimum error is 1.02 %, the maximum error is 8.89 % and the mean-error for all images is 3.2 %.

5.2   Tracking

To evaluate the performance of the proposed approach a MATLAB prototype has been realized. The prototype has been applied on a dataset of worm videos from the Max F. Perutz Laboratories in Vienna. All videos in the dataset are in gray scale with a size of 1280 × 960 pixels and a variable frame rate between 30

Table 1. Evaluation results on whole dataset.

Re-identifications		Head-tail decisions	
# of re-identifications	1.308	# of head-tail decisions	2.962
# of false re-identifications	3	# of false head-tail decisions	42
False re-identifications in %	0.229 %	False head tail decisions in %	1.418 %

and 60 frames per second (43 frames per second in average). To speed up the elaboration time the videos have been re-sized to 640 × 480 pixels. The whole dataset consists of 25 videos that have a total duration of more than 2 h and are composed of 320.630 frames (elaborated videos can be found on our web site[1]).

The ground truth of the worm identities has been generated manually by selecting head and tail position and the worm identifier (male or female) in every frame. Table 1 shows the result of the evaluation of the tracking on the whole dataset. The number of re-identifications is equal to the number of occlusions in which the worms overlap themselves. The number of head-tail decision is bigger than the number of re-identifications, because it is two times the number of re-identifications plus the times a single worm occludes itself (e.g. forms a circle). As can be seen in Table 1, the number of false gender decisions is different in comparison to the number of false head-tail decisions, because these decisions are made independent of each other. Furthermore, it is important to note that false head-tail decisions do not have repercussions on the following head-tail decisions, but false gender decisions can influence the following decisions on gender.

Regarding the gender association, it can be observed that in twenty-two videos there are zero association errors and in three videos, starting from a certain frame to the end of the video, the worms' genders are confused.

Regarding the head and tail association, it can be observed that the majority of false decisions are made when the worms lose their vitality and start to stay quiet or do not move at all. In these situations, the trajectories of the worms are ambiguous and depend on the movement of the water. Instead, when the worms swim in a *natural way*, it is unlikely that false decisions are made.

To illustrate the diversity of the worm features, Table 2 shows the values before and after the occlusion shown in Fig. 4 occurs. This example was chosen because the male worm appears bigger than the female worm before the occlusion, which changes after the occlusion. It shows why it is necessary to consider all features and use a sliding window.

Table 2. Worm features before and after the occlusion shown in Fig. 4 occurs.

Feature	Male worm	Female worm
Area (pixels)	477/384	441/485
Mean gray scale value	50/49	71 / 66
Geodesic length (pixels)	68,63/66,77	71,04/66,61

---

[1] http://www.prip.tuwien.ac.at/research/worms.php.

6   Conclusion

In this paper, a novel approach to track Platynereis dumerilii is proposed. It is able to handle occlusions and maintain the identity of the tracked worms with the help of a novel feature, the *normalized shape*. The normalized shape allows the comparison of the shape of worms independent of their deformation. It is used in conjunction with other features to correctly re-identify the worms after occlusions. Experimental evaluations on more than two hours of video material showed that the proposed approach is able to reliably analyze the nuptial dance of the worms. In 99.8 % of the cases the gender of the worms was correctly re-identified after an occlusion. The head and the tail where correctly labeled in 98.6 % of the cases.

The proposed method for the re-identification and the trajectory analysis to assign head and tail are not limited to the presented application and can be applied to other tracking problems. Especially the normalized shape is a suitable representation for all kinds of non-rigid objects having a main axis.

Acknowledgements. The authors thank Stephanie Bannister from the Max F. Perutz Laboratories GmbH for valuable discussions and providing videos of the spawning worms.

References

1. Aigerman, N., Poranne, R., Lipman, Y.: Lifted bijections for low distortion surface mappings. ACM Trans. Graph **33**(4), 69:1–69:12 (2014)
2. Baek, J.H., Cosman, P., Feng, Z., Silver, J., Baek, J., Schafer, W.: Using machine vision to analyze and classify Caenorhabditis elegans behavioral phenotypes quantitatively. J. Neurosci. Methods **118**, 9–21 (2002)
3. Bentley, M.G., Olive, P.J.W., Last, K.: Sexual satellites, moonlight and the nuptial dances of worms: the influence of the moon on the reproduction of marine animals. Earth, Moon, Planets **85**, 67–84 (1999)
4. Chai, T., Draxler, R.R.: Root mean square error (RMSE) or mean absolute error (MAE)? Arguments against avoiding rmse in the literature. Geosci. Model Dev. **7**(3), 1247–1250 (2014)
5. Coifman, B., Beymer, D., McLauchlan, P., Malik, J.: A real-time computer vision system for vehicle tracking and traffic surveillance. Transp. Res. Part C: Emerg. Technol. **6**(4), 271–288 (1998)
6. Hardaker, L.A., Singer, E., Kerr, R., Schafer, W.R.: Serotonin modulates locomotory behavior and coordinates egg-laying and movement in C. elegans. J. Neurobiol. **49**, 303–313 (2001)
7. Hoshi, K., Shingai, R.: Computer-driven automatic identification of locomotion states in Caenorhabditis elegans. J. Neurosci. Methods **157**(2), 355–363 (2006)
8. Huang, K.M.: Tracking and Analysis of Caenorhabditis Elegans Behavior Using Machine Vision. ProQuest, Ann Arbor (2008)
9. Husson, S.J., Costa, W.S., Schmitt, C., Gottschalk, A.: Keeping track of worm trackers. In: WormBook (2012)

10. Kovvali, N., Banavar, M.K., Spanias, A.: An Introduction to Kalman Filtering with MATLAB Examples. Synthesis Lectures on Signal Processing. Morgan and Claypool Publishers, San Rafael (2013)
11. Lascio, R.D., Foggia, P., Percannella, G., Saggese, A., Vento, M.: A real time algorithm for people tracking using contextual reasoning. CVIU **117**(8), 892–908 (2013)
12. Osborne, J.W.: Best Practices in Quantitative Methods. Sage Publications, Thousand Oaks (2008). Ed. by Osborne, J.W.: Includes bibliographical references and index
13. Pucher, D.: 2D tracking of platynereis dumerilii worms during spawning. Technical report PRIP-TR-135, TU Wien, Austria, April 2016
14. Ramot, D., Johnson, B.E., Berry, T.L., Carnell, L., Goodman, M.B.: The parallel worm tracker: a platform for measuring average speed and drug-induced paralysis in nematodes. PLoS ONE **3**(5), e2208+ (2008)
15. Shingai, R.: Durations and frequencies of free locomotion in wild type and gabaergic mutants of Caenorhabditis elegans. Neurosci. Res. **38**(1), 71–84 (2000)
16. Swierczek, N.A., Giles, A.C., Rankin, C.H., Kerr, R.A.: High-throughput behavioral analysis in C. elegans. Nat. Methods **8**(7), 592–598 (2011)
17. Wang, S.J., Wang, Z.-W.: Track a worm, an open-source system for quantitative assessment of C. elegans locomotory, bending behavior. PLoS ONE **8**(7), 1–10 (2013)
18. Yemini, E.: High-Throughput, Single-worm Tracking and Analysis in Caenorhabditis Elegans. University of Cambridge, Cambridge (2013)
19. Zantke, J., Bannister, S., Rajan, V.B.V., Raible, F., Tessmar-Raible, K.: Genetic, genomic tools for the marine annelid platynereis dumerilii. Genetics **197**, 19–31 (2014). World Polychaeta Database (WPolyDb)

Entity Extraction and Correction Based on Token Structure Model Generation

Najoua Rahal[1]([✉]), Mohamed Benjlaiel[2], and Adel M. Alimi[2]

[1] Tunis el Manar University, FST, Tunis, Tunisia
`najoua.rahal.tn@ieee.org`
[2] Sfax University, ENIS, Sfax, Tunisia
`benjlaiel@yahoo.fr, adel.alimi@ieee.org`

Abstract. The logical and semantic structure analysis is a basic process for invoice understanding. Be able to carry out a robust layout analysis is very difficult due to highly heterogeneous invoice templates. In this paper, we propose a local structure for entity extraction and correction from scanned invoices. It attempts to extract entity in contiguous and noncontiguous structure by automatic finding the local structure of each entity without structure model matching and user intervention. Firstly, the entities are labeled in OCRed invoice. Combining labeled entities with geometric and semantic relations, token structure models are generated. These models are used for entity extraction and mislabeling correction by ignoring some superfluous tokens detected by labeling step. The correction module to the contiguous structure differs from that of the noncontiguous structure. The obtained results with a dataset of real invoices are reported in experimental section.

Keywords: Contextual search · Contiguous and noncontiguous structure · Mislabeling correction · Token structure models

1 Introduction

In accordance with [1], Automatic document processing refers to three main categories; doctype classification, data capture/Functional Role Labeling, and document sets. Doctype classification is to assign a document image to a prestored template. Data capture represents the extraction of relevant human understandable information from document image. The category Document sets relates between documents and their contents depending on business logic. In this paper, we focus on automatic data capture from invoices regardless of their high geometric variations.

Figure 1 shows some examples of entities in contiguous (Fig. 1(a)) and noncontiguous (Fig. 1(b)) structure. It illustrates how closeness, direction and graphical elements may differ in conjunction Reference Words (RWs) e.g., "FACTURE N°", "Date", "Net à payer", etc. with Key Fields (KFs) e.g., "006651", "22/08/2015", "228 276.300", etc. for an entity, in various invoices.

© Springer International Publishing AG 2016
A. Robles-Kelly et al. (Eds.): S+SSPR 2016, LNCS 10029, pp. 401–411, 2016.
DOI: 10.1007/978-3-319-49055-7_36

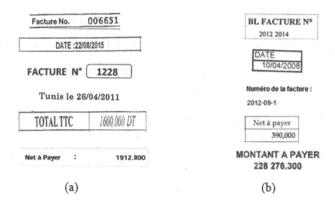

(a)   (b)

Fig. 1. Sample of entities showed the diversity of layout styles used in invoice. (a) Entity in contiguous structure. (b) Entity in noncontiguous structure.

In this context, many initiative works, like [2], learn a local structure layout from training document and reuse it for extracting the fields in the test document. The weakness of such work is that require the human intervention for labeling semantic fields. Authors in [3,4] propose to correct the mislabeling by adding the missing labels. However, they require high regularity of structures and automatic blocks and segments obtained by OCR (Optical Character Recognition). Also, the mislabeling correction is based on matching a structure graph with a model graph.

Experimental studies have shown that the mismatch between unstructured data obtained by the OCR, like Tesseract[1] (OCR without layout analysis), and its physical representation generates another type of mislabeling called superfluous tokens. This problem is caused by mishandling of spaces by OCR. Figure 2 shows sample of entities with superfluous tokens mislabeling. At the bottom of Fig. 2(a), there is the result of labeling applied to the text of OCR to extract the "Balance Due". At the top, there is the physical representation of this labeling.

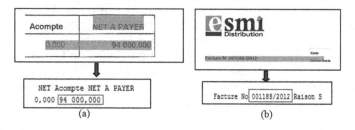

(a)   (b)

Fig. 2. (a) Entity in noncontiguous structure with superfluous tokens. (b) Entity in contiguous structure with superfluous tokens (Color figure online)

---

[1] https://github.com/tesseract-ocr.

The final impact of this mishandling is a wrong extraction of an entity in which "0,000" represents a noisy token.

In an earlier work [5], we have proposed a method for treatment entity in contiguous structure. However, it does not able to extract entity in noncontiguous structure. Also, it does not benefit of physical and logical structure of the entity in the invoice which is the purpose of this work. The ultimate goal of our method is to extract only the relevant tokens from an entity (framed in red color) and increases the accuracy of the extraction process.

Our contributions are: (i) a robust system for entity extraction based on contextual search of local structure of each entity. Then, there is no need to classify contiguous and noncontiguous structure. Our system starts its contextual search in contiguous structure. If no result is found, then, it moves automatically to the treatment of extraction entity in noncontiguous structure. (ii) Adoption of correction step to eliminate the superfluous tokens caused by the labeling step.

In the remainder of our paper, we firstly describe in detail our solution. Next, we discuss obtained experimental results. Finally, the paper is concluded.

2   Proposed Method

The overview of our proposed method is given by Fig. 3. Firstly, invoice image is processed by OCR engine. Secondly, entities are labeled in OCRed invoice image. Once labeled, the local structure of each entity is detected. For each entity structure, a token model is generated which aims to eliminate the superfluous tokens caused by the labeling step. In this model, the KFs represent the tokens and the distances represent the relationships between them. Finally, an incremental algorithm is applied to concatenate each two consecutive tokens of which the distance between them respects a certain threshold.

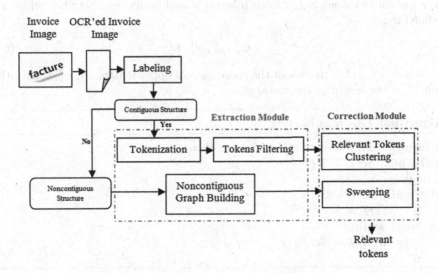

Fig. 3. Global schema of proposed method

2.1   Labeling

Entities are labeled in the invoice using Patterns of Regular Expressions (Regex). Each invoice I is defined as:

$$I = \{L_i\} \tag{1}$$

Where $\{L_i\}$ is a set of labels. Each label is represented by:

$$L_i = \{R_i, F_i\} \tag{2}$$

Where R_i is the Reference Words (RWs) label and F_i is the Key Fields (KFs) label.

2.2   Extraction Entity in Contiguous Structure

Tokenization. The tokenization allows the presentation of a label in the form of a set of tokens, $SetT$, which are separated by whitespace character as:

$$L_i = SetT \tag{3}$$

The tokens of R_i are defined as:

$$SetR = \{T_i^R | SetR \in SetT\} \tag{4}$$

The tokens of F_i are defined as:

$$SetF = \{T_i^F | SetF \in SetT\} \tag{5}$$

Tokens Filtering. In this step, Algorithm 1 is iteratively used to delete $SetR$. This algorithm is stopped when $SetR$ is empty. At this stage, $SetT$ contains only the entire tokens $SetF$. Each token is stored on its bounding box which is defined by:

$$T_i^F \rightarrow [x_i^F, y_i^F, w_i^F, h_i^F] \tag{6}$$

Where x_i^F and y_i^F represent the coordinate of upper left corner, w_i^F the width and h_i^F is the height of the rectangle.

Algorithm 1. Tokens filtering

1: **Input:** $SetT = SetR.SetF$ // Set of tokens
2: **Output:** $SetT = SetF$
3: **begin**
4:    **while** $SetR \neq \emptyset$ **do**
5:        $SetT = SetT/T_i^R$
6:    **end while**
7:    return $SetT = SetF$
8: **end**

Relevant Tokens Clustering. In this step, we propose a correction module for elimination the superfluous tokens. It represents the arrangement of relevant tokens of local entity. The geometric relations of a structure is modeled by distances measuring. The clustering of relevant tokens and eliminating the noisy content require the distances measuring between consecutive tokens T_i^F and T_j^F $(j = i + 1)$. Each distance is calculated as:

$$d_{ij} = x_j^F - (x_i^F + w_i^F) \tag{7}$$

The incremental algorithm, detailed in Algorithm 2, is applied to concatenate relevant tokens. $SetF$ contains at least one token. In this case, the latter represents the relevant token. If $SetF > 1$, then, we need to cluster relevant tokens. To achieve this goal, a threshold S is defined as the maximum distance between two consecutive tokens. This threshold is empirically defined. The measured distance is compared with S. If $d_{ij} \leq S$, then, the tokens are concatenated. If it is not the case, then, the algorithm is stopped and the rest of tokens, $SetN$, are ignored.

Algorithm 2. Incremental algorithm

1: **Input:** $SetF$ // Entity containing at least one token and may contain noisy tokens $SetN = \{T_i^N | SetN \subset SetF\}$
2: **Output:** $RT = SetF/SetN$ // Relevant tokens
3: **begin**
4:     **while** $SetF \neq \emptyset$ **do**
5:       **if** $SetF = 1$ **then**
6:         $RT = SetF$
7:       **elseif** $SetF > 1$ **then**
8:         **for** $j = 1 : SetF - 1$ **do**
9:           **if** $d_{ij} \leq S$ **then**
10:            $T_i^F = concat(T_i^F, T_j^F)$
11:            $T_j^F = []$
12:           **elseif** $d_{ij} > S$ **then**
13:            $RT = SetF/SetN$
14:          **end if**
15:         **end for**
16:         return $RT = SetF/SetN$
17:       **end if**
18:     **end while**
19: **end**

2.3 Extraction Entity in Noncontiguous Structure

Entity in noncontiguous structure means that RWs and KFs appear in the invoice in vertical structure. Since the drawing of relationships between RWs and all the KFs is time consuming and no avail, we propose to filter the labels. This requires the detection of KFs in a given region.

For relevant entity extraction, we build a graph of structural relationships. This graph is called Noncontiguous Graph.

Noncontiguous Graph Building. For noncontiguous entity structure extraction, as detailed in Algorithm 3, a graph is built $G = (N, M, E)$ in which N is a node of the label R_i. M is a set of finite nodes that represent the labels F_j having the centers under the center of N. $E \subseteq N \times M$ is a finite set of arcs which represent a geometric relationships between the node N and the nodes of M. Each arc $e_{ij} \in E$ relating the node N and m_j is represented by Nm_j. We define a feature vector which describes the geometric relationships between N and m_j.

$$a_{ij} = (CN_i, Cm_j, e_{ij}) \tag{8}$$

Where: CN_i is the center of the node N (step 4 in Algorithm 3). Cm_j is the center of each node m_j (step 6 in Algorithm 3). e_{ij} is the distance that separates the bounding boxes of the labels corresponding to N and m_j (step 10 in Algorithm 3), as we can view in Fig. 4. The idea is to detect the nearest m_j to N (step 13 in Algorithm 3). We consider only the nodes having the centers under the center of N. The distances are calculated as:

$$e_{ij} = \begin{cases} 1, & \text{if } Cm_j(2) > CN_i(2) \\ 0, & \text{else} \end{cases} \tag{9}$$

Where: $Cm_j(2)$ is the second coordinate (ordinate) of the center Cm_j. $CN_i(2)$ is the second coordinate of the center CN_i. e_{ij} is calculated to filter the KFs labels i.e., we bethink only the centers having the upright under the center of N.

The centers are calculated to determinate the nearest m_j to N. In Fig. 4, m_4 represents the nearest label KFs node to N i.e., m_4 is the relevant KFs. However, the latter may contain noisy tokens that must be eliminated. So, we need to tokenize the relevant KFs (step 14 in Algorithm 3) for clustering relevant tokens and ignore the noisy one.

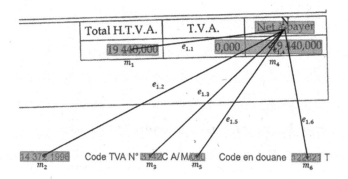

Fig. 4. Noncontiguous graph

The difficulty of detecting the relevant tokens of field in a vertical structure resides in this step. In a horizontal structure, the starting token from which begins the clustering of tokens is known. In addition, noisy tokens are found only on the right side. By cons, in a vertical structure, it is first necessary to determine the starting token. Then, we have to perform a sweeping to eliminate noisy tokens to the left and then to the right. To achieve this goal, a subgraph of relationships is built between the node N and the tokens $K = \{k_j\}$ of the nearest node m_4. The nearest token is the frame used for a sweeping. To determinate the nearest token, we calculate the distance p_{ij} between the node N and each token k_j (step 16 in Algorithm 3). The nearest token possesses the minimum distance with N (step 18 in Algorithm 3). We call this token "ind" as it represents an index from which begins the sweeping. In Fig. 5, the "ind" is k_2.

Sweeping. The sweeping is the exploration token by token of an entity. It is done in both directions to the left (step 20 in Algorithm 3) and then to the right (step 23 in Algorithm 3) for superfluous tokens elimination. The geometrical relationships, provided by the distances measured between tokens inside $Left\_M$, are used to concatenate relevant tokens. This matrix is defined as:

$$Left\_M = (K(1:ind)) \tag{10}$$

Whenever, we calculate two distances between two consecutive tokens. The first distance is calculated as:

$$n_{Z-1,Z} = Left\_M_{(Z,1)} - (Left\_M_{(Z-1,1)} + Left\_M_{(Z-1,3)}) \tag{11}$$

This distance must not exceed the threshold S previously identified (explained in Sect. 2.3).

To ensure the horizontal alignment of consecutive tokens, we need to calculate the distance between their second coordinates. This distance must not exceed certain threshold H and is calculated as:

$$g_{Z-1,Z} = Left\_M_{(Z,2)} - Left\_M_{(Z-1,2)} \tag{12}$$

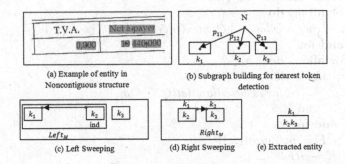

(a) Example of entity in Noncontiguous structure

(b) Subgraph building for nearest token detection

(c) Left Sweeping

(d) Right Sweeping

(e) Extracted entity

Fig. 5. Subgraph of tokens

The left sweeping outcome is $Left\_M$ containing only one element grouping the relevant tokens. This element is added to the beginning of the created matrix $Right\_M$ for the right sweeping. So, all relevant tokens are grouped in $Right\_M$ (step 25 in Algorithm 3) which is defined as:

$$Right\_M = (K(Left\_M + 1 : end)) \qquad (13)$$

The concatenation in the right sweeping is done with the same principles detailed in the left sweeping. Figure 5 shows the process of sweeping for tokens concatenation. In Fig. 5(b), a subgraph of geometric relationships is established between the nodes. The nearest token, ind, having the minimum distance with the node N is detected. The latter is "19". In Fig. 5(c), $Left\_M$ contains two tokens "0,000" and "19". The distance $n_{Z-1,Z}$ between these tokens exceeds the threshold S. For that, the token "0,000" is eliminated. In Fig. 5(d), $Left\_M$, containing one element, is integrated in the start of $Right\_M$ and the right sweeping begins. In Fig. 5(e), the right sweeping allows the concatenating of relevant tokens ("19", "440,000").

3   Experiments

3.1   Dataset

For test, we use a dataset of 930 real invoices obtained from Compagnie des Phosphates de Gafsa (CPG)[2]. The entities are categorized into 7 types: Invoice Number (N°), Invoice Date (DT), Account Identity (AI), Pre-tax Amount (PA), Total Including Tax (IT), Holdback (H) and Balance Due (BA).

Algorithm 3. Noncontiguous entity structure extraction

1: **Input:** N // RWs
M // KFs
K // Tokens
2: **Output:** RT// Relevant tokens
3: **begin**
4:     $CN_i = calculate\_Center(N)$
5:         **for all** m_j **do**
6:             $Cm_j = calculate\_Center(m_j)$
7:         **end for**
8:         **for all** m_j **do**
9:             **if** $Cm_j(2) > CN_i(2)$
10:                $e_{ij} = Euclediandist(CN_i, Cm_j)$
11:            **end if**
12:        **end for**
13:        $RF = m_j(dmin(e_{ij}))$
14:        $K \leftarrow RF$

---

[2] http://www.cpg.com.tn.

```
15:        for all k_j do
16:            p_ij = Euclediandist(CN_i, Ck_j)
17:        end for
18:        ind = k_j(dmin(p_ij))
19:        while Left_M ≠ ∅ do
20:            left_sweep()
21:        end while
22:        while Right_M ≠ ∅ do
23:            right_sweep()
24:        end while
25:        Right_M = RT
26:end
```

It is important to indicate that our system sustains the data extraction from grayscale, color and bi-tonal (black and white) images. Our system is insensitive to the multiplicity of fonts. Although the preprocessing step does not belong of our work, our system is able to manipulate little noisy invoices with a slight skew. These invoices contain graphical elements, logos, vertical and horizontal lines, and tables. Some manipulated invoices are shown in Fig. 6.

We have used our ground truth to evaluate our system's performance. This ground truth was manually prepared.

3.2 Erroneous RWs Correction

In our system, entities are labeled using Regex. The patterns are written to allow some OCR errors in RWs such as confusing zero with capital or lowercase O (e.g., Facture no⇒ Facture n0). This can allow unconstrained input that nearly matches the Regex pattern to be taken in account and significantly improve the performance. The refined Regex has allowed us to detect correctly 62 N°, 13 DT, 53 PA, 17 IT, 153 AI, and 23 BA.

3.3 Structure Correction Evaluation

To capture contiguous and noncontiguous structures of entity, a set of Regex patterns are used in conjunction with geometric relationships between labels. The correction step is integrated for superfluous tokens elimination. The goal is to increase the accuracy of the extraction. The correction step has allowed us to correct to 100 % the superfluous tokens and yielded a growth of the accuracies. Table 1 shows the impact of this step for each entity. For that, we use two options; without correction (W/o C) and with correction (With C). The most interesting result is for the N° entity since it has a countless number of formats. The obtained rates justify the fixed threshold distances between any consecutive tokens. We use two thresholds: S is around 32. It is fixed for concatenating the tokens whatever in contiguous or noncontiguous structure. The second threshold H is around 12. It is fixed only in noncontiguous structure to ensure the alignment and the consecutiveness of the tokens. The thresholds set show the power possessed by

Fig. 6. Sample of invoices in our dataset

our method for correction. To ensure the robustness of our correction method, we propose to strengthen these thresholds by other features such as font size to avoid bad detection that can be generated by using the few thresholds in other models.

Missed entities, as detailed in Table 2, are due to the following issues: errors in RW represent the RWs completely wrong, so, they cannot be identified by the Regex. Errors in KF are the KFs partially corrected or completely not corrected: if one field is not properly extracted, then, the entity was regarded as erroneous. The confusing labels (CL) means that the label is not associated with the correct entity which leads a failed match for another entity. The OCR sometimes missed the text zone (MT) due to: skewed image, noise, degraded characters, bad detection of tabular structure, etc.

Table 1. Impact of correction

Entity	Method (%)	
	W/o C	With C
N°	83.22	98.17
DT	97.63	98.82
BA	95.01	97.04
PA	94.01	96.72
IT	93.42	96.24
H	90.91	93.94

Table 2. Missed entities

Entity	Error Types (%)			
	RW	KF	CL	MT
N°	29.41	29.41	11.76	29.41
DT	9.10	36.36	63.64	9.10
BA	11.76	23.53	29.41	35.29
PA	41.37	20.69	6.90	31.04
IT	41.67	33.33	8.33	16.67
AI	40.48	19.05	0	40.48

Table 3. Rates comparison

	Recall(%)	Precision(%)
Method [3]	88.88	95.23
Method + Module correction [3]	93.37	97.50
Our method	92.49	98.86
Our method + Module correction	96.78	98.91

3.4   Comparison with Existing System

Table 3 synthesizes the obtained Recall and Precision of our system. In this table, we also compare our work with the results obtained by the system proposed in [3]. Recall and Precision are defined as:

$$Recall = \frac{relevant\ extracted\ entities}{relevant\ entities} \qquad (14)$$

$$Precision = \frac{relevant\ extracted\ entities}{extracted\ entities} \qquad (15)$$

4   Conclusion

We have proposed an approach for entity extraction from scanned invoices. We have showed how adopting a local structure of entities is very efficient for data extraction. This represents a powerful tool in dealing with variant layout entities. Our method is reinforced by correction step for superfluous tokens elimination. The experimental results have showed an interesting improvement in the performance and accuracy of the extraction process.

Acknowledgment. We are grateful to CPG Company for providing real invoices for test.

References

1. Saund, E.: Scientific challenges underlying production document processing. In: Document Recognition and Retrieval XVIII, DRR (2011)
2. Rusinol, M., Benkhelfallah, T., Poulain, V.D.: Field extraction from administrative documents by incremental structural templates. In: International Conference on Document Analysis and Recognition, ICDAR (2013)
3. Kooli, N., Belaid, A.: Semantic label and structure model based approach for entity recognition in database context. In: International Conference on Document Analysis and Recognition, ICDAR (2015)
4. Dejean, H.: Extracting structured data from unstructured document with incomplete resources. In: International Conference on Document Analysis and Recognition, ICDAR (2015)
5. Rahal, N., Benjlaiel, M., Alimi, Adel. M.: Incremental structural model for extracting relevant tokens of entity. In: IEEE International Conference on Systems, Man, and Cybernetics (SMC) (2016, to be published)

Detecting Ellipses in Elongated Shapes
Using the Thickness Profile

Aysylu Gabdulkhakova[(✉)] and Walter G. Kropatsch

Pattern Recognition and Image Processing Group,
TU Wien, Vienna, Austria
{aysylu,krw}@prip.tuwien.ac.at

Abstract. This paper presents a method that detects elliptical parts of
a given elongated shape. For this purpose, first, the shape is represented
by its skeleton. In case of branches, the skeleton is partitioned into a
set of lines/curves. Second, the ellipse parameters are estimated using
the thickness profile along each line/curve, and the properties of its first
and second derivatives. The proposed method requires no prior information about the model, number of ellipses and their parameter values.
The detected ellipses are then used in our second proposed approach for
ellipse-based shape description. It can be applied for analysing motion
and deformation of biological objects like roots, worms, and diatoms.

Keywords: Shape analysis · Shape description · Ellipse detection

1 Introduction

The advantages of modelling the structure of the shape with ellipses cover various
aspects. Human perceptual system interprets the significant parts of the shape
or the rigid parts of articulated objects with the ellipses [10]. Also the ellipsoids
and super quadrics provide a strong cue about the orientation of anisotropic
phenomena [11]. Indeed, the major axis of the ellipse, when interpreted as an
orientation, becomes useful for judging the motion direction or for constraining
the motion of articulated parts of the object [29]. The eccentricity of the ellipse, ε,
characterizes its elongation, and is estimated as $(\sqrt{a^2 - b^2})/a$, where a and b are
the lengths of the semi-major and semi-minor axes of the ellipse respectively. It
can be applied for analysis of deformations like shrinking and stretching. Tuning
the ellipse elongation along its orientation can be used to enhance the traditional
skeletonization algorithms for the problem of representing the intersection of two
straight lines (or, alternatively, a cross), compare cases (a) and (d) in Fig. 1.
Another advantage of using ellipses refers back to the paper of Rosenfeld [23].
He defines the ribbon-like shapes that can be represented given a spine and a

A. Gabdulkhakova—Supported by the Austrian Agency for International Cooper-
ation in Education and Research (OeAD) within the OeAD Sonderstipendien program, financed by the Vienna PhD School of Informatics.

A. Robles-Kelly et al. (Eds.): S+SSPR 2016, LNCS 10029, pp. 412–423, 2016.
DOI: 10.1007/978-3-319-49055-7_37

generator (disc or line segment). There are also algorithms that use ellipse as the generator [27]. To the best of our knowledge the ellipse was not assumed as a unified generator that can degenerate into a disc ($\varepsilon = 0$) and/or into a line segment ($\varepsilon = 1$). On one side, this provides a smooth transition between the rectangular parts of the object that have different properties, compare Fig. 1, cases (b) and (e). On the other side, symmetric shapes with a high positive curvature can be represented by a single spine instead of a branched skeleton, compare Fig. 1, cases (c) and (f).

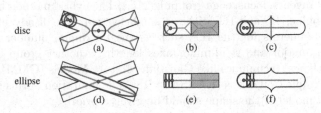

Fig. 1. Advantages of using ellipses as a generator for improving the skeletonization in case of intersection of two straight lines [(a) and (d)]; a union of two distinct rectangles [(b) and (e)]; a shape with a high positive curvature [(c) and (f)]. Blue indicates a type of generator, and the skeleton is shown in pink. (Color figure online)

The goal of this paper is twofold: (1) detect elliptical parts of the given elongated shape, and (2) describe its structure with ellipses and their degenerations (line segments and discs). Existing approaches for ellipse detection either try to minimize the deviation between a single ellipse and the given shape (based on Least-Squares [8,19]), or require an information about a number of ellipses and their parameters (based on Hough Transform [4,22] and Gaussian Mixture Models [14]). In contrast, the proposed method does not have the above constraints. Taking the skeleton of the shape, we, first, decompose it into a set of simple lines/curves w.r.t. branches. Then, for each line/curve we analyse its thickness profile and detect the parameters of ellipses. To obtain invariance to deformation, we find the ellipses on the flattened version of the shape, while preserving the mapping to the original shape [1,13].

The remaining of the paper is structured as follows. Section 2 discusses the state-of-the art in the area of ellipse detection. Then we explain the proposed methods for ellipse detection and shape description in Sects. 3 and 4 respectively. Specific algorithms and the related research are directly described in the corresponding subsections. In Sect. 5 we provide the experimental results, and give the final remarks in Sect. 6.

2   Related Work

Wong et al. [28] distinguished three major categories of approaches that aim at finding ellipses: (1) based on Least-Squares (LS), (2) based on voting scheme, (3) uncategorized (statistical, heuristic, and hybrid techniques).

LS approaches minimize the deviation between the computed ellipse points and the original set of points [8,19]. The disadvantage of the standard LS method is that it finds only a single ellipse and is sensitive to noise and outliers. Voting scheme based methods mainly operate with the Hough Transform (HT) [4,16,20, 22,24]. Ellipse has five parameters - orientation, length of the semi-major axis a, length of the semi-minor axis b, coordinates of a center x and y. The complexity of searching a combination of these parameters is high - $O(n^5)$, where n is the number of input points [7]. In order to optimize the computation and reduce the complexity of the parameter space, the ellipse-specific properties are used (symmetry, tangents, locus, edge grouping, etc.). The hybrid method by Cicconet et al. [5] combines LS and HT within a single system. The disadvantage lies in specifying parameters for the HT, such as the maximum number of ellipses, range of semi-major and semi-minor axes lengths. Another group of methods uses the statistical techniques like Gaussian Mixture Models (GMM) [14,29], or RANdom SAmple Consensus (RANSAC) [18]. The limitation of this idea is that the elliptical model of the shape should be given a priori.

3   Thickness-Based Ellipse Detection (TED)

The *TED* algorithm aims at finding the parameters of ellipse according to the thickness profile along the skeleton of the shape, or, in case of branches, along the lines/curves that form a skeleton. The finest detection is achieved on the symmetric shape that satisfies the following criterion. Consider a symmetric 2D shape, $S1$, in a Cartesian XY plane. Its major symmetry axis coincides with the abscissa. Rotate $S1$ about the abscissa (Fig. 2(a)), project the resultant 3D volume (Fig. 2(b)) orthogonally w.r.t. the original XY plane (Fig. 2(c)). The criterion requires the original shape $S1$ to be identical to the resultant projection $S2$. In general, the present work aims at the above type of symmetric shapes and/or at their combination. To emphasize the fact that points of the skeleton are equidistant from the opposite borders of the shape, we will also refer to it as a symmetry axis.

(a)   (b)   (c)

Fig. 2. Description of the target shape: (a) original shape $S1$ (b) 3D volume, which is obtained by rotating $S1$ about X-axis (c) projection of the 3D volume onto the XY plane - $S2$. For the target shape, $S1$ is identical to $S2$.

The pipeline of the proposed *TED* contains three algorithmic steps: *Step 1* - skeletonization, *Step 2* - computation of the thickness profile along the skeleton,

Step 3 - ellipse parameter estimation w.r.t. the thickness profile. When the shape is deformed, the algorithm has an additional step of flattening the shape after skeletonization. In the scopes of this paper we will call this modification as a *Straightened TED*, or *S-TED*.

3.1   Skeletonization

Numerous skeletonization techniques exist. The majority of them refers to the concept of Medial Axis Transformation (MAT), introduced by Blum [3]. The underlying principle is to fit the circles of maximum radii inside the shape and sequentially connect their centres. The resultant representation does not contain the ridges of the contour, as opposed to *grass-fire* skeleton [3] and the Distance Transform (DT) [26]. By the definition, DT is an operator that yields a distance-valued image by assigning to each pixel of the given binary image its distance to the nearest background pixel.

Another group of methods evolves from Voronoi diagrams [21]. The idea is to use the boundary points of the shape in order to produce the Voronoi tesselation. After that the skeleton is obtained as the set of points and of the Voronoi diagram that belongs to the given shape.

The last approach within this overview is homotopic *thinning* [9] (a recent survey on skeletonization was conducted by Saha et al. [25]). It iteratively peels the surface of the object, while preserving topology. The result is a one-pixel-thick skeleton.

In general, all the above approaches suffer from noise. This impact is reflected by spurious branches that do not correspond to the essential parts of the shape.

3.2   Shape Flattening

One way to flatten the shape is to compute the curvature along the symmetry axis. Starting at the origin, the points of the symmetry axis are sequentially translated towards the alignment direction [12]. Such an approach has a high computational cost, and fails on the spiral shapes.

Our idea is to transform the Cartesian coordinates of each point of the shape in image space to the object-specific space. For this purpose we use Straightened Curved Planar Reformation (CPR) [13]. It takes longitudinal values along one axis and latitudinal values along the other axis.

The *longitudinal axis* reflects the length of the main axis of the object. One of the two end points is used as the origin. The values range from 0 to N, where N equals the number of points that form the axis. The *latitudinal axis* reflects the thickness of the object for each point along the longitudinal axis. The advantages of the above method include (1) the stability to deformations (excluding crossing), (2) the invariance to affine transformations, (3) the reconstruction of the straightened representation of the shape without an additional computation. The method can be extended to 3D case.

3.3   Thickness Profile

Given a symmetry axis of the object, SA, a position function, $pos : \mathbb{R}_+ \rightarrow \mathbb{R}^2$, associates the point $P = (x, y) \in SA$ to the arc length s between the point P and the origin. In this context, thickness profile, $f : \mathbb{R} \rightarrow \mathbb{R}_+$, is a 1D function of arc length s along SA that returns the thickness of the object at $pos(s)$:

$$f(s) = \{lthick(pos(s))|s \in [0, \ldots, length(SA)]\} \tag{1}$$

By *lthick* we denote a distance metric which defines the local thickness of the object at the point P. This metric is chosen such that it satisfies the requirements of the particular problem. Indeed, in *TED* algorithm, the target objects have a straight symmetry axis, and are rotated such that this axis is parallel to abscissa. Let us assume that local thickness equals the width of a cut taken along the normal at the given skeletal point. Then, *lthick* can be defined as a double value of a chamfer DT [26] with a City-Block metric at this skeletal point. The reason is that (1) in this problem statement, the City-Block metric provides a better thickness estimation than the Euclidean metric since having a 4-neighbourhood, (2) as opposed to computing the number of the shape points corresponding to each cut along the skeleton, chamfer DT is less sensitive to noise and local perturbations, (3) chamfer DT has a linear time complexity.

In *S-TED* algorithm *lthick* can be defined by latitudinal values along the flattened shape. Computing the latitudinal values with the normals to the symmetry axis is sensitive to noise. The linear LS algorithm enables to reduce the impact of such local perturbations. Also, the algorithm enables to build a hierarchy of ellipses by smoothing the thickness profile with the Gaussian filter. Starting at a fine level with many ellipses that correspond to every local perturbation (in case of a noisy data), and finishing at a coarse level that represents the whole shape with a single ellipse.

In literature also exist another thickness-estimation algorithms, such as [6], that may improve the performance. Though, comparison of various distance metrics for thickness was not the goal of the present paper.

3.4   Ellipse Parameter Estimation

The idea of detecting the parameters of ellipse is derived from the properties of the thickness change along the elongated shape. The centre of an ellipse-candidate corresponds to the position of the *local maximum* (C in Fig. 3), and/or to the position of the *bending point* (C_1, C_2 in Fig. 3). The *local minimum* corresponds to some point on the ellipse-candidate (P_1, P_2 in Fig. 4).

Position of the Ellipse Centre. The algorithm for detecting the centres of ellipses is described in relation to three possible configurations of the target shapes: (1) a single ellipse, (2) a union of multiple ellipses, and (3) a union of multiple ellipses with a smooth transition.

Single ellipse. Thickness profile of a single ellipse monotonically increases to the length of the semi-minor axis, b, and then monotonically decreases. Thus, the position of the ellipse centre is exactly at the local maximum of the thickness profile. Indeed, consider the canonical ellipse representation (see Fig. 4). Its centre is at the origin, $O = (0,0)$, minor and major axes are correspondingly collinear with the ordinate and abscissa. The canonical ellipse equation is:

$$\frac{x^2}{a^2} + \frac{y^2}{b^2} = 1, \text{where } P = (x,y) \text{ is a point on the ellipse.} \tag{2}$$

Substituting x with s, and assuming that $y = \pm f(s)$ w.r.t. Eq. 1 gives:

$$\frac{s^2}{a^2} + \frac{f^2(s)}{b^2} = 1 \Rightarrow f^2(s) = \frac{b^2}{a^2}(a^2 - s^2) \Rightarrow 2f(s)f'(s) = \frac{b^2}{a^2}(-2s) \tag{3}$$
$$f'(s) = 0, a \neq 0, b \neq 0 \Rightarrow s = a, x = 0$$

Since dealing with the discrete space, $f'(s)$ is approximated by the differences between the neighbouring points of the profile. Therefore, *local maximum* is the point, where $f'(s)$ changes from positive to zero/negative value.

Multiple ellipses. In general, a union of multiple ellipses along the symmetry axis results in multiple *local maxima*. So, the positions of the centres are found w.r.t. $f'(s)$ as for the single ellipse. Specific case is the union of two identical ellipses with the common major or minor axis - there is only one *local maximum*, and, thus, one detected ellipse. Another issue is the union of two ellipses of different sizes, that overlap along the minor axis of the smaller ellipse. For the small ellipse, the local maximum cannot be detected, since the sign of the derivative does not change. This case is discussed in the next paragraph. For the complete enclosure case, estimation of the ellipse centre positions w.r.t. the local maxima is possible, if performed for each object separately.

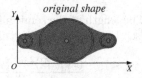

original shape

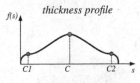

thickness profile

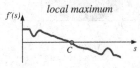

local maximum

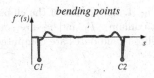

bending points

Fig. 3. Example of multiple ellipses with smooth transitions. The positions of the centres are detected w.r.t. to the local maximum at the point C, and the bending points - C_1 and C_2.

Multiple ellipses with smooth transitions. Consider a shape that consists of ellipses and smooth transitions between them which cover at most a half of each ellipse (see Fig. 3). In this case, the ellipse centre positions C_1 and C_2 w.r.t. the local maxima cannot be estimated. At these points, the sign of $f'(s)$ stays the same, but its value is changing by Δ while connecting two different monotonic functions. As a result, there is a Dirac function like peak at $f''(s)$

with a negative Δ amplitude at the corresponding positions. We call C_1 and C_2 *bending points.*

The algorithm for detecting bending points has 2 steps. First, we find the set of the peak-candidates by comparing the ratio between the maximum and median values of the $f''(s)$. This is done, in order to eliminate the impact of noise. As opposed to mean, median is not distorted by the maximum values.

In the second step we fit ellipses inside the peak-candidates, and compute their eccentricities using the equation $(\sqrt{a^2 - b^2})/a$. For the Dirac function like peaks this ratio converges to 1.

The Lengths of the Semi-major and Semi-minor Axes. Let us consider that $pos(C) = (x_C, y_C)$ is the center of the ellipse-candidate, $pos(P_1) = (x_{P_1}, y_{P_1})$ and $pos(P_2) = (x_{P_2}, y_{P_2})$ are the points on it, detected as the closest local minima to C (Fig. 4). Then the length of the minor axis b equals:

$$b = f(C) \tag{4}$$

The length of the semi-major axis a is then computed w.r.t. the thickness profile (see Fig. 4) and implicit ellipse representation:

$$\frac{(P_i - C)^2}{a^2} + \frac{(f(P_i) - 0)^2}{b^2} = 1 \Rightarrow a = \arg\min_{1 \leq i \leq 2} \frac{|P_i - C|}{\sqrt{1 - \frac{f^2(P_i)}{b^2}}} \tag{5}$$

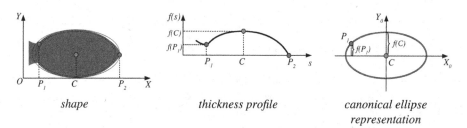

| shape | thickness profile | canonical ellipse representation |

Fig. 4. Estimation of the ellipse parameters. C is the ellipse centre (found as the local maximum); P_1 and P_2 are some points on the ellipse (found as local minima). Two ellipse candidates are computed w.r.t. P_1, P_2 and the implicit ellipse representation.

3.5   Complexity of the Algorithms

TED contains 3 algorithmic steps. *Step 1* - skeletonization of the shape using the chamfer DT - $O(n)$, where n is the number of shape points. Next, in *Step 2*, the thickness profile is computed by taking the maximum DT value along the skeleton. The complexity remains $O(n)$. *Step 3* - finding the local extrema and the bending points has a linear complexity. Since the above steps are performed sequentially, the resultant complexity equals $O(n)$.

S-TED contains shape flattening in addition to *Step 1*. The complexity of this procedure is $O(n)$, where n is the number of shape points. *Steps 2–3* are identical to those described in *TED*. Therefore, the overall complexity of the *S-TED* is $O(n)$. In contrast, the approach proposed by Cicconet et al. [5] has a higher complexity of $O(m \cdot n^2)$, where n is the number of shape contour points, and m is the number of ellipses.

4   Shape Description

Given a shape as a single ellipse, or a union of multiple ellipses along the symmetry axis, the shape description contains a set of detected ellipses. Given a shape as a union of smoothly connected ellipses, the shape description contains a set of detected ellipses and line segments. The centres of these line segments correspond to the points of the skeleton that connect the neighbouring ellipses. The length of each line segment equals to the corresponding local shape thickness.

Our approach enables to describe the shapes which parts have a high positive curvature. If function $f(s)$ at some point P changes from concave to convex, or vice versa, then P is a *point of inflection*. In this case the sign of $f''(s)$ should change from positive to zero/negative value or from negative to zero/positive value. In contrast, axial representations with a single generator are not able to handle the shape with round ends and high positive curvature [23].

5   Experiments and Discussion

The performance of *TED* and *S-TED* was compared to the state-of-the-art approach of Cicconet et al. [5]. As opposed to standard LS-based methods [8,19], [5] can detect multiple overlapping ellipses, and does not require an a priori model, in contrast to GMM-based methods [14,29]. As reported, [5] outperformed the algorithm of Prasad et al. [22], which in turn showed better results than [2,15,17,18,20].

The test dataset contains 80 synthetic images of one and two ellipses, 500 by 500 pixels. As discussed with the authors [5], the test shapes closely resemble ellipses, and contain boundary points of each ellipse. For the proposed method we use the union of binary masks of these ellipses. With the reference to the target shape definition, there are 8 classes of possible configurations, 10 images each (see Table 2). For a single ellipse: (1) different degree of elongation (a varies from 40 to 200 pixels, b - from 4 to 40 pixels), (2) rotation, (3) distortion with a white Gaussian noise, (4) distortion with a "sinusoidal" noise[1]. For two ellipses: (5) a union of ellipses of the same size, (6) a union of ellipses of the different sizes, (7) ellipses with mutually orthogonal major axes, and (8) configurations that approximate an articulated motion. Since targeting the elongated shapes without holes, having more than two ellipses will not reveal additional test cases. The quantitative comparison was performed in Precision/Recall manner, as described in [5] (see Table 1, *Ellipse Detection*).

---

[1] Sinusoidal noise takes every i-th sin wave value, and adds it to the original signal.

Table 1. Interpretation of Precision/Recall methods w.r.t. the experiment type

	True positive (tp)	False positive (fp)	False negative (fn)	Precision (**P**)	Recall (**R**)
Ellipse detection	a, b, x_C, y_C deviate from the ground truth by at most 20 pixels, and rotation angle by at most $5°$	$N-1$ True positives	a, b, x_C, y_C deviate from the ground truth by more than 20 pixels, and rotation angle by more than $5°$	$\frac{tp}{(tp+fp)}$	$\frac{tp}{(tp+fn)}$
Shape description	Pixels that belong to both original and reconstructed shapes	Pixels that belong to the reconstructed shape only	Pixels that belong to the original shape only		

Table 2. On the left: the results of *TED* and *S-TED* as compared to Cicconet et al. [5]. On the right: the examples of ellipse configurations w.r.t. 8 classes, and examples of the input data for the compared algorithms.

1 ellipse

	1		2		3		4	
	P	R	P	R	P	R	P	R
Cicconet et al. [5]	1	0.6	1	0.91	1	0.9	1	1
TED ($\sigma=5$, $n=15, L=30$)	0.91	1	1	1	1	1	1	1
TED ($\sigma=5$, $n=15, L=15$)	0.75	0.75	1	1	1	1	1	1
TED ($\sigma=5$, $n=15, L=5$)	0.7	0.54	0.9	1	0.89	0.61	0.9	0.75
S-TED ($\sigma=5$, $n=15, L=15$)	0.91	1	1	1	1	0.67	1	1
S-TED ($\sigma=5$, $n=15, L=30$)	1	1	1	0.6	1	0.64	1	1

2 ellipses

	5		6		7		8	
	P	R	P	R	P	R	P	R
Cicconet et al. [5]	1	0.5	1	0.55	1	0.15	1	0.2
TED ($\sigma=5$, $n=15, L=30$)	1	0.75	0.93	0.7	1	0.25	-	-
TED ($\sigma=5$, $n=15, L=15$)	1	0.85	0.79	0.75	1	0.3	-	-
TED ($\sigma=5$, $n=15, L=5$)	1	1	0.6	0.75	0.55	0.29	-	-
S-TED ($\sigma=5$, $n=15, L=30$)	1	0.9	1	0.75	-	-	0.88	0.7
S-TED ($\sigma=5$, $n=15, L=15$)	1	0.85	1	0.75	-	-	1	0.65

shape examples from the synthetic dataset | input for TED and S-TED | input for [5]

Table 3. The results of shape reconstruction as tested on a diatom dataset.

	Number of images	tp	fp	fn	**P**	**R**
TED ($\sigma=5, n=15, L=5$)	357	20532911	1266345	291387	**0.942**	**0.986**

The experimental results[2] are summarized in Table 2. The proposed detection of the ellipse parameters relies on $f'(s)$ and $f''(s)$. So, both *TED* and *S-TED*

---

[2] According to the algorithm description, *TED* focuses on shapes with a straight symmetry axis. Thus, it was tested on configurations (1)–(7). The current implementation of *S-TED* considers the symmetry axis to be the longest path of the given skeleton. In case of cross intersection of the ellipses, configuration (7), a different strategy for skeleton decomposition should be considered.

are sensitive to perturbations, and on a single ellipse their Recall rate worsens under the noise impact. Applying Gaussian function, $G(n) = \frac{1}{\sigma\sqrt{2\pi}}e^{-0.5(n/\sigma)^2}$, to the thickness profile enables to decrease the impact of noise, and achieve the highest Precision and Recall rates for TED ($\sigma = 5, n = 15$, and $L = 30$). Here n is the size of the Gaussian window, σ - a standard deviation, L is the number of smoothing iterations. The method of Cicconet et al. [5] assumes the number of ellipses to be given a priori. As a result, it demonstrates a high Precision rate on a single ellipse, as well as on two ellipses (always 1). Though, the Recall rates are lower as compared to TED and S-TED, and decrease sufficiently (from the minimum of 0.6 for a single ellipse to the minimum of 0.15 for two ellipses).

The quality of the shape reconstruction from the proposed shape description was tested on a part of a diatom dataset[3], that contains 357 real images. Figure 5 shows the main steps of processing a diatom image: segmentation (b), shape description (c), and shape reconstruction (d). Elliptical parts of the diatoms were detected w.r.t. TED algorithm. We performed the evaluation using the Precision/Recall method (as described in Table 1, *Shape Description*), and achieved the results as shown in Table 3.

Fig. 5. Example of the shape description and reconstruction of a diatom: (a) original image, (b) binary shape of the diatom, (c) shape description that includes detected ellipses (in yellow) and line segments (in blue), (d) reconstructed shape (Color figure online)

6 Conclusion and Future Work

The paper presents a novel approach that finds the parameters of ellipses by tracking the thickness change along the symmetry axis. The shape does not have to be a combination of fine ellipses. The proposed method enables to build a hierarchy of ellipse-based descriptions by smoothing the thickness profile with the Gaussian filter. The presented shape description approach has a high recoverability rate. In future we plan to use it for motion, growth and deformation analysis of the biological objects like roots, worms, and diatoms. A hierarchy of descriptions will enable to describe the motion at multiple levels of abstraction: from local movements of the parts to global behaviour of the entire object.

---

[3] ADIAC Diatom Dataset. http://rbg-web2.rbge.org.uk/ADIAC/pubdat/downloads/public_images.htm Accessed: 2016-10-22.

References

1. Aigerman, N., Poranne, R., Lipman, Y.: Lifted bijections for low distortion surface mappings. ACM Trans. Graph. **33**(4), 69:1–69:12 (2014)
2. Bai, X., Sun, C., Zhou, F.: Splitting touching cells based on concave points and ellipse fitting. Pattern Recogn. **42**(11), 2434–2446 (2009)
3. Blum, H.: A transformation for extracting new descriptors of shape. In: Models for the Perception of Speech and Visual Form, pp. 362–380. MIT Press (1967)
4. Chien, C.-F., Cheng, Y.-C., Lin, T.-T.: Robust ellipse detection based on hierarchical image pyramid and hough transform. J. Opt. Soc. Am. A **28**(4), 581–589 (2011)
5. Cicconet, M., Gunsalus, K., Geiger, D., Werman, M.: Ellipses from triangles. In: IEEE International Conference on Image Processing, pp. 3626–3630. IEEE (2014)
6. Coeurjolly, D.: Fast and accurate approximation of the Euclidean opening function in arbitrary dimension. In: IEEE International Conference on Pattern Recognition, pp. 229–232. IEEE Computer Society (2010)
7. Davis, L.M., Theobald, B.-J., Bagnall, A.: Automated bone age assessment using feature extraction. In: Yin, H., Costa, J.A.F., Barreto, G. (eds.) IDEAL 2012. LNCS, vol. 7435, pp. 43–51. Springer, Heidelberg (2012). doi:10.1007/978-3-642-32639-4_6
8. Fitzgibbon, A., Pilu, M., Fisher, R.: Direct least square fitting of ellipses. IEEE Trans. Pattern Anal. Mach. Intell. **21**(5), 476–480 (1999)
9. Haralick, R.M., Shapiro, L.G.: Computer and Robot Vision, vol. 1. Addison Wesley, Boston (1992)
10. Hoffman, D., Richards, W.: Parts of recognition. Cognition **18**(1), 65–96 (1984)
11. Jaklic, A., Leonardis, A., Solina, F.: Segmentation and Recovery of Superquadrics, vol. 20. Springer Science & Business Media, Berlin (2000)
12. Jiang, T., Dong, Z., Ma, C., Wang, Y.: Toward perception-based shape decomposition. In: Lee, K.M., Matsushita, Y., Rehg, J.M., Hu, Z. (eds.) ACCV 2012. LNCS, vol. 7725, pp. 188–201. Springer, Heidelberg (2012). doi:10.1007/978-3-642-37444-9_15
13. Kanitsar, A., Fleischmann, D., Wegenkittl, R., Felkel, P., Gröller, E.: CPR-curved planar reformation. In: Proceedings of the IEEE Visualization, pp. 37–44 (2002)
14. Kemp, M., Da Xu, R.Y.: Geometrically-constrained balloon fitting for multiple connected ellipses. Pattern Recogn. **48**(7), 2198–2208 (2015)
15. Kim, E., Haseyama, M., Kitajima, H.: Fast and robust ellipse extraction from complicated images. In: Proceedings of IEEE Information Technology and Applications (2002)
16. Lei, Y., Wong, K.C.: Ellipse detection based on symmetry. Pattern Recogn. Lett. **20**(1), 41–47 (1999)
17. Liu, Z.-Y., Qiao, H.: Multiple ellipses detection in noisy environments: a hierarchical approach. Pattern Recogn. **42**(11), 2421–2433 (2009)
18. Mai, F., Hung, Y.S., Zhong, H., Sze, W.F.: A hierarchical approach for fast and robust ellipse extraction. Pattern Recogn. **41**(8), 2512–2524 (2008)
19. Maini, E.S.: Enhanced direct least square fitting of ellipses. Int. J. Pattern Recogn. Artif. Intell. **20**(06), 939–953 (2006)
20. McLaughlin, R.: Randomized hough transform: improved ellipse detection with comparison. Pattern Recogn. Lett. **19**(3), 299–305 (1998)
21. Ogniewicz, R., Ilg, M.: Voronoi skeletons: theory and applications. In: IEEE Computer Society Conference on Computer Vision and Pattern Recognition, pp. 63–69 (1992)

22. Prasad, D., Leung, M., Cho, S.-Y.: Edge curvature and convexity based ellipse detection method. Pattern Recogn. **45**(9), 3204–3221 (2012)
23. Rosenfeld, A.: Axial representations of shape. Comput. Vis. Graph. Image Process. **33**(2), 156–173 (1986)
24. Rosin, P.: Ellipse fitting by accumulating five-point fits. Pattern Recogn. Lett. **14**(8), 661–669 (1993)
25. Saha, P., Borgefors, G., Sanniti di Baja, G.: A survey on skeletonization algorithms and their applications. Pattern Recogn. Lett. **76**, 3–12 (2016)
26. Soille, P.: Morphological Image Analysis: Principles and Applications. Springer Science & Business Media, Berlin (2013)
27. Talbot, H.: Elliptical distance transforms and applications. In: Kuba, A., Nyúl, L.G., Palágyi, K. (eds.) DGCI 2006. LNCS, vol. 4245, pp. 320–330. Springer, Heidelberg (2006). doi:10.1007/11907350_27
28. Wong, Y., Lin, S., Ren, T., Kwok, N.: A survey on ellipse detection methods. In: IEEE International Symposium on Industrial Electronics, pp. 1105–1110 (2012)
29. Xu, D., Yi, R., Kemp, M.: Fitting multiple connected ellipses to an image silhouette hierarchically. IEEE Trans. Image Process. **19**(7), 1673–1682 (2010)

Spatio-temporal Pattern Recognition

Unsupervised Interpretable Pattern Discovery in Time Series Using Autoencoders

Kevin Bascol[1], Rémi Emonet[1], Elisa Fromont[1(✉)], and Jean-Marc Odobez[2]

[1] Univ Lyon, UJM-Saint-Etienne, CNRS, Institut d'Optique Graduate School,
Laboratoire Hubert Curien UMR 5516, 42023 Saint-Etienne, France
elisa.fromont@univ-st-etienne.fr
[2] Idiap Research Institute, 1920 Martigny, Switzerland

Abstract. We study the use of feed-forward convolutional neural networks for the unsupervised problem of mining recurrent temporal patterns mixed in multivariate time series. Traditional convolutional autoencoders lack interpretability for two main reasons: the number of patterns corresponds to the manually-fixed number of convolution filters, and the patterns are often redundant and correlated. To recover clean patterns, we introduce different elements in the architecture, including an adaptive rectified linear unit function that improves patterns interpretability, and a group-lasso regularizer that helps automatically finding the relevant number of patterns. We illustrate the necessity of these elements on synthetic data and real data in the context of activity mining in videos.

1 Introduction

Unsupervised discovery of patterns in temporal data is an important data mining topic due to numerous application domains like finance, biology or video analysis. In some applications, the patterns are solely used as features for classification and thus the classification accuracy is the only criterion. This paper considers different applications where the patterns can also be used for data analysis, data understanding, and novelty or anomaly detection [4–6,18].

Not all time series are of the same nature. In this work, we consider the difficult case of multivariate time series whose observations are the result of a combination of different recurring phenomena that can overlap. Examples include traffic videos where the activity of multiple cars causes the observed sequence of images [6], or aggregate power consumption where the observed consumption is due to a mixture of appliances [10]. Unlike many techniques from the data mining community, our aim is not to list *all* recurrent patterns in the data with their frequency but to reconstruct the entire temporal documents by means of a *limited and unknown* number of recurring patterns together with their occurrence times in the data. In this view, we want to un-mix multivariate time series to recover how they can be decomposed in terms of recurrent temporally-structured patterns. Following the conventions used in [6], we will call a temporal pattern a *motif*, and an input multivariate time series a *temporal document*.

© Springer International Publishing AG 2016
A. Robles-Kelly et al. (Eds.): S+SSPR 2016, LNCS 10029, pp. 427–438, 2016.
DOI: 10.1007/978-3-319-49055-7_38

Artificial neural networks (or deep learning architectures) have (re)become tremendously popular in the last decade due to their impressive, and so far not beaten, results in image classification, speech recognition and natural language processing. In particular, autoencoders are artificial neural networks used to learn a compressed, distributed representation (encoding) for a set of data, typically for the purpose of dimensionality reduction. It is thus an unsupervised learning method whose (hidden) layers contain representations of the input data sufficiently powerful for compressing (and decompressing) the data while loosing as few information as possible. Given the temporal nature of your data, our pattern discovery task is fundamentally convolutional (the same network is applied at any instant and is thus time-shift invariant) since it needs to identify motifs whatever their time(s) of occurrence. To tackle this task, we will thus focus on a particular type of autoencoders, the convolutional ones. However, while well adapted for discriminative tasks like classification [1], the patterns captured by (convolutional) autoencoders are not fully interpretable and often correlated.

In this paper, we address the discovery of interpretable motifs using convolutional auto-encoders and make the following contributions:

- we show that the interpretability of standard convolutional autoencoders is limited;
- we introduce an adaptive rectified linear unit (AdaReLU) which allows hidden layers to capture clear occurrences of motifs,
- we propose a regularization inspired by group-lasso to automatically select the number of filters in a convolutional neural net,
- we show, through experiments on synthetic and real data, how these elements (and others) allow to recover interpretable motifs[1].

It is important to note that some previous *generative models* [6,21] have obtained very good results on this task. However, their extensions to semi-supervised settings (i.e. with partially labelled data) or hierarchical schemes are cumbursome to achieve. In contrast, in this paper, to solve the same modeling problem we present a radically different method which will lend itself to more flexible and systematic end-to-end training frameworks and extensions.

The paper is organized as follows. In Sect. 2, we clarify the link between our data mining technique and previous work. Section 3 gives the details of our method while Sect. 4 shows experiments both on synthetic and real data. We conclude and draw future directions in Sect. 5.

2   Related Work

Our paper shows how to use a popular method (autoencoders) to tackle a task (pattern discovery in time series) that has seldom been considered for this type of method. We thus briefly review other methods used in this context and then, other works that use neural networks for unsupervised time series modeling.

---

[1] The complete source code will be made available online.

Unsupervised Pattern Discovery in Time Series. Traditional unsupervised approaches that deal with time series do not aim at modeling series but rather at extracting interesting pieces of the series that can be used as high level descriptions for direct analysis or as input features for other algorithms. In this category fall all the event-based (e.g. [7,22,23]), sequence [15] and trajectory mining methods [25]. On the contrary of the previously cited methods, we do not know in advance the occurrence time, type, length or number of (possibly) overlapping patterns that can be used to describe the entire multivariate time series. These methods cannot be directly used in our application context.

The *generative methods* for modeling time series assume an apriori model and estimate its parameters. In the precursor work of [16], the unsupervised problem of finding patterns was decomposed into two steps, a supervised step involving an oracle who identifies patterns and series containing such patterns and an EM-step where a model of the series is generated according to those patterns. In [13], the authors propose a functional independent component analysis method for finding linearly varying patterns of activation in the data. They assume the availability of pre-segmented data where the occurrence time of each possible pattern is known in advance. Authors of [10] address the discovery of overlapping patterns to disaggregate the energy level of electric consumption. They propose to use additive factorial hidden Markov models, assuming that the electrical signal is univariate and that the known devices (each one represented by one HMM) have a finite known number of states. This also imposes that the motif occurrences of one particular device can not overlap. The work of [6] proposes to extract an apriori unknown number of patterns and their possibly overlapping occurrences in documents using Dirichlet processes. The model automatically finds the number of patterns, their length and occurrence times by fitting infinite mixtures of categorical distributions to the data. This approach achieved very good results, but its extensions to semi-supervised settings [19] or hierarchical schemes [2] were either not so effective [19] or more cumbursome [2]. In contrast, the neural network approach of this paper will lend itself to more flexible and systematic end-to-end training frameworks and extensions.

Networks for Time Series Mining. A recent survey [11] reviews the network-based unsupervised feature learning methods for time series modeling. As explained in Sect. 1, autoencoders [17] and also Restricted Boltzmann Machines (RBM) [8] are neural networks designed to be trained from unsupervised data. The two types of networks can achieve similar goals but differ in the objective function and related optimization algorithms. Both methods were extended to handle time series [1,14], but the goal was to minimize a reconstruction error without taking care of the interpretability or of finding the relevant number of patterns. In this paper, we show that convolutional autoencoders can indeed capture the spatio-temporal structure in temporal documents. We build on the above works and propose a model to discover the right number of meaningful patterns in the convolution filters, and to generate sparse activations.

3   Motif Mining with Convolutional Autoencoders (AE)

Convolutional AEs [12] are particular AEs whose connection weights are constrained to be convolution kernels. In practice, this means that most of the learned parameters are shared within the network and that the weight matrices which store the convolution filters can be directly interpreted and visualized. Below, we first present the traditional AE model and then introduce our contributions to enforce at the same time a good interpretability of the convolutional filters and a clean and sparse activation of these filters.

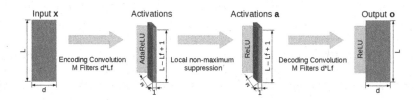

Fig. 1. Autoencoder architecture. Temporal documents of L time steps of d dimensional observations are encoded (here using M convolutional filters of size $d \times L_f$ forming the $^e\mathbf{W}$ weights) to produce an activation layer. A decoding process (symmetric to encoding; parameterized by the weights $^d\mathbf{W}$ of M decoding convolutional filters of size $d \times L_f$) regenerates the data.

3.1   Classical Convolutional Autoencoders

A main difference between an AE and a standard neural network is the loss function used to train the network. In an AE, the loss does not depend on labels, it is the reconstruction error between the input data and the network output. Figure 1 illustrates the main network modeling components of our model. In our case, a training example is a multivariate time series $\mathbf{x}$ whose L time steps are described by a vector $\mathbf{x}_{(:,t)} \in \mathcal{R}^d$, and the network is parameterized by the set of weights $\mathbf{W} = \{^e\mathbf{W}, {}^d\mathbf{W}\}$ involved in the coding and decoding processes. If we denote by $\mathbf{X} = \{\mathbf{x}^b \in \mathcal{R}^{L \times d}, b = 1 \dots N\}$ the set of all training elements, the estimation of these weights is classically conducted by optimizing the cost function $C(\mathbf{W}, \mathbf{X}) = MSE(\mathbf{W}, \mathbf{X}) + R_{reg}(\mathbf{W}, \mathbf{X})$ where the Mean Squared Error (MSE) reconstruction loss can be written as:

$$MSE(\mathbf{W}, \mathbf{X}) = \frac{1}{N} \sum_{b=1}^{N} \sum_{i=1}^{d} \sum_{t=1}^{L} \left(\mathbf{x}_{(i,t)}^b - \mathbf{o}_{(i,t)}^b \right)^2 \qquad (1)$$

where $\mathbf{o}^b$ (which depends on parameters $\mathbf{W}$) is the AE output of the bth input document. To avoid learning trivial and unstable mappings, a regularization term R_{reg} is often added to the MSE and usually comprises two terms. The first one, known as weight decay as it avoids unnecessary high weight values, is a ℓ_2 norm on the matrix weights. The second one (used with binary activations) consists of a Kullback-Leibler divergence $\sum_{j=1}^{M} KL(\rho||\hat{\rho}_j)$ encouraging *all* hidden activation units to have their probability of activation $\hat{\rho}_j$ estimated across samples to be close to a chosen parameter ρ, thus enforcing some activation sparsity when ρ is small. The parameters are typically learned using a stochastic gradient descent algorithm (SGD) with momentum using an appropriate rate scheduling [3].

3.2   Interpretable Pattern Discovery with Autoencoders

In our application, the learned convolution filters should not only minimize the reconstruction error but also be directly interpretable. Ideally, we would like to only extract filters which capture and represent interesting data patterns, as illustrated in Fig. 2-c–d. To achieve this, we add a number of elements in the network architecture and in our optimization cost function to constrain our network appropriately.

Enforcing Non-negative Decoding Filters. As the AE output is somehow defined as a linear combination of the *decoding* filters, then these filters can represent the patterns we are looking for and we can interpret the hidden layers activations **a** (see Fig. 1) as the occurrences of these patterns. Thus, as our input is non-negative (a temporal document), we constraint the decoding filters weights to be non-negative by thresholding them at every SGD iteration. The assumption that the input is non-negative holds in our case and it will also hold in deeper AEs provided that we use ReLU-like activation functions. Note that for encoding, we do not constrain filters so they can have negative values to compensate for the pattern auto-correlation (see below).

Sparsifying the Filters. The traditional ℓ_2 regularization allows many small but non-zero values. To force these values to zero and thus get sparser filters, we replaced the ℓ_2 norm by the sparsity-promoting norm ℓ_1 known as lasso:

$$R_{las}(\mathbf{W}) = \sum_{f=1}^{M}\sum_{i=1}^{d}\sum_{k=1}^{L_f} \left|{}^{e}\mathbf{W}_{(i,k)}^{f}\right| + \sum_{f=1}^{M}\sum_{i=1}^{d}\sum_{k=1}^{L_f} \left|{}^{d}\mathbf{W}_{(i,k)}^{f}\right| \qquad (2)$$

Encouraging Sparse Activations. The traditional KL divergence aims at making *all* hidden units equally useful *on average*, whereas our goal is to have the activation layer to be as sparse as possible for each given input document. We achieve this by encouraging peaky activations, i.e. of low entropy when seen as a document-level probability distribution, as was proposed in [20] when dealing on topic models for motif discovery. This results in an entropy-based regularization expressed on the set $\mathbf{A} = \{\mathbf{a}^b\}$ of document-level activations:

$$R_{ent}(\mathbf{A}) = -\frac{1}{N}\sum_{b=1}^{N}\left(\sum_{f=1}^{M}\sum_{t=1}^{L-L_f+1}\hat{\mathbf{a}}_{f,t}^{b}\log\left(\hat{\mathbf{a}}_{f,t}^{b}\right)\right) \text{ with } \hat{\mathbf{a}}_{f,t}^{b} = \mathbf{a}_{f,t}^{b}\bigg/\sum_{f=1}^{M}\sum_{t=1}^{L-L_f+1}\mathbf{a}_{f,t}^{b} \qquad (3)$$

Local Non-maximum Activation Removal. The previous entropy regularizer encourages peaked activations. However, as the encoding layer remains a convolutional layer, if a filter is correlated in time with itself or another filter, then the activations cannot be sparse. This phenomenon is due to the feed forward nature of the network, where activations depend on the input, not on each others: hence, no activation can inhibit its neighboring activations. To handle this issue we add a local non-maximum suppression layer which, from a network perspective, is obtained by convolving activations with a temporal Gaussian filter, subtracting from the result the activation intensities, and applying a ReLU, focusing in this way spread activations into central peaks.

Handling Distant Filter Correlations with AdaReLU. The Gaussian layer cannot handle non local (in time) correlations. To handle this, we propose to replace the traditional ReLU activation function by a novel one called adaptive ReLU. AdaReLU works on groups of units and sets to 0 all the values that are below a percentage (e.g., 60 %) of the maximal value in the group. In our architecture, AdaReLU is applied separately on each filter activation sequence.

Finding the True Number of Patterns. One main advantage and contribution of our AE-based method compared to methods presented in Sect. 2 is the possibility to discover the "true" number of patterns in the data. One solution to achieve this is to introduce in the network a large set of filters and "hope" that the learning leads to only a few non null filters capturing the interesting patterns. However, in practice, standard regularization terms and optimizations tend to produce networks "using" all or many more filters than the number of true patterns which results in partial and less interpretable patterns. To overcome this problem, we propose to use a group lasso regularization term called $\ell_{2,1}$ norm [24] that constrains the network to "use" as few filters as possible. It can be formulated for our weight matrix as:

$$R_{grp}(\mathbf{W}) = \sum_{f=1}^{M} \sqrt{\sum_{i=1}^{d} \sum_{k=1}^{L_f} \left({}^{\mathsf{e}}\mathbf{W}^f_{(i,k)} \right)^2} + \sum_{f=1}^{M} \sqrt{\sum_{i=1}^{d} \sum_{k=1}^{L_f} \left({}^{\mathsf{d}}\mathbf{W}^f_{(i,k)} \right)^2} \quad (4)$$

Overall Objective Function. Combining Eqs. (1), (2), (3) and (4), we obtain the objective function that is optimized by our network:

$$C(\mathbf{W}, \mathbf{X}) = MSE(\mathbf{W}, \mathbf{X}) + \lambda_{las} R_{las}(\mathbf{W}) + \lambda_{grp} R_{grp}(\mathbf{W}) + \lambda_{ent} R_{ent}(\mathbf{A}(\mathbf{W}, \mathbf{X})) \quad (5)$$

4 Experiments

4.1 Experimental Setting

Datasets. To study the behavior of our approach, we experimented with both synthetic and real video datasets. The synthetic data were obtained using a known generation process: temporal documents were produced by sampling random observations of random linear combinations of motifs along with salt-and-pepper noise whose amount was defined as a percentage of the total document intensities (noise levels: 0 %, 33 %, 66 %). Six motifs (defined as letter sequences for ease of visualization) were used. A document example is shown in Fig. 2-a, where the the feature dimension ($d = 25$) is represented vertically, and time horizontally ($L = 300$). For each experiments, 100 documents were generated using this process and used to train the autoencoders. This controlled environment allowed us to evaluate the importance of modeling elements. In particular, we are interested in (i) the number of patterns discovered (defined as the non empty decoding filters[2]); (ii) the "sharpness" of the activations; and (iii) the robustness of our method according to parameters like $\lambda_{lasso}, \lambda_{grp}, \lambda_{ent}$, the number of filters M, and the noise level.

We also applied our approach on videos recorded from fixed cameras. We used videos from the QMUL [9] and the far-field datasets [21]. The data preprocessing steps from the companion code of [6] were applied. Optical flow features were obtained by estimating, quantifying, and locally collecting optical flow over 1 second periods. Then, temporal documents were obtained by reducing the dimensionality of these to $d = 100$, and by cutting videos into temporal documents of size $L = 300$ time steps.

Architecture Details and Parameter Setting. The proposed architecture is given in Fig. 1. As stated earlier, the goal of this paper is to make the most of a convolutional AE with a *single* layer (corresponding to the activation layer)[3]. Weights are initialized according to a uniform distribution between 0 and $\frac{1}{d*L_f}$.

In general, the filter length L_f should be large enough to capture the longest expected recurring pattern of interest in the data. The filter length has been set to $L_f = 45$ in synthetic experiments, which is beyond the longer motif of the ground-truth. In the video examples, we used $L_f = 11$, corresponding to 10 seconds, and which allows to capture the different traffic activities and phases of our data [21].

4.2 Results on the Synthetic Dataset

Since we know the "true" number of patterns and their expected visualization, we first validate our approach by showing (see Fig. 2-c) that we can find a set of

---

[2] We consider a filter *empty* if the sum of its weights is lower or equal to $\frac{1}{2}$ (the average sum value after initialization).

[3] Note however that the method can be generalized to hierarchical motifs using more layers, but then the interpretation of results would slightly differ.

parameters such that our filters exactly capture our given motifs and the number of non empty filters is exactly the "true" number of motifs in the dataset even when this dataset is noisy (this is also true for a clean dataset). In this case (see Fig. 2-e) the activations for the complete document are, as expected, sparse and "peaky". The output document (see Fig. 2-b) is a good un-noisy reconstruction of the input document shown in Fig. 2-a.

In Fig. 3, we evaluate the influence of the given number of filters M and the noise level on both the number of recovered motifs an the MSE while fixing the parameters as in Fig. 2. We can see that with this set of parameters, the AE is able to recover the true number of filters for the large majority of noise levels and values of M. For all noise levels, we see from the low MSE that the AEs is able to well reconstruct the original document as long as the number of given filters is at least equal to the number of "true" patterns in the document.

Model Selection: Influence of λ_{grp}. Figure 4 shows the number of non zero filters in function of λ_{grp} and of the noise level for the synthetic dataset with 6 known motifs when using 12 filters (left) and 16 filters (right). The light blue area is the area in which the AEs was able to discover the true number of patterns. With no group lasso regularization ($\lambda_{grp} = 0$), the AE systematically uses all the available filters capturing the original patterns (see $2^{nd}, 4^{th}$ or 5th filters in Fig. 2-d), redundant variants of the same pattern (filters 1st and 3rd in Fig. 2-d) or a more difficult to interpret mix of the patterns (filters 6th and 7th in Fig. 2-d). On the contrary, with too high values of λ_{grp}, the AE does not find any patterns (resulting in a high MSE). A good heuristic to set the value of λ_{grp} could thus be to increase it as much as possible until the resulting MSE starts increasing. In the rest of the experiments, λ_{grp} is set equal to 2.

Influence of $\lambda_{ent}, \lambda_{lasso}$, AdaReLU, and Non-local Maxima Suppression. We have conducted the same experiments as in Fig. 2 on clean and noisy datasets (up to 66 % of noise) with $M = 3$, $M = 6$ $M = 12$ to assess the behavior of our system when canceling the parameters: (1) λ_{ent} that controls the entropy of the activation layer, (2) λ_{las}, the lasso regularizer (3) the AdaReLU function (we used a simple ReLU in the encoding layer instead) and (4) the Non-Local Maxima activation suppression layer. In all cases, all parameters but one were fixed according to the best set of values given in Fig. 2. For lack of space, we do not give all the corresponding figures but we comment the main results.

The λ_{ent} is particularly important in the presence of noise. Without noise and when this parameter is set to 0, the patterns are less sharp and smooth and the activations are more spread along time with much smaller intensities. However, the MSE is as low as for the default parameters. In the presence of noise (see Fig. 2-f), the AE is more likely to miss the recovery of some patterns even when the optimal number of filters is given (e.g. in some experiments only 5 out of the 6 filters were not empty) and the MSE increases a lot compared to experiments on clean data. This shows again that the MSE can be a good heuristic to tune the parameters on real data. The λ_{las} has similar effects with and without noise: it helps removing all the small activation values resulting in much sharper (and thus interpretable) patterns.

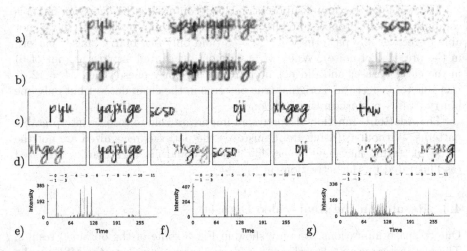

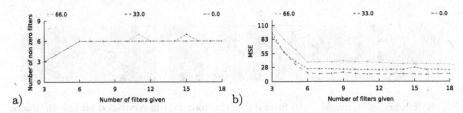

Fig. 2. Results on the synthetic data built from 6 patterns, with 66 % of noise, $M = 12$ filters, and unless stated otherwise $\lambda_{las} = 0.004$, $\lambda_{grp} = 2$, $\lambda_{ent} = 0.2$. (a) Sample document; (b) Obtained output reconstructed document; (c) Weights of seven out of the 12 obtained filters (the 5 remaining filters are empty); (d) Weights of seven filters when not using group lasso, i.e. with $\lambda_{grp} = 0$ (note that the 5 remaining filters are non empty); (e, f, g) Examples of activation intensities (colors correspond to a given filter) with default parameters (e); without the entropy sparsifying term ($\lambda_{ent} = 0$) (f); with ReLU instead of AdaReLU (g). (Color figure online)

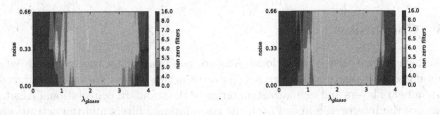

Fig. 3. Influence of the given number of filters M and the noise level (0 %, 33 % and 66 %) on: (a) the number of recovered motifs and (b) the Mean Squared Error. Experiments on the synthetic dataset with $\lambda_{las} = 0.004$, $\lambda_{grp} = 2$, $\lambda_{ent} = 0.2$.

Fig. 4. Evolution of the number of non-zero filters (sparsity) with respect to the noise level when we vary the parameter λ_{grp} (λ_{glasso} in the figure) that controls the group lasso regularization for the synthetic dataset with 6 known motifs when using 12 filters (right) and 16 filters (left). (Color figure online)

The non-local maximum suppression layer (comprising the *Gaussian* filter) is compulsory in our proposed architecture. Indeed, without it, the system was not able to recover any patterns when $M = 3$ (and only one blurry "false" pattern in the presence of noise). When $M = 6$, it only captured 4 patterns (out of 6) in the clean dataset and did not find any in the noisy ones. When $M = 12$, it was able to recover the 6 original true patterns in the clean dataset but only one blurry "false" pattern in the noisy ones.

The AdaReLU function also plays an important role to recover interpretable patterns. Without it (using ReLU instead) the patterns recognized are not the "true" patterns, they have a very low intensity and are highly auto-correlated (as illustrated by the activations in Fig. 2-g).

4.3   Results on the Real Video Dataset

Due to space limitations, we only show in Fig. 5 some of the obtained results. The parameters were selected using grid search by minimizing the MSE on the targeted dataset. For instance, on the Junction 1 dataset, the final parameters used are $\lambda_{las} = 0.2$, $\lambda_{grp} = 50$, $\lambda_{ent} = 5$. Note that this is larger than in the synthetic case but the observation size is also much larger (100 vs 25) and the filters are thus sparser in general. In the Junction 1 dataset, the autoencoder recovers 4 non-empty and meaningful filters capturing the car activities related to the different traffic signal cycles, whereas in the far-field case, the main trajectories of cars were recovered as also reported in [21].

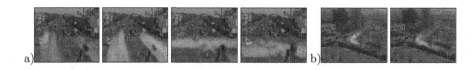

Fig. 5. Traffic patterns. $M = 10$ filters. (a) The four motifs recovered on the Junction 1 dataset, (6 empty ones are not shown). (b) Two filters (out of the five recovered) on the far-field dataset.

5   Conclusion

We have shown that convolutional AEs are good candidate unsupervised data mining tools to discover interpretable patterns in time series. We have introduced a number of layers and regularization terms to the standard convolutional AEs to enforce the interpretability of both the convolutional filters and the activations in the hidden layers of the network. The filters are directly interpretable as spatio-temporal patterns while the activations give the occurrence times of each patterns in the temporal document. This allow us to un-mix multivariate time series. A direct perspective of this work is the use of multi-layer AEs to capture combination of motifs. If this was not the aim of this article, it may help to reduce the number of parameters needed to obtain truly interpretable patterns and capture more complex patterns in data.

Acknowledgement. This work has been supported by the ANR project SoLStiCe (ANR-13-BS02-0002-01).

References

1. Baccouche, M., Mamalet, F., Wolf, C., Garcia, C., Baskurt, A.: Spatio-temporal convolutional sparse auto-encoder for sequence classification. In: British Machine Vision Conference (BMVC) (2012)
2. Chockalingam, T., Emonet, R., Odobez, J.-M.: Localized anomaly detection via hierarchical integrated activity discovery. In: AVSS (2013)
3. Darken, C., Moody, J.E.: Note on learning rate schedules for stochastic optimization. In: NIPS, pp. 832–838 (1990)
4. Du, X., Jin, R., Ding, I., Lee, V.E., Thornton, J.H.: Migration motif: a spatial-temporal pattern mining approach for nancial markets. In: KDD, pp. 1135–1144. ACM (2009)
5. Emonet, R., Varadarajan, J., Odobez, J.-M.: Multi-camera open space human activity discovery for anomaly detection. In: IEEE International Conference on Advanced Video and Signal-Based Surveillance (AVSS), Klagenfurt, Austria, August 2011
6. Emonet, R., Varadarajan, J., Odobez, J.-M.: Temporal analysis of motif mixtures using dirichlet processes. IEEE PAMI **36**(1), 140–156 (2014)
7. Marwah, M., Shao, H., Ramakrishnan, N.: A temporal motif mining approach to unsupervised energy disaggregation: applications to residential and commercial buildings. In: Proceedings of the 27th AAAI Conference (2013)
8. Hinton, G.E., Salakhutdinov, R.R.: Reducing the dimensionality of data with neural networks. Science **313**(5786), 504–507 (2006)
9. Hospedales, T., Gong, S., Xiang, T.: A Markov clustering topic model for mining behavior in video. In: ICCV (2009)
10. Kolter, J.Z., Jaakkola, T.: Approximate inference in additive factorial HMMs with application to energy disaggregation. In: Proceedings of AISTATS Conference (2012)
11. Karlsson, L., Längkvist, M., Loutfi, A.: A review of unsupervised feature learning and deep learning for time-series modeling. Pattern Recognit. Lett. **42**, 11–24 (2014)
12. Masci, J., Meier, U., Cireşan, D., Schmidhuber, J.: Stacked convolutional auto-encoders for hierarchical feature extraction. In: Honkela, T., Duch, W., Girolami, M., Kaski, S. (eds.) ICANN 2011. LNCS, vol. 6791, pp. 52–59. Springer, Heidelberg (2011). doi:10.1007/978-3-642-21735-7_7
13. Mehta, N.A., Gray, A.G.: Funcica for time series pattern discovery. In: Proceedings of the SIAM International Conference on Data Mining, pp. 73–84 (2009)
14. Memisevic, R., Hinton, G.E.: Unsupervised learning of image transformations. In: Computer Vision and Pattern Recognition (CVPR) (2007)
15. Mooney, C.H., Roddick, J.F.: Sequential pattern mining - approaches, algorithms. ACM Comput. Surv. **45**(2), 19:1–19:39 (2013)

16. Oates, T.: PERUSE: an unsupervised algorithm for finding recurring patterns in time series. In: ICDM (2002)
17. Ranzato, M., Poultney, C., Chopra, S., LeCun, Y.: Efficient learning of sparse representations with an energy-based model. In: NIPS. MIT Press (2006)
18. Sallaberry, A., Pecheur, N., Bringay, S., Roche, M., Teisseire, M.: Sequential patterns mining and gene sequence visualization to discover novelty from microarray data. J. Biomed. Inform. **44**(5), 760–774 (2011)
19. Tavenard, R., Emonet, R., Odobez, J.-M.: Time-sensitive topic models for action recognition in videos. In: International Conference on Image Processing (ICIP), Melbourne (2013)
20. Varadarajan, J., Emonet, R., Odobez, J.-M.: A sparsity constraint for topic models - application to temporal activity mining. In: NIPS Workshop on Practical Applications of Sparse Modeling: Open Issues and New Directions (2010)
21. Varadarajan, J., Emonet, R., Odobez, J.-M.: A sequential topic model for mining recurrent activities from long term video logs. Int. J. Comput. Vis. **103**(1), 100–126 (2013)
22. Chu, W.-S., Zhou, F., Torre, F.: Unsupervised temporal commonality discovery. In: Fitzgibbon, A., Lazebnik, S., Perona, P., Sato, Y., Schmid, C. (eds.) ECCV 2012. LNCS, vol. 7575, pp. 373–387. Springer, Heidelberg (2012). doi:10.1007/978-3-642-33765-9_27
23. Peng, W.-C., Chen, Y.-C., Lee, S.-Y.: Mining temporal patterns in time interval-based data. IEEE Trans. Knowl. Data Eng. **27**(12), 3318–3331 (2015)
24. Yuan, M., Lin, Y.: Model selection and estimation in regression with grouped variables. J. R. Stat. Soc. (B) **68**(1), 49–67 (2006)
25. Zheng, Y.: Trajectory data mining: an overview. ACM Trans. Intell. Syst. Technol. **6**(3), 29:1–29:41 (2015)

Gesture Input for GPS Route Search

Radu Mariescu-Istodor[(✉)] and Pasi Fränti

University of Eastern Finland, Joensuu, Finland
{radum, franti}@cs.uef.fi

Abstract. We present a simple and user-friendly tool for an efficient search from a spatial database containing GPS tracks. The input is a sketch of a route drawn by a user on a map by mouse, hand or other means. This type of interaction is useful when a user does not remember the date and time of a specific route, but remembers its shape approximately. We evaluate the efficiency of the retrieval when the shape given by the gesture is simple or complex, and when the area contains either a small or large number of routes. We use the Mopsi2014 route dataset to demonstrate that the search works in real time.

Keywords: GPS · Route · Gesture · Matching · Touchscreen · Draw

1 Introduction

GPS-enabled smartphones allow users to collect large amounts of location-based data such as geo-tagged notes, photos, videos and geographical trajectories hereafter referred to as *routes*. Mobile users track routes for reasons like: recording travel experiences, recommending a certain path and keeping track of personal statistics in sports such as hiking, running, cycling and skiing. A sample route collection is shown in Fig. 1. From a large collection like this, it is difficult to find a specific route unless user remembers the date when it was recorded. Otherwise the amount of data is overwhelming to perform systematic search from among all the records.

Many applications exist that allow users to track their movement; some of these are: *Sports Tracker*[1], *Endomondo*[2], *Strava*[3] and *Mopsi*[4]. Mopsi is a location-based social network created by the School of Computing at the University of Eastern Finland. Mopsi users can find out who or what is around. They can track their movements, share photos and chat with friends. Mopsi includes fast retrieval and visualization of routes [1] using a real-time route reduction technique [2]. Transport mode information is automatically inferred by analyzing the speed variance of the route [3]. Movement is classified as either: walking, running, cycling or car. Route similarity, novelty, inclusion and noteworthiness [4, 5] are computed by using cell representations of the routes created by a grid which covers the planet. Searching for spatially similar routes is done efficiently by indexing these cells.

---

[1] http://www.sports-tracker.com.
[2] https://www.endomondo.com.
[3] https://www.strava.com.
[4] http://cs.uef.fi/mopsi.

© Springer International Publishing AG 2016
A. Robles-Kelly et al. (Eds.): S+SSPR 2016, LNCS 10029, pp. 439–449, 2016.
DOI: 10.1007/978-3-319-49055-7_39

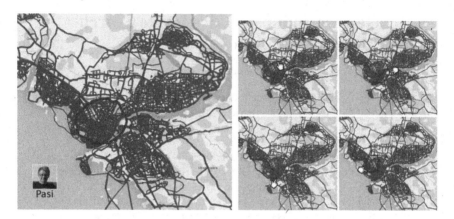

Fig. 1. Route collection of user Pasi over the city of Joensuu, Finland is shown on left. The collection spans from 2008 to 2014. A circle-shaped route that we want to find is emphasized. Four attempts (all failed) to find this route by clicking the map are shown right.

We propose a real-time search for routes in the Mopsi collection by using gestures and pattern matching. The gesture is a hand-drawn input in the form of a free shape done on a map. The shape approximates the locations where the targeted route passes through. According to [6], this gesture can be classified as of the *symbolic* type, implying that it has no meaning when performed in other contexts (not using a map). Referring to the taxonomy in [7] the result of the gesture is to trigger a command: search for route(s) with given spatial characteristics. This search works by computing the similarity between the input gesture and every route in the database. The most similar route candidates are provided to the user.

Gestures have been used as a means to access menu items without the need to traverse large hierarchies. In [8], gestures are continuous pen traces on top of a stylus keyboard. This soft keyboard can be inconvenience as it wastes screen space unnecessarily. In our method, we use the underlying map as a canvas for drawing the gestures. On desktop computers, the gesture mode is explicitly activated by holding a hotkey while drawing the gesture by mouse. On touchscreens, we need to distinguish the gesture from normal map interaction (panning and zooming). In [9], it was discovered that it is possible to distinguish gesture from other touch events such as scrolling or tapping by buffering the touch events and analyzing the queue to determine if the sequence is a gesture or not. We use this method to activate the gesture mode, and neither designated area nor activation button are therefore needed.

Typically, symbolic gesture-based systems require the user to learn a set of symbols [6]. Our method is simpler as no learning of symbols is required. However, the user is expected to understand and be able to read maps because the roads, buildings and terrain elements such as forests, lakes and rivers are the key information used when giving the input. For example, user may draw the input by following a river front, road, or other landmarks visible on map. Users who have a large route collection benefit most from the gesture search. It is therefore fair to assume that these users have also the necessary skills to understand maps.

2   User Interface

2.1   System Overview

Let us assume that Mopsi user Pasi wants to review the statistics of a specific route from his collection but he does not recall the date. Pasi knows that the route is in Joensuu, Finland so he proceeds to move the map to this location. Figure 1 shows that Pasi has a large route collection in Joensuu. Let us further assume that he wishes to find the highlighted circular route. Exhaustive search among all the routes would not be reasonable so best change is to try to distinguish the route on map. In Mopsi, this is possible by clicking any individual route on the map. However, this is also difficult because the targeted route overlaps with many others.

Gesture search enables a user to search routes by drawing the sample shape of the desired route over the map. Figure 2 shows how Pasi's route is found by drawing a circular gesture on the map around the center of Joensuu. The search returns four possible candidates, including the one he was looking for.

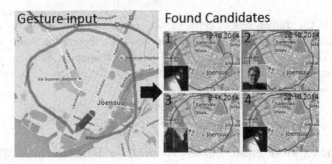

Fig. 2. A circle-shaped gesture surrounding the center of Joensuu reveals four circular route candidates. Pasi's route is number two in the list.

2.2   Map Handling

Mopsi uses Google Maps and OpenStreetMap. They both offer several built in functions for user interaction. A user can pan the map by clicking and dragging it in the desired direction. Zooming in can be done by double left-click and zooming out is done by double right-click. Zooming can be also done using the mouse wheel or by the pinch gesture.

To start the gesture search on a computer, user presses a hotkey (Ctrl). When pressed, the built-in map handling functions are disabled and the gesture input mode is enabled. In this mode, a user can draw on the map by clicking, holding and moving the mouse around while keeping the hotkey pressed. Releasing the hotkey causes two things to happen simultaneously: the input gesture is sent to the server for processing the query, and default map behavior is reactivated.

Majority of touchscreens nowadays do not have a keyboard and existing buttons serve for different purposes such as exiting applications, changing volume levels or enabling the camera. It is possible to implement a soft button on the screen to toggle the gesture input mode however this wastes screen space which makes drawing more difficult, especially on small screens.

Instead, we activate the gesture first by a click (tap) and then, immediately, touch the screen again to draw the shape. We denote this event as *Tap-and-Draw*. The *Draw* event works similarly as panning the map, however, the preceding *Tap* event triggers gesture input mode. When the *Draw* gesture is complete, the input gesture is sent to the server and the search is initiated; default map behavior is reactivated.

2.3   Real-Time Route Search

The search returns the route(s) that are most similar to the shape of the gesture input. For the matching, we use the method in [5]. It computes the spatial similarity between routes by first representing them as cells in a grid and then using the Jaccard similarity coefficient:

$$J(C_A, C_B) = \frac{|C_A \cap C_B|}{|C_A \cup C_B|}, \tag{1}$$

where C_A and C_B are two sets of cells. However, because of the arbitrary division of the grid, route points may end up in different cells even though the points are close to each other. This problem is solved by applying morphological dilation with square structural element and using the additional cells as a buffer region when computing the similarity. The similarity is then formulated as:

$$S(C_A, C_B) = \frac{|C_A \cap C_B| + |C_A \cap C_B^d| + |C_B \cap C_A^d|}{|C_A| + |C_B| - |C_A \cap C_B|}, \tag{2}$$

where C_A^d and C_B^d are the dilated regions of routes C_A and C_B respectively. To make the search efficient we pre-compute the cell representation and use B-tree index [12] on the cell database. With this setup the search works real-time.

To perform the search, the input shape is converted into cells. The similarity between this cell set and all routes is then computed using (2). The result is similarity ranking which often contains a multitude of results with varying levels of similarity to the given shape. To the user we present only the most significant candidates.

2.4   Map Projection and the Grid

Most online maps (*Google Maps, OpenStreetMap, Yahoo! Maps, Bing Maps*) use a variant of the *Mercator* projection [10]. In Mopsi, we use Google Maps or Open-StreetMap as the map interface. *Mercator* is a cylindrical map projection which preserves the angles, however, the linear scale increases with latitude. The parallels and meridians are straight and perpendicular to each other. The meridians are equidistant, but the parallels become sparser as they further themselves from the equator.

Creating a grid by choosing a fixed cell size (in degrees) will cause the cells to appear vertically stretched when viewed on the Mercator projection. The amount cells stretch increases the farther away they are from the equator. In Joensuu, Finland the cells appear twice as tall as they are wide.

2.5 Multi-resolution Grids

The precision of drawing the gesture should be independent on the zoom level of the map. When the zoom level is decreased by one unit the content of the map becomes half of its previous size, and consequently, the regions on the map become twice as difficult to read. We create 10 grids with different resolutions and store the routes at each of these approximation levels (see Table 1).

The finest grid has a cell size of 25×25 meters. Finer grids are not needed because at this level, GPS error becomes already apparent and the route approximations become unreliable. The amount of cells needed increases exponentially when finer grids are produced. Therefore, we do not compute unnecessary levels in vain. Sparsest grid has cell length of 12.5 km. At lower levels (≥ 25 km) the cell size becomes so big that even the longest routes are represented by only a few cells.

Table 1. A mapping from zoom-level to the grid resolution. The statistics are for Mopsi2014 Route dataset using each of the grid resolutions.

Zoom level	≤ 6	7	8	9	10	11	12	13	14	≥ 15
Grid resolution	0	1	2	3	4	5	6	7	8	9
Cell size (km)	12,8	6,4	3,2	1,6	0,8	0,4	0,2	0,1	50 m	25 m
Amount of cells	7×10^4	9×10^4	1×10^5	2×10^5	4×10^5	7×10^5	1×10^6	3×10^6	5×10^6	1×10^7
Memory (MB)	3,5	4,5	6,5	9,5	16,5	30,6	59,6	118,6	238	486
B-tree Index (MB)	8,5	9,5	13,5	21,5	35,6	66,7	131,8	263,1	526	1,1 GB

3 Route Search

We present next our algorithm for performing the gesture-based route search. The algorithm (*GSearch*) first extracts the cells that the input shape passes through using the *Find-Cells* function. This function chooses the correct grid resolution based on the zoom level using the mapping presented in Table 1. Every point is then mapped to the cell it resides in. At the Equator, one degree is roughly 111 km and the smallest cell length we support is 25×25 meters. We dilate the input route C_A with 3×3 square structural element to obtain C_A^d.

GSearch: Searches for route candidates matching a given gesture.

Input: gesture shape G, zoom level z
Output: candidates list L
C, C^d ← **Find-Cells** (G, z)
ranking   ← **RSR** (C, C^d, z)
Top-Cluster   ← **Cluster** (ranking. similarities)

for i ← 1 to size (ranking) **do**
    if ranking [i] is in Top-Cluster **then**
        add ranking [i] to L. append (ranking [i])

Find-Cells: Obtains the cells that a given shape passes through at a specific zoom level.

Input: shape S, zoom level z
Output: cell set C, dilated region C^d
r   ← **Get-Resolution** (z) // according to Table 1
min-cell-size   ← 25 // meters
degree-size   ← 111 // km
dividing-factor ← min-cell-size * degree-size / pow(2, r)

for i ← 1 to size (S) **do**
    x ← round (S [i]. latitude * dividing-factor)
    y ← round (S [i]. longitude * dividing-factor)
    add (x, y) to C
C^d ← **Dilate** (C)

Route Similarity Ranking (RSR) algorithm is then applied to find all similar routes in the database. *RSR* iterates through every cell in C_A and C_A^d, and finds what other routes pass through the same cells. For each found route C_B, it checks if the cell belongs to $C_A \cap C_B$, $C_A \cap C_B^d$ or $C_A^d \cap C_B$. The algorithm maintains counters for each type and uses them for computing the similarity values using (2). Time complexity is $O((|C| + |C^d|)$ $(\log(MQ)) + a(C) + a(C^d))$ where M is the number of routes in the database, Q is the average route size in cells and $a(C) = \sum \left(|C \cap C_i| + |C \cap C_i^d| \right)$, $i = \overline{1,M}$.

The similarity ranking usually results in a large number of routes, of which only few are relevant to the user. It might be possible to filter out routes below a given threshold, but then we might get no result in some cases; the other extreme is when searching for a very common route. Then there can be too many results above the threshold. Therefore, we limit the number of results using clustering as follows.

RSR: Computing the route similarity ranking.

Input: cells C, dilated part C^d, zoom level z
Output: ranking of routes according to similarity values
SC ← initialize SetCounter array; // structure defined below
// process input route
for i ← 1 to size (C) **do**
    R_i, R^d_i ← **Get-Routes** (C [i])
    for j ← 1 to size (R_i) **do**
        SC [R_i[j]]. A ++; SC [R_i[j]]. B ++; SC [R_i[j]]. AB ++;
    for j ← 1 to size (R^d_i) **do**
        SC [R^d_i[j]]. A ++; SC [R^d_i[j]]. B ++; SC [R^d_i[j]]. AB^d ++;
// process dilated part
for i ← 1 to size (C^d) **do**
    R_i, R^d_i ← **Get-Routes** (C^d [i])
    for j ← 1 to size (R_i) **do**
        SC [R_i[j]]. B ++; SC [R_i[j]]. $A^d B$ ++;
    for j ← 1 to size (R^d_i) **do**
        SC [R^d_i[j]]. B ++; SC [R^d_i[j]]. $A^d B^d$ ++;
ranking ← new list;
for each r_{id} in SC **do**
    sim ← (SC [r_{id}]. AB + SC [r_{id}]. $A^d B$ + SC [r_{id}]. AB^d) /
            (SC [r_{id}]. A + SC [r_{id}]. B - SC [r_{id}]. AB)
    add (r_{id}, sim) to ranking
SetCounter { A←0; B←0; AB←0; $A^d B$←0; AB^d←0; }

We cluster the threshold values by *Random Swap* (*RS*) algorithm [11] with 10,000 iterations with 16 clusters. The algorithm alternates between K-Means and random relocation of centroids in order to avoid getting stuck in a local optimum. The algorithm converges to the final result in few hundreds of iterations, on average. However, since *Random Swap* is fast, we can afford to use 10,000 iterations to increase the probability of finding optimal partitioning.

From the clustering result, we take the cluster having the routes of highest similarities. The idea is that the clustering will find the size of this set automatically by fitting the clustering structure to the distribution of the similarities.

Cluster: Limits the ranking to the most likely candidates.

Input: similarities S
Output: cluster with highest similarities
T   ← 10.000 // number of iterations
M   ← 16 // number of clusters
P   ← **RS** (S, T, M) // Random Swap clustering
Top-Partition ← 0; Top-Cluster ← empty set

for i ← 1 to size (S) **do**
   if S [i] = max (S) **then**
      Top-Partition ← P [i]

for i ← 1 to size (S) **do**
   if P [i] = Top-Partition **then**
      add S [i] to Top-Cluster

---

4   Experiments

We perform experiments using the Mopsi2014[5] dataset, which is a subset of all routes from the Mopsi database collected by the end of 2014. It contains 6,779 routes recorded by 51 users who have 10 or more routes. Routes consists of a wide range of activities including walking, cycling, hiking, jogging, orienteering, skiing, driving, traveling by bus, train or boat. Most routes are in Joensuu region, Finland, which creates a very dense area suitable for stressing the method. A summary of the dataset is shown in Table 2. All experiments were performed on Dell R920, 4 x E7-4860 (total 48 cores), 1 TB, 4 TB SAS HD.

Table 2. Statistics of Mopsi2014 route dataset.

Routes	Points	Kilometers	Hours
6,779	7,850,387	87,851	4,504

---

[5] http://cs.uef.fi/mopsi/routes/dataset.

4.1 Efficiency of the Search

The efficiency of the search is proportional to the size of the database, and to the resolution of the grid. The grid to be chosen depends on the zoom level required to view the targeted route on the map: small routes are best viewed using a higher zoom-level. The grid depends also on the size of the screen. To get a better understanding of this we computed the zoom-level for the best-view of each route in the Mopsi2014 dataset. We consider the best-view as the maximum zoom-level that shows the entire route on screen. The results in Fig. 3 show that lowest zoom levels are rarely used. Routes in such zoom levels should span across multiple countries or even continents, and thus, are rare in the dataset. The highest zoom levels (20–21) are also not often used because they cover only very short routes, usually non-movement records.

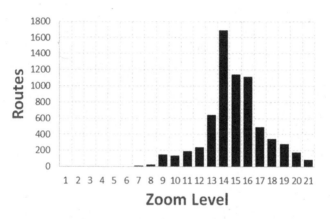

Fig. 3. Histogram showing what zoom-levels are used more often when viewing routes.

When computing the histogram from Fig. 3, we assumed a screen size of 1366 × 768, which, according to the free statistics provided by W3Counter[6], was the most used screen size during March 2016.

We next compute the efficiency of the G-Search algorithm by taking every route in Mopsi2014 as the target route. The best-viewed zoom level for them is first found. A perfect gesture is then simulated for the route by selecting the cells it travels through. Search is then performed using the default screen size of 1366 × 768. The results are summarized in Fig. 4. As expected, the time required is small (0.2–0.8 s) at small zoom levels. At the largest zoom levels the time is also small, but this is against expectations. The reason for the low execution times is the fact that for zoom level 15 and above, the same grid is used and, as a result, the number of cells required to represent each route is lower. Only the middle level routes can take slightly more than 1 s.

This experiment shows that, given a random target route, the expected search time is about 1 s or less, thus, it can be considered real-time. In practice, a smaller zoom level is used by the user than the best-fitting one is selected. Thus, < 1 s result happens

---

[6] https://www.w3counter.com/globalstats.php.

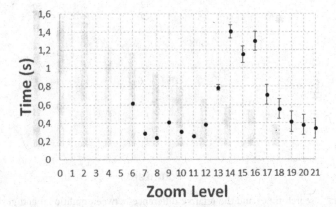

Fig. 4. Times required by G-Search when searching all routes in Mopsi2014 dataset. The results are grouped by the zoom-level and averaged. The average of all searches is 0.9 s.

more often. The reason is that often at zoom levels just below the best-fitting one it is easier to see the landmarks on the map. Furthermore, it is possible that at the best-fitting level the gesture implies drawing on the edges of the map, which are more difficult to target than the central area. Another reason is that the 1366 × 768 screen size is large, and using a smaller screen implies a finer grid will be used. Processing times with default screen size of 320 × 658 yields even smaller processing time of about 0.2 s.

The search time also depends on the density of the routes. In low density areas (< 200 routes), the search time is 0.14 s, on average. In very dense areas (> 1000 routes) the search time is 2.2 s, on average. There is also minor dependency on the size of the gesture. A gesture passing through 50 cells takes 0.7 s time on average, whereas as gesture passing through 200 cells takes 0.7 s, on average. The upper limit is the number of cells that can fit on the screen (3600 with the 1366 × 768 screen size).

4.2 Usability Evaluation

We study next the efficiency of the gesture search from usability point of view. We compare the average time user spends on searching a randomly chose route using the gesture search and using the previous (traditional) system. Eleven volunteers were asked to search randomly selected routes using a tool[7] built for this purpose as follows:

A target route was shown on map but no date, length or duration were shown. User can study and memorize the route as long as wanted.

When user pressed the *Start* button, user was (randomly) directed either to the traditional system or to the new Gesture search. Timer was started.

The task was to find the route and input its date and then press *Stop* button. If the date was correct the timer was stopped. If the user considered the task too difficult he was allowed to press *Give-up* button.

---

[7] http://cs.uef.fi/mopsi/routes/gestureSearch/qual.php.

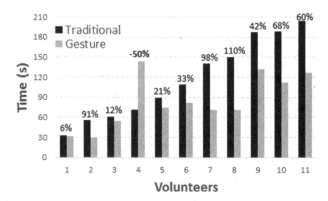

Fig. 5. Average search times and the relative difference between traditional and gesture search.

Each volunteer was asked to repeat the test at least 10 times, or as long as he/she found it fun to do.

In total, 106 routes were searched using the traditional system, and 98 using the gesture search. The searched routes were found 77 % of the time using traditional search compared to 91 % when using gestures. Gesture search was 41 % faster, on average. The individual performance differences are shown in Fig. 5. Traditional search is slower on average than gesture search for all except one user.

The search time is affected also by other factors such as complexity and length of the route, and density of the areas the route passes through. We next group the results by these three factors. The complexity is calculated as the number of points used by the polygonal approximation [2] to represent the route at its best-fit zoom level. Density is calculated as the proportion of cells that are overloaded by other routes; it is the opposite to the noteworthiness value in [5]. Results in Table 3 show that although shorter and less complex routes in low density areas are faster to find, the Gesture search outperforms the traditional approach in all cases.

The volunteers were also asked if they liked the Gesture search and which one they would prefer for such search task. They all rated Gesture search as good (10) or excellent (1). Most (9) preferred Gesture search, none (0) preferred the traditional search, and some (2) would not use either. Written comments included "*I really liked it*" and "*It was fun*".

Table 3. Average search times when grouped by different factors.

	Length		Complexity		Density	
	Short 2.7 km	Long 12.7 km	Low 31 pts	High 128 pts	Low 12 %	High 75 %
Traditional	90 s	116 s	87 s	120 s	90 s	117 s
Gesture	64 s	78 s	65 s	77 s	54 s	88 s
Reduction	30 %	33 %	25 %	36 %	30 %	24 %

5   Conclusion

We showed that gestures can be successfully used as input for searching routes from large data collections. We solved all the components of the search including user input, database optimization, pattern matching, and selecting threshold by clustering to show only the most significant results. The effectiveness of the method was demonstrated by run time analysis showing that it works real time, and by usability experiments showing that it outperforms traditional search.

References

1. Waga, K., Tabarcea, A., Mariescu-Istodor, R., Fränti, P.: Real time access to multiple GPS tracks. In: International Conference on Web Information Systems and Technologies (WEBIST 2013), Aachen, Germany, pp. 293–299 (2013)
2. Chen, M., Xu, M., Fränti, P.: A fast multiresolution polygonal approximation algorithm for GPS trajectory simplification. IEEE Trans. Image Process. **21**(5), 2770–2785 (2012)
3. Waga, K., Tabarcea, A., Chen, M., and Fränti, P.: Detecting movement type by route segmentation and classification. In: IEEE International Conference on Collaborative Computing: Networking, Applications and Worksharing (CollaborateCom 2012), Pittsburgh, USA, pp. 508–513 (2012)
4. Mariescu-Istodor, R., Tabarcea, A., Saeidi, R., Fränti, P.: Low complexity spatial similarity measure of GPS trajectories. In: International Conference on Web Information Systems and Technologies (WEBIST 2014), Barcelona, Spain, pp. 62–69 (2014)
5. Mariescu-Istodor, R., Fränti, P.: Grid-based method for GPS route analysis and retrieval. Manuscript (2016, submitted)
6. Cirelli, M., Nakamura, R.: A survey on multi-touch gesture recognition and multi-touch frameworks. In: ACM Conference on Interactive Tabletops and Surfaces (ITS 2014), Dresden, Germany, pp. 35–44 (2014)
7. Karam, M., Schraefel, M.C.: A taxonomy of Gestures in Human Computer Interaction. ACM Transactions on Computer-Human Interactions (2015, in press)
8. Kristensson, P.O., Zhai, S.: Command strokes with and without preview: using pen gestures on keyboard for command selection. In: SIGCHI Conference on Human Factors in Computing Systems (CHI 2007), New York, USA, pp. 1137–1146 (2007)
9. Li, Y.: Gesture search: a tool for fast mobile data access. In: ACM Symposium on User Interface Software and Technology (UIST 2010), New York, USA, pp. 87–96 (2010)
10. Kennedy, M., Kopp, S.: Understanding Map Projections. ESRI Press, Redlands (2001)
11. Fränti, P., Kivijärvi, J.: Randomized local search algorithm for the clustering problem. Pattern Anal. Appl. **3**(4), 358–369 (2000)
12. Cormen, T.H.: Introduction to Algorithms. MIT press, Cambridge (2009)

GPS Trajectory Biometrics: From Where You Were to How You Move

Sami Sieranoja[⊠], Tomi Kinnunen, and Pasi Fränti

School of Computing, University of Eastern Finland, P.O. Box 111, FIN-80101 Joensuu, Finland
{samisi,tkinnu,franti}@cs.uef.fi

Abstract. In this paper, we study how well GPS data can be used for biometric identification. Previous work has considered only the location and the entire route trajectory pattern. These can reveal the user identity when he repeats his every day moving patterns but not when traveling to new location where no route history is recorded for him. Instead of the absolute location, we model location-independent micro movements measured by speed and direction changes. The resulting short-term trajectory dynamics are modelled by Gaussian mixture model - universal background model (GMM-UBM) classifier from speed and direction change features. The results show that we can indentify users from OpenstreetMap data with an equal error rate (EER) of 19.6 %. Although this is too modest result for user authentication, it indicates that GPS traces do contain identifying cues, which could potentially be used in forensic applications.

1 Introduction

Thanks to smart devices combined with an increasing number of social media applications, collecting and sharing of personal data has never been easier as it is today. Besides photo and video uploads, smart-phones provide direct or processed information of the user's location or behavior via *global positioning system* (GPS), accelerometer or other sensor data. As an example, a sportswoman might upload her running route coordinates along with physical performance data.

GPS coordinate data contains a rich source of information about the user's whereabouts and behavior. This information can be used to provide useful services such as recommending potential friends based on user's trajectories [17]. On the other hand, it also raises a question of privacy [4].

Location-related or *spatial* cues include the most commonly used locations (such as user's home) or routes (such as daily route from home to work). Once combined with *temporal* (time-stamp) information, one is able to, for example, infer the future movements of a user [14] and the most likely times she will be absent from her home [3]. Speed estimates can be used for inferring the most likely means of transport (walking, bicycling, driving) [15] or whether the user respects speed limits.

© Springer International Publishing AG 2016
A. Robles-Kelly et al. (Eds.): S+SSPR 2016, LNCS 10029, pp. 450–460, 2016.
DOI: 10.1007/978-3-319-49055-7_40

Even if a GPS trajectory data would be anonymized by obscuring the obvious identifying information, such as name and home address, the user might still be *re-identifiable* by linking an anonymized GPS coordinate data with non-anonymized data in the user's public profile in a social media application. Such information could be very sensitive; examples might be visit to an abortion clinic, church or premises of a political party [10].

In this study we focus on *user identification* based on GPS trajectory data. Differently from prior work that use user's location history for identification (*where you were*) [10,13], we approach the problem as a biometric identification task: to identify the user based on his or her physical or behavioral characteristics, but independently of the location or absolute timing of the trajectory data. We view the GPS trajectory coordinates of a person as an inaccurate measurement of the physical behavior of the user related to his or her muscle activity, such as gait or the way of steering a bicycle.

Our primary goal is to find out whether and how much of person-identifying traits exists in GPS trajectory data. Unlike [10,13] where the question was approached by the possibility to identify users by only identifying individual routes, we approach it as a statistical pattern recognition problem. That is, we model the distribution of short-term feature vectors derived from a set of GPS routes that reflect user's physical activity, rather than the locations visited. This way we are able to obtain a more accurate picture of how well users could be identified in situations where the training and test routes originate from different locations or dates. We utilize two public datasets to study the question whether person identification is feasible from GPS trajectory data, and if so, how much training and testing data is required.

2   Related Work

2.1   Location Privacy

The topic of location privacy has been a subject of many studies [6]. Of recent work, route uniqueness has been studied in [10] where it was shown that even low resolution mobile traces collected from mobile phone carriers are highly unique. Selecting only three points of a trace was enough to uniquely identify most traces. Similarly in [13] it was shown for GPS data that even when points are sampled out of the routes, they can still be reliably linked with the original routes.

2.2   Spatio-Temporal Similarity

Considerable amount of work has been devoted on *recommendation* based on GPS trajectory. As an example, in [17], potential friends are recommended based on user's trajectories. So-called *stay cells* are created based on detected stops. They are considered important since the user stayed there longer time. Similarity of trajectories is then measured based on *longest common subsequence* (LCS) and giving more importance to longer patterns. In [8], revised version of LCS is applied by partitioning the trajectories based on speed and detected turn points. A similarity score is computed using both geographic and semantic similarity.

In [1], similarity of a person's days is assessed based on the trajectory by discovering their semantic meaning. The data is collected from tracking users' cars and pre-processed by detecting stop points. Most common pairs of stops are assumed to be user's home and work locations. Dynamic time warping of the raw trajectories using geographic distances of the points was reported to work best. In [16], personalized search for similar trajectories is performed by taking into account user preferences of which parts of the query trajectory is more important.

Complete trajectories are not always available and the similarity must then be measured based on sparse location data such as visits, favorite places or check-ins. In [7], user data is hierarchically clustered into geographic regions. A graph is constructed from the clustered locations so that a node is a region user has visited, and an edge between two nodes represents the order of the visits to these regions. This method still relies on the order of the locations visited.

Algorithm 1. Features for Segment

Input: Route segment S, Window width w
Output: Set of feature vectors F
   procedure FEATURESFORSEGMENT(S,w)
      $F \leftarrow \emptyset$
      for all $s_i \in S$ **do**
         $W \leftarrow (s_i, \ldots, s_{i+w})$
         speed $\leftarrow$ SPEED(W,w)
         turns $\leftarrow$ TURNS(W,w)
         $F_{\text{speed}} \leftarrow$ DFTFEATURE(speed,w)
         $F_{\text{turns}} \leftarrow$ DFTFEATURE(turns,w)
         $F_i \leftarrow (F_{\text{speed}}, F_{\text{turns}})$
      end for
   end procedure

Algorithm 2. DFT Feature

Input: Local window W, Window width w
Output: Feature X
   procedure DFTFEATURE(W, w)
      $X \leftarrow (W - \text{mean}(W)) \circ \text{HammingWindow}(w)$
      $X \leftarrow |\text{fft}(X)|$
      $X \leftarrow \text{dct}(\log 10(X))$
      $X \leftarrow X(1:24)$
   end procedure

---

3 Statistical User Characterization Using Short-Term GPS Dynamics

We model GPS user behavior by first calculating *discrete Fourier transform* (DFT) features from local speed and direction changes (Sect. 3.1). Then Gaussian mixture model - universal background model (GMM-UBM) classifier is trained on these features (Sect. 3.2).

3.1 Short-Term GPS Dynamics

DFT features are calculated from speed and turn angle data (see Algorithms 1 and 2). The route is processed using a sliding window of 100 s (100 or 50 points, depending on sample rate). Speed and turn angles are then calculated for each point inside the window. Turn angles are further processed by integration to produce a turn angle measure similar to what speed is to acceleration.

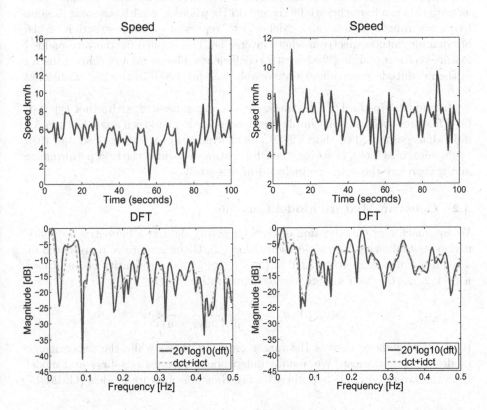

Fig. 1. DFT feature processing [user1]   **Fig. 2.** DFT feature processing [user2]

After that, DFT is calculated separately for speed and turn angle. For each type of feature and each short-term segment, this yields complex-valued spectrum represented in polar form as $X(\omega_k) = |X(\omega_k)|e^{i\theta(\omega_k)}$, where $|X(\omega_k)|$ and

$\theta(\omega_k)$ denote, respectively, the magnitude and phase of the kth frequency component. Similar to short-term speech processing, we discard the phase part. Logarithm of the retained part, magnitude, is then parameterized using discrete cosine transform (DCT) for dimensionality reduction and decorrelation purposes. We retain the first 24 DCT coefficients (including the DC coefficient) and concatenate the speed and turn angle features to yield feature vectors of dimensionality $24 \cdot 2 = 48$.

The DFT feature was designed to model how frequently and by how much the speed and turn angles change in a route segment. Additionally, they also reflect the speed of the user to some extent. The greater the speed, the more frequently there are turns and deacceleration/acceleration. This is expected to shift the part of the spectrum that correlates with road network to the right (to higher frequencies).

Figures 1 and 2 illustrate the features for two different users. In this example, user2 has less low frequency variations in the speed of the segment and this shows as a dip between frequencies 0.05 Hz and 0.1 Hz whereas user1 has a spike in same frequency range. Additionally, "dct + idct" represents a reconstruction of the original magnitude spectrum where inverse DCT is applied on the zero-padded feature vector containing the lowest 24 coefficients. The prominent characteristics of the magnitude spectrum are reasonably well preserved in the 24-dimensional features.

In addition to the DFT features, we also experimented with other types of GPS features such as using simple speed, acceleration and turn angle features for individual points. Also, short (2–20 point) windows of relative speed and turn angle were considered. However, the DFT features provided the best performance and is therefore the only one included in this study.

3.2 Gaussian Mixture Model Classifier

We approach user classification of GPS trajectory data as a biometric authentication task following Bayes' decision theory [2]. Given a route $\mathcal{R}$ represented by a sequence of feature vectors, $\mathcal{X} = \{x_1, \ldots, x_T\}$, and an identity claim of user $u \in \{1, 2, \ldots, U\}$, we evaluate log-likelihood ratio:

$$\text{score}(\mathcal{X}, u) = \log \frac{p(\mathcal{X}|\text{user} = u)}{p(\mathcal{X}|\text{user} \neq u)}. \tag{1}$$

Here the numerator models the target user hypothesis while the denominator models its complement. We assume independent observations $\{x_t\}$ and model both the target and anti-hypotheses using Gaussian mixture models (GMMs),

$$p(\mathcal{X}|\boldsymbol{\lambda}) = \prod_{t=1}^{T} \sum_{m=1}^{M} P_m \mathcal{N}(x_t|\boldsymbol{\mu}_m, \boldsymbol{\Sigma}_m), \tag{2}$$

where M is the number of Gaussians (model order) and $\boldsymbol{\lambda} = \{P_m, \boldsymbol{\mu}_m, \boldsymbol{\Sigma}_m : m = 1, \ldots, M\}$ denotes the model parameters: mixing weights (component priors) P_m, mean vectors $\boldsymbol{\mu}_m$ and covariance matrices $\boldsymbol{\Sigma}_m$. In our implementation the covariance matrices are constrained to be diagonal ones[1]. The number of Gaussians, M, is a control parameter to trade off between precise user modeling and generalization power (this will be explored below).

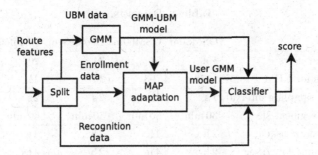

Fig. 3. Modeling and classification system

To train the model, we follow a well-known two-step recipe based on *adapted Gaussian mixture models* or GMM-UBM [12], based on *maximum a posteriori* (MAP) adaptation principle [5] (see Fig. 3). First, a *universal background model* (UBM), used for modeling the anti-hypothesis in (1), is trained by pooling route feature vectors from a large number of off-line users. This is achieved using *expectation-maximization* (EM) algorithm. The UBM is intended for representing common information shared across all the users and is trained once only, using users disjoint from the target users. The UBM is then used in (1) across all the target users for score normalization. To enroll (register) a new user, the UBM parameters are moved towards the enrollment data. We copy the mixing weights and component variance vectors from the UBM and use the adapted mean vectors to represent users. We point the interested reader to [12] for further details.

4   Experimental Setup

4.1   Data Sets

We created four different evaluation protocols based on two publicly available source GPS data sets: **Geolife**[2] and **Openstreetmap**[3]. As the original datasets

---

[1] This assumption, in general, is non-restrictive. As pointed in [11], since the individual Gaussians act together to model the input feature density, full covariance matrices are not needed even for features that are statistically dependent. For a given amount of training data, one can fit either a GMM with full covariance matrices with less Gaussians, or equivalently a larger GMM with diagonal covariances.

[2] http://research.microsoft.com/en-us/downloads/b16d359d-d164-469e-9fd4-daa38f2b2e13/.

[3] http://planet.openstreetmap.org/gps/gpx-planet-2013-04-09.tar.xz.

were originally not designed for biometric tasks, we designed specific evaluation protocols similar to those used in speaker verification and other biometrics. The details of the resulting datasets are presented in Table 1.

Table 1. Data sets.

	OSM30m	OSM60m	Geolife30m	Geolife60m
Target users	156	51	34	20
Users for UBM	1178	793	35	41
GPS sampling interval	1 s	1 s	2 s	2 s
Test segment size	30 min	60 min	30 min	60 min
Test data/user	2 h	4 h	2 h	4 h
Training data/user	2 h	4 h	2 h	4 h
Genuine trials	624	204	136	80
Impostor trials	96720	10200	4488	1520

The Geolife data set was collected by Microsoft Research Asia during a course of three years (from 04/2007 to 08/2012). It contains data from 182 users (which our filtering reduced to 34) logged in varying sample rates, covering a broad range of outdoor activities including life routines, sports activities, shopping, sightseeing, dining, hiking and bicycling.

Openstreetmap (OSM) data is a collection of public traces[4]. A 23 GB compressed dump of these traces was published in 2013[5]. It contains GPS traces uploaded to **openstreetmap.org** by 41413 different users logged in varying sample rates. We select those routes which (1) privacy option was set to "identifiable"[6], and (2) route sampling interval was 1 s. Due to computational limitations, only a subset of these routes were processed.

To reduce the risk of having two almost identical routes (such as from work to home), and thus detecting a route instead of user, each user's routes were filtered by removing any overlapping points (< 30 m). To remove most car routes, we applied a simple heuristic speed filtering by taking the top 4-quantile speed of a segment and discarding the route if the speed exceeded 35 km/h. The route data was then processed to contain only uniform interval (1 s or 2 s) routes. For the 2 s routes, it contains also 1 s routes with every second point removed. Only the OSM data set contained enough data to create a 1 s interval data set.

From the filtered routes, we created a subset where each user has exactly the same amount of training and test data, for example 4 h of testing data split into four 60 min test items for OSM60m. If the desired amount of data (time) was not reached, we excluded the user from the data set. The routes of users

---

[4] https://www.openstreetmap.org/traces.

[5] http://planet.openstreetmap.org/gps/.

[6] http://wiki.openstreetmap.org/wiki/Visibility_of_GPS_traces.

which did not have sufficient training or testing data were retained for training the universal background model.

4.2   Evaluation

Classifier scores were computed for all the possible (usermodel, route) pairs; whenever the user identity of the test route matches the target (model) user, this constitutes a *same user* (genuine) trial, otherwise a *different user* (impostor) trial. We measure the performance using a standard performance measure of biometric systems, *equal error rate* (EER), which is the misclassification rate at the detection threshold where false acceptance and false rejection rates are equal. In practice, we use an implementation in BOSARIS toolkit[7] to compute EERs. In addition to EER, we also report *relative rank* (RRANK) measure, defined as the average rank of the correct user model score and normalized to range from 0 (best) to 1 (worst) by dividing with the number of users.

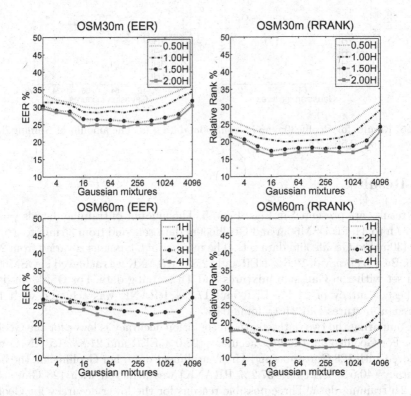

Fig. 4. Results for OSM30m and OSM60m data sets. The amount of training data is varied from 1 to 4 h.

---

[7] https://sites.google.com/site/bosaristoolkit/.

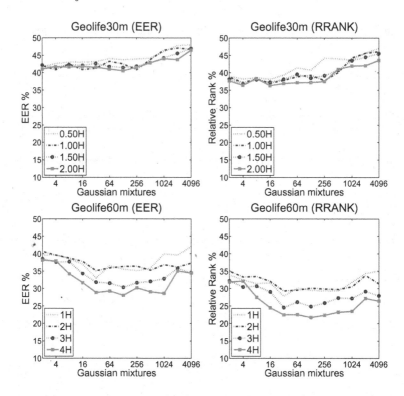

Fig. 5. Results for Geolife30m and Geolife60m data sets. The amount of training data is varied from 0.5 to 2 h.

5   Results

The results are presented in Figs. 4 and 5. The amount of training data is varied from 1 h to 4 h for OSM60m and Geolife60m data sets and from 30 min to 2 h for OSM30m and Geolife30m data sets. The number of Gaussians is varied from 2 to 4096. Best accuracy of 19.6 % EER and 12.8 % RRANK was achieved in OSM60m data set with 256 Gaussian mixtures and 4 h training data. For OSM30m data set best accuracy of 24.4 % EER and 17.2 % RRANK was achieved with 128 Gaussian mixtures and 2 h training data.

Comparing the two datasets, the recognition accuracy is lower for the Geolife data. For Geolife60m, the best accuracy (28.0 % EER and 21.8 % RRANK) was achieved with 128 Gaussians and 4 h of enrollment data. For Geolife30m, the best accuracy (40.5 % EER and 37.2 % RRANK) was achieved with 128 Gaussians and 2 h training data. Three possible reasons for the lower accuracy for Geolife include (1) lower GPS sampling, leading to less discriminative spectral features, and (2) much smaller number of users to train UBM.

Concerning the amount of training data and the number of model parameters, we observe three expected results. Firstly, larger amount of training data generally leads to higher accuracy. Secondly, the optimal number of Gaussians lies in between the tested parameter range. This is expected from the bias-variance trade-off in statistical modeling: too many Gaussians leads to overfitting while too few do not discriminate the users well. Thirdly, for larger amounts of training data, the optimal model size is obtained with a larger number of Gaussian components.

Compared with other user movement based biometrics, the GPS features did not achieve as good recognition accuracy as accelerometer. For example, the gait-based recognition in [9] reached 7% EER, compared to our 19.6% using GPS signal. Although it is only indirect comparison, it is reasonable evidence that even if GPS signal can be used to recognize user, accelerometer probably provides more reliable source—if available. GPS technology is also likely to develop further to make it more reliable. Probably also more accurate in user identification. One potential future idea would be to study the joint use of GPS and accelerometer data.

6   Conclusion

Our experiments indicate that local variations in GPS data possess user specific characteristic that can potentially be used to recognize a person and should therefore be handled with similar care as any other private information such as voice or fingerprints.

Our method achieved an accuracy of 19.6% EER and 12.8% RRANK in the best case. While these error rates are clearly too high to be useful in user authentication applications requiring high level of security and trustworthiness, they do indicate that local GPS trajectory movements contain person-identifying information. This information might be useful for applications such as recommendation systems or to detect sudden changes in user's behavior.

We assume that the accuracy is less than optimal partly due to inherent inaccuracy of GPS data and low sampling frequency on available data sets. However, future development in movement tracking technology is likely to increase accuracy of routes, and consequently improve recognition accuracy of our method. Also, even larger amount and diversity of training and testing data would likely improve the accuracy further.

Due to limitations of available data sets, we were unable to rule out the impact of certain factors which may be user specific while not being characteristic of users. These include properties of GPS tracker, and the area where user lives. Therefore it is still an open question how much the recognition accuracy relates to what user is and how much to user's surroundings.

References

1. Biagioni, J., Krumm, J.: Days of our lives: assessing day similarity from location traces. In: Carberry, S., Weibelzahl, S., Micarelli, A., Semeraro, G. (eds.) UMAP 2013. LNCS, vol. 7899, pp. 89–101. Springer, Heidelberg (2013). doi:10.1007/978-3-642-38844-6_8

2. Duda, R., Hart, P., Stork, D.: Pattern Classification. WIS, New York (2000)
3. Fletcher, D.: Please rob me: The risks of online oversharing. Time Mag. Online (2010)
4. Gambs, S., Killijian, M.O., del Prado Cortez, M.N.: Show me how you move and I will tell you who you are. In: Proceedings of 3rd ACM SIGSPATIAL International Workshop on Security and Privacy in GIS and LBS, SPRINGL 2010, pp. 34–41. ACM, New York (2010)
5. Gauvain, J.L., Lee, C.H.: Maximum a posteriori estimation for multivariate Gaussian mixture observations of Markov chains. IEEE Trans. Speech Audio Process. 2(2), 291–298 (1994)
6. Krumm, J.: A survey of computational location privacy. Pers. Ubiquit. Comput. 13(6), 391–399 (2009)
7. Li, Q., Zheng, Y., Xie, X., Chen, Y., Liu, W., Ma, W.Y.: Mining user similarity based on location history. In: Proceedings of 16th ACM SIGSPATIAL International Conference on Advances in Geographic Information Systems, GIS 2008, pp. 34:1–34:10. ACM, New York (2008)
8. Liu, H., Schneider, M.: Similarity measurement of moving object trajectories. In: Proceedings of 3rd ACM SIGSPATIAL International Workshop on GeoStreaming, IWGS 2012, pp. 19–22. ACM, New York (2012)
9. Mäntyjärvi, J., Lindholm, M., Vildjiounaite, E., Mäkelä, S.M., Ailisto, H.: Identifying users of portable devices from gait pattern with accelerometers. In: Proceedings of IEEE International Conference on Acoustics, Speech, and Signal Processing, vol. 2, pp. ii-973. IEEE (2005)
10. de Montjoye, Y.A., Hidalgo, C.A., Verleysen, M., Blondel, V.D.: Unique in the crowd: the privacy bounds of human mobility. Scientific reports 3 (2013)
11. Reynolds, D.: Gaussian Mixture Models. In: Li, S.Z., Jain, A. (eds.) Encyclopedia of Biometric Recognition. Springer, New York (2008)
12. Reynolds, D., Quatieri, T., Dunn, R.: Speaker verification using adapted gaussian mixture models. Digit. Sig. Process. 10(1), 19–41 (2000)
13. Rossi, L., Walker, J., Musolesi, M.: Spatio-Temporal Techniques for User Identification by Means of GPS Mobility Data. CoRR abs/1501.06814 (2015)
14. Song, C., Qu, Z., Blumm, N., Barabási, A.L.: Limits of predictability in human mobility. Science 327(5968), 1018–1021 (2010)
15. Waga, K., Tabarcea, A., Chen, M., Fränti, P.: Detecting movement type by route segmentation and classification. In: 2012 8th International Conference on Collaborative Computing: Networking, Applications and Worksharing (CollaborateCom), pp. 508–513. IEEE (2012)
16. Wang, H., Liu, K.: User oriented trajectory similarity search. In: Proceedings of ACM SIGKDD International Workshop on Urban Computing, UrbComp 2012, pp. 103–110. ACM, New York (2012)
17. Ying, J.J.C., Lu, E.H.C., Lee, W.C., Weng, T.C., Tseng, V.S.: Mining user similarity from semantic trajectories. In: Proceedings of 2nd ACM SIGSPATIAL International Workshop on Location Based Social Networks, LBSN 2010, pp. 19–26. ACM, New York (2010)

Structural Matching

Enriched Bag of Words for Protein Remote Homology Detection

Andrea Cucci, Pietro Lovato, and Manuele Bicego[⊠]

Dipartimento di Informatica - Ca' Vignal 2, Università degli Studi di Verona,
Strada le Grazie 15, 37134 Verona, Italy
manuele.bicego@univr.it

Abstract. One of the most challenging Pattern Recognition problems in Bioinformatics is to detect if two proteins that show very low sequence similarity are functionally or structurally related – this is the so-called Protein Remote Homology Detection (PRHD) problem. Even if in this context approaches based on the "Bag of Words" (BoW) paradigm showed high potential, there is still room for further refinements, especially by considering the peculiar application context. In this paper we proposed a modified BoW representation for PRHD, which enriches the classic BoW with information derived from the evolutionary history of mutations each protein is subjected to. An experimental comparison on a standard benchmark demonstrates the feasibility of the proposed technique.

Keywords: Bag of words · N-grams · Sequence classification

1 Introduction

In recent years, several Pattern Recognition problems have been successfully faced by approaches based on the "Bag of Words" (BoW) representation [21]. This representation is particularly appropriate when the pattern is characterized (or assumed to be characterized) by the repetition of basic, "constituting" elements called words. By assuming that all possible words are stored in a dictionary, the BoW vector for one particular object is obtained by *counting* the number of times each element of the dictionary occurs in the object. One of the main advantages of this representation is that it can represent in a vector space many types of objects, even ones that are non-vectorial in nature (like documents, strings, sequences), for which less computational tools are available. The success of this paradigm has been demonstrated in many fields [2–4,21]: in particular, in the bioinformatics context, different BoW approaches [6,14–16] have been proposed in recent years – with the name of N-gram methods – to face the so-called Protein Remote Homology Detection (PRHD) problem [1,10,12]. This represents a central problem in bioinformatics aimed at identifying functionally or structurally-related proteins by looking at amino acid sequence similarity – where the term *remote* refers to some very challenging situations where

© Springer International Publishing AG 2016
A. Robles-Kelly et al. (Eds.): S+SSPR 2016, LNCS 10029, pp. 463–473, 2016.
DOI: 10.1007/978-3-319-49055-7_41

homologous proteins exhibit very low sequence similarity. In this context, the BoW paradigm is instantiated by considering as words the so-called N-grams, i.e. short sequences of aminoacids of fixed length (N), extracted from the aminoacidic sequence – in the basic formulation [6] – or even from evolutionary representations, i.e. the profiles [14,15].

In this context, approaches based on the BoW representation achieved state of the art prediction performances. Yet, the potentialities of this representation have not been completely exploited, but can be enriched by using some peculiarities of the specific application scenario. In particular, to solve the PRHD task it is needed to capture the homology between proteins, linked to evolutionary aspects, such as insertions, deletions and mutations incurred between the two sequences. Let us concentrate to this last operation, which represents the case when an aminoacid in the sequence is substituted with another aminoacid during evolution. Biologically, there are mutations which are very likely to happen (due to the similar chemical-physical characteristics of the aminoacids), whereas some others are less likely. A good representation for PRHD should capture this aspect; the BoW approach, in its original formulation for PRHD, does not permit to model this aspect[1]: if there is a mutation, we simply count for a *different* word, independently from the fact that the mutation is highly probable or not to happen in nature. However, the BoW paradigm can be extended to cope with this aspect, and this represents the main goal of this paper. More in detail, here we propose a BoW approach to PRHD which modifies the process of counting words, in order to take into account the evolutionary relations between words. The idea is straightforward: in the classical setting, when we observe a word w, we increment its counter by 1. Here we propose to extend this process and to increment also the counters of words which are "biologically likely" mutations of the word w. More specifically, we propose to increment the counter of all other words w' by a value which is directly proportional to the probability of mutation of w in w'. This information is estimated from the so-called substitution matrices (the most famous example being the BLOSUM [9]), employed in sequence-alignment approaches, which quantitatively measure how likely it is, in nature, to observe particular mutations. In this sense, the BoW vector is *enriched* by evolutionary information derived from the specific application scenario.

The proposed approach has been thoroughly evaluated using the standard SCOP[2] 1.53 superfamily benchmark [12], representing the most widely employed dataset to test PRHD approaches. Obtained results demonstrate that the proposed approach reaches satisfactory results in relation to other N-gram based techniques, as well as in comparison to a broader spectrum of approaches proposed in the recent literature.

The rest of the paper is organized as follows: in Sect. 2 we summarize the classic Bag of Words approaches for Protein Remote Homology Detection, whereas in Sect. 3 we present the proposed approach. The experimental

---

[1] Actually, in computer vision, some approaches dealing with weights have been proposed – e.g. see [17].

[2] http://scop.berkeley.edu/ [7].

evaluation is described in Sect. 4; finally, in Sect. 5, conclusions are drawn and future perspectives are envisaged.

2   BoW Approaches for PRHD

In this section we summarize how a BoW representation can be extracted from a biological sequence – this scheme being at the basis of different PRHD systems [6,14–16]. First, we introduce how "words" and "dictionary" are defined in this context. We consider as words *sequence N-grams*: a N-gram of a sequence $S = s_1 \ldots s_L$ is defined as a subsequence of N consecutive symbols $g_l = s_l \ldots s_{l+N-1}$. Once fixed the length N, we can define a dictionary $\mathbb{D}$ as the set of all possible subsequences of length N built using the alphabet $\mathcal{A}$ (the four symbols A, T, C, G in case of nucleotides, or 20 symbols in case of aminoacids). Therefore the dictionary $\mathbb{D}$ contains $W = \mathcal{A}^N$ words.

Given a sequence S, its Bag of Words representation $BoW(S)$ is obtained by counting how many times each word (N-gram) $v_i \in \mathbb{D}$ occurs in S. Let us introduce more formally the counting process, mainly to fix the notation used to present the proposed approach. In the first step all the N-grams $g_1, \ldots g_G$ present in the sequence S are extracted (where G depends on the length L of the sequence and on the degree of overlap with which the N-grams are extracted from the sequence). Then, each g_i is represented via a vector $\mathbf{w}_i$,

$$g_i \longrightarrow \mathbf{w}_i = [0, 0, \cdots, 1, \cdots 0] \tag{1}$$

This W-dimensional vector encodes the fact that g_i corresponds to the j-th word v_j of the dictionary $\mathbb{D}$ via the "1-of-W" scheme: in the $\mathbf{w}_i$ vector all the elements are zero, except one, which is 1; the position of the non zero element is the position in the dictionary $\mathbb{D}$ of the N-gram g_i extracted from the sequence. Given such representation, the Bag of Words representation of S is obtained by summing element-wise all the vectors $\mathbf{w}_1, \ldots, \mathbf{w}_G$:

$$BoW(S) = \mathbf{w}_1 + \mathbf{w}_2 + \cdots + \mathbf{w}_G \tag{2}$$

See the left part of Fig. 1 for a schematic sketch of the BoW scheme.

This representation has been successfully employed in the case of Protein Remote Homology Detection, typically as direct input to discriminative classifier such as Support Vector Machines [14,15], or after the employment of more sophisticated models, such as topic models [16]. In all these approaches, the BoW representation has been extracted from different kinds of biological sequences: raw sequences (as in [6]), evolutionary representations of the biological sequences – called *profiles* (as in [14,15]), or even in combination with the corresponding 3D structures (as in [16]).

3   The Proposed Approach

The main idea behind the proposed approach stems from the observation that the classic Bag of Words scheme for Protein Remote Homology Detection is

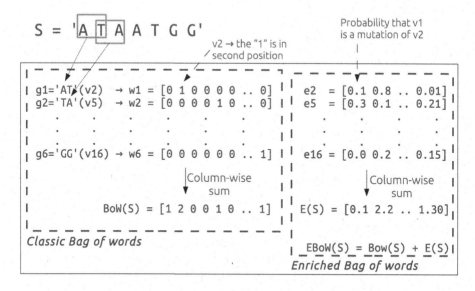

Fig. 1. Sketch of the Bag of Words representation and the proposed Enriched Bag of Words approach. We are considering nucleotidic sequences (therefore the alphabet contains the symbols A,T,C,G); in the specific case, the sequence length is 7 ($L = 7$), and we used N-grams of length 2, with overlap = 1. This means that the number of N-grams extracted from the sequence is 6 ($G = 6$).

not able to encode evolutionary relations which can exist between words: if a substitution occurs (i.e. an aminoacid is replaced by another during evolution), the classic BoW simply counts for another word in the dictionary, independently from how likely is this substitution: actually, in nature, there are definitely different probabilities of mutation between aminoacids (which compose the words in the BoW), which depend on the family, the chemical properties or the structural features. To cope with this aspect, we propose a scheme based on the following idea: if an N-gram g_i in the sequence corresponds to the word v_j, we increment the count of v_j by 1 (as in the classic BoW), but we also increment the counters of the words which are "biologically likely" mutations of the word v_j: such increments are clearly directly proportional to the probability of being mutation of v_j.

More formally, a given N-gram g_i, corresponding to the j-th word of the dictionary, is represented by $\mathbf{w}'_i$, defined as:

$$g_i \longrightarrow \mathbf{w}' = \mathbf{w}_i + \mathbf{e}_j \tag{3}$$

where $\mathbf{w}_i$ is defined as in Eq. (1) (all zeros and a 1 in position j), and

$$\mathbf{e}_j = [e_{j1}, e_{j2}, \cdots, e_{jW}] \tag{4}$$

is the *enrichment vector*, a vector of length W which, in every position k, indicates how much probable is that the k-th word of the dictionary is a mutation of v_j. This vector permits to explicitly encode the biological a priori knowledge on the relation which occurs between the words of the dictionary.

Given this correction, the *Enriched* Bag of Words representation is obtained by following the same scheme of Eq. (2), i.e. by the columwise summation of all $\mathbf{w}_i'$:

$$EBoW(S) = \mathbf{w}_1' + \mathbf{w}_2' + \cdots + \mathbf{w}_G' \tag{5}$$

Rearranging the summation, we obtain:

$$EBoW(S) = BoW(S) + E(S) \tag{6}$$

where $E(S)$ represents the "enrichment" (or correction) made to the Bag of Words representation of sequence S.

3.1 Computing the Enrichment Vectors

The enrichment vectors are obtained by starting from the so-called substitution matrices (the most famous one called BLOSUM [9]), matrices which encode the biological knowledge related to mutations. This matrix is typically used to perform sequence alignments, and a given entry (i, j) encodes the rate at which the aminoacid i is likely to mutate into the aminoacid j (the higher, the more likely it is). Intuitively, this matrix has the highest values on the diagonal; for off-diagonal elements, the matrix should reflect the fact that there are some mutations that are highly improbable, due for example to physical or chemical properties of aminoacids.

To define our enrichment vectors, we start from the approach used to derive the BLOSUM matrix, proposed in [9]. In that paper, the matrix has been built by starting from *blocks* of related sequences[3]. From these blocks (more than 2000 blocks have been used in the original paper of [9]), the *expected probability of occurrence* of each pair of symbols can be computed, and this represents our starting information. For example, in the case of the nucleotidic sequences (i.e. the alphabet is composed by 'A', 'T', 'C', 'G'), this matrix is

$$\mathbf{M}_1 = \begin{bmatrix} P(A \to A) & P(A \to T) & P(A \to C) & P(A \to G) \\ P(T \to A) & \cdots & & P(T \to G) \\ & & \ddots & \\ P(G \to A) & \cdots & \cdots & P(G \to G) \end{bmatrix}$$

---

[3] In biological terms, related sequences are the ones which belong to the same evolutionary family – namely, they share the same biological function.

where $P(x \rightarrow y)$ indicates the probability that the nucleotide 'x' mutates into the nucleotide 'y'. If we use as words 1-grams, namely the entries of the alphabet $\mathcal{A}$, then we can directly employ this matrix to derive the required enrichment vectors. For $N > 1$, however, we should provide a larger matrix, containing the mutation probability for each pair of N-grams. Here we define this probability via the multiplication of the probabilities of pairs of symbols; for example, in the case of 2-grams, we have

$$P(xy \rightarrow kj) = P(x \rightarrow k)P(y \rightarrow j)$$

We are aware that by employing this simple scheme we are assuming that the symbols inside the N-gram are probabilistically independent: however this simplifying assumption is accepted and employed in many applications dealing with biological sequences – e.g. for multiple sequence alignment [18]. In formula, the mutation matrix $\mathbf{M}_N$ for N-grams of length N is obtained inductively by employing the Kronecker tensor product "$\otimes$":

$$\mathbf{M}_N = \mathbf{M}_{N-1} \otimes \mathbf{M}_1 \tag{7}$$

Finally, the matrix of enrichment vectors for N-grams of length N $\mathbf{E}_N = [\mathbf{e}_1; \mathbf{e}_2; ...; \mathbf{e}_W]$ is obtained by normalizing the mutation matrix in order to have a reasonable range.

$$\mathbf{E}_N = \frac{\mathbf{M}_N}{\max_{i,j} \mathbf{M}_N} \tag{8}$$

4   Experimental Evaluation

The experimental evaluation is based on a famous benchmark[4] widely employed to assess the detection capabilities of many protein remote homology detection systems [12], extracted from SCOP version 1.53 and containing 4352 sequences from 54 different families. The protein remote homology detection task is cast into a binary classification problem: to simulate remote homology, 54 different subsets are created: in each of this, an entire target family is left out as positive testing set. Positive training sequences are selected from other families belonging to the same superfamily (i.e. sharing remote homology), whereas negative examples are taken from other super-families. Please note that class labels are very unbalanced, with a vast majority of objects belonging to the negative class (on average the positive class (train + test) is composed by 49 sequences, whereas the negative one is made by 4267).

As in many previous works [5,6,13–15,19], classification is performed using SVM via the public GIST implementation[5], setting the kernel type to radial basis, and keeping the remaining parameters to their default values. Detection accuracies are measured using the receiver operating characteristic (ROC) score and the ROC50 score [8]. In both cases, the larger the value the better the

---

[4] Available at http://noble.gs.washington.edu/proj/svm-pairwise/.

[5] Downloadable from http://www.chibi.ubc.ca/gist/ [12].

detection. In particular, the former represents the usual area under the ROC curve, whereas the latter measures the area under the ROC curve up to the first 50 false positives. A score of 1 indicates perfect separation of positives from negatives, whereas a score of 0 indicates that none of the top 50 sequences selected by the algorithm were positives.

4.1 Results and Discussion

The proposed approach has been compared with the corresponding classic Bag of Words representation in different experimental conditions. In particular, we tested the improvement obtained by the enrichment on BoW representations defined from the raw sequence and from its evolutionary representation (the profile), using different N-grams (1-gram, 2-gram, 3-gram). For what concerns the proposed method, we employed different variants of the BLOSUM matrices. Roughly speaking, a different number after the name "BLOSUM" indicates a more or less strict definition of "similar sequences" in the construction of blocks (see Sect. 3.1).

ROC and ROC50 scores, averaged over all the families of the dataset, are shown in Table 1. To assess statistical significance of our results and demonstrate

Table 1. ROC (top) and ROC50 (bottom) scores. "EBoWX" indicates that the enrichment vectors have been obtained by using the BLOSUMX matrix. In bold we put results for which the p-value of the statistical test is less than 0.05.

Sequence based	1-grams	2-grams	3-grams	Profile based	1-grams	2-grams	3-grams
BoW	0.8601	0.8709	0.8117	BoW	0.9070	0.9290	0.8876
EBoW-45	0.8644	**0.8998**	**0.9131**	EBoW-45	0.9054	**0.9458**	**0.9494**
EBoW-50	0.8638	**0.8996**	**0.9139**	EBoW-50	0.9048	**0.9453**	**0.9494**
EBoW-62	0.8647	**0.8990**	**0.9114**	EBoW-62	0.9042	**0.9453**	**0.9466**
EBoW-80	0.8653	**0.8968**	**0.9061**	EBoW-80	0.9046	**0.9440**	**0.9413**
EBoW-90	0.8652	**0.8950**	**0.9016**	EBoW-90	0.9051	**0.9427**	**0.9384**

(ROC)

Sequence based	1-grams	2-grams	3-grams	Profile based	1-grams	2-grams	3-grams
BoW	0.6054	0.6331	0.5848	BoW	0.6928	0.7741	0.7220
EBoW-45	0.6256	**0.6925**	**0.7175**	EBoW-45	**0.6552**	**0.7914**	0.7832
EBoW-50	0.6270	**0.6888**	**0.7216**	EBoW-50	0.6670	**0.7830**	**0.7944**
EBoW-62	0.6274	0.6763	**0.7052**	EBoW-62	0.6719	0.7863	0.7931
EBoW-80	0.6325	**0.6894**	**0.6776**	EBoW-80	0.6826	0.7829	**0.8007**
EBoW-90	0.6301	**0.6886**	**0.6737**	EBoW-90	0.6731	0.7774	**0.8151**

(ROC50)

that increments in ROC/ROC50 scores gained with the proposed approach are not due to mere chance, we performed a Wilcoxon signed-rank test, reporting in the tables in bold the results for which the corresponding p-value is less than 0.05 (i.e. bold numbers indicate a statistically significant difference). From the tables different observations can be derived. In general, it can be seen that the proposed enrichment is almost always beneficial for 2-grams and 3-grams, with some really important improvements – for example, with 3-grams and BoW based on sequences, the ROC (ROC50) score improves from 0.81 to 0.91 (from 0.58 to 0.72), this representing a remarkable result. This is more evident by looking at the ROC scores. For what concerns the different BLOSUM employed, no differences can be observed in the ROC scores; however, considering the ROC50 scores, it seems evident that this choice has an impact. Unfortunately, a general rule can not be derived: for some configurations a stricter BLOSUM is better, for others the other way around holds. In general, we can say that for 2-grams and 3-grams there is always a configuration for which a statistically significant improvement can be obtained.

For 1-grams such improvement is not so evident (ROC50 results also highlight one case when the biological enrichment results in a worst performance). For what concerns 2- and 3-grams, it seems evident that the proposed enrichment permits to derive a better representation for classification. We think that this is due to a twofold beneficial effect that the approach produces on the representation: from one hand, we are injecting useful information which permits to recover from the uncertainty present in the counting process – a peculiarity of this application. From the other hand, the proposed approach permits to reduce the huge sparsity of the Bag of Words vectors within this application. In fact, within the SCOP datasets the sequences have an average length of 200, thus resulting in around 200 N-grams (if we consider the maximum possible overlap); when using 3-grams, the Bag of Words vector has 8000 entries (the size of the dictionary, 20^3) to be filled with around 200 ones; this implies that most of the entries are zero (this problem is less severe with 2-grams). Even if good classification methods able to deal with sparse representations exist, in this specific case a SVM with the rbf kernel has been used, for fair comparison with state of the art, thus this sparsity problem may have an impact. To provide some empirical support to our intuition, we performed two experiments, focusing on BoW computed from profiles. In the first, we select as Enrichment Matrix a random probability matrix – this solution would in principle alleviate the sparsity problem, but it does not injects any evolutionary information. In the second, we removed from the Enriched BoW low values so that the number of zero-value entries is the same as in the standard Bag of Words representation – this solution only injects evolutionary information, without solving the sparsity problem. ROC values are shown in Table 2: the accuracies obtained in the 3-grams case suggest that there is a beneficial effect both in only the reduction of the sparsity (BoW plus random Enrichment) and in the truncated injection of relevant information (Truncated Effect): however the proposed approach, which combines both effect, obtain the best effect. This is not so evident by looking at the results with 2-grams, where only the complete approach permits to improve the accuracies.

Table 2. Properties of the Enriched BoW representation in the PRHD.

	Classic BoW	BoW + random **E**	Truncated EBoW	Proposed EBoW
2-grams	0.9290	0.9086	0.9213	0.9458
3-grams	0.8876	0.9140	0.9042	0.9494

As a final analysis, we reported in Table 3 some comparative results with other approaches of the literature applied to the SCOP 1.53 benchmark. When compared to other techniques that are based on Bag of Words, the proposed approach behaves very well, outperforming all the alternative techniques; looking at the global picture, the table shows very promising results, also in comparison with other approaches, where satisfactory performances are reached both using the ROC and the ROC50 evaluation measures. Please note that the results can be further improved, for example by deriving the enrichment vectors from more recent and accurate substitution matrices or by tuning the impact of the enrichment (for example by putting a weight α in Eq. (3)).

Table 3. Average ROC scores for the 54 families in the SCOP 1.53 superfamily benchmark for different methods.

Method	ROC	ROC50	Reference
Enriched BoW (3-gram)	0.949	0.815	This paper
Bag of words based methods			
SVM-N-gram	0.826	0.589	[6]
SVM-N-gram-LSA	0.878	0.628	[6]
SVM-Top-N-gram $(n = 1)$	0.907	0.696	[14]
SVM-Top-N-gram $(n = 2)$	0.923	0.713	[14]
SVM-Top-N-gram-combine	0.933	0.767	[14]
SVM-N-gram-p1	0.887	0.726	[15]
SVM-N-gram-KTA	0.892	0.731	[15]
Other methods			
SVM-pairwise	0.908	0.787	[15]
SVM-LA	0.925	0.752	[20]
Profile (5,7.5)	0.980	0.794	[11]
SVM-Pattern-LSA	0.879	0.626	[6]
SVM-Motif-LSA	0.860	0.628	[6]
PSI-BLAST	0.676	0.330	[5]
SVM-Bprofile-LSA	0.921	0.698	[5]
SVM-PDT-profile $(\beta = 8, n = 2)$	0.950	0.740	[13]
SVM-LA-p1	0.958	0.888	[15]

5   Conclusions

In this paper we proposed an enriched BoW representation for Protein Remote Homology Detection, which injects evolutionary information into the counting process, thus resulting in a richer and biologically relevant representation. The proposed scheme has been tested on a standard benchmark, obtaining very promising results. As a future work, we plan to investigate the suitability of the proposed scheme in other domains, such as text processing. Clearly, in this latter case, the main challenge is to define how words are related through similarities in meaning.

References

1. Altschul, S.F., Madden, T.L., Schffer, A.A., Zhang, J., Zhang, Z., Miller, W., Lipman, D.J.: Gapped BLAST and PSI-BLAST: a new generation of protein database search programs. Nucleic Acid Res. **25**(17), 3389–3402 (1997)
2. Bicego, M., Lovato, P., Perina, A., Fasoli, M., Delledonne, M., Pezzotti, M., Polverari, A., Murino, V.: Investigating topic models' capabilities in expression microarray data classification. IEEE/ACM Trans. Comput. Biol. Bioinform. **9**(6), 1831–1836 (2012)
3. Brelstaff, G., Bicego, M., Culeddu, N., Chessa, M.: Bag of peaks: interpretation of nmr spectrometry. Bioinformatics **25**(2), 258–264 (2009)
4. Csurka, G., Dance, C., Fan, L., Willamowski, J., Bray, C.: Visual categorization with bags of keypoints. In: Workshop on Statistical Learning in Computer Vision, ECCV, pp. 1–22 (2004)
5. Dong, Q., Lin, L., Wang, X.: Protein remote homology detection based on binary profiles. In: Hochreiter, S., Wagner, R. (eds.) BIRD 2007. LNCS, vol. 4414, pp. 212–223. Springer, Heidelberg (2007). doi:10.1007/978-3-540-71233-6_17
6. Dong, Q., Wang, X., Lin, L.: Application of latent semantic analysis to protein remote homology detection. Bioinformatics **22**(3), 285–290 (2006)
7. Fox, N.K., Brenner, S.E., Chandonia, J.: SCOPe: structural classification of proteins - extended, integrating SCOP and ASTRAL data and classification of new structures. Nucleic Acids Res. **42**(Database–Issue), 304–309 (2014)
8. Gribskov, M., Robinson, N.L.: Use of receiver operating characteristic (ROC) analysis to evaluate sequence matching. Comput. Chem. **20**(1), 25–33 (1996)
9. Henikoff, S., Henikoff, J.: Amino acid substitution matrices from protein blocks. PNAS **89**(22), 10915–10919 (1992)
10. Karplus, K., Barrett, C., Hughey, R.: Hidden Markov models for detecting remote protein homologies. Bioinformatics **14**, 846–856 (1998)
11. Kuang, R., Ie, E., Wang, K., Wang, K., Siddiqi, M., Freund, Y., Leslie, C.: Profile-based string kernels for remote homology detection and motif extraction. J. Bioinform. Comput. Biol. **3**(03), 527–550 (2005)
12. Liao, L., Noble, W.S.: Combining pairwise sequence similarity and support vector machines for detecting remote protein evolutionary and structural relationships. J. Comput. Biol. **10**(6), 857–868 (2003)
13. Liu, B., Wang, X., Chen, Q., Dong, Q., Lan, X.: Using amino acid physicochemical distance transformation for fast protein remote homology detection. PLoS ONE **7**(9), e46633 (2012)

14. Liu, B., Wang, X., Lin, L., Dong, Q., Wang, X.: A discriminative method for protein remote homology detection and fold recognition combining top-n-grams and latent semantic analysis. BMC Bioinf. **9**(1), 510 (2008)
15. Liu, B., Zhang, D., Xu, R., Xu, J., Wang, X., Chen, Q., Dong, Q., Chou, K.C.: Combining evolutionary information extracted from frequency profiles with sequence-based kernels for protein remote homology detection. Bioinformatics **30**(4), 472–479 (2014)
16. Lovato, P., Giorgetti, A., Bicego, M.: A multimodal approach for protein remote homology detection. IEEE/ACM Trans. Comput. Biol. Bioinform. **12**(5), 1193–1198 (2015)
17. Marszaek, M., Schmid, C.: Spatial weighting for bag-of-features. In: Proceedings of International Conference on Computer Vision and Pattern Recognition, vol. 2, pp. 2118–2125 (2006)
18. Pevsner, J.: Bioinformatics and Functional Genomics. Wiley, Hoboken (2003)
19. Rangwala, H., Karypis, G.: Profile-based direct kernels for remote homology detection and fold recognition. Bioinformatics **21**(23), 4239–4247 (2005)
20. Saigo, H., Vert, J.P., Ueda, N., Akutsu, T.: Protein homology detection using string alignment kernels. Bioinformatics **20**(11), 1682–1689 (2004)
21. Salton, G., McGill, M.J.: Introduction to Modern Information Retrieval. McGraw-Hill Inc., New York (1986)

The Average Mixing Matrix Signature

Luca Rossi[1]([✉]), Simone Severini[2], and Andrea Torsello[3]

[1] School of Engineering and Applied Science, Aston University, Birmingham, UK
l.rossi@aston.ac.uk
[2] Department of Computer Science, University College London, London, UK
[3] DAIS, Università Ca' Foscari Venezia, Venice, Italy

Abstract. Laplacian-based descriptors, such as the Heat Kernel Signature and the Wave Kernel Signature, allow one to embed the vertices of a graph onto a vectorial space, and have been successfully used to find the optimal matching between a pair of input graphs. While the HKS uses a heat diffusion process to probe the local structure of a graph, the WKS attempts to do the same through wave propagation. In this paper, we propose an alternative structural descriptor that is based on continuous-time quantum walks. More specifically, we characterise the structure of a graph using its average mixing matrix. The average mixing matrix is a doubly-stochastic matrix that encodes the time-averaged behaviour of a continuous-time quantum walk on the graph. We propose to use the rows of the average mixing matrix for increasing stopping times to develop a novel signature, the Average Mixing Matrix Signature (AMMS). We perform an extensive range of experiments and we show that the proposed signature is robust under structural perturbations of the original graphs and it outperforms both the HKS and WKS when used as a node descriptor in a graph matching task.

Keywords: Graph characterisation · Structural descriptor · Quantum walks · Average mixing matrix

1 Introduction

Graph-based representations have been used with considerable success in computer vision in the abstraction and recognition of object shape and scene structure [4,8,13]. A fundamental problem in graph-based pattern recognition is that of recovering the set of correspondences (matching) between the vertices of two graphs. In computer vision, graph matching has been applied to a wide range of problems, from object categorisation [1,4] to action recognition [14,15]. More formally, in the graph matching problem the goal is to find a mapping between the nodes of two graphs such that the edge structure is preserved.

While there exists a wide range of methods to solve this problem, many graph matching algorithms greatly benefit from the use of local structural descriptors to maximise their performance. To this end, a structural descriptor or signature is assigned to each node of the graphs, effectively embedding the graphs nodes onto

© Springer International Publishing AG 2016
A. Robles-Kelly et al. (Eds.): S+SSPR 2016, LNCS 10029, pp. 474–484, 2016.
DOI: 10.1007/978-3-319-49055-7_42

a vectorial space. Given an underlying correspondence between the node sets of the graphs, the assumption underpinning these approaches is that corresponding nodes will be mapped to points that are close in the signature space.

Two structural signatures that have proven to be particularly successful are the Heat Kernel Signature (HKS) [12] and the Wave Kernel Signature (WKS) [2]. Both signatures belong to the family of Laplacian-based spectral descriptors and were introduced for the analysis of non-rigid three-dimensional shapes. More specifically, they are similarly based on the idea of using the spectrum of the graph Laplacian (and more in general the Laplace-Beltrami operator on the shape surface) to characterise the points on the shape surface. While the Heat Kernel Signature is based on a heat diffusion process, the Wave Kernel Signature is based on the propagation of a wavefunction.

In this paper, we introduce a novel structural signature based on a quantum mechanical process taking place over the graph. In particular, we propose to probe the graph structure using a continuous-time quantum walk [6]. We make use of the average mixing matrix [5] to define a quantum analogue of the heat kernel, and we propose to take the rows of the average mixing matrix at increasing times t as our vertex descriptor. The average mixing matrix was introduced by Godsil, and it describes the time-averaged probability of a quantum walks starting from node x to visit node y at time t. The motivation behind our work is based on the fact that quantum walks have proven to be very successful in characterising graph structures [3,9–11]. Moreover, using the average mixing matrix allows us to overcome to lack of convergence of quantum walks. We show that the proposed signature is robust to structural noise and outperforms state-of-the-art signatures in a graph matching task.

The remainder of the paper is organised as follows: Sect. 2 reviews the necessary quantum mechanical background and Sect. 3 introduces the proposed signature. The experimental evaluation is presented in Sect. 4 and Sect. 5 concludes the paper.

2   Quantum Walks and Average Mixing Matrix

In this section, we introduce the necessary quantum mechanical background. We start by reviewing the define of continuous-time quantum walks, and then we show how to compute the average mixing matrix of a graph.

2.1   Continuous-Time Quantum Walks

The continuous-time quantum walk represents the quantum analogue of the continuous-time random walk [6]. Let $G = (V, E)$ denote an undirected graph over n nodes, where V is the vertex set and $E \subseteq V \times V$ is the edge set. In a continuous-time random walk, $\boldsymbol{p}(t) \in \mathbb{R}^n$ denotes the state of a walk at time t. The state vectors evolves according to the heat equation

$$\boldsymbol{p}(t) = e^{-Lt}\boldsymbol{p}(0), \tag{1}$$

where the graph Laplacian L is the infinitesimal generator matrix of the underlying continuous-time Markov process.

Similarly to its classical counterpart, the state space of the continuous-time quantum walks is the vertex set of the graph. The classical state vector is replaced by a vector of complex amplitudes over V whose squared norm sums to unity, and as such the state of the system is not constrained to lie in a probability space, thus allowing interference to take place. The general state of the walk at time t is a complex linear combination of the basis states $|u\rangle$, i.e.,

$$|\psi(t)\rangle = \sum_{u \in V} \alpha_u(t)|u\rangle, \tag{2}$$

where the amplitude $\alpha_u(t) \in \mathbb{C}$ and $|\psi(t)\rangle \in \mathbb{C}^{|V|}$ are both complex. Moreover, we have that $\alpha_u(t)\alpha_u^*(t)$ gives the probability that at time t the walker is at the vertex u, and thus $\sum_{u \in V} \alpha_u(t)\alpha_u^*(t) = 1$ and $\alpha_u(t)\alpha_u^*(t) \in [0,1]$, for all $u \in V$, $t \in \mathbb{R}^+$.

The evolution of the walk is governed by the Schrödinger equation

$$\frac{\partial}{\partial t}|\psi(t)\rangle = -iH|\psi(t)\rangle, \tag{3}$$

where we denote the time-independent Hamiltonian as H. Generally speaking, a continuous-time quantum walk is induced whenever the structure of the graphs is reflected by the (0,1) pattern of the Hamiltonian. For example, we could take the adjacency matrix or the Laplacian. In the following we assume $H = A$.

Given an initial state $|\psi(0)\rangle$, solving the Schrödinger equation gives the expression of the state vector at time t,

$$|\psi(t)\rangle = U(t)|\psi(0)\rangle, \tag{4}$$

where $U(t) = e^{-iAt}$ is the unitary operator governing the temporal evolution of the quantum walk. Equation 4 can be conveniently expressed in terms of the spectral decomposition of the adjacency matrix $A = \Phi\Lambda\Phi^\top$, i.e., $|\psi(t)\rangle = \Phi^\top e^{-i\Lambda t}\Phi|\psi(0)\rangle$, where Φ denotes the $n \times n$ matrix $\Phi = (\phi_1|\phi_2|...|\phi_j|...|\phi_n)$ with the ordered eigenvectors ϕ_js of H as columns and $\Lambda = \mathrm{diag}(\lambda_1, \lambda_2, ..., \lambda_j, ..., \lambda_n)$ is the $n \times n$ diagonal matrix with the ordered eigenvalues λ_j of A as elements, and we have made use of the fact that $\exp[-iAt] = \Phi^\top \exp[-i\Lambda t]\Phi$.

2.2   Average Mixing Matrix

Given a graph $G = (V, E)$ with adjacency matrix A and its associated unitary operator $U(t)$, the behaviour of a continuous-time quantum walk at time t is captured by the mixing matrix [5]

$$M_G(t) = e^{iAt} \circ e^{-iAt} = U(-t) \circ U(t), \tag{5}$$

where $A \circ B$ denotes the Schur-Hadamard product of two matrices A and B.

Note that, while in the classical case the probability distribution induced by a random walk converges to a steady state, this does not happen in the quantum case. However, we will show that we can enforce convergence by taking a time-average even if $U(t)$ is norm-preserving. Let us define the average mixing matrix [5] at time T as

$$\widehat{M}_{G;T} = \frac{1}{T} \int_0^T M_G(t) \, dt. \tag{6}$$

The entry uv can be interpreted as the average probability that a walker is found at vertex v when starting the walk from vertex u. Let us define $P_\lambda = \sum_{k=1}^{\mu(\lambda)} \phi_{\lambda,k} \phi_{\lambda,k}^\top$ to be the projection operator on the subspace spanned by the $\mu(\lambda)$ eigenvectors $\phi_{\lambda,k}$ associated with the eigenvalue λ of A. Given this set of projectors, the unitary operator inducing the quantum walk can be rewritten as

$$U(t) = \sum_\lambda e^{-i\lambda t} P_\lambda \tag{7}$$

Given Eq. 7, we can rewrite the equation for the mixing matrix as

$$M_G(t) = \sum_{\lambda_1 \in \Lambda} \sum_{\lambda_2 \in \Lambda} e^{-i(\lambda_1 - \lambda_2)t} P_{\lambda_1} \circ P_{\lambda_2}, \tag{8}$$

and thus we can reformulate Eq. 6 as

$$\widehat{M}_{G;T} = \sum_{\lambda_1 \in \Lambda} \sum_{\lambda_2 \in \Lambda} P_{\lambda_1} \circ P_{\lambda_2} \frac{1}{T} \int_0^T e^{-i(\lambda_1 - \lambda_2)t} \, dt, \tag{9}$$

which has solution

$$\widehat{M}_{G;T} = \sum_{\lambda_1 \in \Lambda} \sum_{\lambda_2 \in \Lambda} P_{\lambda_1} \circ P_{\lambda_2} \frac{i(1 - e^{iT(\lambda_2 - \lambda_1)})}{T(\lambda_2 - \lambda_1)}. \tag{10}$$

In the limit $T \to \infty$, Eq. 10 becomes

$$\widehat{M}_{G;\infty} = \sum_{\lambda \in \tilde{\Lambda}} P_\lambda \circ P_\lambda \tag{11}$$

where $\tilde{\Lambda}$ is the set of distinct eigenvalues of the adjacency matrix.

2.3   Properties of the Average Mixing Matrix

Given a graph $G = (V, E)$, let $\widehat{M}_{G;\infty}(i, j)$ denote the element of $\widehat{M}_{G;\infty}$ at the ith row and jth column. Recall that two nodes $u, v \in V$ are strongly cospectral if and only if $P_\lambda e_u = \pm P_\lambda e_v$ for all eigenvalues λ, where e_u is an element of the standard basis of $\mathbb{R}^{|V|}$. It can be shown that the following properties hold for the matrix $\widehat{M}_{G;\infty}$:

1. $\widehat{M}_{G;\infty}$ is doubly stochastic and rational, and that $\widehat{M}_{G;\infty}$ is positive semidefinite whenever G is connected [5];
2. Let G and H be cospectral, regular, and switching equivalent graphs. Then $\widehat{M}_{G;\infty}(j,j) = \widehat{M}_{H;\infty}(j,j)$, for every $j = 1, 2, ..., n$;
3. Let G and H be cospectral, walk-regular graphs. Then $\widehat{M}_{G;\infty}(j,j) = \widehat{M}_{H;\infty}(j,j)$, for every $j = 1, 2, ..., n$;
4. two vertices i and j are strongly cospectral if and only if the two corresponding rows of the average mixing matrix are the same, i.e., $\widehat{M}_{G;\infty}(i,k) = \widehat{M}_{G;\infty}(j,k)$, for each k [5];
5. suppose G is a regular and connected graph and $\overline{G}$ is the complementary graph. Then $\widehat{M}_{G;\infty} = \widehat{M}_{\overline{G};\infty}$ if and only if $\overline{G}$ is connected.

We omit the proof of $2, 3$ and 5 for lack of space, while the proofs of 1 and 4 can be found in [5]. Note that $2, 3, 4$ and 5 outline the conditions under which some diagonal element of the mixing matrix of can be duplicated, the rows of two mixing matrices are identical, and two mixing matrices have exactly the same elements, respectively. In other words, the above properties give an important indication of the limitations of the matrix $\widehat{M}_{G;\infty}$ to discriminate between different graphs. Although the same properties have not been proved for an arbitrary stopping time T, our experimental analysis in Sect. 4 suggests that some of these properties may still hold for $T < \infty$.

3   The Average Mixing Matrix Signature

Given a graph $G = (V, E)$ and a vertex $x \in V$, we define the Average Mixing Matrix Signature (AMMS) at x at time T as

$$AMMS(x, T) = \mathtt{sort}(\widehat{M}_{G;T}(x, -)), \tag{12}$$

where $\widehat{M}_{G;T}(x, -)$ denotes the row of the average mixing matrix corresponding to x, and $\mathtt{sort}$ is a sorting function. Given a time interval $[T_{min}, T_{max}]$, the AMMS is then created by concatenating the sorted rows for every $T \in [T_{min}, T_{max}]$.

As Eq. 12 shows, for each stopping time T, we decide to take the whole row of the average mixing matrix rather than just the diagonal element $\widehat{M}_{G;T}(x, x)$. Recall from Sect. 2 that under some particular conditions two graphs can have the same diagonal entries of $\widehat{M}_{G;\infty}$. Although the same has not been proved for an arbitrary T, we take the entire row in an attempt to maximise the discriminatory power of the signature. Finally, to ensure the permutational invariance of the signature, for each T we sort the elements of the average mixing matrix rows.

Note that, while there is no specific reason to take all the elements of the sorted row rather than simply the first k, in the experimental evaluation we show that the best performance is usually achieved for large values of k, so we propose to take the entire row. Finally, from Eq. 10 we observe that the computational complexity of the proposed signature is $O(|V|^4)$.

4   Experimental Evaluation

To evaluate the proposed signature, we compare it on a graph matching problem against the HKS as well as the WKS. In the following, we refer to the percentage of correctly matched nodes as accuracy.

4.1   HKS and WKS

Heat Kernel Signature. The HKS is based on the analysis of a heat diffusion process on the shape governed by the heat equation

$$-\frac{\partial u(x,t)}{\partial t} = \Delta u(x,t), \tag{13}$$

where Δ is the Laplace-Beltrami operator. Assuming at time $t = 0$ all the heat is concentrated at a point x, it can be shown that the amount of heat that remains at x at time t is

$$HKS(x,t) = \sum_{i=0}^{\infty} e^{-\lambda_i t}\phi_i(x)^2 \tag{14}$$

where λ_i and ϕ_i denote the ith eigenvalue and the ith eigenfunction of the Laplace-Beltrami operator, respectively. The HKS at a point x is then constructed by evaluating Eq. 14 for different $t \in [t_{min}, t_{max}]$. As suggested in [12], here we sample 100 points logarithmically over the time interval defined by $t_{min} = 4\ln(10)/\lambda_{max}$ and $t_{max} = 4\ln(10)/\lambda_{min}]$, where λ_{min} and λ_{max} denote the minimum and maximum eigenvalue of the graph Laplacian, respectively.

Wave Kernel Signature. The WKS is based on the analysis of wavefunction evolution on the shape surface governed by the Schrödinger equation

$$-\frac{\partial \psi(x,t)}{\partial t} = i\Delta\psi(x,t). \tag{15}$$

Although similar in the definition, note that the dynamics behind the WKS (oscillation) and the HKS (dissipation) are fundamentally different. The WKS evaluates the expected probability of measuring a quantum particle in the point x of the shape surface at any time, i.e.,

$$WKS(x,e) = \sum_{i=0}^{\infty} \phi_i(x)^2 f_e(\lambda_i)^2, \tag{16}$$

where λ_i and ϕ_i denote the ith eigenvalue and the ith eigenfunction of the Laplace-Beltrami operator, respectively, and f_e is a log-normal energy distribution that depends on a parameters e. The WKS for a point x is then constructed by evaluating Eq. 16 for different values of $e \in [e_{min}, e_{max}]$. Here we set all the parameters of the WKS as illustrated in [2].

4.2   Datasets

We use a dataset made of the images from the CMU house sequence, where each image is abstracted as a Delaunay graph over a set of corner feature points. All the resulting graphs are composed of 30 to 32 nodes. Figure 1 shows the ten images with the feature points and the graphs superimposed.

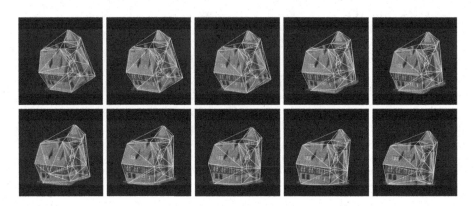

Fig. 1. CMU house sequence with the feature points and the Delaunay graphs superimposed.

In addition, we consider a synthetic datasets consisting of 10 Delaunay graphs over 20 nodes. This dataset is generated as follows. We create a first graph by generating 20 computing the Delaunay triangulation of 20 uniformly randomly scattered two-dimensional points. Then, we generate 9 additional graphs by slightly perturbing the original points and recomputing the Delaunay triangulation.

4.3   Graph Matching Experimental Setup

For a pair of input graphs, we compute the AMMS for each graph node. Here we decide to sample 10 points logarithmically over the time interval $[0.1, 1]$. The choice of $T_{max} = 1$ is experimentally motivated in the next subsection. In general, we observe that on all the datasets considered in this paper the best performance is achieved for low values of T_{max}. Given the nodes signatures, we then compute the matrix of pairwise Euclidean distances between the signatures. Finally, we cast the graph matching problem as an assignment problem which can be solved in polynomial time using the well-known Hungarian method [7].

4.4   Sensitivity to Parameters

We first investigate the sensitivity of the proposed signatures to the number k of row elements that we consider for each stopping time $T \in [T_{min}, T_{max}]$, as well the value of T_{max}. Staring from a seed graph G from the synthetic dataset, we generate 100 noisy copies G' by removing or adding an edge with

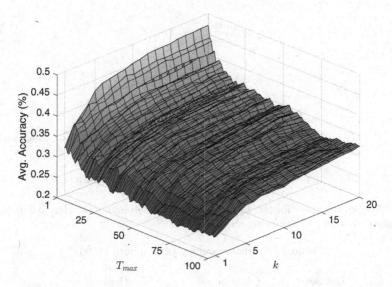

Fig. 2. Evaluation of the sensitivity to the parameters values. Here k denotes the number of row elements used to construct the signature and T_{max} is the maximum stopping time considered.

probability $p = 0.03$. Then, we compute the matching for each pair (G, G') and each choice of the parameters k and T_{max}, and we plot the average accuracy in Fig. 2. Here we let $k = 1, 2, \cdots, n$, where n is the number of nodes of G, and we let $T_{max} = 1, 2, \cdots, 100$. Figure 2 shows that the best accuracy is achieved for k larger than 10 and $T_{max} = 1$.

4.5   Robustness to Erdős-Rényi Structural Noise

In order to evaluate the robustness of the AMMS to structural noise, we perform the following experiment. Given a graph G from the synthetic dataset, we create a series of noisy versions of G, for increasing amounts of Erdős-Rényi noise. To this end, we flip an edge of the graph with probability p and obtain the noisy graph G'. For each value of p, we repeat the noise addition process 100 times to obtain 100 noisy copies of G. Then, we compute the optimal matching using the proposed signature and we compare the results against those obtained using the HKS and the WKS. Figure 3(a) shows the average accuracy for increasing levels of structural noise. Note that the AMMS and the WKS are significantly less susceptible to noise than the HKS. It is also interesting to observe that as p grows over 0.8, the number of correctly matched nodes for the AMMS start to increase (see Fig. 3(b)). We posit that this is evidence that the average mixing matrix of a graph G is equivalent or very similar to that of its complement $\overline{G}$ also for arbitrary $T < \infty$. Note that for $T \to \infty$ this is true if G is regular and connected, and $\overline{G}$ is connected as well. Although this is not the case for the graphs used in these experiments, the geometric nature of the Delaunay triangulation yields indeed fairly regular graphs.

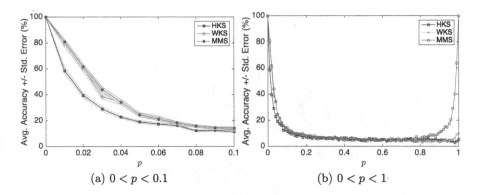

Fig. 3. Average accuracy for increasing levels of Erdős-Rényi noise. Here p denotes the probability of adding or removing an edge.

Table 1. Average accuracy ($\pm$ standard error).

Dataset	CMU house	Synthetic
HKS	30.2963 ± 3.8879	14 ± 1.3162
WKS	44.3704 ± 4.1975	14.5556 ± 1.7467
AMMS(*diag*)	43.3333 ± 3.8461	16.3333 ± 1.6314
AMMS(*row*)	**52.5185 ± 3.8170**	**19.4444 ± 2.1095**

4.6 Graph Matching Accuracy

Table 1 shows the average accuracy on the CMU house and the synthetic datasets. We consider two alternative versions of the proposed signatures: (1) AMMS(*diag*) is the signature where only the diagonal elements $\widehat{M}_{G;T}(x,x)$ are used; (2) AMMS(*row*) is the signature where all the rows elements are used. In both the datasets considered, our signature performs significantly better than both the HKS and the WKS. Moreover, the AMMS(*row*) always outperforms the AMMS(*diag*), highlighting the importance of using all the elements of the row and suggesting once again that the properties listed in Sect. 2 hold for an arbitrary T. Finally, we note that the AMMS(*diag*) always outperforms the HKS but not the WKS.

5 Conclusion

In this paper, we have introduced a novel structural signature for graphs. We probed the structure of the graph using continuous-time quantum walks and we proposed to consider the rows of the average mixing matrix for increasing stopping times T as the node descriptor. The experimental results show that this signature can outperform state-of-the-art signatures like the Heat Kernel Signature and the Wave Kernel Signature in a graph matching task. Future work should focus on the optimisation of the signature parameters for general graphs, as well as reducing the computational complexity associated with the Average Mixing Matrix Signature. Recall that we observed that the best performance is usually achieved for low values of T_{max}. Therefore, one idea may be to reduce the computational complexity of the proposed signature by resorting to a linear approximation of its values.

Acknowledgments. Simone Severini was supported by the Royal Society and EPSRC.

References

1. Albarelli, A., Bergamasco, F., Rossi, L., Vascon, S., Torsello, A.: A stable graph-based representation for object recognition through high-order matching. In: 2012 21st International Conference on Pattern Recognition (ICPR), pp. 3341–3344. IEEE (2012)
2. Aubry, M., Schlickewei, U., Cremers, D.: The wave kernel signature: a quantum mechanical approach to shape analysis. In: 2011 IEEE International Conference on Computer Vision Workshops (ICCV Workshops), pp. 1626–1633. IEEE (2011)
3. Bai, L., Rossi, L., Torsello, A., Hancock, E.R.: A quantum Jensen-Shannon graph kernel for unattributed graphs. Pattern Recogn. **48**(2), 344–355 (2015)
4. Duchenne, O., Joulin, A., Ponce, J.: A graph-matching kernel for object categorization. In: 2011 IEEE International Conference on Computer Vision (ICCV), pp. 1792–1799. IEEE (2011)
5. Godsil, C.: Average mixing of continuous quantum walks. J. Comb. Theor. Ser. A **120**(7), 1649–1662 (2013)
6. Kempe, J.: Quantum random walks: an introductory overview. Contemp. Phys. **44**(4), 307–327 (2003)
7. Munkres, J.: Algorithms for the assignment and transportation problems. J. Soc. Ind. Appl. Math. **5**(1), 32–38 (1957)
8. Riesen, K., Bunke, H.: Graph Classification and Clustering Based on Vector Space Embedding. World Scientific Publishing Co., Inc., Hackensack (2010)
9. Rossi, L., Torsello, A., Hancock, E.R.: Measuring graph similarity through continuous-time quantum walks and the quantum Jensen-Shannon divergence. Phys. Rev. E **91**(2), 022815 (2015)
10. Rossi, L., Torsello, A., Hancock, E.R., Wilson, R.C.: Characterizing graph symmetries through quantum Jensen-Shannon divergence. Phys. Rev. E **88**(3), 032806 (2013)
11. Suau, P., Hancock, E.R., Escolano, F.: Analysis of the schrödinger operator in the context of graph characterization. In: Hancock, E., Pelillo, M. (eds.) SIMBAD 2013. LNCS, vol. 7953, pp. 190–203. Springer, Heidelberg (2013)
12. Sun, J., Ovsjanikov, M., Guibas, L.: A concise and provably informative multiscale signature based on heat diffusion. In: Computer graphics forum, vol. 28, pp. 1383–1392. Wiley Online Library (2009)
13. Torsello, A., Rossi, L.: Supervised learning of graph structure. In: Pelillo, M., Hancock, E.R. (eds.) SIMBAD 2011. LNCS, vol. 7005, pp. 117–132. Springer, Heidelberg (2011). doi:10.1007/978-3-642-24471-1_9

14. Yao, B., Fei-Fei, L.: Action recognition with exemplar based 2.5D graph matching. In: Fitzgibbon, A., Lazebnik, S., Perona, P., Sato, Y., Schmid, C. (eds.) ECCV 2012. LNCS, vol. 7575, pp. 173–186. Springer, Heidelberg (2012). doi:10.1007/978-3-642-33765-9_13

15. Yi, Y., Lin, M.: Human action recognition with graph-based multiple-instance learning. Pattern Recogn. **53**(C), 148–162 (2016)

Exact Graph Edit Distance Computation Using a Binary Linear Program

Julien Lerouge[1], Zeina Abu-Aisheh[2], Romain Raveaux[2], Pierre Héroux[1], and Sébastien Adam[1(✉)]

[1] Normandie Univ, UNIROUEN, UNIHAVRE, INSA Rouen, LITIS,
76000 Rouen, France
Sebastien.Adam@univ-rouen.fr
[2] LI Tours, Avenue Jean Portalis, Tours, France

Abstract. This paper presents a binary linear program which computes the exact graph edit distance between two richly attributed graphs (i.e. with attributes on both vertices and edges). Without solving graph edit distance for large graphs, the proposed program enables to process richer and larger graphs than existing approaches based on mathematical programming and the A^* algorithm. Experiments are led on 7 standard graph datasets and the proposed approach is compared with two state-of-the-art algorithms.

Keywords: Graph edit distance · Binary linear program

1 Introduction

Computing the dissimilarity between two graphs is a crucial issue for graph-based pattern recognition problems, e.g. malware detection [10], chemoinformatics [5], or document analysis [2]. A large number of algorithms are proposed in the literature to compute graph dissimilarity. Among existing approaches, Graph Edit Distance (GED) has retained a lot of attention during the two last decades. Using GED, graph dissimilarity computation is directly linked to a matching process through the introduction of a set of graph edit operations (e.g. vertex insertion, vertex deletion). Each edit operation being characterized by a cost, the GED is the total cost of the least expensive sequence of edit operations that transforms one graph into the other one. A major theoretical advantage of GED is that it is a dissimilarity measure for arbitrarily structured and arbitrarily attributed graphs. Moreover, together with the dissimilarity, it provides the corresponding sequence of edit operations, which can be itself useful for application purpose. A limitation for the use of GED is its computational complexity. Indeed, as stated in [22], the GED problem is NP-hard. This explains why many recent contributions focus on the computation of GED approximations which can be applied on large graphs. Among existing approximations, some are based on the proposition of new heuristics to improve the performance of exact approaches [3,21] whereas others propose faster but suboptimal methods which approximate the

A. Robles-Kelly et al. (Eds.): S+SSPR 2016, LNCS 10029, pp. 485–495, 2016.
DOI: 10.1007/978-3-319-49055-7_43

exact GED (e.g. [1,4,14–16,18,20]). These latter are faster, but do not guarantee to find the optimal solution.

In this paper, we are interested in the exact computation of the GED between two graphs, for a given set of edit costs. In this context, the main family of existing approaches is based on the widely known A* algorithm [3,21]. This algorithm relies on the exploration of the tree of solutions. In this tree, a node corresponds to a partial edition of the input graph towards the target one. A leaf of the tree corresponds to a complete edit path which transforms one graph into the other. The exploration of the tree is led by developing the most promising ways on the basis of an estimation of GED. For each node, this estimation is the sum of the cost associated to the partial edit path and an estimation of the cost for the remaining path. The latter is given by a heuristic. Provided that the estimation of the future cost is lower than or equal to the real cost, an optimal path from the root node to a leaf node is guaranteed to be found [7]. The different A*-based methods published in the literature mainly differ in the implemented heuristics for the future cost estimation which correspond to different tradeoffs between approximation quality and their computation time [3,21].

A second family of algorithms consists in using Binary Linear Programing (BLP) for computing the GED. Justice and Hero [9] proposed a BLP formulation of the GED problem. The proposed program searches for the permutation matrix which minimizes the cost of transforming G_1 into G_2, with G_1 and G_2 two unweighted and undirected attributed graphs. The criterion to be minimized takes into account costs for matching vertices, but the formulation does not integrate the ability to process graphs with labels on their edges.

In some previous works, we also have investigated the use of Integer Linear Programing for graph matching problems. In [11,13], a formulation and a toolbox are proposed for substitution-tolerant subgraph isomorphism and its use for symbol spotting in technical drawings. In [12], a BLP-based minimum cost subgraph matching is described. It has been used for document analysis purpose in [6]. In this paper, we propose an extension of [12] for GED computation. The proposed framework can be used to compute the distance between richly attributed graphs (i.e. with attributes on both vertices and edges) which can not be tackled using the BLP formulation proposed in [9]. Experiments led on reference datasets show that the new BLP formulation can compute an exact GED on larger graphs than A* based algorithm.

This paper is organized as follows: Sect. 2 presents the important definitions necessary for introducing our formulations of the GED. Then, Sect. 3 describes the proposed binary linear programming formulation. Section 4 presents the experiments and analyses the obtained results. Section 5 provides some concluding remarks.

2   Definitions

In this paper, we are interested in computing GED between attributed graphs. This section is dedicated to the introduction of the notations and definitions used in the remaining of the paper.

Definition 1. *An attributed graph G is a 4-tuple $G = (V, E, \mu, \xi)$, where :*

- *V is a set of vertices,*
- *E is a set of edges, such that $\forall e = ij \in E, i \in V$ and $j \in V$,*
- *$\mu : V \rightarrow L_V$ is a vertex labeling function which associates the label $\mu(v)$ to all vertices v of V, where L_V is the set of possible labels for the vertices,*
- *$\xi : E \rightarrow L_E$ is an edge labeling function which associates the label $\xi(e)$ to all edges e of E, where L_E is the set of possible labels for the edges.*

The vertices (resp. edges) label space L_V (resp. L_E) may be composed of any combination of numeric, symbolic or string attributes.

Definition 2. *The graph edit distance $d(.,.)$ is a function*

$$d : \mathcal{G} \times \mathcal{G} \rightarrow \mathbb{R}^+$$

$$(G_1, G_2) \mapsto d(G_1, G_2) = \min_{o = (o_1, \ldots, o_k) \in \Gamma(G_1, G_2)} \sum_{i=1}^{k} c(o_i)$$

where $G_1 = (V_1, E_1, \mu_1, \xi_1)$ and $G_2 = (V_2, E_2, \mu_2, \xi_2)$ are two graphs from the set $\mathcal{G}$ and $\Gamma(G_1, G_2)$ is the set of all edit paths $o = (o_1, \ldots, o_k)$ allowing to transform G_1 into G_2. An elementary edit operation o_i is one of vertex substitution ($v_1 \rightarrow v_2$), edge substitution ($e_1 \rightarrow e_2$), vertex deletion ($v_1 \rightarrow \epsilon$), edge deletion: ($e_1 \rightarrow \epsilon$), vertex insertion ($\epsilon \rightarrow v_2$) and edge insertion ($\epsilon \rightarrow e_2$) with $v_1 \in V_1$, $v_2 \in V_2$, $e_1 \in E_1$ and $e_2 \in E_2$. ϵ is a dummy vertex or edge which is used to model insertion or deletion. $c(.)$ is a function which associates a cost to each elementary edit operation o_i.

In the more general case where GED is computed between attributed graphs, edit costs are generally defined as functions of vertices (resp. edges) attributes. More precisely, substitution costs are defined as a function of the attributes of the substituted vertices (resp. edges), whereas insertion and deletion are penalized with a value linked to the attributes of the inserted/deleted vertex (resp. edge):

$$c(v_1 \rightarrow v_2) = c(v_2 \rightarrow v_1) = f_v(\mu_1(v_1), \mu_2(v_2))$$
$$c(e_1 \rightarrow e_2) = c(e_2 \rightarrow e_1) = f_e(\xi_1(e_1), \xi_2(e_2))$$
$$c(v \rightarrow \epsilon) = c(\epsilon \rightarrow v) = g_v(\mu(v))$$
$$c(e \rightarrow \epsilon) = c(\epsilon \rightarrow e) = g_e(\xi(e))$$

3   BLP Formulation of GED

Binary Linear Programing is a restriction of integer linear programming (ILP) with binary variables. Its general form is :

$$\min_{x} c^T x \tag{1a}$$

$$\text{subject to } Ax \leq b \tag{1b}$$

$$x \in \{0, 1\}^n \tag{1c}$$

where $c \in \mathbb{R}^n, A \in \mathbb{R}^{n \times m}$ and $b \in \mathbb{R}^m$ are data of the problem. A feasible solution is a vector x of n binary variables (1c) which respects linear inequality constraints (1b). The constraint (1c) which enumerates the admissible values of variables is called a domain constraint. If the program has at least a feasible solution, then the optimal solutions are the ones that minimize the objective function (1a) which is a linear combination of variables of x weighted by the components of the vector c.

In this section, we successively define the variables, the objective function and the linear constraint of the BLP used for formulating GED as a BLP.

3.1   BLP-GED Variables

Our goal is to compute GED between two graphs $G_1 = (V_1, E_1, \mu_1, \xi_1)$ and $G_2 = (V_2, E_2, \mu_2, \xi_2)$. In the rest of this section, for the sake of simplicity of notations, we consider that the graphs G_1 and G_2 are simple directed graphs. However, the formulations given in this section can be applied directly without modification to multigraphs. Besides, the extension to the undirected case only needs some slight modifications.

In the GED definition provided in Sect. 2, the edit operations that are allowed to match the graphs G_1 and G_2 are (i) the substitution of a vertex (respectively an edge) of G_1 with a vertex (resp. an edge) of G_2, (ii) the deletion of a vertex (or an edge) from G_1 and (iii) the insertion of a vertex (or an edge) of G_2 in G_1. For each type of edit operation, we define a set of corresponding binary variables:

$$- \forall (i, k) \in V_1 \times V_2, x_{i,k} = \begin{cases} 1 \text{ if } i \text{ is substituted with } k, \\ 0 \text{ otherwise.} \end{cases}$$

$$- \forall (ij, kl) \in E_1 \times E_2, y_{ij,kl} = \begin{cases} 1 \text{ if } ij \text{ is substituted with } kl, \\ 0 \text{ otherwise.} \end{cases}$$

$$- \forall i \in V_1, u_i = \begin{cases} 1 \text{ if } i \text{ is deleted from } G_1 \\ 0 \text{ otherwise.} \end{cases}$$

$$- \forall ij \in E_1, e_{ij} = \begin{cases} 1 \text{ if } ij \text{ is deleted from } G_1 \\ 0 \text{ otherwise.} \end{cases}$$

$$- \forall k \in V_2, v_k = \begin{cases} 1 \text{ if } k \text{ is inserted in } G_1 \\ 0 \text{ otherwise.} \end{cases}$$

$$- \forall kl \in E_2, f_{kl} = \begin{cases} 1 \text{ if } kl \text{ is inserted in } G_1 \\ 0 \text{ otherwise.} \end{cases}$$

Using these notations, we define an edit path between G_1 and G_2 as a 6-tuple $(\mathbf{x}, \mathbf{y}, \mathbf{u}, \mathbf{v}, \mathbf{e}, \mathbf{f})$ where $\mathbf{x} = (x_{i,k})_{(i,k) \in V_1 \times V_2}$, $\mathbf{y} = (y_{ij,kl})_{(ij,kl) \in E_1 \times E_2}$, $\mathbf{u} = (u_i)_{i \in V_1}$, $\mathbf{e} = (e_{ij})_{ij \in E_1}$, $\mathbf{v} = (v_k)_{k \in V_2}$ and $\mathbf{f} = (f_{kl})_{kl \in E_2}$.

In order to evaluate the global cost of an edit path, elementary costs for each edit operation must be defined. We adopt the following notations for these costs:

$- \forall (i, k) \in V_1 \times V_2, c(i \rightarrow k)$ is the cost of substituting the vertex i with k,
$- \forall (ij, kl) \in E_1 \times E_2, c(ij \rightarrow kl)$ is the cost of substituting the edge ij with kl,

- $\forall i \in V_1, c(i \to \epsilon)$ is the cost of deleting the vertex i from G_1,
- $\forall ij \in E_1, c(ij \to \epsilon)$ is the cost of deleting the edge ij from G_1,
- $\forall k \in V_2, c(\epsilon \to k)$ is the cost of inserting the vertex k in G_1,
- $\forall kl \in E_2, c(\epsilon \to kl)$ is the cost of inserting the edge kl in G_1.

3.2   Objective Function

The objective function is the overall cost induced by applying an edit path $(\mathbf{x}, \mathbf{y}, \mathbf{u}, \mathbf{v}, \mathbf{e}, \mathbf{f})$ that transforms a graph G_1 into G_2, using the elementary costs defined above. GED between G_1 and G_2 is the minimum value of the objective function (2) over $(\mathbf{x}, \mathbf{y}, \mathbf{u}, \mathbf{v}, \mathbf{e}, \mathbf{f})$ subject to constraints detailed in Sect. 3.3.

$$
\begin{aligned}
d(G_1, G_2) = \min_{\mathbf{x}, \mathbf{y}, \mathbf{u}, \mathbf{v}, \mathbf{e}, \mathbf{f}} \Bigg(&\sum_{i \in V_1} \sum_{k \in V_2} c(i \to k) \cdot x_{i,k} + \sum_{ij \in E_1} \sum_{kl \in E_2} c(ij \to kl) \cdot y_{ij,kl} \\
&+ \sum_{i \in V_1} c(i \to \epsilon) \cdot u_i + \sum_{k \in V_2} c(\epsilon \to k) \cdot v_k \\
&+ \sum_{ij \in E_1} c(ij \to \epsilon) \cdot e_{ij} + \sum_{kl \in E_2} c(\epsilon \to kl) \cdot f_{kl} \Bigg)
\end{aligned}
\tag{2}
$$

3.3   Constraints

The constraints presented in this part are designed to guarantee that the admissible solutions of the BLP are edit paths that transform G_1 in a graph which is isomorphic to G_2. An edit path is considered as admissible if and only if the following conditions are respected:

1. It provides a one-to-one mapping between a subset of the vertices of G_1 and a subset of the vertices of G_2. The remaining vertices are either deleted or inserted,.
2. It provides a one-to-one mapping between a subset of the edges of G_1 and a subset of the edges of G_2. The remaining edges are either deleted or inserted,
3. The vertices matchings and the edges matchings are consistent, i.e. the graph topology is respected.

The following paragraphs describe the linear constraints used to integrate these conditions into the BLP.

1 - Vertices Matching Constraints. The constraint (3) ensures that each vertex of G_1 is either matched to exactly one vertex of G_2 or deleted from G_1, while the constraint (4) ensures that each vertex of G_2 is either matched to exactly one vertex of G_1 or inserted in G_1:

$$u_i + \sum_{k \in V_2} x_{i,k} = 1 \quad \forall i \in V_1 \tag{3}$$

$$v_k + \sum_{i \in V_1} x_{i,k} = 1 \quad \forall k \in V_2 \tag{4}$$

2 - Edges Matching Constraints. Similar to the vertex matching constraints, the constraints (5) and (6) guarantee a valid mapping between the edges:

$$e_{ij} + \sum_{kl \in E_2} y_{ij,kl} = 1 \quad \forall ij \in E_1 \tag{5}$$

$$f_{kl} + \sum_{ij \in E_1} y_{ij,kl} = 1 \quad \forall kl \in E_2 \tag{6}$$

3 - Topological Constraints. In order to respect the graph topology in the matching, an edge $ij \in E_1$ can be matched to an edge $kl \in E_2$ only if the head vertices $i \in V_1$ and $k \in V_2$, on the one hand, and if the tail vertices $j \in V_1$ and $l \in V_2$, on the other hand, are respectively matched. This quadratic constraint can be expressed linearly with the following constraints (7) and (8):

– ij and kl can be matched if and only if their head vertices are matched:

$$y_{ij,kl} \le x_{i,k} \quad \forall (ij, kl) \in E_1 \times E_2 \tag{7}$$

– ij and kl can be matched if and only if their tail vertices are matched:

$$y_{ij,kl} \le x_{j,l} \quad \forall (ij, kl) \in E_1 \times E_2 \tag{8}$$

Equations 2 to 8, coupled with domain constraints which ensure that the solution is binary leads to our BLP formulation :

(BLP-GED)

$$\min_{\mathbf{x,y,u,v,f}} \left(\sum_{i \in V_1} \sum_{k \in V_2} c(i \to k) \cdot x_{i,k} \right.$$
$$+ \sum_{ij \in E_1} \sum_{kl \in E_2} c(ij \to kl) \cdot y_{ij,kl}$$
$$+ \sum_{i \in V_1} c(i \to \epsilon) \cdot u_i + \sum_{k \in V_2} c(\epsilon \to k) \cdot v_k \tag{9a}$$
$$\left. + \sum_{ij \in E_1} c(ij \to \epsilon) \cdot e_{ij} + \sum_{kl \in E_2} c(\epsilon \to kl) \cdot f_{kl} \right)$$

$$\text{subject to}\quad u_i + \sum_{k \in V_2} x_{i,k} = 1 \quad \forall i \in V_1 \tag{9b}$$

$$v_k + \sum_{i \in V_1} x_{i,k} = 1 \quad \forall k \in V_2 \tag{9c}$$

$$e_{ij} + \sum_{kl \in E_2} y_{ij,kl} = 1 \quad \forall ij \in E_1 \tag{9d}$$

$$f_{kl} + \sum_{ij \in E_1} y_{ij,kl} = 1 \quad \forall kl \in E_2 \tag{9e}$$

$$y_{ij,kl} \leq x_{i,k} \quad \forall (ij, kl) \in E_1 \times E_2 \tag{9f}$$

$$y_{ij,kl} \leq x_{j,l} \quad \forall (ij, kl) \in E_1 \times E_2 \tag{9g}$$

$$\text{with}\quad x_{i,k} \in \{0,1\} \quad \forall (i,k) \in V_1 \times V_2 \tag{9h}$$

$$y_{ij,kl} \in \{0,1\} \quad \forall (ij, kl) \in E_1 \times E_2 \tag{9i}$$

$$u_i \in \{0,1\} \quad \forall i \in V_1 \tag{9j}$$

$$v_k \in \{0,1\} \quad \forall k \in V_2 \tag{9k}$$

$$e_{ij} \in \{0,1\} \quad \forall ij \in E_1 \tag{9l}$$

$$f_{kl} \in \{0,1\} \quad \forall kl \in E_2 \tag{9m}$$

4 Experiments

In this section, we present some experimental results obtained using the proposed formulation, compared with those of two reference methods. The first one is based on the A* algorithm with a bipartite heuristic [21]. This method is the most well-known exact GED method and is often used to evaluate the accuracy of approximate methods. The second exact method is the BLP proposed by Justice and Hero in [9]. This method, called JH in the paper, is directly linked to our proposal. Since this method cannot deal with edge attributes, we could not perform JH on all our datasets.

In this practical work, our method has been implemented in $C\#$ using CPLEX Concert Technology. All the methods were run on a 2.6 GHz quad-core computer with 8 GB RAM. For the sake of comparison, none of the methods were parallelized and CPLEX was set up in a deterministic manner.

The 3 methods are evaluated on 7 reference datasets:

- **GREC** [17] is composed of undirected graphs of rather small size (i.e. up to 20 vertices in our experiments). In addition, continuous attributes on vertices and edges play an important role in the matching procedure. Such graphs are representative of pattern recognition problems where graphs are involved in a classification stage.
- **PROTEIN** [17] is a molecule dataset. In similar datasets, vertices are labeled with atomic chemical elements, which imposes a stringent constraint of label equality in the matching process. In this particular dataset, vertices are labeled with chemical element sequences. The stringent constraint can be relaxed

thanks to the string edit distance. So the matching process can be tolerant and accommodate with differences on labels.

- **ILPISO** [8] stands apart from the others in the sense that this dataset hold directed graphs. The aim is to illustrate the flexibility of our proposal that can handle different types of graphs.
- **LETTER** [17] is broken down into three parts (LOW, MED, HIGH) which corresponds to distortion levels. The LETTER dataset is useful because it holds graphs of rather small size (maximum of 9 vertices), what makes feasible the computations of all GED methods.
- **PAH**[1] is a purely structural database with no labels at all. PAH is a hard dataset gathering graphs of rather large size (≥ 20 *vertices*). Graphs are complex and hard to match in cases where neighborhoods and attributes do not allow easily to differentiate between vertices.

All these datasets are publicly available on IAPR TC15 website[2]. In order to evaluate the algorithms behaviours when the size of the problem grows, we have built subsets where all graphs have the same number of vertices for GREC, PROTEIN, ILPISO datasets. Concerning edit costs, we borrow the setting from [19] for GREC, PROT, and LETTER databases.

In order to evaluate an exact GED computation method, the natural criterion is the computation time. However, from a practical point of view, exact GED solving can be infeasible in a reasonable time or can overstep memory capacity. In our framework, if a time limit of **300** s is reached, the instance is considered as unsolved. The same situation arises when the memory limit of 1 Gb is reached.

Obtained results when comparing the 3 methods are presented in Table 1. Two values are given to qualify each method on each dataset: the percentage of instances solved to optimality (without time or memory problems), called "Opt" and the average running time in milliseconds called "Time". In the objective of a fair comparison, the average running time is calculated only from instances solved to optimality by all the methods. If the memory limit is reached, "OM" appears in the table. The "NA" values mean that the algorithm can not be applied to the dataset. It occurs for the JH method on GREC, PROT and ILPISO dataset since JH formulation does not take into account edge labels.

Some observations can be made from these results. First, as expected, A* is the worst method. It reaches the optimal solution only for datasets composed of very small graphs (LETTER and GREC10). A* is the fastest method for small graphs. When graphs are larger than 10 vertices, all the instances are not optimally solved. When graphs hold 10 vertices, the execution time increase considerably by a factor 2000 compared to graphs of 5 vertices. The combinatorial explosion is not prevented by the pruning strategy. Beyond 10 vertices, A* cannot converge to the optimality because of memory saturation phenomenon. The size of the list OPEN containing pending solutions grows exponentially according to the graph size and the bipartite heuristic fails to prune the search tree efficiently.

---

[1] https://brunl01.users.greyc.fr/CHEMISTRY/.

[2] https://iapr-tc15.greyc.fr/links.html.

Table 1. Results

	BLP-GED		A*		JH	
	Opt	Time	Opt	Time	Opt	Time
LETTER HIGH	100	43	100	3	100	13
LETTER MED	100	8	100	1	100	10
LETTER LOW	100	8	100	1	100	10
GREC 5	100	4	100	3	NA	NA
GREC 10	100	60	90.1	7980	NA	NA
GREC 20	99.5	11016	OM	OM	NA	NA
PROT 20	83.1	35376	OM	OM	NA	NA
PROT 30	37.9	81689	OM	OM	NA	NA
ILPISO 10	100	23	OM	OM	NA	NA
ILPISO 25	91.6	6545	OM	OM	NA	NA
ILPISO 50	56.9	17045	OM	OM	NA	NA
PAH	51	92991	OM	OM	100	948

Second, when analysing BLP-GED results, one can see that for datasets composed of smaller graphs (LETTER, ILPISO10, GREC5-10), BLP-GED converges to the optimality. For larger graphs (GREC20, PROT and ILPISO25-50), the optimal solution is not always reached, because of the time restriction but BLP-GED is the only method to solve instances and to output solutions. An interesting comment comes from the GREC dataset where nearly all instances are solved to optimality however increasing the graph size of only 5 vertices can lead to an important increase of time. The relation between graph size and solving time is not linear at all and it recalls us humbly how hard is the GED problem.

Finally, concerning JH, the method performs well for the PAH dataset solving 100 % of instances optimally against 51 % for BLP-GED. JH achieved better results because it is not generic but dedicated to compare unweighted graphs. However, on a smaller graphs like in LETTER, JH is the slowest method due to a higher number of variables and constraints than BLP-GED formulation. Finally, JH is not flexible and the method is unable to consider edge labels and cannot be applied to GREC, PROT and ILPISO.

5 Conclusion

In this paper, an exact binary linear programming formulation of the GED problem has been presented. One of the major advantage of our proposal against another binary program (JH) is that it can deal with a wide range of attributed relational graphs: directed or undirected graphs, simple graphs or multigraphs, with a combination of symbolic, numeric and/or string attributes on vertices and edges. Moreover, obtained results show that our BLP is about 100 times faster than $A*$ for medium-size graphs while being more memory efficient. Our future works concern the optimization and the approximation of our formulation of GED computation.

References

1. Fankhauser, S., Riesen, K., Bunke, H., Dickinson, P.J.: Suboptimal graph isomorphism using bipartite matching. IJPRAI **26**(6), 1250013 (2012)
2. Fischer, A., Bunke, H.: Character prototype selection for handwriting recognition in historical documents. In: 2011 19th European Signal Processing Conference, pp. 1435–1439, August 2011
3. Fischer, A., Plamondon, R., Savaria, Y., Riesen, K., Bunke, H.: A hausdorff heuristic for efficient computation of graph edit distance. In: Fränti, P., Brown, G., Loog, M., Escolano, F., Pelillo, M. (eds.) S+SSPR 2014. LNCS, vol. 8621, pp. 83–92. Springer, Heidelberg (2014). doi:10.1007/978-3-662-44415-3_9
4. Fischer, A., Suen, C.Y., Frinken, V., Riesen, K., Bunke, H.: Approximation of graph edit distance based on Hausdorff matching. Pattern Recogn. **48**(2), 331–343 (2015)
5. Gaüzère, B., Brun, L., Villemin, D.: Two new graphs kernels in chemoinformatics. Pattern Recogn. Lett. **33**(15), 2038–2047 (2012). http://www.science direct.com/science/article/pii/S016786551200102X, Graph-Based Representations in Pattern Recognition
6. Hammami, M., Héroux, P., Adam, S., d'Andecy, V.P.: One-shot field spotting on colored forms using subgraph isomorphism. In: 2015 13th International Conference on Document Analysis and Recognition (ICDAR), pp. 586–590, August 2015
7. Hart, P., Nilsson, N., Raphael, B.: A formal basis for the heuristic determination of minimum cost paths. IEEE Trans. Syst. Sci. Cybern. **4**(2), 100–107 (1968)
8. Héroux, P., Bodic, P., Adam, S.: Datasets for the evaluation of substitution-tolerant subgraph isomorphism. In: Lamiroy, B., Ogier, J.-M. (eds.) GREC 2013. LNCS, vol. 8746, pp. 240–251. Springer, Heidelberg (2014). doi:10.1007/978-3-662-44854-0_19
9. Justice, D., Hero, A.: A binary linear programming formulation of the graph edit distance. IEEE Trans. Pattern Anal. Mach. Intell. **28**(8), 1200–1214 (2006)
10. Kostakis, O.: Classy: fast clustering streams of call-graphs. Data Min. Knowl. Discov. **28**(5), 1554–1585 (2014). doi:10.1007/s10618-014-0367-9
11. Le Bodic, P., Héroux, P., Adam, S., Lecourtier, Y.: An integer linear program for substitution-tolerant subgraph isomorphism and its use for symbol spotting in technical drawings. Pattern Recogn. **45**(12), 4214–4224 (2012)
12. Lerouge, J., Hammami, M., Héroux, P., Adam, S.: Minimum cost subgraph matching using a binary linear program. Pattern Recogn. Lett. **71**, 45–51 (2016)
13. Lerouge, J., Bodic, P., Héroux, P., Adam, S.: GEM++: a tool for solving substitution-tolerant subgraph isomorphism. In: Liu, C.-L., Luo, B., Kropatsch, W.G., Cheng, J. (eds.) GbRPR 2015. LNCS, vol. 9069, pp. 128–137. Springer, Heidelberg (2015). doi:10.1007/978-3-319-18224-7_13
14. Myers, R., Wilson, R.C., Hancock, E.R.: Bayesian graph edit distance. IEEE Trans. Pattern Anal. Mach. Intell. **22**(6), 628–635 (2000)
15. Neuhaus, M., Riesen, K., Bunke, H.: Fast suboptimal algorithms for the computation of graph edit distance. In: Yeung, D.-Y., Kwok, J.T., Fred, A., Roli, F., Ridder, D. (eds.) SSPR /SPR 2006. LNCS, vol. 4109, pp. 163–172. Springer, Heidelberg (2006). doi:10.1007/11815921_17

16. Raveaux, R., Burie, J.C., Ogier, J.M.: A graph matching method and a graph matching distance based on subgraph assignments. Pattern Recogn. Lett. **31**(5), 394–406 (2010)

17. Riesen, K., Bunke, H.: IAM graph database repository for graph based pattern recognition and machine learning. In: da Vitoria Lobo, N., Kasparis, T., Roli, F., Kwok, J.T., Georgiopoulos, M., Anagnostopoulos, G.C., Loog, M. (eds.) Structural, Syntactic, and Statistical Pattern Recognition. LNCS, vol. 5342, pp. 287–297. Springer, Heidelberg (2008)

18. Riesen, K., Bunke, H.: Approximate graph edit distance computation by means of bipartite graph matching. Image Vis. Comput. **27**(7), 950–959 (2009)

19. Riesen, K., Bunke, H.: Graph Classification and Clustering Based on Vector Space Embedding. World Scientific Publishing Co. Inc., River Edge (2010)

20. Riesen, K., Bunke, H.: Improving bipartite graph edit distance approximation using various search strategies. Pattern Recogn. **48**(4), 1349–1363 (2015)

21. Riesen, K., Fankhauser, S., Bunke, H.: Speeding up graph edit distance computation with a bipartite heuristic. In: Frasconi, P., Kersting, K., Tsuda, K. (eds.) Mining and Learning with Graphs, MLG 2007, Firence, Italy, 1–3 August 2007, Proceedings (2007)

22. Zeng, Z., Tung, A.K.H., Wang, J., Feng, J., Zhou, L.: Comparing stars: on approximating graph edit distance. Proc. VLDB Endowment. **2**, 25–36 (2009)

Approximating Graph Edit Distance Using GNCCP

Benoît Gaüzère[1](✉), Sébastien Bougleux[2], and Luc Brun[3]

[1] Normandie Univ, INSA Rouen, LITIS, Rouen, France
benoit.gauzere@insa-rouen.fr
[2] Normandie Univ, UNICAEN, CNRS, GREYC, Rouen, France
[3] Normandie Univ, ENSICAEN, CNRS, GREYC, Rouen, France

Abstract. The graph edit distance (GED) is a flexible and widely used dissimilarity measure between graphs. Computing the GED between two graphs can be performed by solving a quadratic assignment problem (QAP). However, the problem is NP complete hence forbidding the computation of the optimal GED on large graphs. To tackle this drawback, recent heuristics are based on a linear approximation of the initial QAP formulation. In this paper, we propose a method providing a better local minimum of the QAP formulation than our previous proposition based on IPFP. We adapt a convex concave regularization scheme initially designed for graph matching which allows to reach better local minimum and avoids the need of an initialization step. Several experiments demonstrate that our method outperforms previous methods in terms of accuracy, with a time still much lower than the computation of a GED.

1 Introduction

Graphs provide a flexible framework to represent data including relationships between elements. In addition, graphs come with an underlying powerful theory which allows to infer a lot of information from this representation. However, graph's space is not an euclidean space. This last point avoids the use of classic machine learning methods mainly designed to operate in euclidean spaces. Several approaches aim to bridge the gap between graph space and euclidean spaces in order to combine machine learning and graphs. A first historical approach consists in embedding graphs onto euclidean spaces by computing a set of descriptors describing graphs. Even if this method is straightforward and can be easily controlled by choosing the information to keep, the loss of structural information induced by the euclidean embedding may constitute a major drawback for some applications. An extension of this approach is based on the well known *kernel trick*. Kernel methods implicitly define a scalar product between embedding of graphs in some Hilbert spaces, without requiring an explicit formulation of the Reproducing Kernel Hilbert Space (RKHS) associated to the kernel. The use of this trick allows to combine graphs and powerful kernel methods such as Kernel Ridge Regression or Support Vector Machines. However, constraints on the design of graph kernels often complexify the consideration of fine similarities or dissimilarities between graphs.

© Springer International Publishing AG 2016
A. Robles-Kelly et al. (Eds.): S+SSPR 2016, LNCS 10029, pp. 496–506, 2016.
DOI: 10.1007/978-3-319-49055-7_44

Another strategy consists in operating directly in the graph space. One of the most used measure is the graph edit distance (GED) [2,14]. The GED of two graphs may be understood as the minimal amount of distortion required to transform a source graph into a target one. This distortion is encoded by an edit path, defined as a sequence of edit operations which includes nodes and edges substitutions, removals and insertions. Depending on the context, each edit operation e included in an edit path γ is associated to a non-negative cost $c(e)$. The sum of all edit operation costs included within γ defines the cost $A(\gamma)$ associated to this edit path. The minimal cost $A(\gamma^\star)$ among all edit paths $\Gamma(G_1, G_2)$ defines the GED between G_1 and G_2:

$$\mathrm{GED}(G_1, G_2) = A(\gamma^\star) = \min_{\gamma \in \Gamma(G_1, G_2)} A(\gamma) \tag{1}$$

with $A(\gamma) = \sum_{e \in \gamma} c(e)$. The edit path $\gamma^\star$ corresponds to an optimal edit path. GED has been widely used by the structural pattern recognition community [4,11,12,15] despite the fact that such distance comes along with several drawbacks. First of all, computing the GED of two graphs requires to find a path having a minimal cost among all possible paths, which is a NP-complete problem [8]. Computing an exact GED is generally done using $A^\star$ algorithm. In practice, due to its high complexity, the computation of an exact GED is intractable for graphs having more than 10 nodes [4,11]. Such a limitation restricts applications of GED on real datasets, hence motivating the graph community to focus on heuristics providing suboptimal solutions of Eq. 1.

In this paper, we propose a method to compute an accurate GED approximation. In Sect. 2, we first show the close relationship between GED and nodes' mappings. This relationship leads to the formulation of the graph edit distance as a quadratic assignment problem (QAP). Then, Sect. 3 first reviews the method used in [13] to find a local minimum of the QAP associated to GED. Then, we propose a more accurate and reliable optimizer of our QAP formulation. Section 4 shows the effectiveness of our proposal on chemoinformatics problems.

2   GED as a Quadratic Assignment Problem

Exact GED computation is based on a tree search algorithm which finds an optimal edit path among possible ones. The resulting algorithm follows the intuition given by the formal definition of the GED. However, this approach has an exponential complexity with the number of nodes.

Another formulation of the GED is based on its relationship with nodes' mapping. First, let us consider two sets of nodes $V_1 = \{v_1, \ldots, v_n\}$ and $V_2 = \{u_1, \ldots, u_m\}$ of two graphs G_1 and G_2, with $n = |V_1|$ and $m = |V_2|$. The substitution of node $v_i \in V_1$ to node $u_j \in V_2$ can be encoded by mapping v_i to u_j. Insertions/removals of nodes can not be encoded by a node to node mapping since a removed node will no longer appear in the target graph. Therefore, we augment the two sets V_1 and V_2 by adding enough null elements ϵ to encode the removal or insertion of any node: $V_1^\epsilon = V_1 \cup \{\epsilon_1, \ldots, \epsilon_m\}$ and

$V_2^\epsilon = V_2 \cup \{\epsilon_1, \ldots, \epsilon_n\}$. Note that $|V_1^\epsilon| = |V_2^\epsilon| = n + m$. Insertion of a node $u_j \in V_2^\epsilon$ can now be represented by mapping an ϵ element to u_j. In the same way, a mapping $v_i \to \epsilon$ encodes the removal of node v_i. It has be shown [1,10] that, under mild assumptions on edit paths, each mapping between the two sets V_1^ϵ and V_2^ϵ of two graphs G_1 and G_2 corresponds to one and only one edit path transforming G_1 into G_2. Considering this bijective relationship, finding the optimal edit path relies on finding an optimal mapping between the two sets V_1^ϵ and V_2^ϵ which minimizes the total mapping cost of nodes and edges.

Let us consider a bijective function $\phi : V_1^\epsilon \to V_2^\epsilon$. The mapping cost induced by this mapping can be defined as a sum of two terms:

$$S(V_1^\epsilon, V_2^\epsilon, \phi) = L_v(V_1^\epsilon, V_2^\epsilon, \phi) + Q_e(V_1^\epsilon, V_2^\epsilon, \phi) \qquad (2)$$

The first term $L_v(V_1^\epsilon, V_2^\epsilon, \phi)$ encodes the cost induced by nodes' mappings and the second term $Q_e(V_1^\epsilon, V_2^\epsilon, \phi)$ the cost induced by edges' mappings. Let us consider the matrix $\mathbf{X} \in \Pi$, with Π representing the set of binary doubly stochastic matrices: $\mathbf{X}$ encodes a mapping function Φ iff $\mathbf{X}_{i,j} = 1$ with $\phi(i) = j$ and $\mathbf{X}_{i,j} = 0$ otherwise. $\mathbf{X}$ corresponds then to a mapping or permutation matrix. The cost induced by nodes mapping can then be defined as:

$$L_v(V_1^\epsilon, V_2^\epsilon, \phi) = \sum_{i=1}^{n} c(v_i \to \phi(v_i)) + \sum_{i=1}^{m} c(\epsilon_i \to \phi(\epsilon_i)) = \sum_{i=1}^{n+m} \sum_{j=1}^{n+m} \mathbf{C}_{i,j} \mathbf{X}_{i,j} \qquad (3)$$

with $\mathbf{C} \in \mathbb{R}_+^{(n+m) \times (n+m)}$ encoding costs associated to edit operations on nodes.

The bipartite GED, proposed by [11], can be understood as an approximation of the problem expressed in Eq. 2. In bipartite approach, only the linear term $L_v(V_1^\epsilon, V_2^\epsilon, \phi)$ is considered. Finding the optimal mapping w.r.t. $L_v(V_1^\epsilon, V_2^\epsilon, \phi)$ corresponds to a Linear Sum Assignment Problem (LSAP) defined as:

$$\mathbf{X}^\star = \underset{\mathbf{X} \in \Pi}{\operatorname{argmin}} \operatorname{vec}(\mathbf{C})^T \operatorname{vec}(\mathbf{X}) \qquad (4)$$

where $\operatorname{vec}(\mathbf{X})$ encodes the vectorization of matrix $\mathbf{X}$, i.e. the concatenation of its rows into one single vector. For a sake of clarity, $\operatorname{vec}(\mathbf{X})$ will be denoted $\mathbf{x}$.

Equation 4 can be resolved using well known algorithms such as Hungarian [6] in $O((n+m)^3)$. The optimal mapping encoded by $\mathbf{X}^\star$ encodes a set of nodes edit operations. Using simple graphs, edges' edit operations are inferred from the nodes' mapping to complete the edit path. Finally, the sum of costs associated to this edit path is taken as the approximation of GED. Note that since this edit path may not be optimal, the computed GED may be an overestimation of the exact one.

Each entry $\mathbf{C}_{i,j}$ of matrix $\mathbf{C}$ encodes the cost of the edit operation induced by mapping the i-th element of V_1^ϵ to the j-th element of V_2^ϵ. Table 1 summarizes the general term $\mathbf{C}_{i,j}$. Computing an optimal mapping using only $\mathbf{C}$ as defined in Table 1 will only take into account information about nodes, ignoring all the structural information encoded by edges. However, since the quality of the GED

Table 1. Edit operations and costs encoded in matrix $\mathbf{C}$

i	j	Edit operation	Cost $\mathbf{C}_{i,j}$
$\in V_1$	$\in V_2$	Substitution of v_i by u_j	$c(v_i \to u_j)$
$\in V_1$	$\notin V_2$	Removal of v_i	$c(v_i \to \epsilon)$
$\notin V_1$	$\in V_2$	Insertion of u_j	$c(\epsilon \to u_j))$
$\notin V_1$	$\notin V_2$	None	0

approximation depends directly from the computed mapping, costs encoded in matrix $\mathbf{C}$ may also estimate costs induced by edit operations on edges. To include this information, methods based on bipartite graph matching [3,4,11] define a cost matrix $\mathbf{C}$ augmented with the costs induced by the mapping of nodes' neighborhoods. The differences between different methods are mainly based on the size of the radius considered around each node. Results presented by [3] have shown that, as expected, the quality of the approximation increases as long as we take into account a larger radius around each node, and thus more structural information. However, we also observe an asymptotic gain for radius greater than 3, hence showing the limit of this method. Obviously, computational times also increase as we increase the radius.

Cost induced by edges' edit operations are inferred from edges' mapping. However, mapping ϕ is only defined between two extended sets of nodes and an edge is defined by a couple of nodes. Therefore, edge mappings can be directly deduced from the mapping of their incident nodes. Let $e = (v_i, v_j) \in E_1$, its mapped edge in V_2^ϵ corresponds to $e' = (\phi(v_i), \phi(v_j))$. If $e' \in E_2$, then the mapping corresponds to an edge substitution, else if $e' \notin E_2$, then the edge has been removed. Reciprocally, if $e' = (u_i, u_j) \in E_2$ and $e = (\phi^{-1}(u_i), \phi^{-1}(u_j)) \notin E_1$, then this edge has been inserted. The cost associated to edges' operations is encoded by a sum of three terms, one for each kind of edit operation:

$$Q_e(V_1^\epsilon, V_2^\epsilon, \phi) = \underbrace{\sum_{\substack{(i,j)\in E_1, \phi(i)=k, \\ \phi(j)=l, (k,l)\in E_2}} c((i,j) \to (k,l))}_{\text{substitutions}} + \underbrace{\sum_{\substack{(i,j)\in E_1, \phi(i)=k, \\ \phi(j)=l, (k,l)\notin E_2}} c((i,j) \to \epsilon)}_{\text{removals}}$$

$$+ \underbrace{\sum_{\substack{(k,l)\in E_2, i=\phi^{-1}(k), \\ j=\phi^{-1}(l), (i,j)\notin E_1}} c(\epsilon \to (k,l))}_{\text{insertions}} \tag{5}$$

Following the quadratic assignment formulation given by [1], the term $Q_e(V_1^\epsilon, V_2^\epsilon, \phi)$ can be expressed as a quadratic term depending on a matrix $\mathbf{D} \in \mathbb{R}^{(n+m)^2 \times (n+m)^2}$ encoding edges' mapping costs :

$$Q_e(V_1^\epsilon, V_2^\epsilon, \phi) = \mathbf{x}^T \mathbf{D} \mathbf{x} \tag{6}$$

Table 2. General term of matrix $\mathbf{D}$

(i,j)	(k,l)	Edit operation	$\mathbf{D}_{ik,jl}$
$\in E_1$	$\in E_2$	Substitution of (i,j) by (k,l)	$c((i,j) \rightarrow (k,l))$
$\in E_1$	$\notin E_2$	Removal of (i,j)	$c((i,j) \rightarrow \epsilon)$
$\notin E_1$	$\in E_2$	Insertion of (k,l) into E_1	$c(\epsilon \rightarrow (k,l))$
$\notin E_1$	$\notin E_2$	None	0

Table 2 summarizes all mapping costs encoded within matrix $\mathbf{D}$. Note that the computation of $\mathbf{x}^T \mathbf{D} \mathbf{x}$ has, in a naive computation, a $\mathcal{O}((n+m)^4)$ computational and memory complexity. However, the vector $\mathbf{x} \in \{0,1\}^{(n+m)^2}$ is sparse since it encodes a mapping and thus only $(n+m)$ terms differs from 0, which reduces the computational complexity if we only process them (Algorithm 2, lines 1 and 3). Therefore, with a proper approach, the computation of $(\mathbf{x}^T \mathbf{D} \mathbf{x})$ has thus a $\mathcal{O}((n+m)^2)$ complexity. Note that this complexity is given for $\mathbf{x}$ encoding a permutation matrix. If more than $n+m$ elements of $\mathbf{x}$ are different from 0, the associated complexity will be higher. We also avoid the storage of $\mathbf{D}$ since we can efficiently compute each term trough a function d (Algorithm 2, line 7) which computes $\mathbf{D}_{ij,kl}$ according to Table 2. The overall memory complexity can thus be reduced to the storage of $n+m$ mappings.

Plugging Eqs. 3 and 6 into Eq. 2, the cost of transforming G_1 into G_2 including both nodes and edge costs according to a mapping ϕ can then be written as:

$$S(V_1^\epsilon, V_2^\epsilon, \phi) = Q_e(V_1^\epsilon, V_2^\epsilon, \phi) + L_v(V_1^\epsilon, V_2^\epsilon, \phi) = \mathbf{x}^T \mathbf{D} \mathbf{x} + \mathbf{c}^T \mathbf{x} \qquad (7)$$

with $\mathbf{D}$ and $\mathbf{c} = \mathrm{vec}(\mathbf{C})$ being fully dependant on the two input graphs G_1 and G_2 and $\mathbf{x}$ encoding the mapping ϕ. Given a pair of graphs, Eq. 7 can be rewritten as only depending on $\mathbf{x}$ as $S(\mathbf{x}) = \mathbf{x}^T \mathbf{D} \mathbf{x} + \mathbf{c}^T \mathbf{x}$.

Given the Algorithm 2, each edge is processed twice, i.e. once as (i,j) and once as (j,i). In case of undirected graphs, it will thus count each edit operation twice.

Algorithm 1. Computation of $\mathbf{x}^T \mathbf{D} \mathbf{x}(\phi)$

1: **for all** $j,l \mid \mathsf{x}_{jl} \neq 0$ **do** // $\mathcal{O}(n+m)$ if $x \in \Pi$
2:     $(\mathbf{x}^T \mathbf{D})_{jl} = 0$
3:     **for all** $i,k \mid \mathsf{x}_{ik} \neq 0$ **do** // $\mathcal{O}(n+m)$ if $x \in \Pi$
4:        $\delta_{ij} \leftarrow (i,j) \in E_1$
5:        $\delta_{kl} \leftarrow (k,l) \in E_2$
6:        $(\mathbf{x}^T \mathbf{D})_{jl} \leftarrow (\mathbf{x}^T \mathbf{D})_{jl} + \mathrm{d}(\delta_{ij}, \delta_{kl}, c(\cdot \rightarrow \cdot))$ // see Table 2
7:     **end for**
8: **end for**
9: **return** x_{k+1}

We have $\mathbf{x}^T \mathbf{D}\mathbf{x} = \frac{1}{2}\mathbf{x}^T(\mathbf{D} + \mathbf{D}^T)\mathbf{x}$ and to handle both directed and undirected graphs, we introduce a matrix Δ :

$$S(\mathbf{x}) = \frac{1}{2}\mathbf{x}^T \Delta \mathbf{x} + \mathbf{c}^T \mathbf{x} \quad \text{with } \Delta = \begin{cases} \mathbf{D} & \text{if } G_1 \text{ and } G_2 \text{ are undirected} \\ \mathbf{D} + \mathbf{D}^T & \text{else} \end{cases}$$
(8)

Note that $\mathbf{D}$ is symmetric if both G_1 and G_2 are undirected [1]. Hence, Δ is symmetric in both cases. Including the linear part into the quadratic one leads to:

$$S(\mathbf{x}) = \mathbf{x}^T \widehat{\Delta} \mathbf{x}, \text{ with } \widehat{\Delta} = \frac{1}{2}\Delta + \text{diag}(\mathbf{c}) \tag{9}$$

Then, computing the GED of two graphs leads to minimize the quadratic function given in Eq. 9 under the constraint that $\mathbf{x}$ is a permutation matrix. This problem corresponds thus to the QAP:

$$\text{GED}(G_1, G_2) = \min_{\mathbf{x} \in \Pi} \mathbf{x}^T \widehat{\Delta} \mathbf{x} \tag{10}$$

Note that quadratic formulation of GED have already been proposed: [5] proposes a binary quadratic program transformed into a binary linear program but restricted to undirected and unlabeled edges. Slightly after, [10] proposed another quadratic program but where optimization is only focused on node substitutions.

3   Resolution of the QAP

As stated in Introduction, computing an exact GED is intractable for most applications. Reducing the complexity relies thus on finding the best possible mapping, i.e. the one having the lowest mapping cost, in an acceptable computational time. Rather than finding an exact solution of a linear approximation of the problem, another strategy consists in optimizing the QAP associated to GED (Eq. 10). However, since Δ is neither positive nor negative definite, finding a global minimum of Eq. 10 is, conversely to LSAP, a NP-complete problem. Since this problem is NP-complete, most of algorithms find thus approximate solutions by relaxing the original problem to the set of doubly stochastic matrices. As we resolve a QAP and not a generic quadratic problem, the solution should be a permutation matrix encoding a mapping. Since most of classic approaches used to resolve quadratic problems do not take this constraint into account, the final projection of the continuous solution to a permutation matrix may alter the approximation of the GED.

Interger-Projected Fixed Point (IPFP) method [7] has originally been proposed to resolve graph matching problems formalized as a maximization of a QAP. Conversely to classical methods used to resolve quadratic problems, IPFP allows to approximate the optimal solution by a gradient descent approach taking into account the nature of the solution. The algorithm mainly iterates over

---

Algorithm 2. IPFP($\mathbf{x_0}, \hat{\Delta}, \varepsilon$)

1: $k = 0$
2: **repeat**
3: $\quad \mathbf{b}^\star \leftarrow \text{argmin}_{\mathbf{b} \in \Pi} \ (\mathbf{x_k}^T \hat{\Delta})$
4: $\quad \alpha \leftarrow (\mathbf{x_k}^T \hat{\Delta}) \mathbf{b}^\star - 2S(\mathbf{x_k}) + \mathbf{c}^T \mathbf{x_k}$
5: $\quad \beta \leftarrow S(\mathbf{b}^\star) + S(\mathbf{x_k}) - (\mathbf{x_k}^T \hat{\Delta}) \mathbf{b}^\star + \mathbf{c}^T \mathbf{x_k}$
6: $\quad t^\star \leftarrow -\alpha/2\beta$
7: $\quad \mathbf{x_{k+1}} \leftarrow \mathbf{x_k} + t^\star (\mathbf{b}^\star - \mathbf{x_k})$
8: $\quad k \leftarrow k + 1$
9: **until** $\|\mathbf{x_k} - \mathbf{x_{k-1}}\| < \varepsilon$
10: **return** $\text{argmax}_{\mathbf{b} \in \Pi} \ \mathbf{x}_k^T \mathbf{b}$

---

two steps: (i) compute a gradient direction which maximizes a linear approximation of the QAP in discrete domain and (ii) maximize the QAP in the continuous domain between the previous solution and the one found at step (i).

IPFP has been adapted from its original definition to GED computation [1,13] by considering a minimization rather than a maximization and by considering the matrix $\hat{\Delta}$. The algorithm corresponding to this adaptation is shown in Algorithm 3. Line 3 corresponds to the computation of the gradient direction. This step corresponds to a LSAP and can be efficiently computed using an Hungarian algorithm such as in bipartite GED framework. Then, a line search is performed to minimize the quadratic objective function in continuous domain. This optimization relies on the computation of a step size $t^\star$ (line 6) which can be computed analytically. Finally, the solution at convergence is projected onto the set of mapping matrices (line 10). Note that the stochastic matrix $\mathbf{x_k}$ is projected by line 10 onto its closest permutation matrix. However, the distance betwenn $\mathbf{x_k}$ and its projection can not be bounded a priori. In [1,13], we reported the results obtained by using this approach and, as expected, it reaches a better accuracy than LSAP based methods. However, due to the non convexity of the objective function, the accuracy of the approximation is strongly dependant on the initialization $\mathbf{x_0}$.

Graduated NonConvexity and Concativity Procedure (GNCCP) [9,16] is a path following algorithm which consists in approximating the solution of a QAP by using a convex-concave relaxation. This approach brings several advantages over IPFP. First, no initial mapping is required. This may avoid the accuracy variations induced by the initialization step observed with IPFP [13]. Second, this algorithm converges towards a mapping matrix [9]. This second property allows to avoid the projection step required at the end of IPFP algorithm which may alter the accuracy of the approximation. GNCCP algorithm is based on a weighted sum of a convex and a concave relaxation of Eq. 9:

$$S_\zeta(\mathbf{x}) = (1 - |\zeta|)S(\mathbf{x}) + \zeta \mathbf{x}^T \mathbf{x} \tag{11}$$

For $\zeta = 1$, $S_\zeta(\mathbf{x})$ is equal to $\mathbf{x}^T \mathbf{x}$ which corresponds to a fully convex objective function. Conversely, for $\zeta = -1$, $S_\zeta(\mathbf{x}) = -\mathbf{x}^T \mathbf{x}$ which now defines a concave

---

Algorithm 3. $\mathrm{GNCCP}(\mathbf{x}_0, \mathbf{c}, \Delta, k_{\max})$

1: $\zeta = 1, d = 0.1, \mathbf{x} = \mathbf{0}$
2: **while** $\zeta > -1$ & $\mathbf{x} \notin \mathbb{A}$ **do**
3:     // Resolution of relaxed QAP according to ζ using Frank-Wolfe like algorithm
4:     $\mathbf{x} \leftarrow \mathrm{argmin}_{\mathbf{x} \in \varPi} (1 - |\zeta|)(\frac{1}{2}\mathbf{x}^T \hat{\boldsymbol{\Delta}} \mathbf{x}) + \zeta \mathbf{x}^T \mathbf{x}$
5:     $\zeta \leftarrow \zeta - d$
6: **end while**
7: **return** $\mathbf{x}_{k+1}$

---

function. GNCCP algorithm starts with $\zeta = 1$, and since the problem is convex, the initialization does not alter the result and will not influence the quality of the approximation as in [13]. Then, the algorithm, detailed in Algorithm 3, smoothly interpolates convex and concave relaxations by passing from $\zeta = 1$ to $\zeta = -1$ with steps of size d until convergence is reached by having a mapping matrix or $\zeta = -1$. Note that the minimum of the concave relaxation is a mapping matrix [16,17]. Hence, conversely to IPFP, we do not need to perform a final projection step which may alter the approximation (end of Algorithm 3).

For a given ζ, each iteration of GNCCP minimizes the quadratic functional defined in Eq. 11 using a Frank-Wolfe like algorithm. In this paper, we adapt IPFP algorithm used in [13] to perform this step and we remove last projection step (Algorithm 3, line 10). However, since we optimize the relaxed objective function (Eq. 11), we have to update the linear subproblem and line search step. In our modified version of Algorithm 3, the gradient direction $\mathbf{b}^\star$ has now to minimize $S_\zeta(\mathbf{x})$:

$$\frac{\partial S_\zeta(\mathbf{x})}{\partial \mathbf{x}} = (1 - |\zeta|)\frac{\partial S(\mathbf{x})}{\partial \mathbf{x}} + \zeta 2\mathbf{x}^T = (1 - |\zeta|)(\mathbf{x}^T \hat{\Delta}) + \zeta 2\mathbf{x}^T \qquad (12)$$

Then, line 3 of Algorithm 3 is updated with:

$$\mathbf{b}^\star \leftarrow \mathop{\mathrm{argmin}}_{\mathbf{b} \in \varPi} \ [(1 - |\zeta|)(\mathbf{x}^T \hat{\Delta}) + 2\zeta \mathbf{x}^T]\mathbf{b} \qquad (13)$$

Given the gradient direction $\mathbf{b}^\star$, line search step has also to be updated to minimize $S_\zeta(\mathbf{x})$ rather than $S(\mathbf{x})$. After some calculus, lines 4 and 5 are updated with:

$$\alpha_\zeta \leftarrow [(1 - |\zeta|)(\mathbf{x_k}^T \hat{\Delta}) + 2\zeta \mathbf{x_k}^T]\mathbf{b}^\star - 2S_\zeta(\mathbf{x_k}) + (1 - |\zeta|)\mathbf{c}^T \mathbf{x_k} \qquad (14)$$

$$\beta_\zeta \leftarrow S_\zeta(\mathbf{b}^\star) + S_\zeta(\mathbf{x_k}) - [(1 - |\zeta|)(\mathbf{x}^T \hat{\Delta}) + 2\zeta \mathbf{x_k}^T]\mathbf{b}^\star + (1 - |\zeta|)\mathbf{c}^T \mathbf{x_k} \qquad (15)$$

Note that some of the terms of Eqs. 14 and 15 have been already computed: $[(1 - |\zeta|)(\mathbf{x}^T \hat{\Delta}) + 2\zeta \mathbf{x_k}^T]\mathbf{b}^\star$ corresponds to the optimal cost computed by LSAP resolver on line 3 of Algorithm 3, $\mathbf{c}^T \mathbf{x_k}$ to the linear term included into $\hat{\Delta}$ and computed for the linear subproblem and $S_\zeta(\mathbf{x_k})$ corresponds to the score of objective function computed at previous iteration. Finally, step size $t^\star$ is computed in the same way as in Algorithm 3 with $t^\star = -\alpha_\zeta/2\beta_\zeta$ (line 6).

Table 3. Accuracy and complexity scores. d is the average edit distance, e the average error and t the average computational time.

Algorithm	Alkane			Acyclic			PAH	
	d	e	t	d	e	t	d	t
$A^\star$	15	-	1.29	17	-	6.02	-	-
LSAP [11]	35	18	$\simeq 10^{-3}$	35	18	$\simeq 10^{-3}$	138	$\simeq 10^{-3}$
LSAP random walks [4]	33	18	$\simeq 10^{-3}$	31	14	$\simeq 10^{-2}$	120	$\simeq 10^{-2}$
LSAP K-graphs [3]	26	11	2.27	28	9	0.73	129	2.01
$IPFP_{\text{Random init}}$	22.6	7.1	0.007	23.4	6.1	0.006	63	0.04
$IPFP_{\text{Init}}$ [4]	20.5	5	0.006	20.7	3.4	0.005	52.5	0.037
$GNCCP$	16.7	1.2	0.46	18.8	1.5	0.33	41.8	6.24

4   Experiments

Our method based on GNCCP is tested on 3 real world chemical datasets[1]. Similarly to our previous papers using these datasets [3,4,13], edge and node substitutions costs have been set to 1 and 3 for insertions and deletions. To avoid bias due to arbitrary order of nodes in data files, each graph adjacency matrix is randomly permuted. We compare our method to others methods computing GED. First, we used an exact GED computed using $A^\star$ but only available for small graphs (line 1). Second, we show results obtained by LSAP based methods using different information around each node: incident edges and adjacent nodes [11] (line 2), bag of random walks up to 3 edges (line 3) [4], subgraphs up to a radius 3 (line 4) [3]. We also show results obtained by IPFP approach, detailed in Sect. 3, with two different initializations: a random one (line 5) and one using [4] (line 6). Finally, results obtained by our method with $d = 0.1$ are displayed (line 7). First of all, Table 3 shows clearly that our method allows to reach the best accuracy. Considering the two first datasets Alkane and Acyclic for which we have a ground truth computed using $A^\star$ (line 1), average error of approximation is divided by over 2 for acyclic and by over 4 for alkane, hence showing a very good approximation of GED. These improvements allow to reduce the relative error to about 10 %. Concerning PAH dataset, the quality of the approximation must be deduced from average distance since we can not compute the exact GED. Since the edit distance is always overestimated, the lowest edit distance may correspond to a better approximation. The quality of our approximation is also validated on PAH dataset since the average edit distance is reduced by about 20 %. Despite the good approximation offered by GNCCP, we also observe an important increase of computationnal times. This phenomenon is not surprising since GNCCP iterates over IPFP. For $d = 0.1$, we may perform about 20 iterations in the worst case (Algorithm 3). In practice, we observe an average of 11 iterations before reaching convergence. Nonetheless, computational times

---

[1] Datasets are available at https://iapr-tc15.greyc.fr/links.html.

increase by a factor 40. This increase is due to the computation of quadratic term $x^T \Delta x$. Indeed, for ζ far from 0, the matrix x may contains a lot of entries different from 0. Therefore, the complexity stated from Algorithm 2 no longer holds.

5 Conclusion

Considering the relationship between graph edit distance and nodes' mapping, we present in this paper a quadratic assignment problem formalizing the graph edit distance with substitutions, insertions and removals of nodes and edges. Following a previous optimization scheme based on a Frank Wolfe algorithm, we adapt a convex concave relaxation framework to the resolution of the QAP associated to edit distance. This approach does not require any initialization step and converges towards a mapping matrix. These two properties allow to compute a more reliable and robust approximation of the graph edit distance. The experiments conducted on real world chemical datasets valid our hypothesis.

References

1. Bougleux, S., Brun, L., Carletti, V., Foggia, P., Gaüzère, B., Vento, M.: A quadratic assignment formulation of the graph edit distance. Technical report, Normandie Univ, NormaSTIC FR 3638, France (2015)
2. Bunke, H., Allermann, G.: Inexact graph matching for structural pattern recognition. Pattern Recogn. Lett. **1**(4), 245–253 (1983)
3. Carletti, V., Gaüzère, B., Brun, L., Vento, M.: Approximate graph edit distance computation combining bipartite matching and exact neighborhood substructure distance. In: Liu, C.-L., Luo, B., Kropatsch, W.G., Cheng, J. (eds.) GbRPR 2015. LNCS, vol. 9069, pp. 188–197. Springer, Heidelberg (2015). doi:10.1007/978-3-319-18224-7_19
4. Gaüzère, B., Bougleux, S., Riesen, K., Brun, L.: Approximate graph edit distance guided by bipartite matching of bags of walks. In: Fränti, P., Brown, G., Loog, M., Escolano, F., Pelillo, M. (eds.) S+SSPR 2014. LNCS, vol. 8621, pp. 73–82. Springer, Heidelberg (2014). doi:10.1007/978-3-662-44415-3_8
5. Justice, D., Hero, A.: A binary linear programming formulation of the graph edit distance. IEEE Trans. Pattern Anal. Mach. Intell. **28**(8), 1200–1214 (2006)
6. Kuhn, H.W.: The hungarian method for the assignment problem. Naval Res. Logist. Quat. **2**, 83–97 (1955)
7. Leordeanu, M., Hebert, M., Sukthankar, R.: An integer projected fixed point method for graph matching and map inference. In: Advances in Neural Information Processing Systems, pp. 1114–1122 (2009)
8. Lin, C.-L.: Hardness of approximating graph transformation problem. In: Du, D.-Z., Zhang, X.-S. (eds.) ISAAC 1994. LNCS, vol. 834, pp. 74–82. Springer, Heidelberg (1994). doi:10.1007/3-540-58325-4_168
9. Liu, Z.-Y., Qiao, H.: GNCCP-graduated NonConvexity and concavity procedure. Pattern Anal. Mach. Intell. **36**(6), 1258–1267 (2014)
10. Neuhaus, M., Bunke, H.: A quadratic programming approach to the graph edit distance problem. In: Escolano, F., Vento, M. (eds.) GbRPR 2007. LNCS, vol. 4538, pp. 92–102. Springer, Heidelberg (2007)

11. Riesen, K., Bunke, H.: Approximate graph edit distance computation by means of bipartite graph matching. Image Vis. Comput. **27**, 950–959 (2009)

12. Riesen, K., Emmenegger, S., Bunke, H.: A novel software toolkit for graph edit ·distance computation. In: Kropatsch, W.G., Artner, N.M., Haxhimusa, Y., Jiang, X. (eds.) GbRPR 2013. LNCS, vol. 7877, pp. 142–151. Springer, Heidelberg (2013). doi:10.1007/978-3-642-38221-5_15

13. Brun, L., Bougleux, S., Gaüzère, B.: Graph edit distance as a quadratic program. In: ICPR (2016, submitted)

14. Sanfeliu, A., Fu, K.-S.: A distance measure between attributed relational graphs for pattern recognition. Syst. Man Cybern. **13**(3), 353–362 (1983)

15. Serratosa, S.: Speeding up fast bipartite graph matching through a new cost matrix. Int. J. Pattern Recogn. **29**(2), 1550010 (2015)

16. Zaslavskiy, M., Bach, F., Vert, J.-P.: A path following algorithm for the graph matching problem. Pattern Anal. Mach. Intell. **31**(12), 2227–2242 (2009)

17. Zhou, F., De la Torre, F.: Factorized graph matching. In: CVPR, pp. 127–134 (2012)

Generalised Median of a Set of Correspondences Based on the Hamming Distance

Carlos Francisco Moreno-García[✉], Francesc Serratosa,
and Xavier Cortés

Departament d'Enginyeria Informàtica i Matemàtiques,
Universitat Rovira I Virgili, Tarragona, Spain
carlosfrancisco.moreno@estudiants.urv.cat,
{xavier.cortes,francesc.serratosa}@urv.cat

Abstract. A correspondence is a set of mappings that establishes a relation between the elements of two data structures (i.e. sets of points, strings, trees or graphs). If we consider several correspondences between the same two structures, one option to define a representative of them is through the generalised median correspondence. In general, the computation of the generalised median is an NP-complete task. In this paper, we present two methods to calculate the generalised median correspondence of multiple correspondences. The first one obtains the optimal solution in cubic time, but it is restricted to the Hamming distance. The second one obtains a sub-optimal solution through an iterative approach, but does not have any restrictions with respect to the used distance. We compare both proposals in terms of the distance to the true generalised median and runtime.

Keywords: Correspondence · Mappings · Hamming distance · Generalised median · Linear assignment problem

1 Introduction

In several pattern recognition applications, there is a need to define an element-to-element relation between two objects. This process, commonly referred as "matching", has been applied on data structures such as sets of points [1], strings [2], trees [3] and most notably, graphs [4–8]. While this previous work demonstrates that there has been a long standing effort to increase the quality of the methods that perform structural matching, it may also derive in scenarios where we encounter two or more parties which, having applied different matching algorithms, have produced several matching solutions. These solutions, onwards referred as "correspondences", may be the result of the several existing methodologies or different parameterisations of these

This research is supported by projects DPI2013-42458-P, TIN2013-47245-C2-2-R, and by Consejo Nacional de Ciencia y Tecnología (CONACyT México).

A. Robles-Kelly et al. (Eds.): S+SSPR 2016, LNCS 10029, pp. 507–518, 2016.
DOI: 10.1007/978-3-319-49055-7_45

methodologies, which generate each time a different set of mappings between the elements of an output data structure and the elements of an input data structure.

Given a set of objects, their median is defined as the object that has the smallest sum of distances (SOD) [9, 10] to all objects in the set [11]. From this definition, we are able to identify the generalised median (GM) and the set median, which difference lies in the space where each median is searched for. In the first case there are no restrictions for the search, while on the second case the exploration space is restricted to the elements in the set.

Due to its robustness, the concept of GM has been implemented to deduce the representative prototype of a set of data structures [12] and for clustering ensemble purposes [13] on data structures such as strings [14], graphs [15] and data clusters [16]. For these data structures (and for correspondences as well), the GM computation turns out to be an NP-complete problem. This drawback has led to a variety of methods solely developed for the GM approximation, such as a genetic search method [11] or approximations through the weighted mean of a pair of strings [17], graphs [18] or data clusters [19]. Most recently, a method known as the Evolutionary method [20] has been presented, offering a good trade-off between accuracy and runtime. This makes the Evolutionary method one of the most viable options for the GM approximation on most domains.

In this paper, we present two methodologies to obtain the GM of a set of correspondences. The first one is based on a voting and minimisation process, and the second one is based on the Evolutionary method adapted to the correspondence case. Notice that the calculation of a representative prototype of a set of correspondence has been studied before in what we called the "correspondence consensus frameworks" [21–25]. Nonetheless, there are some differences between approximating towards the GM and these consensus frameworks; the most important one being the function to be minimised. As commented before, the GM computation aims to minimise the SOD, whereas in the consensus framework, the function could also include the reduction of some other restrictions, such as the cost defined on the correspondences or the structural information of the data structures mapped.

The rest of the paper is structured as follows. In Sect. 2, we introduce the basic definitions. In Sects. 3 and 4, we present both methods. In Sect. 5, we compare them and evaluate the results in terms of the distance to the ground truth GM and the runtime. Finally, Sect. 6 is reserved for conclusions and further work.

2 Basic Definitions

Let us represent any kind of data structure as $G = (\Sigma, \gamma)$, where $v_i \in \Sigma$ is an element inside the structure (elements can be, for instance, characters for strings, or nodes and edges for trees and graphs), and γ is a function that maps each element to a set of attributes. To allow maximum flexibility in the matching process, these structures may have been extended with null elements (represented as Φ), which have a set of attributes that differentiate them from the rest of the elements.

Given $G = (\Sigma, \gamma)$ and $G' = (\Sigma', \gamma')$ of the same order N (naturally or due to the aforementioned null element extension), we define the set of all possible correspondences $T_{G,G'}$ such that each correspondence in this set maps elements of G to elements of G', $f : \Sigma \rightarrow \Sigma'$ in a bijective manner. Onwards, we refer to T instead of $T_{G,G'}$. Moreover, consider a subset of correspondences $S \in T$ between the same pair of structures that have been produced using different matching approaches.

Let f^1 and f^2 denote two different correspondences within S. We can deduct how dissimilar these two correspondences are through the Hamming distance (HD) function, which calculates the distance (number of different mappings) between f^1 and f^2. More formally, the HD is defined as:

$$Dist_{HD}\left(f^1, f^2\right) = \sum_{i=1}^{N} \left(1 - \partial\left(v'_x, v'_y\right)\right) \tag{1}$$

being x and y such that $f^1(v_i) = v'_x$ and $f^2(v_i) = v'_y$, and ∂ being the well-known Kronecker Delta function.

$$\partial(a, b) = \begin{cases} 0 \text{ if } a \neq b \\ 1 \text{ if } a = b \end{cases} \tag{2}$$

Given a set of M input correspondences, the GM is the correspondence that has the smallest SOD to all objects in such set.

$$\hat{f} = \underset{\forall f \in T}{argmin} \sum_{i=1}^{M} Dist_{HD}\left(f, f^i\right) \tag{3}$$

If the minimisation to find $\hat{f}$ is restricted to be within the elements of S, then the solution is called the set median. Conversely, a search of all elements within T is known as the GM, which is considered a more attractive but more computationally demanding option. As noticed, the calculation of a median is closely related to the distance between the objects involved, and thus, the importance of defining the HD for the correspondence case.

3   Minimisation Method

The first method presented in this paper is called Minimisation method. The name is related to the minimisation of the sum of the linear assignment problem (SLAP). To introduce it, consider the following example. Suppose that three separate entities have proposed correspondences as shown in Fig. 1, depicted as f^1 (red lines), f^2 (blue lines) and f^3 (green lines). Notice that as commented in Sect. 2, the input set has been extended with a null element (Φ) to make correspondences mutually bijective.

Fig. 1. Example of three correspondences. (Color figure online)

We are able to represent these correspondences as correspondence matrices F^1, F^2 and F^3 as shown in Fig. 2. These matrices are defined as follows. $F^k[x, y] = 1$ if $f^k(v_x) = v'_y$ and $F^k[x, y] = 0$ otherwise.

$$F^1{}_\phi \qquad F^2{}_\phi \qquad F^3{}_\phi$$

	1'	2'	3'	4'	5'	6'		1'	2'	3'	4'	5'	6'		1'	2'	3'	4'	5'	6'
1	0	1	0	0	0	0		0	0	0	0	0	1		0	1	0	0	0	0
2	1	0	0	0	0	0		0	1	0	0	0	0		1	0	0	0	0	0
3	0	0	1	0	0	0		0	0	1	0	0	0		0	0	1	0	0	0
4	0	0	0	0	1	0		0	0	0	1	0	0		0	0	0	1	0	0
5	0	0	0	1	0	0		0	0	0	0	1	0		0	0	0	0	0	1
6	0	0	0	0	0	1		1	0	0	0	0	0		0	0	0	0	1	0

Fig. 2. Correspondences matrices F^1, F^2 and F^3.

Our method minimises the following expression:

$$\hat{f} = \underset{\forall f \in T}{\operatorname{argmin}} \left\{ \sum_{x,y=1}^{N} [H \circ F]\{x, y\} \right\} \qquad (4)$$

where $\{x, y\}$ is a specific cell and H is the following matrix:

$$H = \sum_{k=1}^{M} \mathbf{1} - F^k \qquad (5)$$

with $\mathbf{1}$ being a matrix of all ones, F being the correspondence matrix of $f \in T$ (if $f(v_x) = v'_y$ then $F\{x, y\} = 1$, otherwise $F\{x, y\} = 0$) and $\circ$ being the Hadamard product.

We deduct $\hat{f}$ through Eq. 4 but we wish to minimise the SOD of all correspondences to obtain the true $\hat{f}$ (Eq. 3). Therefore, we have to demonstrate that the obtained correspondence $\hat{f}$ in Eq. 3 is the same than the one in Eq. 4. For this reason, we have to demonstrate that Eq. 6 holds:

$$\sum_{k=1}^{M} Dist_{HD}(f,f^k) = \sum_{x,y=1}^{N} \left[\left[\sum_{k=1}^{M} 1 - F^k \right] \circ F \right] \{x,y\} \qquad (6)$$

Applying the associative property of Hadamard product, the following expression is obtained:

$$\sum_{k=1}^{M} Dist_{HD}(f,f^k) = \sum_{k=1}^{M} \left(\sum_{x,y=1}^{N} \left[[1 - F^k] \circ F \right] \{x,y\} \right) \qquad (7)$$

Then, if we demonstrate that each individual term holds the equality $Dist_{HD}(f,f^k) = \sum_{x,y=1}^{N} \left[[1 - F^k] \circ F \right] \{x,y\}$, then for sure Eq. 7 holds. As shown in its definition, the HD counts the number of mappings that are different between the two correspondences and similarly, expression $\sum_{x,y=1}^{N} \left[[1 - F^k] \circ F \right] \{x,y\}$ does, since this last expression counts the number of times $F\{x,y\} = 1$ and that simultaneously $F^k\{x,y\} = 0$.

Notice that by adding all correspondence matrices in Eq. 5, we create a structure similar to a voting matrix [26]. This method is based on minimising the linear assignment problem applied to this voting matrix using any solver such as the Hungarian method [27], the Munkres algorithm [28] or the Jonker-Volgenant solver [29] as shown in Algorithm 1:

Algorithm 1: Minimisation
    Input: A set of correspondences
    Output: GM correspondence $\hat{f}$
        Compute matrix H (equation 5)
        $\hat{f} = linear\ solver\ (H)$ ([27], [28] or [29])
End Algorithm

Since the three solvers have all been demonstrated to obtain an optimal value, it is possible to guarantee that this method obtains the exact GM, given that there is only first order information involved and no second order information is considered. That is, we do not take into account relations between the mapped elements inside their set. Figure 3 shows the GM correspondence obtained for this particular practical example.

Fig. 3. GM correspondence of the three correspondences in Fig. 1.

4 Evolutionary Method

The second option explored in this paper for the GM correspondence computation is the use of the meta algorithm presented in [20] called Evolutionary method. This proposal relies on the concept of the weighted mean of a pair of correspondences, which is defined as follows. Given f^1, f^2 and a distance $Dist$ between them (for instance $Dist_{HD}$), the mean correspondence of f^1 and f^2 is a correspondence $\bar{f} \in T$ such that:

$$Dist(f^1,\bar{f}) = Dist(\bar{f},f^2)$$
$$Dist(f^1,f^2) = Dist(f^1,\bar{f}) + Dist(\bar{f},f^2) \tag{8}$$

Additionally, the weighted mean correspondence $\bar{f}_\alpha \in T$ is defined as a correspondence in T that holds:

$$Dist(f^1,\bar{f}_\alpha) = \alpha$$
$$Dist(f^1,f^2) = \alpha + Dist(\bar{f}_\alpha,f^2) \tag{9}$$
$$\text{where } \alpha \text{ is a constant}: \ 0 \le \alpha \le Dist(f^1,f^2)$$

Clearly, $\bar{f}_0 = f^1$ and $\bar{f}_{Dist(f^1,f^2)} = f^2$. Nevertheless, both the mean correspondence and the weighted mean correspondence given a specific α are usually not unique. The concept of the weighted mean has been previously defined for strings [17], graphs [18] and data clusters [19].

As proven in [11], the GM of some elements in any space can be estimated through an optimal partition of the pairs of these elements. This is because they demonstrated that by computing the weighted mean of such optimal pairs of elements, all of those weighted means tend to match in one element that can be considered a good estimation of the GM of the set. Since in some cases the GM can be far away from the deducted element, an iterative algorithm is proposed in [20] which tends to achieve the true GM. This algorithm, applied to the correspondence domain, consists on the steps shown in Algorithm 2:

Algorithm 2: Evolutionary
Input: A set of correspondences
Output: GM correspondence $\hat{f}$
While convergence
1. Deduct the optimal pairs of correspondences.
2. Estimate some weighted means per each pair.
3. Add the weighted means to the current set of correspondences.
4. Select the optimal correspondences in the current set.
End Algorithm

We proceed to detail steps 1, 2 and 4 in the correspondence domain. Notice the third step is simply adding the obtained weighted mean correspondences to the current set of correspondences.

4.1   Optimal Pairs of Correspondences

We generate the distance matrix given the whole correspondences, where any distance between these correspondences can be used. Then, the optimal pairs of elements are considered the ones that generate the minimum SOD between them [11]. Thus, we simply obtain the pairs of correspondences by applying a SLAP solver such as the Hungarian method [27], the Munkres algorithm [28] or the Jonker-Volgenant solver [29]. Note that we do not want one correspondence to be assigned as the optimal pair of itself and for this reason, instead of filling the diagonal of the distance matrix with zeros, we impose a high value. Nevertheless, if there is an odd number of correspondences, for sure the solver returns a correspondence mapped to itself. In this case, this correspondence is stored until the third step.

4.2   Weighted Means of Pairs of Correspondences

The aim of the second step is to estimate Ω equidistant weighted means per each pair of correspondences. Thus, we generate $\bar{f}_{\alpha_1}, \ldots, \bar{f}_{\alpha_\Omega}$ such that $\alpha_i = \frac{i}{\Omega+1}$ (Eq. 9). The order of Ω is usually set from 1 to 3. This is because, through the practical validation, we have seen that restricting the process to calculate only the mean correspondence (that is $\Omega = 1$) makes the process converge slower than when having for instance three equidistant weighted means, even though these are obtained in a sub-optimal form. Moreover, experimentation has shown that if $\Omega > 3$, the computational cost is also increased without gaining in accuracy.

The weighted mean search strategy we used is inspired by the "Moving Elements Uniformly" strategy presented in [19] for the domain of data clusters. In that case, they were able to generate multiple weighted mean data clusters from two initial ones. To do so, authors defined an initial weighted mean as one of the data clusters, and then they systematically swap elements that belong to two different clusters in the weighted data cluster in such a way that the weighted mean data clusters formed tends to move from one of the initial data clusters into the other one.

Our proposal initially defines the weighted mean correspondence as one of the correspondences. Then, it simply swaps pairs of element-to-element mappings in the proposed weighted mean $\bar{f}_\alpha$. Note, every time a swap is performed, the value $Dist(f^1, \bar{f}_\alpha)$ is increased by two, but we cannot guarantee that $Dist(f^2, \bar{f}_\alpha)$ is also decreased by two. For this reason, the strategy checks if the current correspondence is a true weighted mean (Eq. 9 holds). If it is the case, a weighted mean has been formed and the swapping process continues until finding all required weighted means. If it is not the case, the process is reset and repeated until finding weighted means. This method has its base on a theorem presented in [21], where it was shown that a weighted mean correspondence has to hold that $\bar{f}_\alpha(v_x) = f^1(v_x)$ or $\bar{f}_\alpha(v_x) = f^2(v_x)$ for all elements in the weighted mean correspondence.

4.3 Selecting the Optimal Correspondences

Once the current correspondences are put together with the new weighted mean correspondences to enlarge the set (step 3), the method could return to the first step of Algorithm 2 with this newly enlarged set without running the fourth step. Nevertheless, the computational cost and memory space needed in each iteration would exponentially increase. For this reason, the aim of this fourth step is to discard the correspondences that are believed not to be a good choice for the GM. To that aim, a distance matrix is computed between the whole correspondences. Then, the ones that have a larger SOD from themselves to the rest are discarded. Note that this methodology is in line of the GM (Eq. 3).

When the fourth step finishes, Algorithm 2 iterates again until one of three options happens: (1) The sum of the minimum SOD of the whole correspondence in the set is lower than a threshold. (2) A maximum number of iterations is achieved. (3) A minimum difference on the total SOD between the previous iteration and the current one is achieved. Independently of the terminating option, Algorithm 2 returns the correspondence in the set that has at the moment the minimum SOD to the set as the GM correspondence. Convergence is assured since the SOD, in each iteration, is equal or lower than the previous iteration. Moreover, in case the SOD is kept equal, Algorithm 2 stops.

5 Experimental Validation

Two methods have been presented. The first one obtains the exact GM correspondence, but is restricted to the use of the HD. The second one deducts an approximation of the GM correspondence, but is not restricted to any distance between correspondences. In this section, we show how close the suboptimal method is with respect to the optimal one, as well as the runtime of both methods. To have a fair comparison, we have used the HD in both cases.

We performed three tests, all of them executed the same way but using correspondences with $N = 5$, $N = 10$ and $N = 30$ mapped elements in each case. Each test
was prepared as follows. We randomly generated 100 sets of $M = 2, 3 \ldots, 50$ correspondences. For each set, both methods to find the GM correspondence are executed.
In the Minimisation method, the Hungarian method [27] was used to solve the SLAP.

Figure 4 shows the normalised difference on the SOD that the GM correspondences
generated by the Evolutionary method obtained with respect to the ones from the
Minimisation method (x-axis) in the first test ($N = 5$), second test ($N = 10$) and third
test ($N = 30$) respectively. Each dot in the plot represents the average of the 100
executions. For the Evolutionary method, we show results using the number of iterations $I_{max} = 1$ and $I_{max} = 2$. Results for larger values of I_{max} are not shown since they
deliver exactly the same values that the ones of $I_{max} = 2$.

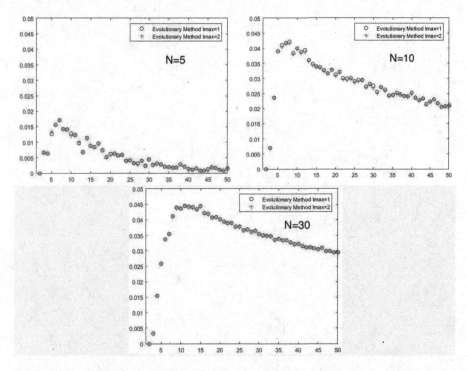

Fig. 4. Average difference of SOD (HD) between the Evolutionary method and Minimisation
method (x-axis, optimal method) for $N = 5$, $N = 10$ and $N = 30$.

In the three cases for a set of two correspondences, $M = 2$, the Evolutionary
method obtains optimal GM correspondences since the method only has to deal with
the mean calculation. Nonetheless as the number of correspondences M increases, this
overestimation has a peak maximum value, and then it decreases until lowering down
again towards the optimal value of the GM correspondence. This leads us to think that

the Evolutionary method has an optimal number of correspondences to be used, since certain values of M lead more overestimation than others. Finally, from these plots we conclude that the Evolutionary method, regardless of the I_{max} value used, obtains values that are really close to the optimal ones. In fact, the worst case overestimates the SOD in 4.5 % with respect to the optimal SOD.

Figure 5 shows the runtime difference between the Evolutionary method and the Minimisation method (x-axis) in seconds. In the case of the Evolutionary method, it is clear that the time spent in each iteration is constant. Comparing both methods, the minimisation one is clearly faster than the evolutionary one, although both have a polynomial computational cost with respect to the number of correspondences used to deduct the GM. Finally, comparing the three plots, we realise the number of elements N in the sets seems to have almost no influence on the runtime.

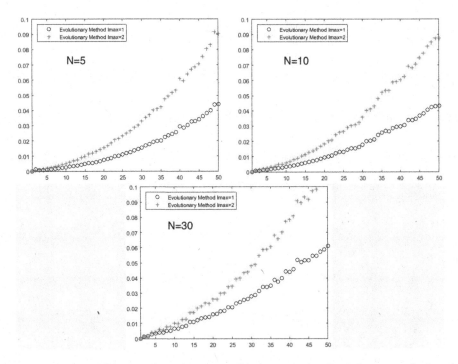

Fig. 5. Average difference of runtime (seconds) between the Evolutionary method and Minimisation method (x-axis) for $N = 5$, $N = 10$ and $N = 30$.

6   Conclusions and Future Work

We have presented two methods to deduct the GM correspondence. The first one, called Minimisation method, computes the exact GM in a reduced runtime, but it is bounded to the use of the HD. Since it is based on the solution of the SLAP, it is limited in scalability for the cases where correspondences have a large size. The second one, called Evolutionary method, obtains a fair approximation of the GM, and may be used with any type of distance between correspondences. This method has better scalability, given that although there is a SLAP solution required (step 1), it only takes into consideration the distance between correspondences, and not the correspondences themselves as in the Minimisation method. Notice that the Evolutionary method has been a viable solution for the approximation of the GM of strings and graphs [20], since these structures imply second order relations and therefore finding their exact GM has an exponential cost.

In the concrete case that the aim is to find the GM of correspondences and the HD can be used, we have shown that this problem can be solved in cubic time (the computational cost of the SLAP) using the Minimisation method. Nevertheless, we consider this paper as a first step towards future research in which other distances between correspondences will be explored, and thus the Evolutionary method should not be discarded for future uses.

We believe that other distances between correspondences which take into consideration not only the element-to-element mapping, but also the structure and attributes of the related elements, could produce more interesting GM correspondences from the application point of view. For instance, in the situation that the correspondences relate attributed graphs, the mapping is defined as node-to-node. In this case, we could consider the local structure of the nodes (its adjacent edges and their terminal nodes) to penalise the cost of the mapping. Then, the Minimisation method would not produce an exact GM and therefore, we would need to compare both algorithms, not only from the runtime point of view, but also in terms of accuracy to deduct the best approximation to the GM correspondence.

References

1. Zitová, B., Flusser, J.: Image registration methods: a survey. Image Vis. Comput. **21**(11), 977–1000 (2003)
2. Navarro, G.: A guided tour to approximate string matching. ACM Comput. Surv. **33**(1), 31–88 (2001)
3. Bille, P.: A survey on tree edit distance and related problems. Theoret. Comput. Sci. **337**(9), 217–239 (2005)
4. Conte, D., Foggia, P., Sansone, C., Vento, M.: Thirty years of graph matching in pattern recognition. Int. J. Pattern Recognit. Artif Intell. **18**(3), 265–298 (2004)
5. Foggia, P., Percannella, G., Vento, M.: Graph matching and learning in pattern recognition in the last ten years. Int. J. Pattern Recognit. Artif Intell. **28**(1), 1450001 (2014)
6. Serratosa, F.: Fast computation of bipartite graph matching. Pattern Recogn. Lett. **45**, 244–250 (2014)
7. Serratosa, F.: Computation of graph edit distance: reasoning about optimality and speed-up. Image Vis. Comput. **40**, 38–48 (2015)
8. Serratosa, F.: Speeding up fast bipartite graph matching through a new cost matrix. Int. J. Pattern Recognit. Artif. Intell. **29**(2), 1550010 (2015)

9. Solé, A., Serratosa, F., Sanfeliu, A.: On the graph edit distance cost: properties and applications. Int. J. Pattern Recognit. Artif Intell. **26**(5), 1260004 (2012)
10. Serratosa, F., Alquézar, R., Sanfeliu, A.: Efficient algorithms for matching attributed graphs and function-described graphs. In: Proceedings of the 15th International Conference on Pattern Recognition, (ICPR), vol. 2, pp. 871–876 (2000)
11. Jiang, X., Munger, A., Bunke, H.: On median graphs: properties, algorithms, and applications. IEEE Trans. Pattern Anal. Mach. Intell. **23**(10), 1144–1151 (2001)
12. Jiang, X., Bunke H.: Learning by generalized median concept. In: Wang, P. (ed) Pattern Recognition and Machine Vision, Chap. 15, pp. 231–246. River Publisher (2010)
13. Vega-Pons, S., Ruiz-Shulcloper, J.: A survey of clustering ensemble algorithms. Int. J. Pattern Recognit. Artif Intell. **25**(3), 337–372 (2011)
14. Jiang, X., Wentker, J., Ferrer, M.: Generalized median string computation by means of string embedding in vector spaces. Pattern Recogn. Lett. **33**, 842–852 (2012)
15. Ferrer, M., Valveny, E., Serratosa, F., Riesen, K., Bunke, H.: Generalized median graph computation by means of graph embedding in vector spaces. Patter Recognit. **43**, 1642–1655 (2010)
16. Franek, L., Jiang, X.: Ensemble clustering by means of clustering embedding in vector spaces. Pattern Recogn. **47**(2), 833–842 (2014)
17. Bunke, H., Jiang, X., Abegglen, K., Kandel, A.: On the weighted mean of a pair of strings. Pattern Anal. Appl. **5**, 23–30 (2002)
18. Bunke, H., Gunter, S.: Weighted mean of a pair of graphs. Computing **67**, 209–224 (2001)
19. Franek, L., Jiang, X., He, C.: Weighted mean of a pair of clusterings. Pattern Anal. Appl. **17**, 153–166 (2014)
20. Franek, L., Jiang, X.: Evolutionary weighted mean based framework for generalized median computation with application to strings. In: Gimel, G., et al. (eds.) SSPR & SPR 2012. LNCS, vol. 7626, pp. 70–78. Springer, Heidelberg (2012)
21. Moreno-García, C.F., Serratosa, F.: Consensus of two sets of correspondences through optimisation functions. Pattern Anal. Appl., 1–13 (2015)
22. Moreno-García, C.F., Serratosa, F.: Consensus of multiple correspondences between sets of elements. Comput. Vis. Image Underst. **142**, 50–64 (2015)
23. Moreno-García, C.F., Serratosa, F., Cortés, X.: Consensus of Two Graph Correspondences Through a Generalisation of the Bipartite Graph Matching. In: Liu, C.-L., Luo, B., Kropatsch, W.G., Cheng, J. (eds.) GbRPR 2015. LNCS, vol. 9069, pp. 87–97. Springer, Heidelberg (2015)
24. Moreno-García, C.F., Serratosa, F.: Online learning the consensus of multiple correspondences between sets. Knowl.-Based Syst. **90**, 49–57 (2015)
25. Moreno-García, C.F., Serratosa, F.: Obtaining the Consensus of Multiple Correspondences between Graphs through Online Learning. Pattern Recognit. Letters. (2016). doi:10.1016/j.patrec.2016.09.003
26. Saha, S., Ekbal, A.: Combining multiple classifiers using vote based classifier ensemble technique for named entity recognition. Data Knowl. Eng. **85**, 15–39 (2013)
27. Kuhn, H.W.: The Hungarian method for the assignment problem export. Naval Res. Logistics Q. **2**(1–2), 83–97 (1955)
28. Munkres, J.: Algorithms for the assignment and transportation problems. J. Soc. of Ind. Appl. Math. **5**(1), 32–38 (1957)
29. Jonker, R., Volgenant, T.: Improving the hungarian assignment algorithm. Oper. Res. Lett. **5**(4), 171–175 (1986)

A Graph Repository for Learning Error-Tolerant Graph Matching

Carlos Francisco Moreno-García[(✉)], Xavier Cortés,
and Francesc Serratosa

Departament d'Enginyeria Informàtica i Matemàtiques,
Universitat Rovira I Virgili, Tarragona, Spain
carlosfrancisco.moreno@estudiants.urv.cat,
{xavier.cortes,francesc.serratosa}@urv.cat

Abstract. In the last years, efforts in the pattern recognition field have been especially focused on developing systems that use graph based representations. To that aim, some graph repositories have been presented to test graph-matching algorithms or to learn some parameters needed on such algorithms. The aim of these tests has always been to increase the recognition ratio in a classification framework. Nevertheless, some graph-matching applications are not solely intended for classification purposes, but to detect similarities between the local parts of the objects that they represent. Thus, current state of the art repositories provide insufficient information. We present a graph repository structure such that each register is not only composed of a graph and its class, but also of a pair of graphs and a ground-truth correspondence between them, as well as their class. This repository structure is useful to analyse and develop graph-matching algorithms and to learn their parameters in a broadly manner. We present seven different databases, which are publicly available, with these structure and present some quality measures experimented on them.

Keywords: Graph database · Graph-matching algorithm · Graph-learning algorithm

1 Introduction

In pattern recognition, benchmarking is the process of measuring the quality of the representation of the objects, or the quality of the algorithms involved on comparing, classifying or clustering these objects. The objective of benchmarking is to improve performance of the involved object representations and pattern recognition algorithms. Pattern recognition, through graph-based representations, has been developed through the last forty years with great success and acknowledgement. Interesting surveys about this subject are [1, 2] or [3]. The first error-tolerant graph matching algorithms were published in 1983, [4, 5], and since then, several new algorithms have been presented.

This research is supported by projects DPI2013-42458-P TIN2013-47245-C2-2-R and by Consejo Nacional de Ciencia y Tecnología (CONACyT México).

© Springer International Publishing AG 2016
A. Robles-Kelly et al. (Eds.): S+SSPR 2016, LNCS 10029, pp. 519–529, 2016.
DOI: 10.1007/978-3-319-49055-7_46

For this reason, in 2008, a specific database to perform benchmarking on graph databases was published for the first time [6]. As authors reported, they presented such database and published its paper with the aim of providing to the scientific community a public and general framework to evaluate graph representations and graph algorithms [7–9], such as error-tolerant graph matching, [10–15] learning the consensus of several correspondences, [16–20], image registration based on graphs, [21, 22], learning graph-matching parameters [23, 24], and so on. Note that a huge amount of methods has been presented, and the previous list is simply a small sample of them. For a detailed list of methods, we refer to the aforementioned surveys [1–3]. This database, called IAM [25], has been largely cited and used to develop new algorithms. It is composed of twelve datasets containing diverse attributed graphs, for instance, proteins, fingerprints, hand written characters, among others.

With the same idea, another graph database had been previously published in 2001 [26, 27]. Nevertheless, the aim of this database [28] is to perform exact isomorphism benchmarking and cannot be used to test error-tolerant graph matching since nodes and edges are unattributed. It contains 166'000 graphs with very diverse graph sizes. Most recently in 2015 [29], a new graph repository [30] was presented in order to compare exact graph edit distance (GED) calculation methods, where data from [26, 31] was collected and enhanced using low-level information.

Note that other papers have presented with new graph-based methodologies and, with the aim of experimental reproducibility, reported their self-made databases and made them public. This is the case of the one first presented in 2006 [32, 33]. It is composed of attributed graphs extracted from image sequences taken from the CMU repository [34]. Graph nodes represent salient points of some images and graph edges have been generated through Delaunay triangulation or represent shape edges.

Registers of the aforementioned databases are composed of a graph and its class (except for the one in [29] that incorporates some additional information). Thus, the only quality measures that we can extract from the algorithms applied to these databases are related on classification purposes. For instance, the usual measures are the false positives, the false negatives and the recognition ratio.

In this paper, we present a new graph-database structure. Registers on this database are composed of a pair of graphs, a ground-truth correspondence between them as well as the class of these graphs. This ground-truth is independent of the graph-matching algorithm and also on their specific parameters, since it has been imposed by a human or an optimal automatic technique. Therefore, the quality measures that we can extract not only are the ones related on classification, but also the ones related on the ground-truth correspondence, such as the Hamming distance (HD) between the obtained correspondence and the ground-truth correspondence. Moreover, some graph-matching learning algorithms that need a given ground-truth correspondence [19, 33, 35–37] could be applied and evaluated. We concretise this structure on seven different databases, and we present some quality measures experimented on them.

Similar to the case of the IAM graph database repository [25], we divide the databases in three sets, viz. learning, test and validation. In machine learning applications, the learning set is used to learn the database knowledge that is usually materialised on the algorithms' input parameters. The validation set is used for regularisation purposes, that is, to tune the over-fitting or under-fitting of the learned parameters. Finally, the test set is used to test the quality measures of the methods learned through the learning and the validation sets.

The rest of the paper is structured in two other sections. In the first one, we present the graph repository and its benchmarks. In the second one, we conclude the paper.

2 The Graph Repository

The "Tarragona Repository" (publicly available at [38]) is described in this section, which is divided into three sub-sections. In the first one, the general structure of the whole databases is described. In the second one, we describe the current databases in the repository. Note the aim of this paper is to define a new method to structure graph databases and therefore, other databases could be included by the authors or other researches in a near future. In the third sub-section, we summarise the main features of each database and we present some experimental results performed on them.

2.1 General Structure

Databases in the "Tarragona repository" are composed of registers with a format (G^i, G'^i, f^i, C^i). Attributed graphs G^i and G'^i need to be defined in the same attribute domain, but may have different orders. The ground-truth correspondence f^i between the nodes of G^i and G'^i may have some nodes of G^i mapped to nodes of G'^i, and other ones mapped to a null node. Nevertheless, two nodes of G^i cannot be mapped to the same node of G'^i. The null node is a mechanism to represent that a node of G^i do not have to be mapped to any node of G'^i [10]. Note some nodes of G'^i may not have been mapped to any node of G^i through f^i. Moreover, we impose both graphs to belong to the same class. This is because we consider it has no sense to map local parts of objects that belong to different classes. For instance, if graphs represent hand-written characters, there is no ground-truth correspondence between an "A" and a "J".

Our databases are composed of five terms: Name, Description, Learning, Test and Validation. Name and Description are obvious, and Learning, Test and Validation are the three common datasets to perform benchmarking.

We present in [38], together with these databases, the following Matlab functions:

- *Load_Register*(*Database, Set, Register*): Returns the register *Register* in the database *Database* and the set *Set* that accepts three values: *Learning, Test* or *Validation*. The output has the format $(G^i, G'^i, C^i, f^i, I^i, I'^i)$. G^i and G'^i are both graphs with their class C^i, f^i is the ground-truth correspondence, and values I^i and I'^i are the indices of graphs G^i and G'^i respectively. These indices are useful to know which graphs have been mapped to other ones since any given graph can appear in several registers although each time has to be mapped to a different graph.
- *Load_Graph*(*Database, Set, Index*): it returns the graph in position *Index*. This function is useful to test the classification ratio.
- *Classification*(*Database, Set1, Set2, K_v, K_e*): Returns the classification ratio and the average Hamming distance given sets *Set1* and *Set2* in *Database*. The fast bipartite

graph matching (FBP) [13] has been used to compute the GED [10] and the correspondences. Parameter K_v is the insertion and deletion costs on the nodes, and parameter K_e is the insertion and deletion cost on the edges.

- *Plot_Graph(Graph, Image)*: Plots the graph over the image where it was extracted from, in the case that the graph represents an object on an image. This function assumes that the first two node attributes are the image coordinates (x, y).

With the aim of reducing the memory space, the *Learning*, *Test* and *Validation* sets of each database have been logically structured as shown in Fig. 1. There is a main vector, where each cell is composed of a structure of three elements. The first one contains a graph, the second one assigns a class to this graph, and the third one describes the correspondences from this graph to the rest of graphs. Considering the graphs, the set of nodes and edges are defined as numerical matrices. The order of each graph is N and nodes have A attributes. Graphs can have different orders N, but they have the same number of attributes A given the whole database. Edges do not have attributes. The existence of an edge is represented by a 1, and the non-existence is represented by a 0. Classes are defined as string of characters. Each correspondence cell $f^{i,a}$ maps the original graph G^i to another graph G^a and it is composed of a structure of two elements that are the index of the input graph and the node-to-node mapping vector. In the node-to-node mapping vector, there are natural numbers representing the index node, and the value -1, which can appear in several positions of the correspondence, represents a mapping to a null node.

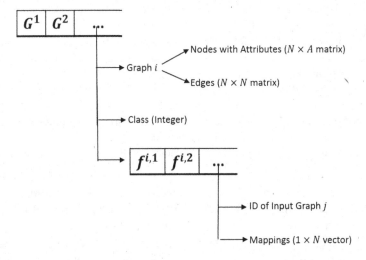

Fig. 1. Scheme representing the distribution of the information contained in each set (learning, validation or test) of a database.

2.2 Databases

The databases that are currently available are:

2.2.1 Rotation Zoom

This database contains graphs that have been extracted from 5 classes that have 10 images of outdoors scenes. Per each class, images were taken from different angles and positions. We were able to generate a correspondence between all the generated graphs by using the image homography, which was provided on the original image database [39]. Each node represents a salient point of the image. It is attributed with the position of the salient point in the image (x, y) and also a 64-size feature vector obtained by the SIFT extractor [40]. Edges are conformed using the Delaunay triangulation and do not have attributes. An example with a graph of each class is shown in Fig. 2.

BOAT   EAST_PARK   EAST_SOUTH

ENSIMAG   RESID

Fig. 2. The first image of each of the 5 classes and their graphs.

2.2.2 Palmprint

In order to construct this database, we used palmprint images contained in the Tsinghua 500 DPI Database [41], which currently has more than 150 subjects whose right and left palm has been scanned a total of 8 times each. Using the first 20 palms of the original database (10 right hands and 10 left hands), this database is constituted by a total of 20 classes of 8 graphs each. Minutiae were extracted using the algorithm proposed in [42] and graphs were constructed with each node representing a minutia. Node attributes contain information such as the minutiae position, angle, type (termination or bifurcation) and quality (good or poor). Edges are conformed using the Delaunay triangulation and do not have attributes. Finally, a correspondence between all graphs of the same class is generated using a greedy matching algorithm based on the Hough transform [43]. An example of a palmprint image and its graph is provided in Fig. 3.

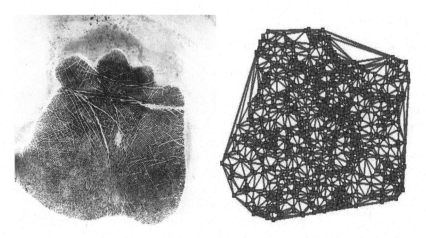

Fig. 3. A palmprint and its graph.

2.2.3 Letters

The *Letters* graph database originally presented in [6] consists on a set of graphs that represent artificially distorted letters of the Latin alphabet. For each class, a prototype line drawing was manually constructed. These prototype drawings are then converted into prototype graphs by representing the lines through undirected edges, and the ending points of such lines through nodes. Attributes on nodes are only the bi-dimensional position of the junctions and edges do not have attributes. Figure 4 shows four samples of letter A.

Fig. 4. Different instances of letter A.

There are three variants of the database depending on the degree of distortion with respect to the original prototype (adding, deleting and moving nodes and edges), viz. low, medium and high. The ground-truth correspondence between the nodes is well-known, because graphs of each class are generated from an original prototype.

2.2.4 Sagrada Familia 3D

The *Sagrada Familia 3D* database consist of a set of graphs, where each one represents a cloud of 3D points with structural relations between them. Nodes represent 3D points and their attributes are the 3D position. Edges represent proximity and do not have attributes. These points have been extracted as follows. First, a sequence of 473 photos were taken from different positions around the Sagrada Familia church in Barcelona (Catalonia, Spain), pointing the camera at the centre of it. Using the whole sequence of 2D images, a 3D model of the monument was built through the Bundler method [44, 45]. This method deducts a global cloud of 3D points of a central object using the salient points of the set of 2D images. Moreover, it also returns the correspondence between the 3D points of the resultant model and the salient points of the 2D images. Each graph in the database represents the 3D information of the salient points that appear in each image. Figure 5 shows the process to generate the graphs. Red points are the 3D model of Sagrada Familia, blue points are the different poses of the camera that has captured the images of the model and black points represent the salient points of images.

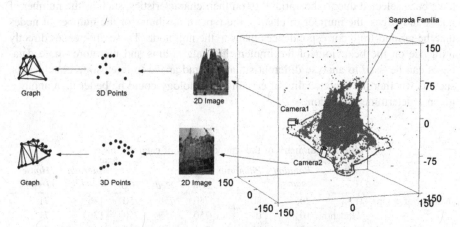

Fig. 5. The process to generate *Sagrada Familia 3D* database. (Color figure online)

2.2.5 House-Hotel

The original CMU "house" and "hotel" databases consist of 111 graphs corresponding to a toy house and 101 graphs corresponding to a hotel [46]. Each frame of these sequences has the same 30 hand-marked salient points identified and labelled with some attributes. Therefore, nodes in the graphs represent the salient points, with their position in the image plus a 60-size feature vector using Context Shape (CS) as attributes. Edges are unattributed and were constructed using the Delaunay triangulation. In this database there are three sets of pairs of frames, considering as baseline the number of frames of separation in the video sequence (Fig. 6).

Fig. 6. Different images of each of the two classes and their graphs.

2.3 Repository Summary

Table 1 summarises the main characteristics of the repository. The databases contained have been selected due to the variability on their characteristics, such as the number of nodes and edges, the number of classes, the type of attributes or the number of nodes that the ground-truth correspondences maps to the null node. These differences directly influence on the behaviour of the implemented algorithms and therefore, these databases can be used to analyse different situations and arrive to interesting conclusions, such as whether the functionality of certain methodology could be better than another, given a determined situation.

Table 1. Summary of the characteristics of each database.

Database		Rotation zoom	Palmprint	Letter			Sagrada Familia	House-Hotel
				Low	Med	High		
Number of graphs	Train	20	80	750	750	750	136	71
	Validation	10	0	750	750	750	136	71
	Test	20	80	750	750	750	135	70
Number of correspondences	Train	80	320	37500	37500	37500	18496	2627
	Validation	40	0	37500	37500	37500	18255	2627
	Test	80	320	37500	37500	37500	18255	2590
Number of classes		5	20	15	15	15	1	2
Number of node attributes		66	5	2	2	2	3	62
Attributes' description		(x,y) 64 SIFT	(x,y) 1 Angle 1 Type 1 Quality	(x,y)			(x,y,z)	(x,y) 60 CS
Avg. nodes		50	836.3	4.6	4.6	4.6	39.3	30
Avg. edges		277.4	4971.2	6.2	6.4	9	456.5	154.4
Avg. null correspondences		31.6	152.1	0.4	0.4	0.4	30.1	0
Max. nodes		50	1505	8	9	9	141	30
Max. edges		284	8962	12	14	18	1918	158
Max. null correspondences		50	619	4	5	5	139	0

Table 2 shows the classification ratio and the average Hamming distance between the computed correspondences and the ground-truth correspondences. It is the result of running the Matlab function *Classification*(*Database, Test, Reference, K_v, K_e*) available in [36] (explained in Sect. 2.1). As commented, the FBP [13] has been used to compute the GEDs [10] and the correspondences. Insertion and deletion cost on nodes, K_v, and insertion and deletion cost on edges, K_e, have been deducted through the learning algorithm presented in [37]. The aim of this table is not to report the best achieved results but simply to show an example of a specific graph-matching algorithm and learning algorithm. We encourage other researches to share their results, while showing these ones as a starting point.

Table 2. Classification ratio and HD obtained with the FBP [13] given edit costs K_v and K_e, which have been learned by a correspondence-based learning algorithm [37].

Database		Edit costs		Classification	Hamming
		K_v	K_e	ratio	distance
Rotation zoom		0.0325	−0.0027	1	0.8598
Palmprint		210	5	0.85	0.4763
Letter	Low	1	1	0.9453	0.9096
	Med	1	1	0.8667	0.8382
	High	1	1	0.8080	0.8303
Sagrada Familia		0.05	0.05	–	0.7439
House-Hotel		1000	1	1	0.8598

3   Conclusions

We have presented a publicly available graph repository to perform benchmarking on graph algorithms such as graph matching, graph clustering, leaning consensus corre-spondence or parameter learning. The main feature of this repository is that registers of these databases do not have the classical structure composed of a graph and its class, but are composed of a pair of graphs, their class and the ground-truth correspondence. We want this repository not to be seen as a concluded project, but a dynamic one, in which other researches contribute with more graph databases. Moreover, we have presented some classification ratios and Hamming distance on these databases, given some specific algorithms and parameterisations. For this aspect as well, we invite other researches to contribute with more results and therefore, to extend and disseminate the results obtained so far.

References

1. Conte, D., Foggia, P., Sansone, C., Vento, M.: Thirty years of graph matching in pattern recognition. Int. J. Pattern Recogn. Artif. Intell. **18**(3), 265–298 (2004)
2. Foggia, P., Percannella, G., Vento, M.: Graph matching and learning in pattern recognition in the last ten years. Int. J. Pattern Recogn. Artif. Intell. **28**(1), 1450001 (2014)
3. Vento, M.: A long trip in the charming world of graphs for pattern recognition. Pattern Recogn. **48**(2), 291–301 (2015)
4. Sanfeliu, A., Fu, K.S.: A distance measure between attributed relational graphs for pattern recognition. IEEE Trans. Syst. Man Cybern. **13**(3), 353–362 (1983)
5. Bunke, H.: Inexact graph matching for structural pattern recognition. Pattern Recogn. Lett. **1**, 245–253 (1983)
6. Riesen, K., Bunke, H.: IAM graph database repository for graph based pattern recognition and machine learning. In: da Vitoria Lobo, N., Kasparis, T., Roli, F., Kwok, J.T., Georgiopoulos, M., Anagnostopoulos, G.C., Loog, M. (eds.) SSPR & SPR 2008. LNCS, vol. 5342, pp. 287–297. Springer, Heidelberg (2008)
7. Wong, A., You, M.: Entropy and distance of random graphs with application to structural pattern recognition. IEEE Trans. Pattern Anal. Mach. Intell. **7**(5), 599–609 (1985)
8. Serratosa, F., Alquézar, R., Sanfeliu, A.: Function-described graphs for modelling objects represented by attributed graphs. Pattern Recogn. **36**(3), 781–798 (2003)
9. Sanfeliu, A., Serratosa, F., Alquézar, R.: Second-order random graphs for modelling sets of attributed graphs and their application to object learning and recognition. Int. J. Pattern Recogn. Artif. Intell. **18**(3), 375–396 (2004)
10. Solé, A., Serratosa, F., Sanfeliu, A.: On the graph edit distance cost: properties and applications. Int. J. Pattern Recogn. Artif. Intell. **26**(5), 1260004 (2012)
11. Lladós, J., Martí, E., Villanueva, J.: Symbol recognition by error-tolerant subgraph matching between region adjacency graphs. Trans. Pattern Anal. Mach. Intell. **23**(10), 1137–1143 (2001)
12. Riesen, K., Bunke, H.: Approximate graph edit distance computation by means of bipartite graph matching. Image Vis. Comput. **27**, 950–959 (2009)
13. Serratosa, F.: Fast computation of bipartite graph matching. Pattern Recogn. Lett. **45**, 244–250 (2014)
14. Serratosa, F.: Computation of graph edit distance: reasoning about optimality and speed-up. Image Vis. Comput. **40**, 38–48 (2015)
15. Serratosa, F.: Speeding up fast bipartite graph matching through a new cost matrix. Int. J. Pattern Recogn. Artif. Intell. **29**(2), 1550010 (2015)
16. Moreno-García, C.F., Serratosa, F.: Consensus of two sets of correspondences through optimisation functions. Pattern Anal. Appl. **2015**, 1–13 (2015)
17. Moreno-García, C.F., Serratosa, F.: Consensus of multiple correspondences between sets of elements. Comput. Vis. Image Underst. **142**, 50–64 (2015)
18. Moreno-García, C.F., Serratosa, F., Cortés, X.: Consensus of two graph correspondences through a generalization of the bipartite graph matching. In: Liu, C.-L., Luo, B., Kropatsch, Walter G., Cheng, J. (eds.) GbRPR 2015. LNCS, vol. 9069, pp. 87–97. Springer, Heidelberg (2015)
19. Moreno-García, C.F., Serratosa, F.: Online learning the consensus of multiple correspondences between sets. Knowl. Based Syst. **90**, 49–57 (2015)
20. Moreno-García, C.F., Serratosa, F.: Obtaining the consensus of multiple correspondences between graphs through online learning. Pattern Recogn. Lett. (2016). doi:10.1016/j.patrec.2016.09.003

21. Sanromà, G., Alquézar, R., Serratosa, F., Herrera, B.: Smooth point-set registration using neighbouring constraints. Pattern Recogn. Lett. **33**, 2029–2037 (2012)
22. Sanromà, G., Alquézar, R., Serratosa, F.: A new graph matching method for point-set correspondence using the EM algorithm and softassign. Comput. Vis. Image Underst. **116** (2), 292–304 (2012)
23. Neuhaus, M., Bunke, H.: Automatic learning of cost functions for graph edit distance. Inf. Sci. **177**(1), 239–247 (2006)
24. Neuhaus, M., Bunke, H.: Self-organizing maps for learning the edit costs in graph matching. IEEE Trans. Syst. Man Cybern. Part B **35**(3), 503–514 (2005)
25. http://www.iam.unibe.ch/fki/databases/iam-graph-database
26. Foggia, P., Sansone, C., Vento, M.: A database of graphs for isomorphism and subgraph isomorphism benchmarking. In: Proceedings of 3rd International Workshop on Graph Based Representations in Pattern Recognition, pp. 176–187 (2001)
27. De Santo, M., Foggia, P., Sansone, C., Vento, M.: A large database of graphs and its use for benchmarking graph isomorphism algorithms. Pattern Recogn. Lett. **24**, 1067–1079 (2003)
28. http://iapr-tc15.greyc.fr/links.html
29. Abu-Aisheh, Z., Raveaux, R., Ramel, J.Y.: A graph database repository and performance evaluation metrics for graph edit distance. In: IAPR International Workshop on Graph Based Representation (2015)
30. http://www.rfai.li.univ-tours.fr/PublicData/GDR4GED/home.html
31. Gao, X., Xiao, B., Tao, D., Li, X.: A survey of graph edit distance. Pattern Anal. Appl. **13** (1), 113–129 (2010)
32. Caetano, T., Caelli, T., Schuurmans, D., Barone, D.: Graphical models and point pattern matching. IEEE Trans. Pattern Anal. Mach. Intell. **28**(10), 1646–1663 (2006)
33. Caetano, T., et al.: Learning graph matching. Trans. Pattern Anal. Mach. Intell. **31**(6), 1048–1058 (2009)
34. http://www.cs.cmu.edu/afs/cs/project/vision/vasc/idb/www/html_permanent//index.html
35. Cortés, X., Serratosa, F.: Learning graph matching substitution weights based on the ground-truth node correspondence. Int. J. Pattern Recogn. Artif. Intell. **30**(2), 1650005 (2016)
36. Leordeanu, M., Sukthankar, R., Hebert, M.: Unsupervised learning for graph matching. Int. J. Comput. Vis. **96**(1), 28–45 (2012)
37. Cortés, X., Serratosa, F.: Learning graph-matching edit-costs based on the optimality of the oracle's node correspondences. Pattern Recogn. Lett. **56**, 22–29 (2015)
38. http://deim.urv.cat/~francesc.serratosa/databases/
39. http://www.featurespace.org
40. Lowe, D.G.: Distinctive image features from scale-invariant keypoints. Int. J. Comput. Vis. **60**(2), 91–110 (2004)
41. Dai, J., Feng, J., Zhou, J.: Robust and efficient ridge based palmprint matching. IEEE Trans. Pattern Anal. Mach. Intell. **34**(8), 1618–1632 (2012)
42. Dai, J., Zhou, J.: Multi-feature based high-resolution palmprint recognition. IEEE Trans. Pattern Anal. Mach. Intell. **33**(5), 945–957 (2011)
43. Ratha, N.K., Karu, K., Chen, S., Jain, A.K.: A real-time matching system for large fingerprint databases. IEEE Trans. Pattern Anal. Mach. Intell. **18**(8), 799–813 (1996)
44. http://www.cs.cornell.edu/~snavely/bundler/
45. Snavely, N., Todorovic, S.: From contours to 3D object detection and pose estimation. In: International Congress on Computer Vision (ICVV), pp. 983–990 (2011)
46. http://vasc.ri.cmu.edu/idb/html/motion/

Graph Edit Distance or Graph Edit Pseudo-Distance?

Francesc Serratosa$^{(\boxtimes)}$, Xavier Cortés, and Carlos-Francisco Moreno

Universitat Rovira i Virgili, Tarragona, Catalonia, Spain
{francesc.serratosa,xavier.cortes,
carlsofrancisco.moreno}@urv.cat

Abstract. Graph Edit Distance has been intensively used since its appearance in 1983. This distance is very appropriate if we want to compare a pair of attributed graphs from any domain and obtain not only a distance, but also the best correspondence between nodes of the involved graphs. In this paper, we want to analyse if the Graph Edit Distance can be really considered a distance or a pseudo-distance, since some restrictions of the distance function are not ful-filled. Distinguishing between both cases is important because the use of a distance is a restriction in some methods to return exact instead of approximate results. This occurs, for instance, in some graph retrieval techniques. Experimental validation shows that in most of the cases, it is not appropriate to denominate the Graph Edit Distance as a distance, but a pseudo-distance instead, since the triangle inequality is not fulfilled. Therefore, in these cases, the graph retrieval techniques not always return the optimal graph.

Keywords: Graph Edit Distance · Edit cost · Distance function

1 Introduction

Attributed graphs have been of crucial importance in pattern recognition throughout more than four decades [1, 2] since they have been used to model several kinds of problems. Interesting reviews of techniques and applications are [3–5]. If elements in pattern recognition are modelled through attributed graphs, error-tolerant graph-matching algorithms are needed that aim to compute a matching between nodes of two attributed graphs that minimizes some kind of objective function. To that aim, one of the most widely used methods to evaluate an error correcting graph isomorphism is the Graph Edit Distance [1, 2, 6].

Graph Edit Distance needs two main input parameters, which are the pair of attributed graphs to be compared and also other calibration parameters. These parameters have to be tuned in order to maximise a recognition ratio in a classification scenario or simply to minimise the Hamming distance between a ground-truth correspondence between nodes of both graphs and the obtained correspondence.

This research is supported by projects DPI2013-42458-P, TIN2013-47245-C2-2-R, and by Consejo Nacional de Ciencia y Tecnología (CONACyT México).

© Springer International Publishing AG 2016
A. Robles-Kelly et al. (Eds.): S+SSPR 2016, LNCS 10029, pp. 530–540, 2016.
DOI: 10.1007/978-3-319-49055-7_47

Unfortunately, little research has been done to analyse if really the Graph Edit Distance is a distance or simply a dissimilarity function that could be classified as a pseudo-distance, since some distance restrictions are not fulfilled. Reference [7] is the only paper related on this idea, and it shows in which conditions of these calibration parameters the Graph Edit Distance is really a distance.

The importance of Graph Edit Distance being indeed a distance has an influence on some applications. As an example, in [8–10], authors present methods to retrieve graphs in a database. They suppose that given three graphs, the triangle inequality is fulfilled and thanks to this assumption, some comparisons are not needed to be performed. It turns out that if the Graph Edit Distance is not a distance, then the triangle inequality is not guaranteed, and then some graphs that would have to be explored are not considered, making the methods sub-optimal.

The aim of this paper is to empirically analyse if the distance definition is hold when the recognition ratio is maximised or the obtained correspondence is close to the ground truth. The outline of the paper is as follows; in Sect. 2, we define the attributed graphs and the Graph Edit Distance. In Sects. 3 and 4, we explain the restrictions that a distance needs to fulfil and we relate these restrictions on the specific case of the Graph Edit Distance. In Sects. 5 and 6, we show the experimental validation and conclude the paper.

2 Graphs and Graph Edit Distance

Let Δ_v and Δ_e denote the domains of possible values for attributed vertices and arcs, respectively. An attributed graph (over Δ_v and Δ_e) is defined by a tuple $G = (\Sigma_v, \Sigma_e, \gamma_v, \gamma_e)$, where $\Sigma_v = \{v_k \mid k = 1, \ldots, R\}$ is the set of vertices (or nodes), $\Sigma_e = \{e_{ij} \mid i, j \in \{1, \ldots, R\}\}$ is the set of edges (or arcs), $\gamma_v : \Sigma_v \to \Delta_v$ assigns attribute values to vertices and $\gamma_e : \Sigma_e \to \Delta_e$ assigns attribute values to edges.

Let $G^p = (\Sigma_v^p, \Sigma_e^p, \gamma_v^p, \gamma_e^p)$ and $G^q = (\Sigma_v^q, \Sigma_e^q, \gamma_v^q, \gamma_e^q)$ be two attributed graphs of order R^p and R^q. To allow maximum flexibility in the matching process, graphs can be extended with null nodes [1] to be of order $R^p + R^q$. We refer to null nodes of G^p and G^q by $\hat{\Sigma}_v^p \subseteq \Sigma_v^p$ and $\hat{\Sigma}_v^q \subseteq \Sigma_v^q$ respectively. Let T be a set of all possible correspondences between two node sets Σ_v^p and Σ_v^q. Correspondence $f^{p,q} : \Sigma_v^p \to \Sigma_v^q$, assigns each node of G^p to only one node of G^q. The correspondence between edges, denoted by $f_e^{p,q}$, is defined accordingly to the correspondence of their terminal nodes.

$$f_e^{p,q}\left(e_{ab}^p\right) = e_{ij}^q \Rightarrow f^{p,q}\left(v_a^p\right) = v_i^q \wedge f^{p,q}\left(v_b^p\right) = v_j^q$$
$$v_a^p, v_b^p \in \Sigma_v^p - \hat{\Sigma}_v^p \text{ and } v_i^q, v_j^q \in \Sigma_v^q - \hat{\Sigma}_v^q \tag{1}$$

The basic idea behind the Graph Edit Distance is to define a dissimilarity measure between two graphs. This dissimilarity is defined as the minimum amount of distortion required to transform a graph into another. To this end, a number of distortion or edit operations, consisting of insertion, deletion and substitution of both nodes and edges are defined. Then, for every pair of graphs (G^p and G^q), there is a sequence of edit operations that transforms G^p into G^q. In general, several edit paths may exist between two given graphs and to evaluate which edit path is the best, edit cost functions are introduced. The basic idea is to assign a penalty cost to each edit operation according to the amount of distortion that it introduces into the transformation.

Each edit path can be related to an univocal correspondence $f^{p,q} \in T$ between the involved graphs. This way, each edit operation assigns a node of the first graph to a node of the second graph. Deletion and insertion operations are transformed to assignments of a non-null node of the first or second graph to a null node of the second or first graph. Substitutions simply indicate node-to-node assignments. Using this transformation, given two graphs, G^p and G^q, and a correspondence between their nodes, $f^{p,q}$, the graph edit cost is given by [1]:

$$
EditCost(G^p, G^q, f^{p,q}) =
$$

$$
\sum_{\substack{v_a^p \in \Sigma_v^p - \hat{\Sigma}_v^p \\ v_i^q \in \Sigma_v^q - \hat{\Sigma}_v^q}} C_{ns}\left(v_a^p, v_i^q\right) + \sum_{\substack{e_{ab}^p \in \Sigma_e^p - \hat{\Sigma}_e^p \\ e_{ij}^q \in \Sigma_e^q - \hat{\Sigma}_e^q}} C_{es}\left(e_{ab}^p, e_{ij}^q\right) + \sum_{\substack{v_a^p \in \Sigma_v^p - \hat{\Sigma}_v^p \\ v_i^q \in \hat{\Sigma}_v^q}} C_{nd}\left(v_a^p, v_i^q\right) +
$$

$$
\sum_{\substack{v_a^p \in \hat{\Sigma}_v^p \\ v_i^q \in \Sigma_v^q - \hat{\Sigma}_v^q}} C_{ni}\left(v_a^p, v_i^q\right) + \sum_{\substack{e_{ab}^p \in \Sigma_e^p - \hat{\Sigma}_e^p \\ e_{ij}^q \in \hat{\Sigma}_e^q}} C_{ed}\left(e_{ab}^p, e_{ij}^q\right) + \sum_{\substack{e_{ab}^p \in \hat{\Sigma}_e^p \\ e_{ij}^q \in \Sigma_e^q - \hat{\Sigma}_e^q}} C_{ei}\left(e_{ab}^p, e_{ij}^q\right)
$$

being $f^{p,q}\left(v_a^p\right) = v_i^q$ and $f_e^{p,q}\left(e_{ai}^p\right) = e_{ij}^q$

$$(2)$$

where C_{ns} is the cost of substituting node v_a^p of G^p by node $f^{p,q}\left(v_a^p\right)$ of G^q, C_{nd} is the cost of deleting node v_a^p of G^p and C_{ni} is the cost of inserting node v_i^q of G^q. Equivalently for edges, C_{es} is the cost of substituting edge e_{ab}^p of graph G^p by edge $f_e^{p,q}\left(e_{ab}^p\right)$ of G^q, C_{ed} is the cost of assigning edge e_{ab}^p of G^p to a non-existing edge of G^q and C_{ei} is the cost of assigning edge e_{ab}^q of G^q to a non-existing edge of G^p.

Finally, the Graph Edit Distance is defined as the minimum cost under any correspondence in T:

$$
GED(G^p, G^q) = \min_{f^{p,q} \in T} EditCost(G^p, G^q, f^{p,q})
\tag{3}
$$

Using this definition, the Graph Edit Distance essentially depends on $C_{ns}, C_{nd}, C_{ni}, C_{es}, C_{ed}$ and C_{ei} functions. Several definitions of these functions exist. Table 1 summarises the five different configurations that have been presented so far.

The first options [11–16] are the ones where the whole costs are defined as functions that depend on the involved attributes and also on either learned or general knowledge. Attributes are density functions instead of vectors of attributes. The second option makes the Graph Edit Distance to be directly related to the maximal common sub-graph. That is, in [17], authors demonstrate that computing the Graph Edit Distance is exactly the same than deducting the maximal common sub-graph. In the third

Table 1. Examples of Graph Edit Costs in the literature.

Reference	C_{ns}	C_{nd}	C_{ni}	C_{es}	C_{ed}	C_{ei}
[11–13, 15]	$d_n(v_a^p, v_i^q)$	$f_{nd}(v_a^p)$	$f_{nd}(v_i^q)$	$d_e(e_{ab}^p, e_{ij}^q)$	$f_{ed}(e_{ab}^p)$	$f_{ei}(e_{ij}^q)$
[17]	$0, \infty$	1	1	$0, \infty$	0	0
[18]	$d_n(v_a^p, v_i^q)$	K_n	K_n	$d_e(e_{ab}^p, e_{ij}^q)$	0	0
[1, 19, 20]	$d_n(v_a^p, v_i^q)$	K_n	K_n	$d_e(e_{ab}^p, e_{ij}^q)$	K_e	K_e
[23]	$0, K_{ns}$	K_n	K_n	$0, K_{es}$	K_e	K_e

option, [18], authors assume that the graphs are complete, and a non-existing edge is an edge with a "null" attribute. In this case, the cost of deleting and inserting an edge is encoded in the edge substitution cost. Inserting and deleting nodes have a constant cost, K_n. With this definition, authors describe several classes of costs that Eq. 3 deducts the same correspondence. The fourth option might be the most used one [1, 19, 20]. Substitution costs are defined as distances between vectors of attributes, usually the Euclidean distance. Insertion and deletion costs are constants, K_n and K_e, that have been manually tested or automatically learned [21, 22]. Finally, the last option is used in fingerprint recognition [23]. It is similar to the previous option, except from the substitution costs that are constants. Nodes represent minutiae and edges are the relations between them. If a specific distance between minutiae is lower than a threshold, then a zero is imposed as a substitution cost. Otherwise, this cost takes a constant value K_{ns}. The same happens with the edges that take a constant value K_{es}.

It is worth noting that for all of the cases except for the first one, the insertion and deletion costs on nodes are considered to be the same, K_n. The same happens for edges, K_e. Nevertheless, in the string edit distance, also known as Levenshtein distance [24], insertion and deletion costs might be considered different depending on the application. The most usual application is an automatic writing correction scenario, in which the possibility of missing a character is different than accidentally adding an extra character [25].

The optimal computation of the Graph Edit Distance is usually carried out by means of a tree search algorithm, which explores the space of all possible mappings of the nodes and edges of the first graph to the nodes and edges of the second graph. A widely used method is based on the A* algorithm, for instance [18]. Unfortunately, the computational complexity of this algorithm is exponential in the number of nodes of the involved graphs. This means that the running time may be non-admissible in some applications, even for reasonably small graphs. This is why bipartite graph matching [26–29] has appeared.

3 Restrictions on the Graph Edit Distance

A distance, also called a metric, is a function that defines a dissimilarity between elements of a set, such as x, y or z. The domain is $[0, \infty)$ and it holds the following restrictions for all elements in the set [30]:

(1) Non-negativity: $dist(x, y) \geq 0$.

(2) Identity of indiscernible elements: $dist(x, y) = 0 \Leftrightarrow x = y$.

(3) Symmetry: $dist(x, y) = dist(y, x)$

(4) Triangle inequality: $dist(x, y) \leq dist(x, z) + dist(z, y)$

(4)

In some cases, there is a need to relax these restrictions and thus the resulting functions are not called distance but pseudo-distance, quasi-distance, meta-distance or semi-distance, depending on which restriction is violated and how it is violated [30]. In the case of the Graph Edit Distance, this relaxation comes from the imposition of a ground truth correspondence. This ground truth is deducted from an independent method and it is not related on the edit costs.

If we wish the Graph Edit Distance to be defined as a true distance function, it is needed to assure the whole edit operations in the ground truth correspondence fulfil the four properties in the following Eq. 5. In these equations, we suppose that the ground truth correspondence is $f^{p,q}$ such that $f^{p,q}\left(v_a^p\right) = v_i^q$ and $f^{p,q}\left(v_b^p\right) = v_j^q$.

(1) Non-negativity: $C_{ns} \geq 0$ and $C_{es} \geq 0$.

(2) Identity of indiscernible elements:

$$C_{ns}\left(v_a^p, v_i^q\right) = 0 \Leftrightarrow \gamma_v\left(v_a^p\right) = \gamma_v(v_i^q)$$

$$C_{es}\left(e_{ab}^p, e_{ij}^q\right) = 0 \Leftrightarrow \gamma_e\left(e_{ab}^p\right) = \gamma_e\left(e_{ij}^q\right)$$

(3) Symmetry:

$$C_{nd}\left(v_a^p, v_i^q\right) = C_{ni}\left(v_{a'}^p, v_{i'}^q\right) \Leftrightarrow \gamma_v\left(v_a^p\right) = \gamma_v\left(v_i^q\right)$$

where $v_a^p \in \Sigma_v^p - \hat{\Sigma}_v^p$, $v_i^q \in \hat{\Sigma}_v^q$, $v_{a'}^p \in \hat{\Sigma}_v^p$ and $v_{i'}^q \in \Sigma_v^q - \hat{\Sigma}_v^q$

$$C_{ed}\left(e_{ab}^p, e_{ij}^q\right) = C_{ei}\left(e_{a'b'}^p, e_{i'j'}^q\right) \Leftrightarrow \gamma_e\left(e_{ab}^p\right) = \gamma_e\left(e_{i'j'}^q\right)$$

where $e_{ab}^p \in \Sigma_e^p - \hat{\Sigma}_e^p$, $e_{ij}^q \in \hat{\Sigma}_e^q$, $e_{a'b'}^p \in \hat{\Sigma}_e^p$ and $e_{i'j'}^q \in \Sigma_e^q - \hat{\Sigma}_e^q$

(4) Triangle inequality:

$$C_{ns}\left(v_a^p, v_i^q\right) \leq C_{nd}\left(v_a^p, v_i^q\right) + C_{ni}\left(v_{a'}^p, v_i^q\right)$$

where $v_a^p \in \Sigma_v^p - \hat{\Sigma}_v^p$, $v_i^q \in \Sigma_v^q - \hat{\Sigma}_v^q$ $v_{i'}^q \in \hat{\Sigma}_v^q$ and $v_{a'}^p \in \hat{\Sigma}_v^p$

$$C_{es}\left(e_{ab}^p, e_{ij}^q\right) \leq C_{ed}\left(e_{ab}^p, e_{i'j'}^q\right) + C_{ei}\left(e_{a'b'}^q, e_{ij}^q\right)$$

where $e_{ab}^p \in \Sigma_e^p - \hat{\Sigma}_e^p$, $e_{ij}^q \in \Sigma_e^q - \hat{\Sigma}_e^q$ $e_{i'j'}^q \in \hat{\Sigma}_e^q$ and $e_{a'b'}^p \in \hat{\Sigma}_e^p$

(5)

For all cited references, functions in Table 1 are defined as distances, and constants as real positive numbers. For this reason, if the Graph Edit Distance cannot be defined as a true distance, it is due to the relations between these functions and constants. Considering the five options proposed in Table 1, it is really difficult to analyse the first option since being a distance or not depends on the specific distance values. We realise that the second and third ones do not hold the triangle inequality and therefore cannot be considered as distances. The fourth option is a distance only if it is guaranteed that the whole substitution operations in the edit path hold:

$$d_n\left(v_a^p, v_i^q\right) \leq 2 \cdot K_n \text{ and } d_e(e_{ab}^p, e_{ij}^q) \leq 2 \cdot K_e \tag{6}$$

That is, we only have to analyse if the triangle inequality of Eq. 5 is fulfilled. Finally, the last option is almost the same than the third one and it is a true distance if constant costs are defined such that,

$$K_{ns} \leq 2 \cdot K_n \text{ and } K_{es} \leq 2 \cdot K_e \tag{7}$$

Since the fourth option is both the most used and the one that can be defined as distance or not, depending on the costs, from now on, we concretise on this specific case.

4 Defining the Graph Edit Distance as a True Distance

Note that given a pair of graphs and an optimal correspondence (the one that minimise *EditCost* in Eq. 3), we can analyse if the used edit costs make the Graph Edit Distance to be a true distance or not. Moreover, each combination of edit costs generates a different optimal correspondence and a Graph Edit Distance value. For this reason, the problem of knowing which are the edit costs that make the Graph Edit Distance to be a true distance is a chicken egg problem. Given some edit costs, we need to compute the optimal correspondence to deduct if the four distance restrictions are violated (Eq. 5), but to deduct the proper edit costs, we need the optimal correspondence.

To solve this problem, we propose to use a ground truth correspondence. That is, given a pair of attributed graphs, and independently of the edit costs, a human or another method deducts which is the "best" correspondence. Thus, we consider that the Graph Edit Distance is a true distance if the four properties in Eq. 5 are fulfilled assuming $f^{p,q}$ is the ground truth correspondence.

Given an application that involves an attributed graph database of M graphs in which the computation of the Graph Edit Distance is needed, the same edit costs have to be used in the whole process and graphs. Thus, we generalise Eq. 6 considering that we have several graphs and also introducing the ground truth concept. We conclude that the Graph Edit Distance is a true distance given some specific insertion and deletion costs for nodes if the following equation holds,

$$\begin{aligned}
&\forall v_a^p \in \Sigma_v^p - \hat{\Sigma}_v^p \text{ given } p: 1..M \text{ and } a: 1..(R^p + R^q) \\
&\text{such that } f^{p,q}\left(v_a^p\right) = v_i^q \text{ \& } v_i^q \in \Sigma_v^q - \hat{\Sigma}_v^q \\
&\text{leads to } d_n\left(v_a^p, v_i^q\right) \leq 2 \cdot K_n \\
&\text{being } f^{p,q} \text{ the ground-truth correspondence}
\end{aligned} \tag{8}$$

Similarly for the edges,

$$\forall\, e^p_{ab} \in \Sigma^p_e - \hat{\Sigma}^p_e \text{ given } p : 1..M \text{ and } a,b : 1..(R^p + R^q)$$

$$\text{such that } f^{p,q}_e\left(e^p_{ab}\right) = e^q_{ij} \ \& \ e^{jq}_i \in \Sigma^q_e - \hat{\Sigma}^q_e$$

$$\text{leads to } d_e\left(e^p_{ab}, e^q_{ij}\right) \leq 2 \cdot K_e \tag{9}$$

being $f^{p,q}_e$ the edge correspondence deducted from the ground-truth correspondence $f^{p,q}$.

5 Experimental Validation

We used four graph databases that are organised in registers such that each register is composed of a pair of graphs and a ground truth correspondence between their nodes. These databases were initially used to automatically learn insertion and deletion edit costs in [21, 22], and are publically available in [31]. These databases do not have attributes on the edges and therefore, we only analyse the insertion and deletion costs on nodes. Nonetheless, what can be deducted on nodes could be easily extrapolated to edges. Graphs in the first two databases, *Letter Low* and *Letter High*, represent hand written characters, which nodes have as only attribute the (x,y) position of the junctions of strokes in the character, and edges being the strokes. Graphs in *House Hotel* database and *Tarragona Rotation Zoom* database have been extracted from images. Their nodes represent salient points in the images with their attributes being the features obtained by the point extractor. Edges have been deducted by the Delaunay triangulation.

Table 2 shows the position of the quartiles, the mean and also half of the maximum values of the node substitution costs $d_n\left(v^p_a, v^q_i\right)$ given the whole correspondences. Clearly, if we want Eq. 8 to hold, the insertion and deletion costs have to be defined such that $K_n \geq \frac{1}{2} Max$.

Table 2. Average node substitution costs given the ground truth correspondences.

	Q1	Q2	Q3	Mean	½ Max
Letter low	0.08	0.12	0.17	0.20	1.68
Letter high	0.28	0.48	0.71	0.55	1.98
House-Hotel	2.82	4.00	5.29	4.08	5.75
Rotation Zoom	0	0	0.002	0.014	0.5

For the sake of clarification, Fig. 1 shows the histograms of $d_n\left(v^p_a, v^q_i\right)$ given all databases with the quartiles and the mean values.

We have used an error-tolerant graph-matching algorithm called Fast Bipartite [27] available in [32] to compute the automatically-deducted correspondence and the distance between the attributed graphs.

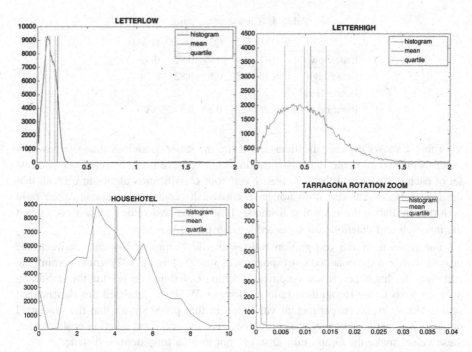

Fig. 1. Histogram of node substitution costs given the ground truth correspondences in the four databases. In green: the first three quartiles. In red: the mean values. (Color figure online)

Table 3 shows the average Hamming distance between the ground-truth correspondence and the automatically obtained correspondence when $K_n = Q1$, $K_n = Q2$, $K_n = Q3$, $K_n = Mean$, and $K_n = \frac{1}{2}Max$. Specific values are shown in Table 2. The Hamming distance is computed as the number of node mappings that are different between both correspondences. Therefore, the lower these values, the better the performance.

Table 3. Average Hamming distance.

	Q1	Q2	Q3	Mean	½ Max
Letter low	0.6	0.6	0.6	0.6	0.7
Letter high	0.9	0.8	0.9	0.9	1.2
House Hotel	0.61	0.71	0.78	0.72	0.80
Rotation Zoom	0.46	0.46	0.27	0.34	0.39

We realise that the lowest Hamming distances are achieved in the positions of the insertion and deletion edit costs such that the triangle inequality is not hold, since these lowest Hamming distances are achieved in the first three quartiles, which are always smaller than $\frac{1}{2}Max$.

Table 4. Classification ratio.

	Q1	Q2	Q3	Mean	½ Max
Letter low	0.97	0.97	0.97	0.97	0.93
Letter high	0.74	0.82	0.83	0.82	0.74
House Hotel	-	-	-	-	-
Rotation Zoom	0.2	0.2	0.35	0.3	0.1

Table 4 shows the classification ratio using the same conditions than the previous experiments. To compute the classification ratio, we have used the reference and test set of each database and the 1-Nearest Neighbour classification algorithm. Recall that the *House Hotel* database does not have classes. It seems as the classification ratio performs similar to the Hamming distance. That is, the best values are achieved when the insertion and deletion edit costs are smaller than $\frac{1}{2} Max$.

The relation of the recognition ratio with the Hamming distance between the ground truth and the obtained correspondences was explored in [22] while learning the edit costs. In that paper, it was empirically demonstrated that decreasing the Hamming distance leads to the recognition ratio to increase. We have validated this dependence again. Moreover, the experimental validation in that paper shows that the presented optimisation method converged to some negative insertion and deletion costs. Again, these values make the Graph Edit Distance not to be a truly defined distance.

Finally, in Table 5 we show the average runtime (in milliseconds) to compute one graph-to-graph comparison. We have used a Windows, I7 and Matlab. We appreciate there is no relation, in general, between the insertion and deletion edit costs and the runtime.

Table 5. Average runtime to match a pair of graphs.

	Q1	Q2	Q3	Mean	½ Max
Letter low	0.61	0.60	0.59	0.58	0.60
Letter high	0.63	0.59	0.59	0.59	0.64
House Hotel	4.8	5.1	5.4	5.1	5.5
Rotation Zoom	15	15	10	8	7

6 Conclusions

Graph Edit Distance is nowadays the most widely used function to compare two graphs and to obtain both a distance and a graph correspondence. This function does not only depend on a pair of graphs, but also on the definition of the insertion and deletion edit costs on nodes and edges. These costs are usually defined as constants, and depending on their definition, we can consider the Graph Edit Distance to be a true distance or not. The fact of not being a true distance can influence on the performance in some applications. Experimental validation has shown that the insertion and deletion costs that obtain the lowest Hamming distances between the ground truth correspondence and the optimal correspondence and also the highest classification ratios are the ones where the triangle inequality does not hold when considering the ground truth correspondence. For instance, the assumption that $EditCost(G^p, G^q, f^{p,q}) \leq GED(G^p, G^t) + GED(G^t, G^q)$, where $f^{p,q}$ is the ground truth correspondence, a fact which is commonly assumed on some applications such as graph retrieval or learning processes. This is because in a properly tuned system, $EditCost(G^p, G^q, f^{p,q}) = GED(G^p, G^q)$.

References

1. Solé, A., Serratosa, F., Sanfeliu, A.: On the graph edit distance cost: properties and applications. Int. J. Pattern Recogn. Artif. Intell. **26**(5), 1260004 (2012). (21 pages)
2. Sanfeliu, A., Fu, K.-S.: A distance measure between attributed relational graphs for pattern recognition. IEEE Trans. Syst. Man Cybern. **13**(3), 353–362 (1983)
3. Conte, D., Foggia, P., Sansone, C., Vento, M.: Thirty years of graph matching in pattern recognition. Int. J. Pattern Recogn. Artif. Intell. **18**(3), 265–298 (2004)
4. Vento, M.: A long trip in the charming world of graphs for pattern recognition. Pattern Recogn. **48**, 291–301 (2015)
5. Foggia, P., Percannella, G., Vento, M.: Graph matching and learning in pattern recognition in the last 10 years. Int. J. Pattern Recogn. Artif. Intell. **28**, 1450001 (2013)
6. Gao, X., et al.: A survey of graph edit distance. Pattern Anal. Appl. **13**(1), 113–129 (2010)
7. Bunke, H., Allermann, G.: Inexact graph matching for structural pattern recognition. Pattern Recogn. Lett. **1**(4), 245–253 (1983)
8. He, L., et al.: Graph matching for object recognition and recovery. Pattern Recogn. Lett. **37** (7), 1557–1560 (2004)
9. Serratosa, F., Cortés, X., Solé, A.: Component retrieval based on a database of graphs for hand-written electronic-scheme digitalisation. Expert Syst. Appl. **40**, 2493–2502 (2013)
10. Berretti, S., Del Bimbo, A., Vicario, E.: Efficient matching and indexing of graph models in content-based retrieval. IEEE Trans. Pattern Anal. Mach. Intell. **23**(10), 1089–1105 (2001)
11. Wong, A., You, M.: Entropy and distance of random graphs with application to structural pattern recognition. Trans. Pattern Anal. Mach. Intell. **7**(5), 599–609 (1985)
12. Neuhaus, M., Bunke, H.: Automatic learning of cost functions for graph edit distance. Inf. Sci. **177**(1), 239–247 (2006)
13. Serratosa, F., Alquézar, R., Sanfeliu, A.: Function-described graphs for modelling objects represented by attributed graphs. Pattern Recogn. **36**(3), 781–798 (2003)
14. Serratosa, F., Alquézar, R., Sanfeliu, A.: Efficient algorithms for matching attributed graphs and function-described graphs. In: Proceedings of 15th International Conference on Pattern Recognition, ICPR 2000, Barcelona, Spain, vol. 2, pp. 871–876 (2000)
15. Sanfeliu, A., Serratosa, F., Alquézar, R.: Second-order random graphs for modelling sets of attributed graphs and their application to object learning and recognition. Int. J. Pattern Recogn. Artif. Intell. **18**(3), 375–396 (2004)
16. Alquézar, R., Sanfeliu, A., Serratosa, F.: Synthesis of function-described graphs. In: Amin, A., Dori, D., Pudil, P., Freeman, H. (eds.) SSPR'98 and SPR'98. LNCS, vol. 1451, pp. 112–121. Springer, Heidelberg (1998)
17. Bunke, H.: On a relation between graph edit distance and maximum common subgraph. Pattern Recogn. Lett. **18**(8), 689–694 (1998)

18. Bunke, H.: Error correcting graph matching: on the influence of the underlying cost function. Trans. Pattern Anal. Mach. Intell. **21**(9), 917–922 (1999)

19. Caetano, T., et al.: Learning graph matching. Trans. Pattern Anal. Mach. Intell. **31**(6), 1048–1058 (2009)

20. Ferrer, M., Serratosa, F., Riesen, K.: Improving bipartite graph matching by assessing the assignment confidence. Pattern Recogn. Lett. **65**, 29–36 (2015)

21. Cortés, X., Serratosa, F.: Learning graph matching substitution weights based on the ground truth node correspondence. Int. J. Pattern Recogn. Artif. Intell. **30**(2), 1650005 (2016). (22 pages)

22. Cortés, X., Serratosa, F.: Learning graph-matching edit-costs based on the optimality of the Oracle's node correspondences. Pattern Recogn. Lett. **56**, 22–29 (2015)

23. Jain, A.K., Maltoni, D.: Handbook of Fingerprint Recognition. Springer, New York (2003)

24. Levenshtein, V.I.: Binary codes capable of correcting deletions, insertions and reversals. Sov. Phys. Dokl. Cybern. Control Theory **10**, 707–710 (1966)

25. Serratosa, F., Sanfeliu, A.: Signatures versus histograms: definitions, distances and algorithms. Pattern Recogn. **39**(5), 921–934 (2006)

26. Riesen, K., Bunke, H.: Approximate graph edit distance computation by means of bipartite graph matching. Image Vis. Comput. **27**(7), 950–959 (2009)

27. Serratosa, F.: Fast computation of bipartite graph matching. Pattern Recogn. Lett. **45**, 244–250 (2014)

28. Serratosa, F.: Computation of graph edit distance: reasoning about optimality and speed-up. Image Vis. Comput. **40**, 38–48 (2015)

29. Serratosa, F.: Speeding up fast bipartite graph matching through a new cost matrix. Int. J. Pattern Recogn. Artif. Intell. **29**(2), 1550010 (2015)

30. Arkhangel'skii, A.V., Pontryagin, L.S.: General Topology I: Basic Concepts and Constructions Dimension Theory. Encyclopaedia of Mathematical Sciences. Springer, Heidelberg (1990)

31. http://deim.urv.cat/ ~ francesc.serratosa/databases/

32. http://deim.urv.cat/ ~ francesc.serratosa/SW

Text and Document Analysis

Handwritten Word Image Categorization with Convolutional Neural Networks and Spatial Pyramid Pooling

J. Ignacio Toledo[1(✉)], Sebastian Sudholt[3], Alicia Fornés[2], Jordi Cucurull[1], Gernot A. Fink[3], and Josep Lladós[2]

[1] Scytl Secure Electronic Voting, Barcelona, Spain
{JuanIgnacio.Toledo,Jordi.Cucurull}@scytl.com
[2] Computer Vision Center, Universitat Autònoma de Barcelona, Barcelona, Spain
{afornes,josep}@cvc.uab.es
[3] Department of Computer Science, TU Dortmund University, Dortmund, Germany
{sebastian.sudholt,gernot.fink}@tu-dortmund.de

Abstract. The extraction of relevant information from historical document collections is one of the key steps in order to make these documents available for access and searches. The usual approach combines transcription and grammars in order to extract semantically meaningful entities. In this paper, we describe a new method to obtain word categories directly from non-preprocessed handwritten word images. The method can be used to directly extract information, being an alternative to the transcription. Thus it can be used as a first step in any kind of syntactical analysis. The approach is based on Convolutional Neural Networks with a Spatial Pyramid Pooling layer to deal with the different shapes of the input images. We performed the experiments on a historical marriage record dataset, obtaining promising results.

Keywords: Document image analysis · Word image categorization · Convolutional neural networks · Named entity detection

1 Introduction

Document Image Analysis and Recognition (DIAR) is the pattern recognition research field devoted to the analysis, recognition and understanding of images of documents. Within this field, one of the most challenging tasks is handwriting recognition [3,6], defined as the task of converting the text contained in a document image into a machine readable format. Indeed, after decades of research, this task is still considered an open problem, specially when dealing with historical manuscripts. The main difficulties are: paper degradation, differences in the handwriting style across centuries, and old vocabulary and syntax.

Generally speaking, handwriting recognition relies on the combination of two models, the optical model and the linguistic model. The former is able to recognize the visual shape of characters or graphemes, and the second interprets

© Springer International Publishing AG 2016
A. Robles-Kelly et al. (Eds.): S+SSPR 2016, LNCS 10029, pp. 543–552, 2016.
DOI: 10.1007/978-3-319-49055-7_48

them in their context based on some structural rules. The linguistic model can range from simple n-grams (probabilities of character or word sequences), to sophisticated syntactic formalisms enriched with semantic information. In this paper we focus in this last concept. Our proposed hipothesis is that in certain conditions where the text can be roughly described by a grammatical structure, the identification of named entities can boost the recognition in a parsing process. Named entity recognition is an information extraction problem consisting in detecting and classifying the text terms into pre-defined categories such as the names of people, streets, organizations, dates, etc. It can also be seen as the semantic annotation of text elements.

However, in many cases, a mere transcription is not the final goal, but more a means to achieve the understanding of the manuscript. Therefore, the aim is to understand the documents and extract the relevant information that these documents contain. For instance, for document collections in archives, museums and libraries, there is a growing interest in making the information available for accessing, searching, browsing, etc. A typical example can be demographic documents containing people's names, birthplaces, occupations, etc. In this application scenario, the extraction of the key contents and its storace in structured databases allows to envision innovative services based in genealogical, social or demographic searches.

A traditional approach to information extraction would be to first transcribe the text, and then use dictionaries, grammars or some other NLP (Natural Language Processing) techniques to detect named entities. Named entity detection [13] has its own caveats dealing with words that have not been seen during training (namely OOV- Out of Vocabulary Words), specially if, like in our case, one wants to detect entities that do not start with a capital letter (e.g. occupations). Moreover in historical handwritten documents, handwriting recognition struggles to produce an accurate transcription further reducing the accuracy of the whole system.

Another option is to transcribe and detect the named entities at the same time. The method described in [16] uses Hidden Markov Models and category n-grams to transcribe and detect categories in demographic documents, obtaining a quite good accuracy. However, the method is following a handwriting recognition architecture, and thus it depends on the performance of the optical model, it needs sufficient training data, and it is unable to detect or recognize OOV words.

A third alternative is to directly detect the named entities from the document image, avoiding the transcription step was recently published [1]. They use a traditional handwriting recognition approach, composed of a preprocessing step for binarization and slant normalization, and then extracting handcrafted features that are then fed into a BLSTM [9] (Bi-directional Long Short-Term Memory Blocks) neural network classifier. Afterwards, they use some post-processing heuristics to reduce false positives. For example, discarding short words or words starting by "Wh" or "Th" because they are more likely to be capitalized because they are the first word in a sentence. The performance of the method is quite good, but its goal is only detecting named entities in uppercase and not categorizing these words. Moreover the post-processing heuristics of this method are specific for the English language.

Another interesting recent work in a related area was proposed by Gordo et al. in [5]. In this work, the authors show that it is possible to extract semantic word embeddings directly from artificially generated word images. They show that the network can even learn possible semantic categories of OOV words by reusing information from prefixes or suffixes of known words. However the training in this dataset required datasets of several millions of synthetically generated word images, that is a very different scenario from the typical handwritten dataset where the annotations are scarce.

In this paper we propose a method able to detect and semantically categorize entities from the word image, without requiring any handwriting recognition system. Our approach is based on the recently popularized Convolutional Neural Networks, with a special Spatial Pyramid Pooling layer to deal with the characteristic variability in aspect ratio of word images.

Our approach has several advantages. First, it is able to detect entities no matter if they start with an uppercase or lowercase letters. Secondly, it can categorize these entities semantically. This means that the detected entity is also classified as belonging to a semantic category, such as name, surname, occupation, etc. The information of the semantic category of a word is a useful information in the parsing process. Third, the effort in the creation of training data is lower than the one needed for handwriting recognition (the word is not transcribed, just classified in several categories). Finally, the method does not have any problem with OOV words because it is not based on transcription or dictionaries. Even in scenarios where transcription would later be required, our method can be helpful by allowing us to use category specific models or dictionaries [16]. It can also be used to simply reduce the transcription cost by using the categorization as a way to select only relevant words to transcribe.

This paper is organized as follows: In Sect. 2 we will describe our method and the architecture of the neural network we built, explaining the function of each of the different layers. In Sect. 3 we will explain the technical details of the dataset used, the training of our neural network and also discuss the results of the experiments. Finally we will draw some conclusions and outline possible ideas for future work.

2   A CNN Based Word Image Categorization Method

In order to classify word images into semantic categories, we propose a CNN based method, inspired by [19] (See Fig. 1). The network is divided into three different parts: the convolutional layers, that can be seen as feature extractors; the fully connected layers that act as a classifier and the Spatial Pyramid Pooling layer that serves as a bridge between features and the classifier, by ensuring a fixed size representation. We will describe each of these different parts in this section.

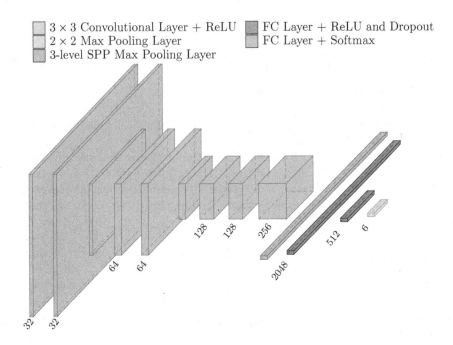

Fig. 1. Outline of our CNN architecture

2.1 Convolutional Neural Networks

Although Convolutional Neural Networks (CNN) were known since at least the early 1990's [12], it has only been recently that they gained major attention due to their high performance in virtually all fields of computer vision. The main building block of these artificial neural networks are the convolutional layers. These layers can be seen as a certain amount of filters. The output of a convolutional layer is generated by a discrete convolution of these filters with the input to the layer. Furthermore, an activation function is applied to the result of the convolution in order to make the layer able to learn non-linear functions. Compared to a standard perceptron, the filters allow sharing weights for different spatial locations thus considerably reducing the number of parameters in general [12].

Convolutional layers serve as feature detectors where each individual filter learns to detect certain features of the input image. In order to introduce a certain amount of translation invariance with respect to these detected features, CNNs usually make use of so called Pooling Layers. In these layers, activations across a certain receptive field are pooled and a single activation is forwarded to the next layer. In most cases this pooling is performed by taking the maximum value seen in the receptive field [10,17].

When stacking layers of convolutional and pooling layers, the filters in the individual convolutional layers learn edge features in the lower layers and more abstract features such as textures and object parts in the higher layers [20].

However, stacking a large amount of layers results in the so called Vanishing Gradient Problem [14] when using traditional activations such as sigmoid or hyperbolic tangent functions. Thus up until the early 2010's, neural network architectures where still fairly shallow [11]. The Vanishing Gradient Problem could first be tackled with the advent of using Rectified Linear Units (ReLU) as activation function [4]. This function is defined as the truncated linear function $f(x) = \max(0, x)$. Using the ReLU, deep CNN architectures are effectively trainable which was first successfully demonstrated in [10].

All of the convolutional layers in our architecture are a set of 3×3 Rectified Linear Units. The size of the filter was chosen to be 3×3 because they have shown to achieve better results compared to those with a bigger receptive field as they impose a regularization on the filter kernels [17]. Similar to the design presented in [17,19], we select a low number of filters in the lower layers and an increasing number in the higher layers. This leads to the neural network learning fewer low-level features for smaller receptive fields that gradually combine into more diverse high-level abstract features.

2.2   Fully Connected Layers

The general layout of CNNs can be split up in a convolutional and a fully connected part. While the convolutional and max pooling layers constitute the former, the latter is a standard Multilayer Perceptron (MLP). Thus, the convolutional part can be seen as a feature extractor while the MLP serves as a classifier. The layers of the MLP are often referred to as Fully Connected Layers (FC) in this context. Just as convolutional layers, the use of ReLU as activation function has shown itself to be effective across various architectures [10,17].

The large amount of free parameters in fully connected layers leads to the problem of the MLP learning the training set "by heart" if the amount of training samples is low. But even for larger training sets, co-adaptation is a common problem in the fully connected layers [8].

In order to counter this, various regularization measures have been proposed with Dropout [18] being one of the most prominent. Here, the output of a neuron has a probability (usually 0.5) to be set to 0 during training. A neuron in the following layer can now no longer rely on a specific neuron in the preceding layer to always be active for the same input image. Thus, the CNN has to learn multiple paths through the neural network for a single input image. This leads to more robust representations and can be seen as an ensemble within the CNN model.

The size of the different layers is a hyperparameter to tune experimentally, except for the final layer whose size has to match the number of classes we want to classify. This final layer usually uses a "softmax" activation function that outputs a probability distribution over the possible semantic categories in our experiment for each input image.

2.3 Spatial Pyramid Pooling

In general, the input to a CNN has to be of a fixed size (defined before training the network). For input images bigger or smaller than this defined size, the usual approach is to perform a (potentially anisotropic) rescale or crop from the image. For word images, with an important degree of variability in size and aspect ratio, cropping is of course not an option and resizing might introduce too strong artificial distortions in character shapes and stroke width. Thus it is important in our case that we allow our CNN to accept differently sized input images.

The key observation is that, while convolutional layers can deal with inputs of arbitrary shape and produce an output of variable shape, the fully connected layers demand a fixed size representation. Thus, the critical part is the connection between the convolutional and the fully connected part. In order to alleviate this problem, the authors in [7] propose a pooling strategy reminiscent of the spatial pyramid paradigm.

The pooling strategy performed over the last layer in the convolutional part is a pyramidal pooling over the entire receptive field. This way, the output of this Spatial Pyramid Pooling layer (SPP) is a representation with fixed dimension which can then serve as input for the ensuing MLP. It was also shown by the authors that this pooling strategy not only enables the CNN to accept differently sized input images, but it also increases the overall performance. In our method, we use a 3-level Spatial Pyramid max pooling with 4×4, 2×2 and 1×1 bin sizes. This allows us to capture meaningful features at different locations and scales whithin the word image.

3 Experimental Validation

In this section we will describe the experimental validation of our proposal. We will first explain iin detail the dataset used as well as some practical details relative to our training. We will then show the results achieved and discuss them.

3.1 Esposalles Dataset

For our experiments we used the Esposalles dataset [2,15]. This dataset consists of historical handwritten marriages records stored in the archives of Barcelona cathedral. The data we used corresponds to the volume 69, which contains 174 handwritten pages. This book was written between 1617 and 1619 by a single writer in old Catalan.

For our purpose the datasets consist of 55632 word images tagged with six different categories: *"male name"*, *"female name"*, *"surname"*, *"location"*, *"occupation"* and *"other"*. From this total we reserve 300 images of each class for testing, up to a total of 1800 images. After discarding word images smaller than 30×30pixels we end up with 53568 training examples for training and 1791 for test. In the training dataset there is a big class imbalance, with 31077 examples

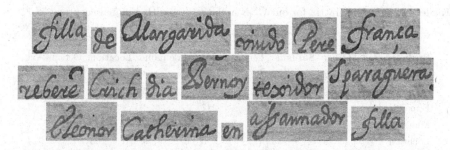

Fig. 2. Several examples of word images in the Esposalles Dataset. The big degree of variability both in size and aspect ratio of the images makes impractical the common approach of resizing images to a common size.

Table 1. Comparative with other methods.

	Precision	Recall	F_1-Measure	Classification accuracy
Adak et al. [1]*	68.42	**92.66**	78.61	-
Romero et al. [16]**	69.1	69.2	69.15	-
Our approach	**84.23**	75.48	**79.61**	78.11

of the class *"other"*, 3636 *"female name"*, 4565 *"male name"*, 2854 *"surname"*, 6581 *"location"* and 4855 *"occupation"*. No normalization or preprocessing was done to word images besides remapping them to grayscale in the interval [0–1] (0: background, 1 foreground). It is worth noting that there are words with the same transcription that could potentially belong to different classes. This is specially significant for the *"surname"* class, since it is quite common for surnames to be related to a location (i.e. a city name), an occupation or even a male name. We can see several examples of word images in Fig. 2. The dataset is available upon request to the authors.

3.2 Experiments and Results

The network was trained using standard backpropagation with stochastic gradient descent with a learning rate of 10^{-4} Nesterov momentum 0.9 and a decay rate of 10^{-6} for 100 epochs, which proved enough to obtain a training accuracy of over 99 % in all the experiments. Since we are working with images of different size, each example had to be processed individually (batch size 1), that is the reason for the low value for the decay rate.

We used the standard categorical cross entropy as loss function. In order to deal with the class imbalance problem, we introduced a "class weight" parameter in the loss function to relatively increase the impact of misclassifying the classes

Table 2. Confusion Matrix for our CNN architecture, with a global accuracy 78.11 %.

Predicted class	True class					
	Other	Surname	Female name	Male name	Location	Occupation
Other	272	52	21	9	61	38
Surname	5	153	19	2	21	3
Female name	2	34	247	9	4	4
Male name	2	14	5	274	7	1
Location	14	33	6	4	203	4
Occupation	3	10	1	0	4	250

with less examples. The weight for each class is calculated by dividing the number of samples of the most populated class by the number of samples of each class.

$$w_i = \frac{max(n_i)}{n_i}$$

We performed several experiments with slightly different network architectures, to empirically calibrate the hyperparameters. After having fixed the learning and decay rates and the number of epochs we noticed that even drastically changing the number of parameters in our architecture the results were similar.

The proposed network architecture, as depicted in Fig. 1 achieved an accuracy of 78.11 % in our dataset. An alternative network with the similar architecture but, halving the number of parameters of all the layers of the network (half of the channels in each of the convolutional layers and half of the neurons in each of the fully connected layers, and keeping the same 3-level pyramid pooling) produced an accuracy of 77.33 %.

In Table 2 we see that the errors are not evenly distributed. Despite the introduced class weights, the network is still more likely to mistakenly assign the class *"other"*. We can also see that examples corresponding to the *"surname"* class are harder to classify. This may be due to several reasons since, as discussed earlier, surnames are usually derived from names, places or occupations. It is also worth noting that the *"surname"* class is the one with fewer samples, thus having seen less examples the model is more likely to overfit this particular class.

The comparison of our method with similar methods in the literature is not an easy task, because this is a relatively recent area of research and there are few publications addressing similar issues. Even the most similar methods have big differences, for instance none of the methods provides a classification accuracy metric. In [16] they address the classification as an aid to transcription, and they work with the Esposalles dataset but with a different labeling with a different number of classes. In the case of [1] the aim is a named entity detection, with a binary output. They provide results with different datasets, so we selected the best result they achieved, usingn the IAM dataset.

We calculated our precision/recall metrics following the approach described in [16], and defined as: *"Let R be the number of relevant words contained in the* *document, let D be the number of relevant words that the system has detected, and let C be the number of the relevant words correctly detected by the system. Precision (π) and recall (ρ) are computed as":*

$$\pi = \frac{C}{D} \qquad \rho = \frac{C}{R}$$

In order to compare our results with the state of the art we can see the class "*Other*" as non-relevant words. Then, for "relevant words" we understand words with a "True Class" other than "*Other*" and for "relevant detected word" we understand word-images assigned with a label other than "*Other*". Finally for "correctly detected" we understand examples where the system correctly assigned a label other than "*Other*". That means we do not consider a word as "correctly detected" unless it is also assigned to the correct category (Table 1).

4   Conclusions

We have presented a simple approach to word categorization using convolutional neural networks. The spatial pyramid pooling layer allows us to deal with the important variability in aspect ratio of word images without artificially distorting our image. We believe that the results are specially promising given that we are classifying just isolated words images with no transcription, context information or language model of any kind. Thus, the addition of context information or simple language models should significantly boost the performance, specially in the mentioned case of surnames, and is probably the next step in this research. It would also be interesting to perform more experiments in order to determine if the network learns heuristic similar to what a human would use. For instance, names, surnames and locations usually start with a capital letter whereas occupations and other words usually do not and some word endings have a much higher likelihood on a particular class.

Finally, we also believe that this model can improve the performance of word spotting methods by reducing the search space or making a more semantic search (for example, one might specifically search for the surname Shepherd or the occupation shepherd).

Acknowledgements. This work has been partially supported by the Spanish project TIN2015-70924-C2-2-R, the European project ERC-2010-AdG-20100407-269796, the grant 2013-DI-067 from the Secretaria d'Universitats i Recerca del Departament d'Economia i Coneixement de la Generalitat de Catalunya and the Ramon y Cajal Fellowship RYC-2014-16831.

References

1. Adak, C., Chaudhuri, B.B., Blumenstein, M.: Named entity recognition from unstructured handwritten document images. In: Doermann, D.S., Govindaraju, V., Lopresti, D.P., Natarajan, P. (eds.) Document Analysis Systems, pp. 375–380 (2016)

2. Fernández-Mota, D., Almazán, J., Cirera, N., Fornés, A., Lladós, J.: BH2M: the Barcelona historical, handwritten marriages database. In: 22nd International Conference on Pattern Recognition (ICPR), pp. 256–261. IEEE (2014)

3. Frinken, V., Bunke, H.: Continuous handwritten script recognition. In: Doermann, D., Tombre, K. (eds.) Handbook of Document Image Processing and Recognition, pp. 391–425. Springer, London (2014)

4. Glorot, X., Bordes, A., Bengio, Y.: Deep sparse rectifier neural networks. In: International Conference on Artificial Intelligence and Statistics, pp. 315–323 (2011)

5. Gordo, A., Almazan, J., Murray, N., Perronin, F.: LEWIS: latent embeddings for word images and their semantics. In: Proceedings of the IEEE International Conference on Computer Vision, pp. 1242–1250 (2015)

6. Graves, A., Liwicki, M., Fernández, S., Bertolami, R., Bunke, H., Schmidhuber, J.: A novel connectionist system for unconstrained handwriting recognition. IEEE Trans. Pattern Anal. Mach. Intell. (PAMI) 31(5), 855–868 (2009)

7. He, K., Zhang, X., Ren, S., Sun, J.: Spatial pyramid pooling in deep convolutional networks for visual recognition. IEEE Trans. Pattern Anal. Mach. Intell. (PAMI) 37(9), 1904–1916 (2015)

8. Hinton, G.E., Srivastava, N., Krizhevsky, A., Sutskever, I., Salakhutdinov, R.R.: Improving neural networks by preventing co-adaptation of feature detectors. arXiv preprint arXiv:1207.0580 (2012)

9. Hochreiter, S., Schmidhuber, J.: Long short-term memory. Neural Comput. 9(8), 1735–1780 (1997)

10. Krizhevsky, A., Sutskever, I., Hinton, G.E.: ImageNet classification with deep convolutional neural networks. In: Pereira, F., Burges, C.J.C., Bottou, L., Weinberger, K.Q. (eds.) Advances in Neural Information Processing Systems 25, pp. 1097–1105 (2012)

11. LeCun, Y., Bottou, L., Bengio, Y., Haffner, P.: Gradient based learning applied to document recognition. Proc. IEEE 86(11), 2278–2324 (1998)

12. LeCun, Y., Boser, B., Denker, J.S., Henderson, D., Howard, R.E., Hubbard, W., Jackel, L.D.: Handwritten digit recognition with a back-propagation network. In: NIPS, pp. 396–404 (1990)

13. Nadeau, D., Sekine, S.: A survey of named entity recognition and classification. Lingvisticae Investigationes 30(1), 3–26 (2007)

14. Pascanu, R., Mikolov, T., Bengio, Y.: On the difficulty of training recurrent neural networks. In: International Conference on Machine Learning, no. 2, pp. 1310–1318 (2013)

15. Romero, V., Fornés, A., Serrano, N., SáNchez, J.A., Toselli, A.H., Frinken, V., Vidal, E., Lladós, J.: The esposalles database: an ancient marriage license corpus for off-line handwriting recognition. Pattern Recogn. 46(6), 1658–1669 (2013)

16. Romero, V., Sánchez, J.A.: Category-based language models for handwriting recognition of marriage license books. In: 12th International Conference on Document Analysis and Recognition (ICDAR), pp. 788–792. IEEE (2013)

17. Simonyan, K., Zisserman, A.: Very deep convolutional networks for large-scale image recognition. arXiv preprint arXiv:1409.1556 (2014)

18. Srivastava, N., Hinton, G., Krizhevsky, A., Sutskever, I., Salakhutdinov, R.: Dropout: a simple way to prevent neural networks from overfitting. J. Mach. Learn. Res. 15, 1929–1958 (2014)

19. Sudholt, S., Fink, G.A.: PHOCNet: a deep convolutional neural network for word spotting in handwritten documents. arXiv preprint arXiv:1604.00187 (2016)

20. Zeiler, M.D., Fergus, R.: Visualizing and understanding convolutional networks. In: Fleet, D., Pajdla, T., Schiele, B., Tuytelaars, T. (eds.) ECCV 2014. LNCS, vol. 8689, pp. 818–833. Springer, Heidelberg (2014). doi:10.1007/978-3-319-10590-1_53

A Novel Graph Database for Handwritten Word Images

Michael Stauffer[1,3](✉), Andreas Fischer[2], and Kaspar Riesen[1]

[1] Institute for Information Systems, University of Applied Sciences and Arts
Northwestern Switzerland, Riggenbachstr. 16, 4600 Olten, Switzerland
{michael.stauffer,kaspar.riesen}@fhnw.ch
[2] University of Fribourg and HES-SO, 1700 Fribourg, Switzerland
andreas.fischer@unifr.ch
[3] Department of Informatics, University of Pretoria, Pretoria, South Africa

Abstract. For several decades graphs act as a powerful and flexible representation formalism in pattern recognition and related fields. For instance, graphs have been employed for specific tasks in image and video analysis, bioinformatics, or network analysis. Yet, graphs are only rarely used when it comes to handwriting recognition. One possible reason for this observation might be the increased complexity of many algorithmic procedures that take graphs, rather than feature vectors, as their input. However, with the rise of efficient graph kernels and fast approximative graph matching algorithms, graph-based handwriting representation could become a versatile alternative to traditional methods. This paper aims at making a seminal step towards promoting graphs in the field of handwriting recognition. In particular, we introduce a set of six different graph formalisms that can be employed to represent handwritten word images. The different graph representations for words, are analysed in a classification experiment (using a distance based classifier). The results of this word classifier provide a benchmark for further investigations.

Keywords: Graph benchmarking dataset · Graph repository · Graph representation for handwritten words

1 Introduction

Structural pattern recognition is based on sophisticated data structures for pattern representation such as strings, trees, or graphs[1]. Graphs are, in contrast with feature vectors, flexible enough to adapt their size to the complexity of individual patterns. Furthermore, graphs are capable to represent structural relationships that might exist between subparts of the underlying pattern (by means of edges). These two benefits turn graphs into a powerful and flexible representation formalism, which is actually used in diverse fields [1,2].

The computation of a dissimilarity between pairs of graphs, termed graph matching, is a basic requirement for pattern recognition. In the last four decades

---

[1] Strings and trees can be seen as special cases of graphs.

© Springer International Publishing AG 2016
A. Robles-Kelly et al. (Eds.): S+SSPR 2016, LNCS 10029, pp. 553–563, 2016.
DOI: 10.1007/978-3-319-49055-7_49

quite an arsenal of algorithms has been proposed for the task of graph matching [1,2]. Moreover, also different benchmarking datasets for graph-based pattern recognition have been made available such as ARG [3], IAM [4], or ILPIso [5]. These dataset repositories consist of synthetically generated graphs as well as graphs that represent real world objects.

Recently, graphs have gained some attention in the field of handwritten document analysis [4] like for instance handwriting recognition [6], keyword spotting [7–9], or signature verification [10,11]. However, we still observe a lack of publicly available graph datasets that are based on handwritten word images. The present paper tries to close this gap and presents a twofold contribution. First, we introduce six novel graph extraction algorithms applicable to handwritten word images. Second, we provide a benchmark database for word classification that is based on the George Washington letters [12,13].

The remainder of this paper is organised as follows. In Sect. 2, the proposed graph representation formalisms are introduced. In Sect. 3, an experimental evaluation of the novel graph representation formalisms is given on the George Washington dataset. Section 4 concludes the paper and outlines possible further research activities.

2   Graph-Based Representation of Word Images

A graph g is formally defined as a four-tuple $g = (V, E, \mu, \nu)$ where V and E are finite sets of nodes and edges, and $\mu : V \to L_V$ as well as $\nu : E \to L_E$ are labelling functions for nodes and edges, respectively. Graphs can either be *undirected* or *directed*, depending on whether pairs of nodes are connected by undirected or directed edges. Additionally, graphs are often divided into *unlabelled* and *labelled* graphs. In the former case we assume empty label alphabets (i.e. $L_v = L_e = \{\}$), and in the latter case, nodes and/or edges can be labelled with an arbitrary numerical, vectorial, or symbolic label.

Different processing steps are necessary for the extraction of graphs from word images. In the framework presented in this paper the document images are first preprocessed by means of *Difference of Gaussian (DoG)*-filtering and binarisation to reduce the influence of noise [14]. On the basis of these preprocessed document images, single word images are automatically segmented from the document and labelled with a ground truth[2]. Next, word images are skeletonised by a 3×3 thinning operator [15]. We denote segmented word images that are binarised and filtered by B. If the image is additionally skeletonised we use the term S.

Graph-based word representations aim at extracting the inherent characteristic of these preprocessed word images. Figure 1 presents an overview of the six different graph representations, which are thoroughly described in the next three subsections. All of the proposed extraction methods result in graphs where the

---

[2] The automatic segmentation is visually inspected and - if necessary - manually corrected. Hence, in our application we assume perfectly segmented words.

nodes are labelled with two-dimensional attributes, i.e. $L_v = \mathbb{R}^2$, while edges remain unlabelled, i.e. $L_e = \{\}$.

In any of the six cases, graphs are normalised in order to reduce the variation in the node labels $(x, y) \in \mathbb{R}^2$ that is due to different word image sizes. Formally, we apply the following transformation to the coordinate pairs (x, y) that occur on all nodes of the current graph.

$$\hat{x} = \frac{x - \mu_x}{\sigma_x} \text{ and } \hat{y} = \frac{y - \mu_y}{\sigma_y}$$

where μ_x, μ_y and σ_x, σ_y denote the mean values and the standard deviations of all node labels in the current graph (in x- and y-direction, respectively).

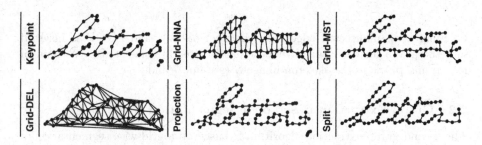

Fig. 1. Different graph representations of the word "Letters"

2.1 Graph Extraction Based on Keypoints

The first graph extraction algorithm is based on the detection of specific key-points in the word images. Keypoints are characteristic points in a word image, such as for instance end- and intersection-points of strokes. The proposed app-roach is inspired by [16] and is actually used for keyword spotting in [17]. In the following, Algorithm 1 and its description are taken from [17].

Graphs are created on the basis of filtered, binarised, and skeletonised word images S (see Algorithm 1 denoted by `Keypoint` from now on). First, end points and junction points are identified for each *Connected Component* (CC) of the skeleton image (see line 2 of Algorithm 1). For circular structures, such as for instance the letter 'O', the upper left point is selected as junction point. Note that the skeletons based on [15] may contain several neighbouring end- or junction points. We apply a local search procedure to select only one point at each ending and junction (this step is not explicitly formalised in Algorithm 1). Both end points and junction points are added to the graph as nodes, labelled with their image coordinates (x, y) (see line 3).

Next, junction points are removed from the skeleton, dividing it into *Con-nected Subcomponents* (CC_{sub}) (see line 4). Afterwards, for each connected sub-component intermediate points $(x, y) \in CC_{sub}$ are converted to nodes and added to the graph in equidistant intervals of size D (see line 5 and 6).

Algorithm 1. Graph Extraction Based on Keypoints

Input: Skeleton image S, Distance threshold D
Output: Graph $g = (V, E)$ with nodes V and edges E
1: **function** KEYPOINT(S,D)
2:     **for** Each connected component $CC \in S$ **do**
3:         $V = V \cup \{(x, y) \in CC \mid (x, y)$ are end- or junction points$\}$
4:         Remove junction points from CC
5:         **for** Each connected subcomponent $CC_{sub} \in CC$ **do**
6:             $V = V \cup \{(x, y) \in CC_{sub} \mid (x, y)$ are points in equidistant intervals $D\}$
7:     **for** Each pair of nodes $(u, v) \in V \times V$ **do**
8:         $E = E \cup (u, v)$ if the corresponding points are connected in S
9:     **return** g

Finally, an undirected edge (u, v) between $u \in V$ and $v \in V$ is inserted into the graph for each pair of nodes that is directly connected by a chain of foreground pixels in the skeleton image S (see line 7 and 8).

2.2 Graph Extraction Based on a Segmentation Grid

The second graph extraction algorithm is based on a grid-wise segmentation of word images. Grids have been used to describe features of word images like *Local Gradient Histogram (LGH)* [18] or *Histogram of Oriented Gradients (HOG)* [19]. However, to the best of our knowledge grids have not been used to represent word images by graphs.

Graphs are created on the basis of binarised and filtered, yet not skeletonised, word images B (see Algorithm 2). First, the dimension of the segmentation grid, basically defined by the number of columns C and rows R, is derived (see line 2 and 3 of Algorithm 2). Formally, we compute

$$C = \frac{\text{Width of } B}{w} \text{ and } R = \frac{\text{Height of } B}{h},$$

where w and h denote the user defined width and height of the resulting segments.

Next, a word image B is divided into $C \times R$ segments of equal size. For each segment s_{ij} $(i = 1, \ldots, C; j = 1, \ldots, R)$ a node is inserted into the resulting graph and labelled by the (x, y)-coordinates of the centre of mass (x_m, y_m) (see line 4). Formally, we compute

$$x_m = \frac{1}{n} \sum_{w=1}^{n} x_w \text{ and } y_m = \frac{1}{n} \sum_{w=1}^{n} y_w, \tag{1}$$

where n denotes the number of foreground pixel in segment s_{ij}, while x_w and y_w denote the x- and y-coordinates of the foreground pixels in s_{ij}. If a segment does not contain any foreground pixel, no centre of mass can be determined and thus no node is created for this segment.

Finally, undirected edges (u, v) are inserted into the graph according to one out of three edge insertion algorithms, viz. *Node Neighbourhood Analysis (NNA)*, *Minimal Spanning Tree (MST)*, or *Delaunay Triangulation (DEL)*. The first algorithm analyses the four neighbouring segments on top, left, right, and bottom of a node $u \in V$. In case a neighbouring segment of u is also represented by a node $v \in V$, an undirected edge (u, v) between u and v is inserted into the graph. The second algorithm reduces the edges inserted by the *Node Neighbourhood Analysis* by means of a *Minimal Spanning Tree* algorithm. Hence, in this case the graphs are actually transformed into trees. Finally, the third algorithm is based on a *Delaunay Triangulation* of all nodes $u \in V$. We denote this algorithmic procedure by `Grid-NNA`, `Grid-MST`, and `Grid-DEL` (depending on which edge insertion algorithm is employed).

Algorithm 2. Graph Extraction Based on a Segmentation Grid

Input: Binary image B, Grid width w, Grid height h
Output: Graph $g = (V, E)$ with nodes V and edges E
1: **function** GRID(B,w,h)
2:     **for** $i \leftarrow 1$ to number of columns $C = \frac{\text{Width of } B}{w}$ **do**
3:         **for** $j \leftarrow 1$ to number of rows $R = \frac{\text{Height of } B}{h}$ **do**
4:             $V = V \cup \{(x_m, y_m) \mid (x_m, y_m)$ is the centre of mass of segment $s_{ij}\}$
5:     **for** Each pair of nodes $(u, v) \in V \times V$ **do**
6:         $E = E \cup (u, v)$ if associated segments are connected by *NNA*, *MST*, or *DEL*
7:     **return** g

2.3 Graph Extraction Based on Projection Profiles

The third graph extraction algorithm is based on an adaptive rather than a fixed segmentation of word images. That is, the individual word segment sizes are adapted to respect to projection profiles. Projection profiles have been used for skew correction [20] and feature vectors of word images [21], to name just two examples. However, to the best of our knowledge projection profiles have not been used to represent word images by graphs.

Graphs are created on the basis of binarised and filtered word images B (see Algorithm 3, denoted by `Projection` from now on). First, a histogram of the vertical projection profile $P_v = \{p_1, \ldots, p_{max}\}$ is computed, where p_i represents the frequency of foreground pixels in column i of B and max is the width of B (see line 2 of Algorithm 3). Next, we split B vertically by searching so called white spaces, i.e. subsequences $\{p_i, \ldots, p_{i+k}\}$ with $p_i = \ldots = p_{i+k} = 0$. To this end, we split B in the middle of white spaces, i.e. position $p = \lfloor (p_i + p_{i+k})/2 \rfloor$, into n segments $\{s_1, \ldots, s_n\}$ (see line 3). In the best possible case a segment encloses word parts that semantically belong together (e.g. characters). Next, further segments are created in equidistant intervals D_v when the width of a segment $s \in B$ is greater than D_v (see line 4 and 5).

The same procedure as described above is then applied to each (vertical) segment $s \in B$ (rather than whole word image B) based on the projection profile of rows (rather than columns) (see lines 6 to 10). Thus, each segment s is individually divided into horizontal segments $\{s_1, \ldots, s_n\}$ (Note that a user defined parameter D_h controls the number of additional segmentation points (similar to D_v)). Subsequently, for each segment $s \in B$ a node is inserted into the resulting graph and labelled by the (x, y)-coordinates of the centre of mass (x_m, y_m) (see (1) as well as line 11 and 12). If a segment consists of background pixels only, no centre of mass can be determined and thus no node is created for this segment.

Finally, an undirected edge (u, v) between $u \in V$ and $v \in V$ is inserted into the graph for each pair of nodes, if the corresponding pair of segments is directly connected by a chain of foreground pixels in the skeletonised word image S (see line 13 and 14).

Algorithm 3. Graph Extraction Based on Projection Profiles

Input: Binary image B, Skeleton image S, Ver. threshold D_v, Hor. threshold D_h
Output: Graph $g = (V, E)$ with nodes V and edges E
1: **function** PROJECTION(B,S,D_v,D_h)
2:     Compute vertical projection profile P_v of B
3:     Split B vertically at middle of white spaces of P_v into $\{s_1, \ldots, s_n\}$
4:     **for** Each segment $s \in B$ with width larger D_v **do**
5:         Split s vertically in equidistant intervals D_v into $\{s_1, \ldots, s_n\}$
6:     **for** Each segment $s \in B$ **do**
7:         Compute horizontal projection profile P_h of s
8:         Split s horizontally at middle of white spaces of P_h into $\{s_1, \ldots, s_n\}$
9:         **for** Each segment $s \in \{s_1, \ldots, s_n\}$ with height larger D_h **do**
10:             Split s horizontally in equidistant intervals D_h into $\{s_1, \ldots, s_n\}$
11:     **for** Each segment $s \in B$ **do**
12:         $V = V \cup \{(x_m, y_m) \mid (x_m, y_m)$ is the centre of mass of segment $s\}$
13:     **for** Each pair of nodes $(u, v) \in V \times V$ **do**
14:         $E = E \cup (u, v)$ if the corresponding segments are connected in S
15:     **return** g

2.4   Graph Extraction Based on Splittings

The fourth graph extraction algorithm is based on an adaptive and iterative segmentation of word images by means of horizontal and vertical splittings. Similar to Projection, the segmentation is based on projection profiles of word images. Yet, their algorithmic procedures clearly distinguishes from each other. To the best of our knowledge such a split-based segmentation has not been used to represent word images by graphs.

Graphs are created on the basis of binarised and filtered word images B (see Algorithm 4, denoted by Split from now on). Thus, each segment $s \in B$ (initially B is regarded as one segment) is iteratively split into smaller subsegments until the width and height of each segment in $s \in B$ is below a certain threshold D_w and D_h, respectively (see lines 2 to 12). Formally, each segment $s \in S$ (with width greater than threshold D_w) is vertically subdivided into subsegments $\{s_1, \ldots, s_n\}$ by means of the projection profile P_v of s (for further details we refer to Sect. 2.3). If the histogram P_v contains no white spaces, i.e. $\forall h_i \in P \neq 0$, the segment s is split in its vertical centre into $\{s_1, s_2\}$ (see lines 3 to 7). Next, the same procedure as described above is applied to each segment $s \in B$ (with height greater than threshold D_h) in the horizontal, rather than vertical, direction (see lines 8 to 12).

Once no segment from $s \in B$ can further be split, the centre of mass (x_m, y_m) (see (1)) is computed for each segment $s \in B$ and a node is inserted into the graph labelled by the (x, y)-coordinates of the closest point on the skeletonised word image S to (x_m, y_m) (see line 13 and 14). If a segment consists of background pixels only, no centre of mass can be determined and thus no node is created for this segment.

Finally, an undirected edge (u, v) between $u \in V$ and $v \in V$ is inserted into the graph for each pair of nodes, if the corresponding pair of segments is directly connected by a chain of foreground pixels in the skeletonised word image S (see line 15 and 16).

Algorithm 4. Graph Extraction Based on Splittings

Input: Binary image B, Skeleton image S, Width threshold D_w, Height threshold D_h
Output: Graph $g = (V, E)$ with nodes V and edges E
1: **function** SPLIT(B,S,D_w,D_h)
2:     **while** Any segment $s \in B$ has a width larger D_w or height larger D_h **do**
3:         **for** Each segment $s \in B$ with width larger D_w **do**
4:             **if** s contains white spaces in vertical projection profile P_v **then**
5:                 Split s vertically at middle of white spaces of P_v into $\{s_1, \ldots, s_n\}$
6:             **else**
7:                 Split s vertically at vertical centre of s into $\{s_1, s_2\}$
8:         **for** Each segment $s \in B$ with height larger D_h **do**
9:             **if** s contains white spaces in horizontal projection profile P_h **then**
10:                 Split s horizontally at middle of white spaces of P_h into $\{s_1, \ldots, s_n\}$
11:             **else**
12:                 Split s horizontally at horizontal centre of s into $\{s_1, s_2\}$
13:     **for** Each segment $s \in B$ **do**
14:         $V = V \cup \{(x_m, y_m) \mid (x_m, y_m) \text{ is the centre of mass of segment } s\}$
15:     **for** Each pair of nodes $(u, v) \in V \times V$ **do**
16:         $E = E \cup (u, v)$ if the corresponding segments are connected in S
17:     **return** g

---

3 Experimental Evaluation

The proposed graph extraction algorithms are evaluated on preprocessed word images of the George Washington (GW) dataset, which consists of twenty different multi-writer letters with only minor variations in the writing style[3]. The same documents have been used in [12,13].

For our benchmark dataset a number of perfectly segmented word images is divided into three independent subsets, viz. a training set (90 words), a validation set (60 words), and a test set (143 words)[4]. Each set contains instances of thirty different words. The validation and training set contain two and three instances per word, respectively, while the test contains at most five and at least three instances per word. For each word image, one graph is created by means of the six different graph extraction algorithms (using different parameterisations).

In Table 1 an overview of the validated meta-parameters for each graph extraction method is given. Roughly speaking, small meta-parameter values result in graphs with a higher number of nodes and edges, while large meta-parameter values result in graph with a smaller number of nodes and edges.

Table 1. Validated meta-parameters of each graph extraction algorithm

Graph extraction algorithm	Validated meta-parameter values
Keypoint	$D = \{2,3,4,5,6\}$
Grid-NNA	$w = \{7,9,11,13,15\} \times h = \{7,9,11,13,15\}$
Grid-MST	$w = \{7,9,11,13,15\} \times h = \{7,9,11,13,15\}$
Grid-DEL	$w = \{7,9,11,13,15\} \times h = \{7,9,11,13,15\}$
Projection	$D_v = \{5,7,9,11\} \times D_h = \{4,6,8,10\}$
Split	$D_w = \{3,5,7,9\} \times D_h = \{7,9,11,13\}$

The quality of the different graph representation formalisms is evaluated by means of the accuracy of a kNN-classifier[5] that operates on approximated *Graph Edit Distances (GED)* [22]. The meta-parameters are optimised with respect to the accuracy of the kNN on the validation set[6]. Then, the accuracy of the kNN-classifier is measured on the test set using the optimal meta-parameters for each graph extraction method.

In Table 2, the optimal meta-parameters, the median number of nodes $|\bar{V}|$ and edges $|\bar{E}|$ (defined over training, validation and test set) as well as the accuracy of

---

[3] George Washington Papers at the Library of Congress, 1741–1799: Series 2, Letter-book 1, pp. 270–279 & 300–309, http://memory.loc.gov/ammem/gwhtml/gwseries2.html.

[4] Available at http://www.histograph.ch.

[5] We define $k = 5$ for all of our evaluations.

[6] If different meta-parameter settings lead to the same accuracy, the setting with the lower average number of nodes and edges is used.

Table 2. Classification accuracy of each graph representation formalism

Graph extraction algorithm	Optimal meta-parameter	$|\bar{V}|$	$|\bar{E}|$	Acc.
Keypoint	D =4,	73	67	0.7762
Grid-NNA	w =13, h=9	39	55	0.6503
Grid-MST	w =9, h=11	46	44	0.7413
Grid-DEL	w =9, h=9	52	138	0.6294
Projection	D_v =9 D_h =6	44	41	0.8182
Split	D_w =7 D_h =9	51	48	0.8042

the kNN on the test set are shown for each extraction method. We observe that the Projection and Split extraction methods clearly perform the best among all algorithms with a classification accuracy of about 82 % and 80 % on the test set, respectively. Keypoint achieves the third best classification result with about 77 %. However, the average number of nodes and edges of both Projection and Split are substantially lower than those of Keypoint. Grid-MST achieves an accuracy which is virtually the same as with Keypoint. Yet, with substantially less nodes and edges than Keypoint. The worst classification accuracies are obtained with Grid-NNA and Grid-DEL (about 65 % and 63 %, respectively). Note especially the large number of edges which are produced with the Delaunay triangulation.

4   Conclusion and Outlook

The novel graph database presented in this paper is based on graph extraction methods. These methods aim at extracting the inherent characteristics of handwritten word images and represent these characteristics by means of graphs. The Keypoint extraction method is based on the representation of nodes by characteristic points on the handwritten stroke, while edges represent strokes between these keypoints. Three of our extraction methods, viz. Grid-NNA, Grid-MST and Grid-DEL, are based on a grid-wise segmentation of a word image. Each segment of this grid is represented by a node which is then labelled by the centre of mass of the segment. Finally, the Projection and Split extraction methods are based on vertical and horizontal segmentations by means of projection profiles. An empirical evaluation of the six extraction algorithm is carried out on the George Washington letters. The achieved accuracy can be seen as a first benchmark to be used for future experiments. Moreover, the experimental results clearly indicate that both Projection and Split are well suited for extracting meaningful graphs from handwritten words.

In future work we plan to extend our graph database to further documents using the presented extraction methods and make them all publicly available. Thus, we will be able to provide a more comparative and thorough study against other state-of-the-art representation formalisms at a later stage.

Acknowledgments. This work has been supported by the Hasler Foundation Switzerland.

References

1. Conte, D., Foggia, P., Sansone, C., Vento, M.: Thirty years of graph matching in pattern recognition. Int. J. Pattern Recogn. Artif. Intell. **18**(03), 265–298 (2004)
2. Foggia, P., Percannella, G., Vento, M.: Graph matching and learning in pattern recognition in the last 10 years. Int. J. Pattern Recogn. Artif. Intell. **28**(01), 1450001 (2014)
3. De Santo, M., Foggia, P., Sansone, C., Vento, M.: A large database of graphs and its use for benchmarking graph isomorphism algorithms. Pattern Recogn. Lett. **24**(8), 1067–1079 (2003)
4. Bunke, H., Riesen, K.: Recent advances in graph-based pattern recognition with applications in document analysis. Pattern Recogn. **44**(5), 1057–1067 (2011)
5. Le Bodic, P., Héroux, P., Adam, S., Lecourtier, Y.: An integer linear program for substitution-tolerant subgraph isomorphism and its use for symbol spotting in technical drawings. Pattern Recogn. **45**(12), 4214–4224 (2012)
6. Fischer, A., Suen, C.Y., Frinken, V., Riesen, K., Bunke, H.: A fast matching algorithm for graph-based handwriting recognition. In: Kropatsch, W.G., Artner, N.M., Haxhimusa, Y., Jiang, X. (eds.) GbRPR 2013. LNCS, vol. 7877, pp. 194–203. Springer, Heidelberg (2013). doi:10.1007/978-3-642-38221-5_21
7. Riba, P., Llados, J., Fornes, A.: Handwritten word spotting by inexact matching of grapheme graphs. In: International Conference on Document Analysis and Recognition, pp. 781–785 (2015)
8. Wang, P., Eglin, V., Garcia, C., Largeron, C., Llados, J., Fornes, A.: A novel learning-free word spotting approach based on graph representation. In: International Workshop on Document Analysis Systems, pp. 207–211 (2014)
9. Bui, Q.A., Visani, M., Mullot, R.: Unsupervised word spotting using a graph representation based on invariants. In: International Conference on Document Analysis and Recognition, pp. 616–620 (2015)
10. Wang, K., Wang, Y., Zhang, Z.: On-line signature verification using segment-to-segment graph matching. In: International Conference on Document Analysis and Recognition, pp. 804–808 (2011)
11. Fotak, T., Bača, M., Koruga, P.: Handwritten signature identification using basic concepts of graph theory. Trans. Sig. Process. **7**(4), 117–129 (2011)
12. Lavrenko, V., Rath, T., Manmatha, R.: Holistic word recognition for handwritten historical documents. In: International Workshop on Document Image Analysis for Libraries, pp. 278–287 (2004)
13. Fischer, A., Keller, A., Frinken, V., Bunke, H.: Lexicon-free handwritten word spotting using character HMMs. Pattern Recogn. Lett. **33**(7), 934–942 (2012)
14. Fischer, A., Indermühle, E., Bunke, H., Viehhauser, G., Stolz, M.: Ground truth creation for handwriting recognition in historical documents. In: International Workshop on Document Analysis Systems, New York, USA, pp. 3–10 (2010)
15. Guo, Z., Hall, R.W.: Parallel thinning with two-subiteration algorithms. Commun. ACM **32**(3), 359–373 (1989)
16. Fischer, A., Riesen, K., Bunke, H.: Graph similarity features for HMM-based handwriting recognition in historical documents. In: International Conference on Frontiers in Handwriting Recognition, pp. 253–258 (2010)

17. Stauffer, M., Fischer, A., Riesen, K.: Graph-based keyword spotting in historical handwritten documents. In: International Workshop on Structural and Syntactic Pattern Recognition (2016)
18. Rodriguez, J.A., Perronnin, F.: Local gradient histogram features for word spotting in unconstrained handwritten documents. In: International Conference on Frontiers in Handwriting Recognition, pp. 7–12 (2008)
19. Almazán, J., Gordo, A., Fornés, A., Valveny, E.: Segmentation-free word spotting with exemplar SVMs. Pattern Recogn. 47(12), 3967–3978 (2014)
20. Hull, J.: Survey and annotated bibliography. Series in Mach. Percept. Artif. Intell. 29, 40–64 (1998)
21. Rath, T., Manmatha, R.: Word image matching using dynamic time warping. In: Computer Vision and Pattern Recognition, vol. 2, pp. II-521–II-527 (2003)
22. Riesen, K., Bunke, H.: Approximate graph edit distance computation by means of bipartite graph matching. Image Vis. Comput. 27(7), 950–959 (2009)

Graph-Based Keyword Spotting in Historical Handwritten Documents

Michael Stauffer[1,3(✉)], Andreas Fischer[2], and Kaspar Riesen[1]

[1] Institute for Information Systems, University of Applied Sciences and Arts
Northwestern Switzerland, Riggenbachstr. 16, 4600 Olten, Switzerland
{michael.stauffer,kaspar.riesen}@fhnw.ch
[2] University of Fribourg and HES-SO, 1700 Fribourg, Switzerland
andreas.fischer@unifr.ch
[3] Department of Informatics, University of Pretoria, Pretoria, South Africa

Abstract. The amount of handwritten documents that is digitally available is rapidly increasing. However, we observe a certain lack of accessibility to these documents especially with respect to searching and browsing. This paper aims at closing this gap by means of a novel method for keyword spotting in ancient handwritten documents. The proposed system relies on a keypoint-based graph representation for individual words. Keypoints are characteristic points in a word image that are represented by nodes, while edges are employed to represent strokes between two keypoints. The basic task of keyword spotting is then conducted by a recent approximation algorithm for graph edit distance. The novel framework for graph-based keyword spotting is tested on the George Washington dataset on which a state-of-the-art reference system is clearly outperformed.

Keywords: Handwritten keyword spotting · Bipartite graph matching · Graph representation for words

1 Introduction

Keyword Spotting (KWS) is the task of retrieving any instance of a given query word in speech recordings or text images [1–3]. Textual KWS can be roughly divided into *online* and *offline* KWS. For online KWS temporal information of the handwriting is available recorded by an electronic input device such as, for instance, a digital pen or a tablet computer. On the other hand side, offline KWS is based on scanned image only, and thus, offline KWS is regarded as the more difficult task than its online counterpart. The focus of this paper is on KWS in historical handwritten documents. Therefore, offline KWS, referred to as KWS from now on, can be applied only.

Most of the KWS methodologies available are either based on *template-based* or *learning-based* matching algorithms. Early approaches of template-based KWS are based on a pixel-by-pixel matching of word images [1]. More elaborated approaches to template-based KWS are based on the matching of feature

A. Robles-Kelly et al. (Eds.): S+SSPR 2016, LNCS 10029, pp. 564–573, 2016.
DOI: 10.1007/978-3-319-49055-7_50

vectors by means of *Dynamic Time Warping (DTW)* [4]. A recent and promising approach to template-based KWS is given by the matching of *Local Binary Pattern (LBP)* histograms [5]. One of the main advantages of template-based KWS is its independence from the actual representation formalism as well as the underlying language (and alphabet) of the document. However, template-based KWS does not generalise well to different writing styles. Learning-based KWS on the other side is based on statistical models like *Hidden Markov Models (HMM)* [6, 7], *Neural Networks (NN)* [3] or *Support Vector Machines (SVM)* [8]. These models have to be trained a priori on a (relatively large) set of training words. An advantage of the learning-based approach, when compared with the template-based approach, is its higher generalisability. Yet, this advantage is accompanied by a loss of flexibility, which is due to the need for learning the parameters of the actual model on a specific training set.

The vast majority of KWS algorithms available are based on statistical representations of word images by certain numerical features. To the best of our knowledge only few graph-based KWS approaches have been proposed so far [9–12]. However, a graph-based representation is particularly interesting for KWS as graphs, in contrast with feature vectors, offer a more natural and comprehensive formalism for the representation of word images.

A first approach for graph-based KWS has been proposed in [10]. The nodes of the employed graphs represent keypoints that are extracted on connected components of the skeletonised word images, while the edges are used to represent the strokes between the keypoints. The majority of the words consists of more than only one connected component, and thus, a word is in general represented by more than one graph. The matching of words is thus based on two separate procedures. First, the individual costs of assignments of all pairs of connected components (represented by graphs) are computed via bipartite graph matching [13]. Second, an optimal assignment between the connected components has to be found. To this end, a DTW algorithm is employed that operates on the costs produced in the first step. This matching procedure is further improved by a so-called *coarse-to-fine approach* in [11].

Another idea for graph-based KWS has been introduced in [12]. In this paper a graph represents a set of prototype strokes (called invariants). First, a word image is segmented into strokes. Eventually, the most similar invariant is defined for every stroke in the word. The nodes of the graph are used to represent these invariants, while edges are inserted between all pairs of nodes. Edges are labelled with the information whether or not strokes of the corresponding nodes are stemming from the same connected component. Finally, for KWS the graph edit distance is computed by means of the bipartite graph matching algorithm [13].

A third approach for graph-based KWS has been proposed in [9] where complete text lines are represented by a *grapheme graph*. Graphemes are sets of prototype convexity paths, similar to invariants, that are defined a priori in a codebook. The nodes of the particular graphs represent the individual graphemes of the text line, while edges are inserted into the graph whenever two graphemes are directly connected to each other. The matching itself is based on a coarse-to-fine approach. Formally, potential subgraphs of the query graph are determined first. These subgraphs are subsequently matched against a query graph by means of the bipartite graph matching algorithm [13].

In the present paper we introduce a novel approach for graph representation of individual words that is based on the detection of keypoints. In contrast with [10] our approach results in a single graph per word. Hence, no additional assignment between graphs of different connected components is necessary during the matching process. Furthermore, in our approach the edges are detected by a novel method based on both the skeleton of connected components and their connected subcomponents. Last but not least, also the graph matching procedure actually employed for KWS has been substantially extended when compared to the previous contributions in the field. In particular, we introduce different types of linear and non-linear cost functions for the edit operations used in [13].

The remainder of this paper is organised as follows. In Sect. 2, the basic concept of graph edit distance is briefly reviewed. In Sect. 3, the proposed graph-based KWS approach is introduced. An experimental evaluation of the proposed framework is given in Sect. 4. Section 5 concludes the paper and outlines possible further research activities.

2   Graph Edit Distance

A graph g is formally defined as a four-tuple $g = (V, E, \mu, \nu)$ where V and E are finite sets of nodes and edges, and $\mu : V \to L_V$ as well as $\nu : E \to L_E$ are labelling functions for nodes and edges, respectively. Graphs can be divided into *undirected* and *directed* graphs, where pairs of nodes are either connected by undirected or directed edges, respectively. Additionally, graphs are often distinguished into *unlabelled* and *labelled* graphs. In the latter case, both nodes and edges can be labelled with an arbitrary numerical, vectorial, or symbolic label from L_v or L_e, respectively. In the former case we assume empty label alphabets, i.e. $L_v = L_e = \{\}$.

Graphs can be matched with exact and inexact methods [14,15]. Inexact graph matching, in contrast to exact graph matching, allows matchings between two non-identical graphs by endowing the matching with a certain error-tolerance with respect to labelling and structure. Several approaches for inexact graph matching have been proposed. Yet, *Graph Edit Distance (GED)* is widely accepted as one of the most flexible and powerful paradigms available [16]. The GED between two graphs g_1 and g_2 is defined as the least costly series of edit operations to be applied to g_1 in order to make it *isomorphic* to g_2. Formally,

$$GED(g_1, g_2) = \min_{(e_1,\ldots,e_k) \in \gamma(g_1,g_2)} \sum_{i=1}^{k} c(e_i)$$

where e_i denotes an edit operation, $(e_1, \ldots, e_k)$ an edit path, $\gamma(g_1, g_2)$ the set of all edit paths that transform g_1 into g_2, and $c(e_i)$ the cost for a certain edit operation e_i. Different types of edit operations are allowed such as substitutions, insertions, deletions, splittings, and mergings of both nodes and edges. Commonly, the cost function $c(e_i)$ considers domain-specific knowledge and reflects the strength of edit operation e_i.

The computation of the exact GED is commonly based on an A*-algorithm that explores all possible edit paths $\gamma(g_1, g_2)$ [17]. However, this exhaustive search is exponential with respect to the number of nodes of the involved graphs.

In order to make the concept of GED applicable to large graphs and/or large graph sets, several fast but approximative algorithms have been proposed [13,18]. In the present paper we make use of the well-known bipartite graph matching algorithm for approximating the GED in cubic time complexity [13]. This algorithm is based on an optimal match between nodes and their local structure (i.e. their adjacent edges). That is, the suboptimal computation of the GED is based on a reduction of the GED problem to a *Linear Sum Assignment Problem (LSAP)*, which can be optimally solved by, for instance, Munkres' algorithm [19]. In case of scalability limitations, one could also make use of the graph matching algorithm for approximating the GED in quadratic, rather than cubic, time complexity [18].

3 Graph-Based Keyword Spotting

The proposed graph-based KWS solution is based on four different processing steps as shown in Fig. 1. First, document images are preprocessed and segmented into words (1). Based on the segmented word images, graphs are extracted by means of a novel keypoint-based method (2) and eventually normalised (3). Finally, the graphs of query words are matched against graphs from the document to create a retrieval index (4). In the following four subsections these four steps are described in greater detail.

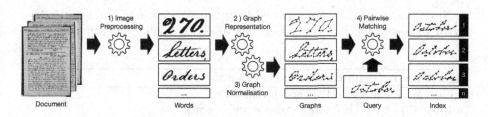

Fig. 1. Process of graph-based keyword spotting of the word "October"

3.1 Image Preprocessing

Image preprocessing aims at reducing variations on document images that are caused, for instance, by noisy background, skewed scanning, or document degradation. In our particular framework, document images are first filtered by a *Difference of Gaussian (DoG)* and binarised by a global threshold [20]. Single word images are then manually segmented. That is, we build our framework on perfectly segmented words in order to focus on the task of KWS. The skew, i.e. the inclination of the document, is removed by a hierarchical rotation of the complete document image such that the horizontal projection profile is step-wise maximised [21]. Note that the skew angle is estimated on complete document images first and then corrected on single word images. Finally, each word image is skeletonised by a 3×3 thinning operator [22].

3.2   Graph Representation

For a graph-based KWS system to succeed, the variations among graphs of the same word have to be minimised, while variations of graphs of different words should remain large. Hence, a graph representation has to represent the inherent characteristic of a word. In the present paper the graph extraction algorithm is based on the detection of keypoints. Keypoints are characteristic points in a word image, such as for instance end- and intersection-points of strokes. The proposed approach is inspired by [7]. However, in contrast with [7] the proposed graph representation makes use of both nodes and edges. Additionally, the keypoint detection is further refined by a local search algorithm.

Graphs are created on the basis of filtered, binarised, and skeletonised word images S (see Algorithm 1). First, end points and junction points are identified for each *Connected Component* (*CC*) of the skeleton image (see line 2 of Algorithm 1). For circular structures, such as for instance the letter 'O', the upper left point is selected as junction point. Note that the skeletons based on [22] may contain several neighbouring end- or junction points. We apply a local search procedure to select only one point at each ending and junction (this step is not explicitly formalised in Algorithm 1). Both end points and junction points are added to the graph as nodes, labelled with their image coordinates (x, y) (see line 3).

Next, junction points are removed from the skeleton, dividing it into *Connected Subcomponents* (CC_{sub}) (see line 4). Afterwards, for each connected

---

Algorithm 1. Graph Extraction Based on Keypoints

Input: Skeleton image S, Distance threshold D
Output: Graph $g = (V, E)$ with nodes V and edges E
 1: **function** KEYPOINT(S,D)
 2:     **for** Each connected component $CC \in S$ **do**
 3:         $V = V \cup \{(x, y) \in CC \mid (x, y)$ are end- or junction points$\}$
 4:         Remove junction points from CC
 5:         **for** Each connected subcomponent $CC_{sub} \in CC$ **do**
 6:             $V = V \cup \{(x, y) \in CC_{sub} \mid (x, y)$ are points in equidistant intervals $D\}$
 7:     **for** Each pair of nodes $(u, v) \in V \times V$ **do**
 8:         $E = E \cup (u, v)$ if the corresponding points are connected in S
 9:     **return** g

subcomponent intermediate points $(x, y) \in CC_{sub}$ are converted to nodes and added to the graph in equidistant intervals of size D (see line 5 and 6).

Finally, an undirected edge (u, v) between $u \in V$ and $v \in V$ is inserted into the graph for each pair of nodes that is directly connected by a chain of foreground pixels in the skeleton image S (see line 7 and 8).

3.3   Graph Normalisation

In order to improve the comparability between graphs of the same word class, the labels $\mu(v)$ of the nodes $v \in V$ are normalised. In our case the node label alphabet is defined by $L_v = \mathbb{R}^2$. A first graph normalisation is based on a centralisation of each node label $\mu(v) = (x, y) \in \mathbb{R}^2$ by

$$\hat{x} = x - \mu_x \text{ and } \hat{y} = y - \mu_y, \tag{1}$$

where $\hat{x}$ and $\hat{y}$ denote the new node coordinates, x and y the original node position, μ_x, and μ_y represent the mean values of all (x, y)-coordinates in the graph under consideration.

The second graph normalisation centralises the node labels and reduces variations of node positions that might occur due to different word image sizes. Formally,

$$\hat{x} = \frac{x - \mu_x}{\sigma_x} \text{ and } \hat{y} = \frac{y - \mu_y}{\sigma_y}, \tag{2}$$

where σ_x and σ_y denote the standard deviation of all node coordinates in the current graph.

3.4   Pairwise Matching

The actual KWS is based on a pairwise matching of a query graph g against all graphs of a set of word graphs $G = \{g_1, \ldots, g_n\}$ stemming from the underlying document. We make use of the bipartite graph matching algorithm [13]. In our system the resulting GED between g and $g_i \in G$ is normalised by using the cost of the maximum cost edit path between g and g_i, viz. the edit path that results from deleting all nodes and edges of g and inserting all nodes and edges of g_i. We refer to this maximum cost as $Max\text{-}GED$ from now on. By means of this procedure a retrieval index $r_i(g) \in [0, 1]$ can be created for every word graph $g_i \in G$ given a certain query graph g. Formally,

$$r_i(g) = \frac{GED(g, g_i)}{Max\text{-}GED(g, g_i)}$$

The effectiveness of edit distance based KWS relies on an adequate definition of cost functions for the basic edit operations. In general, the cost $c(e)$ of a particular edit operation e is defined with respect to the underlying label alphabets L_V and L_E. In our framework the nodes are labelled with two-dimensional numerical labels while edges remain unlabelled, i.e. $L_V = \mathbb{R}^2$ and $L_E = \{\}$. In the present section four cost functions are defined for this particular labelling.

For all of our cost models a constant cost $\tau_v \in \mathbb{R}^+$ for node deletion and insertion is used. Formally, the cost for the node deletions and insertions is defined by $c(u \rightarrow \varepsilon) = c(\varepsilon \rightarrow v) = \tau_v$. For edges a similar cost with another constant cost $\tau_e \in \mathbb{R}^+$ is defined. The cost models to be used in our framework differ in the definition of the cost for node substitutions. The basic intuition behind all approaches is that the more dissimilar two labels are, the stronger is the distortion associated with the corresponding substitution.

The first cost model is based on a weighted Euclidean distance of the two corresponding labels. Formally, given two graphs $g_1 = (V_1, E_1, \mu_1, \nu_1)$ and $g_2 = (V_2, E_2, \mu_2, \nu_2)$, where $\mu_1, \mu_2 : V_1, V_2 \rightarrow \mathbb{R}^2$ the cost for a node substitution $(u \rightarrow v)$ with $\mu_1(u) = (x_i, y_i)$ and $\mu_2(v) = (x_j, y_j)$ is defined by

$$c_E(u \rightarrow v) = \sqrt{\alpha(x_i - x_j)^2 + (1 - \alpha)(y_i - y_j)^2},$$

where $\alpha \in [0, 1]$ is a weighting parameter to define whether the x- or the y-coordinate is more important for the resulting substitution cost.

For graphs with scaled node labels (see Sect. 3.3) the standard deviation σ of the node labels of a query graph might be additionally included in the cost model by defining

$$c_{E_\sigma}(u \rightarrow v) = \sqrt{\alpha\sigma_x(x_i - x_j)^2 + (1 - \alpha)\sigma_y(y_i - y_j)^2},$$

where σ_x and σ_y denote the standard deviation of all node coordinates in the query graph.

The third and fourth cost function are based on the weighted Euclidean distance that is additionally scaled by means of a Sigmoidal function to $[0, 2\tau_v]$. Formally,

$$c_S(u \rightarrow v) = \frac{2\tau_v}{1 + e^{(kc_E(u \rightarrow v) - \gamma)}} \text{ and } c_{S_\sigma}(u \rightarrow v) = \frac{2\tau_v}{1 + e^{(kc_{E_\sigma}(u \rightarrow v) - \gamma)}},$$

where k is the steepness and γ the threshold of the Sigmoidal function. For both cost functions c_S and c_{S_σ} the maximal substitution cost is equal to the sum of cost of a node deletion and node insertion.

4   Experimental Evaluation

The experimental evaluation of the proposed KWS system is carried out on the *George Washington (GW)* dataset, which consists of twenty pages of handwritten letters with only minor variations in the writing style[1]. The same dataset has

---

[1] George Washington Papers at the Library of Congress, 1741–1799: Series 2, Letterbook 1, pp. 270–279 & 300–309, http://memory.loc.gov/ammem/gwhtml/gwseries2.html.

already been used in [6,23,24]. For our KWS framework individual graphs are created for each of the 4893 words of the dataset by means of the keypoint-based graph representation algorithm described in Sect. 3.2. We use a threshold of $D = 5$ for all of our evaluations.

The performance of KWS is measured by the *mean Average Precision (mAP)* in two subsequent experiments. First, the meta-parameters and the different image and graph normalisations are optimised for all cost functions. To this end, the mAP is computed on a small validation set, consisting of ten different query words with a frequency of at least 10 as well as a reduced training set based on 1000 different words including all instances of the query word.

In Table 1 the results of this validation phase are shown. We distinguish between graphs that are based on word images where the skew is corrected or not. For both variants we use graphs where the node labels remain unnormalised (denoted by U in Table 1), and graphs where the labels are normalised by using (1) and (2) (denoted by N_1 and N_2, respectively). Note that the cost models c_E and c_S can be applied to graph normalisation with N_2 only.

Table 1. mAP of Euclidean and Sigmoidal cost functions for different preprocessing

Preprocessing Cost function	Skew not corrected			Skew corrected		
	U	N_1	N_2	U	N_1	N_2
c_E	50.17	72.87	-	47.08	72.24	-
c_{E_σ}	-	-	**76.53**	-	-	75.59
c_S	49.71	72.72	-	50.60	73.53	-
c_{S_σ}	-	-	**76.24**	-	-	75.24

We observe that graphs based on not skew corrected word images in combination with scaled and centralised node labels (N_2) is optimal for both the Euclidean and the Sigmoidal cost functions. These two models are further optimised by means of the node label weighting factor α. By using this weighting parameter, the mAP can be increased from 76.53 to 80.94 and from 76.24 to 79.32 with c_{E_σ} and c_{S_σ}, respectively.

Using this optimal parameter settings, the proposed KWS system is compared with a reference system based on DTW [23,24] with optimised Sakoe-Chiba band. This evaluation is conducted in a four-fold cross-validation, where each fold consists of a test set (avg. 2447 words) that is tested with a training set (avg. 1223.5 words).

In Table 2 the results of our novel graph-based KWS system (using both c_{E_σ} and c_{S_σ}) and the reference DTW system are given. Our graph-based system outperforms the DTW-based KWS system in both cases. The Euclidean and Sigmoidal cost models improve the mAP of the reference system by 2.31 % and 5.62 %, respectively.

Table 2. Graph-based vs. DTW-based KWS

System	mAP	Improvement
DTW	54.08	
Proposed c_{E_σ}	55.33	+2.31%
Proposed c_{S_σ}	57.12	+5.62%

5 Conclusion and Outlook

The novel KWS system proposed in this paper is based on a keypoint-based graph representation of individual words. Keypoints are characteristic points in a word image that are represented by nodes, while edges are represented by strokes between two keypoints. The actual KWS is based on a bipartite matching of pairs of graphs. Four different cost functions have been introduced to quantify the substitution cost of nodes that are matched. These cost functions in combination with different image and graph normalisations are optimised on the George Washington dataset. The optimal system clearly outperforms the reference DTW algorithm.

In future work, we aim at extending our word-based approach to a line-based approach. The actual KWS would therefore be based on finding a subgraph isomorphism of a query graph in the larger line graph. Moreover, other graph representation formalisms as well as more powerful labelling functions could be a rewarding avenue to be pursued. Thus, we will be able to conduct a more thorough comparison against other state-of-the-art systems using further graph representations and documents.

Acknowledgments. This work has been supported by the Hasler Foundation Switzerland.

References

1. Manmatha, R., Han, C., Riseman, E.: Word spotting: a new approach to indexing handwriting. In: Computer Vision and Pattern Recognition, pp. 631–637 (1996)
2. Rath, T., Manmatha, R.: Word image matching using dynamic time warping. In: Computer Vision and Pattern Recognition, vol. 2, pp. II-521–II-527 (2003)
3. Frinken, V., Fischer, A., Manmatha, R., Bunke, H.: A novel word spotting method based on recurrent neural networks. IEEE Trans. PAMI **34**(2), 211–224 (2012)
4. Kolcz, A., Alspector, J., Augusteijn, M., Carlson, R., Viorel Popescu, G.: A line-oriented approach to word spotting in handwritten documents. Pattern Anal. Appl. **3**(2), 153–168 (2000)
5. Dey, S., Nicolaou, A., Llados, J., Pal, U.: Local binary pattern for word spotting in handwritten historical document. Comput. Res. Repository (2016)
6. Lavrenko, V., Rath, T., Manmatha, R.: Holistic word recognition for handwritten historical documents. In: Proceedings of the International Workshop on Document Image Analysis for Libraries, pp. 278–287 (2004)

7. Fischer, A., Riesen, K., Bunke, H.: Graph similarity features for HMM-based handwriting recognition in historical documents. In: International Conference on Frontiers in Handwriting Recognition, pp. 253–258 (2010)

8. Huang, L., Yin, F., Chen, Q.H., Liu, C.L.: Keyword spotting in offline chinese handwritten documents using a statistical model. In: International Conference on Document Analysis and Recognition, pp. 78–82 (2011)

9. Riba, P., Llados, J., Fornes, A.: Handwritten word spotting by inexact matching of grapheme graphs. In: International Conference on Document Analysis and Recognition, pp. 781–785 (2015)

10. Wang, P., Eglin, V., Garcia, C., Largeron, C., Llados, J., Fornes, A.: A novel learning-free word spotting approach based on graph representation. In: Proceedings of the International Workshop on Document Analysis for Libraries, pp. 207–211 (2014)

11. Wang, P., Eglin, V., Garcia, C., Largeron, C., Llados, J., Fornes, A.: A coarse-to-fine word spotting approach for historical handwritten documents based on graph embedding and graph edit distance. In: International Conference on Pattern Recognition, pp. 3074–3079 (2014)

12. Bui, Q.A., Visani, M., Mullot, R.: Unsupervised word spotting using a graph representation based on invariants. In: International Conference on Document Analysis and Recognition, pp. 616–620 (2015)

13. Riesen, K., Bunke, H.: Approximate graph edit distance computation by means of bipartite graph matching. Image Vis. Comput. **27**(7), 950–959 (2009)

14. Conte, D., Foggia, P., Sansone, C., Vento, M.: Thirty years of graph matching in pattern recognition. Int. J. Pattern Rec. Artif. Intell. **18**(03), 265–298 (2004)

15. Foggia, P., Percannella, G., Vento, M.: Graph matching and learning in pattern recognition in the last 10 years. Int. J. Pattern Rec. Artif. Intell. **28**(01), 9–42 (2014)

16. Bunke, H., Allermann, G.: Inexact graph matching for structural pattern recognition. Pattern Rec. Lett. **1**(4), 245–253 (1983)

17. Hart, P., Nilsson, N., Raphael, B.: A formal basis for the heuristic determination of minimum cost paths. IEEE Trans. Syst. Sci. Cybern. **4**(2), 100–107 (1968)

18. Fischer, A., Suen, C.Y., Frinken, V., Riesen, K., Bunke, H.: Approximation of graph edit distance based on Hausdorff matching. Pattern Rec. **48**(2), 331–343 (2015)

19. Munkres, J.: Algorithms for the assignment and transportation problems. J. Soc. Ind. Appl. Math. **5**(1), 32–38 (1957)

20. Fischer, A., Indermühle, E., Bunke, H., Viehhauser, G., Stolz, M.: Ground truth creation for handwriting recognition in historical documents. In: International Workshop on Document Analysis Systems, New York, USA, pp. 3–10 (2010)

21. Hull, J.: Document image skew detection: survey and annotated bibliography. In: Series in Machine Perception and Artificial Intelligence, vol. 29, pp. 40–64 (1998)

22. Guo, Z., Hall, R.W.: Parallel thinning with two-subiteration algorithms. Commun. ACM **32**(3), 359–373 (1989)

23. Rath, T.M., Manmatha, R.: Word spotting for historical documents. Int. J. Doc. Anal. Rec. **9**(2–4), 139–152 (2007)

24. Fischer, A., Keller, A., Frinken, V., Bunke, H.: Lexicon-free handwritten word spotting using character HMMs. Pattern Rec. Lett. **33**(7), 934–942 (2012)

Local Binary Pattern for Word Spotting in Handwritten Historical Document

Sounak Dey[1(✉)], Anguelos Nicolaou[1], Josep Llados[1], and Umapada Pal[2]

[1] Computer Vision Center, Universitat Autonoma de Barcelona, Bellaterra, Spain
{sdey,anguelos,josep}@cvc.uab.es
[2] CVPR Unit, Indian Statistical Institute, Kolkata, India
umapada@isical.ac.in

Abstract. Digital libraries store images which can be highly degraded and to index this kind of images we resort to word spotting as our information retrieval system. Information retrieval for handwritten document images is more challenging due to the difficulties in complex layout analysis, large variations of writing styles, and degradation or low quality of historical manuscripts. This paper presents a simple innovative learning-free method for word spotting from large scale historical documents combining Local Binary Pattern (LBP) and spatial sampling. This method offers three advantages: firstly, it operates in completely learning free paradigm which is very different from unsupervised learning methods, secondly, the computational time is significantly low because of the LBP features, which are very fast to compute, and thirdly, the method can be used in scenarios where annotations are not available. Finally, we compare the results of our proposed retrieval method with other methods in the literature and we obtain the best results in the learning free paradigm.

Keywords: Local binary patterns · Spatial sampling · Learning-free · Word spotting · Handwritten historical document analysis · Large-scale data

1 Introduction

A lot of initiatives has been taken to convert the paper scriptures to digitized media for preservation in digital libraries. Digital libraries store different types of scanned images of documents such as historical manuscripts, documents, obituary and handwritten notes. The challenges in this area become diverse as more and more types of images are considered as input for archival and retrieval.

Documents of different languages are also been archived which is an another challenge. Traditional Optical Character Reader (OCR) techniques cannot be applied generally to all types of imagery due to several reasons. In this context, it is advantageous to explore techniques for direct characterization and manipulation of image features in order to retrieve document images containing textual and other non-textual components. A document image retrieval system

© Springer International Publishing AG 2016
A. Robles-Kelly et al. (Eds.): S+SSPR 2016, LNCS 10029, pp. 574–583, 2016.
DOI: 10.1007/978-3-319-49055-7_51

asks whether an imaged document contains particular words, which are of interest to the user, ignoring other unrelated words. This is also known as **keyword spotting** or simply 'word-spotting' with no need for correct and complete character recognition. Word spotting technique in terms of pattern recognition can be defined as classification of word images.

The problem of word spotting, especially in the setting of large-scale datasets, is the balancing engineering trade-offs between number of documents indexed, queries per second, update rate, query latency, information kept about each document and retrieval algorithm. In order to handle such large scale data, computational efficiency and dimensionality are critical aspects which are effectively taken care by the use of LBP in word spotting.

Word spotting is fundamentally based on appearance based features. In this work we would like to explore the textural features as an alternative representation offering a richest description with minimal computational cost. Moreover the nature of the handwritten words suggests that there is a stable structural pattern due to the ascender and descenders in the words. In this paper, our aim is to propose an end-to-end method which can improve the performance for word spotting in handwritten historical document images. The specific objectives are: (1) To develop a word spotting method for large scale un-anotated handwritten historical data. (2) Apply texture feature like LBP to capture the fine grained information about the handwritten words which is computationally cheap. Converting the text to meta-information. (3) Combine the spatial knowledge using a Quad tree spatial structure [19] for pooling.

We use LBP as a generic low level texture classification features, that don't incorporate any assumptions specific to a task. Out here we consider every text-block as a bi-modal oriented texture.

2   State of the Art

2.1   Taxonomy of Methods

The state of the art word spotting techniques can be classified based on various criteria: (1) Depending on whether segmentation is needed or not i.e. segmentation-free or segmentation-based. (2) Based on possibility on learning: learning-free or learning-based, supervised/unsupervised. (3) Based on usability: Query-By-Example (QBE) or Query-By-String (QBS).

Methods on Segmentation-Free or Segmentation-Base: In the segmentation-based approach, there is a tremendous effort towards solving the word segmentation problem [8,15]. One of the main challenges of keyword spotting methods, either learning-free or learning-based, is that they usually need to segment the document images into words [10,15] or text lines [6] using a layout analysis step. In critical scenarios, dealing with handwritten text and highly degraded documents [11] segmentation is highly crucial. Any segmentation errors have a cumulative effect on subsequent word representations and matching steps.

The work of Rusinol et al. [18] avoids segmentation by representing regions with a fixed-length descriptor based on the well-known bag of visual words (BoW) framework [3]. The recent works of Rodriguez et al. [17] propose methods that relax the segmentation problem by requiring only segmentation at the text line level. In [7], Gatos and Pratikakis perform a fast and very coarse segmentation of the page to detect salient text regions. The represented queries are in the form of a descriptor based on the density of the image patches. Then, a sliding-window search is performed only over the salient regions of the documents using an expensive template-based matching.

Methods on Learning-Free or Learning-Based: Learning-based methods, such as [5,6,17], use supervised machine learning techniques to train models of the query words. On the contrary, learning-free methods, use dedicated matching scheme based on image sample comparison without any necessary training process [9,15]. Learning-based methods are preferred for applications where the keywords to spot are a priori known and fixed. If the training set is large enough they are usually able to deal with multiple writers. However, the cost of having a useful amount of annotated data available might be unbearable in most scenarios. In that sense methods running with few or none training data are preferred. Learning-based methods [14] employ statistical learning methods to train a keyword model that is then used to score query images. A very general approach was recently introduced in [14], where the learning-based approach is applied at word level based on Hidden Markov Models (HMMs). The trained word models are expected to show a better generalization capability than template images. However, the word models still need a considerable amount of templates for training and the system is not able to spot out-of-vocabulary keywords. In the above work holistic word features in conjunction with a probabilistic annotation model is also used. In [5] Fischer et al. used nine features. The first three were the features regarding the cropped window (height, width and center of gravity) and the rest were the geometric features of the contours of the writing. Peronin et al. [14] presented a very general learning-based approach at word level based on local gradient features. In our case we use LBP which is much faster and can be calculated at the run-time. In this paper we use the LBP for the first time to do a fast learning free word spotting schematic. The learning free method, unlike unsupervised methods, can be used without any kind of tuning to any database.

Methods Based on Query-By-Example (QBE) or Query-By-String (QBS). The query can be either an example image (QBE) or a string containing the word to be searched (QBS). In QBS approaches, character models typically HMMs have been pre-trained. At query time the models of the characters forming the string are concatenated into a word-model. Both approaches have their advantages and disadvantages. QBE requires examples of the word to be spotted, whereas, QBS approaches require large amounts of labeled data to train character models. The work of Almazan [1] and Rusinol et al. [18], where the word images are represented with HOG and SIFT descriptors aggregated respectively, can successfully be applied in a retrieval scenario. Most of the popular methods either work on QBE or in QBS and the success in one paradigm cannot be replicated in another, as comparison between images and texts is not well defined. We will focus on the QBE scenario.

The LBP explained in the Sec. 3.2 uses the uniformity to reduce the dimensionality to speed up the process of matching the feature vectors in the learning free paradigm unlike other state of the art methods.

3   Proposed Method

In this section the oriented gradient property of the LBP has been used to develop a fast learning free method for information spotting for large scale document database where annotated data is unavailable.

3.1   End-to-end Pipeline Overview

In the pipeline given below we consider segmented words. Flow-diagram of our proposed approach is shown in Fig. 1. We use a median filtering preprocessing technique to reduce noise.

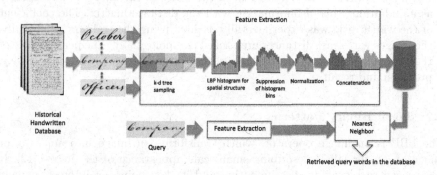

Fig. 1. Flow-diagram of oour proposed approach.

- **LBP Transform:** LBP is a scale sensitive operator whose scaling depends on the sampling rate. In our case this rate is fixed to 8. For each sub-window zone obtained in the spatial sampling state, a uniform compressed LBP histogram is generated. Each histogram is then normalized and weighted by a edge pixel ratio in the sub-window. This was perceived in this way, because the uniform LBP transform contains information regarding the sign transition of gradients which is prominent in case of the edges of the stroke width. We gave importance to the sub-windows with more edge information. The non-uniform pattern is suppressed to reduce the dimensionality of the final feature vector and also reduce its effect on normalization. It can be seen in purple almost on the medial axis in the Fig. 2(d). The information lost by this exclusion is very less compared to that of the uniform patterns. The final feature is the concatenation of the histogram of each sub-window. Though the dimensionality increases with respect to the number of level in spatial sampling, the texture information becomes more distinctive for that space.

- **Quad Tree Spatial Sampling:** The gray level word images are then used to compute the spatial sub windows zones. This is done based on the center of mass of the image which divides the image in four quadrants. Each quadrant was further subdivided based on the center of mass of those quadrants. This brought about twenty such sub windows for the first two level. The levels are experimentally fixed using the train set. The spatial information is embedded in the final feature vector using this technique. LBP histogram is pooled over the zone created by this sampling technique as it gives more weightage to the zones having the pen strokes.

- **Nearest Neighbor:** The feature thus obtained is compared to that of the query using Bray-Curtis (BC) dissimilarity matching as shown in Eq. 1.

$$BC(a,b) = (\sum_i \mid a_i - b_i \mid)/(\sum_i a_i + b_i) \tag{1}$$

where a_i and b_i are the i-th elements of the histograms. We then use the width ratio which is the ratio between the width of the query and the images as an additional bit of information with the distance matrix. The coefficient of the width ratio was experimentally decided using the training set. Finally, the images with least distances are ranked chronologically. The performance of the system was measured by well established mean average precision, accuracy, precision and recall.

3.2   Local Binary Patterns

The LBP is an image operator, which transforms an image into an array or image of integer labels describing small-scale appearance of the image [12]. It has proven to be highly discriminating and its key points of interest, namely its in-variance to monotonic gray level changes and computational proficiency, make it suitable for demanding image analysis tasks. The basic LBP operator, introduced by Ojala et al. [13], was based on the assumption that texture has locally two complementary aspects, a pattern and its strength. LBP feature extraction consists of two principal steps: the LBP transform, and the pooling of LBP into histogram representation of an image. As explained in [13] gray scale in-variance is achieved because of the difference of the intensity of the neighboring pixel to that of the central pixel. It also encapsulates the local geometry at each pixel by encoding binarized differences with pixels of its local neighborhood:

$$LBP_{P,R,t} = \sum_{p=0}^{P-1} s_t(g_p - g_c) \times 2^P, \tag{2}$$

where g_c is the central pixel being encoded, g_p are P symmetrically and uniformly sampled points on the periphery of the circular area of radius R around g_c, and s_t is a binarization function parameter by t. The sampling of g_p is performed with bi-linear interpolation. t, which in the standard definition is considered zero, is a parameter that determines when local differences are big enough for consideration. In our version, the LBP operator works in a 3×3 pixel block of an image. The pixels in this block are threshold by its center pixel value, multiplied by powers of two and then summed to obtain a label for the center pixel. As the neighborhood consists of 8 pixels, a total of $2^8 = 256$ different labels can be obtained depending on the relative gray values of the center and the pixels in the neighborhood.

(a)   (b)   (c)   (d)

Fig. 2. (a) Input Image (b) LBP image (c) LBP with median filtering (d) LBP uniformity with median filter.

In [13] a class of patterns called uniform is described as all patterns having at most two bit-value transitions on a clock-wise traverse. These uniform patterns provide the vast majority, over 90 %, of all the LBP patterns occuring in the analysed historical documents. By binning all non-uniform patterns into a single bin, the histogram representation goes to 59 from originally 256 bins. The final texture feature employed in texture analysis is the histogram of the operator outputs (i.e. pattern labels) accumulated over a texture sample. The reason why the histogram of 'uniform' patterns provides better discrimination in comparison to the histogram of all individual patterns comes down to differences in their statistical properties as shown in Fig. 2(d). The relative proportion of 'non-uniform' patterns of all patterns accumulated into a histogram is so small that their probabilities can not be estimated reliably.

3.3 Spatial Sampling

Since LBP histograms disregard all information about the spatial layout of the features, they have severely limited descriptive ability. We consider the spatial sampling as sub-windows on the whole image. These are obtained from the spatial pyramid with two levels. The Quad tree image sampling is based on the center of mass which yields much better results as shown in the Fig. 3. The intuition was that small sub-windows have more discriminating power than others because of their high black pixel density. The black pixel concentration suggests several handwritten letters together. To determine the sub windows, the gray images were considered for center of mass which was calculated on its binarized image obtained using Ostu's technique. On this center point the image was

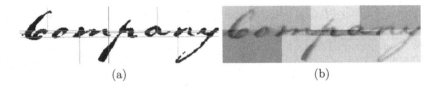

(a)   (b)

Fig. 3. Spatial sampling using Quad tree technique. (a) Quad tree applied based on the center of mass where the red lines is the first level of sampling with blue point as center using sub optimal binarization. The green lines are second level of sampling with blue points as center respectively. (b) The different zones of the sampling is shown by different colors.

divided into four quadrants. The number of hierarchical levels determines the total number of sub-window used. The number of hierarchical levels determines the total amount of sub-windows used, for level one being 4 sub-windows and for level 2, 16 (for each quadrant, 4 more quadrants were generated and so on). In our case, we just used level 2 which was experimentally fixed.

3.4   Nearest Neighbour K-NN

Nearest neighbor search (NNS) is an optimization problem for finding closest (or most similar) points. Closeness is typically expressed in terms of a dissimilarity function: the less similar the objects, the larger the function values. In our case it is the Bray-curtis Dissimilarity as defined in Eq. 1. We made a very naive NN technique for our pipeline.

4   Experiments

4.1   Experimental Framework

Our approach was evaluated on two public datasets (The George Washington (GW) dataset [5] and the Barcelona Historical Handwritten Marriages Database (BHHMD) [4]) which are available online. The proposed algorithm was only evaluated on segmentation-based word spotting scenarios. We have used a set of pre-segmented words to compare our approach with other methods in the literature with the aim of testing the descriptor in terms of speed, compacity and learning independence. The results for all the methods considered all the words in the test pages as queries. The used performance evaluation measures are mean average precision (mAP), precision and rPrecision. Given a query, we label the set of relevant objects with regard to the query as *rel* and the set of retrieved elements from the database as *ret*. The precision is defined in terms of *ret* and *rel* in Eq. 3. rPrecision is the precision at rank *rel*. For a given query, $r(n)$ is a binary function on the relevance of the n-th item returned in the ranked list. The used performance evaluation measures is mean average precision (mAP) which defined in Eq. 3.

$$Precision(P) = \frac{\mid ret \cap rel \mid}{\mid ret \mid}, \ mAP = \frac{\sum\limits_{n=1}^{\mid ret \mid}(P@n \times r(n))}{\mid rel \mid} \qquad (3)$$

4.2 Results on George Washington Dataset

The George Washington database was created from the George Washington Papers at the Library of Congress. The dataset was divided in 15 pages for training and validation and the last 5 pages for testing. Table 1 shows the performance of our method compared to others. The best results are highlighted in each category. Here, Quad Tree method is an adaptation of [19].

Table 1. Retrieval results on the George Washington dataset

Method	Learning	mAP	Accuracy (P@1)	rPrecision	Speed(secs) (on Test)
Quad-Tree	Standardization	15.5 %	30.14	15.32	44.41
BoVW [18]	Unsupervised, codebook	68.26 %	85.87	62.56	NA
FisherCCA [2]	Supervised	**93.11 %**	**95.44**	**90.08**	137.63
DTW [16]	No	20.94 %	41.34	20.3	78095.89
HOG pooled Quad-Tree	No	48.22 %	64.96	43.04	45.34
Proposed method	No	**54.44 %**	**72.86**	**48.87**	**43.14**

Some qualitative results are shown in Fig. 4. It is interesting to see that most of the words have been retrieved. This example takes the image *Company* as a query. The system correctly retrieves the first 15 words, whereas the 16th is *observing*, which is very similar to the query word in length and pattern. The next word is again a correct retrieval.

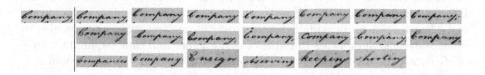

Fig. 4. Qualitative result examples. Query (estreme left) and First 20 retrieved words.

4.3 Results on BHHMD Dataset

This collection consists of marriage registers in the Barcelona area between the 15th and 20th centuries. The ground truth-ed subset contains 40 images. The dataset was divided in 30 pages for train and validation and the last 10 pages for test. Table 2 shows the results. Our method is the second best performing, with best computational time in this dataset. We have also performed tests for cross-dataset evaluation and our method in all categories is just 2 % behind the best state of the art method.

Table 2. Retrieval results on the BHHMD dataset

Method	Learning	mAP	Accuracy(P@1)	rPrecision
Quad-Tree	Standardization	38.4 %	61.92	48.47
FisherCCA [2]	Supervised	**95.40 %**	**95.49**	**94.27**
HOG pooled Quad-Tree	No	66.66 %	80.59	62.35
DTW [16]	No	7.36 %	4.69	2.99
Proposed method (LBP)	No	**70.84 %**	**84.13**	**70.44**

5   Conclusion

We have proposed a fast learning free word spotting method based on LBP-representations and a k-d tree sampling approach. The most important contribution is that the proposed word spotting approach has been shown to be the best among the learning free ones in terms of performance. The computational speed measured with the same benchmark for the proposed method is best compared to other state of the art methods. We have shown that LBP based on uniformity can be stable under the deformations of handwriting. For the pooling approach, the main contribution of the proposed framework is the pooling of the LBP based on the Quad Tree zones. The LBP has been defined as the textural feature which we use as oriented texture recognition. A sampling architecture has been designed to maximize the usage of the pen strokes and preserve the LBP patterns specific to the region. In terms of a retrieval problems, competitive mAP was obtained. The time complexity of this indexation is linear with the number of words in the database. This was reduced to the order of $logN$ by using a k-NN approach. It leads us to conclude that a feature extraction scheme as it is proposed here is very useful to compute inexact matching in large-scale scenarios. We have demonstrated that compact textural descriptors are useful information for handwritten word spotting, despite the variability of handwriting. The experimental results demonstrates that our approach is comparable to other statistical approaches in terms of performance and time requirements.

Future work will focus on the evaluation of the stability of LBP-based representations in large multi-writer document collections.

Acknowledgment. This work has been partially supported by the Spanish project TIN2015-70924-C2-2-R, and RecerCaixa, a research program from ObraSocial La Caixa. We are thankful to the authors of the compared methods for their support.

References

1. Almazán, J., Gordo, A., Fornés, A., Valveny, E.: Efficient exemplar word spotting. In: BMVC, vol. 1, p. 3 (2012)
2. Almazán, J., Gordo, A., Fornés, A., Valveny, E.: Segmentation-free word spotting with exemplar svms. Pattern Recogn. **47**(12), 3967–3978 (2014)

3. Csurka, G., Dance, C., Fan, L., Willamowski, J., Bray, C.: Visual categorization with bags of keypoints. In: Workshop on Statistical Learning in Computer Vision, ECCV, Prague, vol. 1, pp. 1–2 (2004)

4. Fernández-Mota, D., Almazán, J., Cirera, N., Fornés, A., Lladós, J.: BH2M: the barcelona historical, handwritten marriages database. In: 2014 22nd International Conference on Pattern Recognition (ICPR), pp. 256–261. IEEE (2014)

5. Fischer, A., Keller, A., Frinken, V., Bunke, H.: Lexicon-free handwritten word spotting using character hmms. Pattern Recogn. Lett. **33**(7), 934–942 (2012)

6. Frinken, V., Fischer, A., Manmatha, R., Bunke, H.: A novel word spotting method based on recurrent neural networks. IEEE Trans. Pattern Anal. Mach. Intell. **34**(2), 211–224 (2012)

7. Gatos, B., Pratikakis, I.: Segmentation-free word spotting in historical printed documents. In: 10th International Conference on Document Analysis and Recognition, ICDAR 2009, pp. 271–275. IEEE (2009)

8. Ghosh, S.K., Valveny, E.: A sliding window framework for word spotting based on word attributes. In: Paredes, R., Cardoso, J.S., Pardo, X.M. (eds.) IbPRIA 2015. LNCS, vol. 9117, pp. 652–661. Springer, Heidelberg (2015). doi:10.1007/978-3-319-19390-8_73

9. Leydier, Y., Ouji, A., LeBourgeois, F., Emptoz, H.: Towards an omnilingual word retrieval system for ancient manuscripts. Pattern Recogn. **42**(9), 2089–2105 (2009)

10. Liang, Y., Fairhurst, M.C., Guest, R.M.: A synthesised word approach to word retrieval in handwritten documents. Pattern Recogn. **45**(12), 4225–4236 (2012)

11. Louloudis, G., Gatos, B., Pratikakis, I., Halatsis, C.: Text line and word segmentation of handwritten documents. Pattern Recogn. **42**(12), 3169–3183 (2009)

12. Nicolaou, A., Bagdanov, A.D., Liwicki, M., Karatzas, D.: Sparse radial sampling lbp for writer identification. arXiv preprint arXiv:1504.06133 (2015)

13. Ojala, T., Pietikainen, M., Maenpaa, T.: Multiresolution gray-scale and rotation invariant texture classification with local binary patterns. IEEE Trans. Pattern Anal. Mach. Intell. **24**(7), 971–987 (2002)

14. Perronnin, F., Rodriguez-Serrano, J., et al.: Fisher kernels for handwritten word-spotting. In: 10th International Conference on Document Analysis and Recognition, ICDAR 2009, pp. 106–110. IEEE (2009)

15. Rath, T.M., Manmatha, R.: Features for word spotting in historical manuscripts. In: Proceedings of the Seventh International Conference on Document Analysis and Recognition, pp. 218–222. IEEE (2003)

16. Rath, T.M., Manmatha, R.: Word image matching using dynamic time warping. In: Proceedings of the 2003 IEEE Computer Society Conference on Computer Vision and Pattern Recognition, vol. 2, pp. 521–527. IEEE (2003)

17. Rodriguez-Serrano, J., Perronnin, F., et al.: A model-based sequence similarity with application to handwritten word spotting. IEEE Trans. Pattern Anal. Mach. Intell. **34**(11), 2108–2120 (2012)

18. Rusinol, M., Aldavert, D., Toledo, R., Lladós, J.: Browsing heterogeneous document collections by a segmentation-free word spotting method. In: 2011 International Conference on Document Analysis and Recognition (ICDAR), pp. 63–67. IEEE (2011)

19. Sidiropoulos, P., Vrochidis, S., Kompatsiaris, I.: Content-based binary image retrieval using the adaptive hierarchical density histogram. Pattern Recogn. **44**(4), 739–750 (2011)

Learning Robust Features for Gait Recognition by Maximum Margin Criterion

(Extended Abstract)

Michal Balazia[✉] and Petr Sojka

Faculty of Informatics, Masaryk University, Botanická 68a, 602 00 Brno,
Czech Republic
xbalazia@mail.muni.cz, sojka@fi.muni.cz

In the field of gait recognition from motion capture (MoCap) data, designing human-interpretable gait features is a common practice of many fellow
researchers. To refrain from ad-hoc schemes and to find maximally discriminative
features we may need to explore beyond the limits of human interpretability. This
paper contributes to the state-of-the-art with a machine learning approach for
extracting robust gait features directly from raw joint coordinates. The features
are learned by a modification of Linear Discriminant Analysis with Maximum
Margin Criterion (MMC) so that the identities are maximally separated and, in
combination with an appropriate classifier, used for gait recognition.

Recognition of a person involves capturing their raw walk sample, extracting
gait features to compose a template that serves as the walker's signature, and
finally querying a central database for a set of similar templates to report the
most likely identity.

The goal of the MMC-based learning is to find a linear discriminant that
maximizes the misclassification margin. We discriminate classes by projecting
high-dimensional input data onto low-dimensional sub-spaces by linear transformations with the goal of maximizing the class separability. We are interested
in finding an optimal feature space where a gait template is close to those of
the same walker and far from those of different walkers. A solution to this optimization problem can be obtained by eigendecomposition of the between-class
scatter matrix minus the within-class scatter matrix. Obtaining the eigenvectors
involves a fast two-step algorithm in virtue of the Singular Value Decomposition.
Similarity of two templates is expressed in the Mahalanobis distance.

Extensive simulations of the proposed method and eight state-of-the-art
methods used a CMU MoCap sub-database of 54 walking subjects that performed 3,843 gait cycles in total, which makes an average of about 71 samples
per subject. A variety of class-separability coefficients and classification metrics
allows insights from different statistical perspectives. Results indicate that the
proposed method is a leading concept for rank-based classifier systems: lowest
Davies-Bouldin Index, highest Dunn Index, highest (and exclusively positive)

The full research paper **Learning Robust Features for Gait Recognition by
Maximum Margin Criterion** has been accepted for publication at the 23rd
International Conference on Pattern Recognition (ICPR 2016), Cancun, Mexico,
December 2016. arXiv:1609.04392.

© Springer International Publishing AG 2016
A. Robles-Kelly et al. (Eds.): S+SSPR 2016, LNCS 10029, pp. 585–586, 2016.
DOI: 10.1007/978-3-319-49055-7

Silhouette Coefficient, second highest Fisher's Discriminant Ratio and, combined with rank-based classifier, the best Cumulative Match Characteristic, False Accept Rate and False Reject Rate trade-off, Receiver Operating Characteristic (ROC) and recall-precision trade-off scores along with Correct Classification Rate, Equal Error Rate, Area Under ROC Curve and Mean Average Precision. We interpret the high scores as a sign of robustness. Apart from performance merits, the MMC method is also efficient: low-dimensional templates (dimension $\leq \#\text{classes} - 1 = 53$) and Mahalanobis distance ensure fast distance computations and thus contribute to high scalability.

Author Index

Printed in the United States
By Bookmasters